NUMERICAL METHODS IN FLUID DYNAMICS

NUMERICAL METHODS IN FLUID DYNAMICS:

INITIAL AND INITIAL BOUNDARY-VALUE PROBLEMS

GARY A. SOD
Department of Mathematics,
Tulane University

The right of the University of Cambridge to print and sell all manner of books was granted by Henry VIII in 1534. The University has printed and published continuously since 1584.

CAMBRIDGE UNIVERSITY PRESS

Cambridge
London New York New Rochelle
Melbourne Sydney

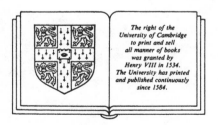

CAMBRIDGE UNIVERSITY PRESS
Cambridge, New York, Melbourne, Madrid, Cape Town, Singapore, São Paulo, Delhi

Cambridge University Press
The Edinburgh Building, Cambridge CB2 8RU, UK

Published in the United States of America by Cambridge University Press, New York

www.cambridge.org
Information on this title: www.cambridge.org/9780521105521

First published 1985
This digitally printed version 2009

A catalogue record for this publication is available from the British Library

Library of Congress Cataloguing in Publication data

Sod, Gary A. 1950–
Numerical methods in fluid dynamics.
Includes bibliographical references and index.
1. Fluid dynamics–Mathematics. 2. Numerical
analysis. I. Title.
TA357.S536 1984 620.1′064 83-25169

ISBN 978-0-521-25924-8 hardback
ISBN 978-0-521-10552-1 paperback

To the woman in my life, Jo

CONTENTS

Chapter IV. Hyperbolic Conservation Laws

Chapter V. Stability in the Presence of Boundaries

Appendices

PREFACE

This first book of a two-volume series on numerical fluid dynamics is concerned with finite difference methods for initial boundary-value problems. The intention is to make the field of numerical fluid dynamics accessible by emphasizing concepts along with the underlying theory rather than presenting a collection of recipes and by building on the classical methods to develop the numerical methods on the leading edge of research.

This book, which is essentially self-contained, is intended to be a text directed at first-year graduate students in engineering and the physical sciences and as a reference to practitioners in the field. It assumes a basic knowledge of partial differential equations. The proofs of some of the classical theorems requiring a greater knowledge of analysis are presented in appendices. This first volume is designed for a course in numerical methods for initial boundary-value problems. The numerical methods and techniques for their analysis contained in Volume I provide the foundation for Volume II, which deals exclusively with the equations governing fluid motion.

I am indebted to many friends and colleagues for their support and cooperation. The formost of these, taught me numerical analysis, was my advisor, and is my friend, Alexandre Chorin. This book is in no small way a reflection of his influence on my way of viewing numerical fluid dynamics. I would also like to express my appreciation to Philip Colella, F. Alberto Grünbaum, Ole Hald, Amiram Harten, J. Mac Hyman, Robert Miller, Joseph Oliger, and Olof Widlund who have read drafts of the book and made useful suggestions.

I wish to thank Meredith Mickel and P.Q. Lam who typed and retyped the manuscript. I wish to thank David Tranah and the staff of the Cambridge University Press for their patience and assistance in the preparation of the book.

Finally, I wish to express my appreciation to my wife, Jo, for her patience, understanding, and her never-ending support, without which this book would not have been completed.

January, 1985

Gary A. Sod

I. INTRODUCTION

I.1. Introductory Example.

Consider the one-dimensional heat (diffusion) equation on $t > 0$

(1.1.1a) $$\partial_t v = a^2 \partial_x^2 v, \quad -\infty < x < +\infty, \quad t > 0$$

with initial condition

(1.1.1b) $$v(x,0) = f(x), \quad -\infty < x < +\infty,$$

where $f(x)$ is uniformly bounded by some constant M, that is, $|f(x)| < M$ for all x.

The derivatives shall be approximated by finite differences. Let k, called the time step, and h called the grid spacing, be small positive numbers. Using the definition of the derivative,

$$\partial_x v = \lim_{h \to 0} \frac{v(x+h,t) - v(x,t)}{h} = \lim_{h \to 0} \frac{v(x,t) - v(x-h,t)}{h} .$$

This leads to two approximations to $\partial_x v$,

$$\partial_x v \sim \frac{v(x+h,t) - v(x,t)}{h} \quad \text{or} \quad \frac{v(x,t) - v(x-h,t)}{h} .$$

Similarly, for $\partial_t v$,

$$\partial_t v \sim \frac{v(x,t+h) - v(x,t)}{k} .$$

Also, for $\partial_x^2 v$, using the approximation above for $\partial_x v$,

$$\partial_x^2 v \sim \frac{\dfrac{v(x+h,t) - v(x,t)}{h} - \dfrac{v(x,t) - v(x-h,t)}{h}}{h}$$

$$= \frac{v(x+h,t) - 2v(x,t) + v(x-h,t)}{h^2}$$

Let $u^n \equiv u(ih,nk)$, where i is an integer and n is a non-negative integer, denote a grid function which approximates the exact solution of $v(x,t)$ at the point (ih,nk). u_i^n is determined by replacing the initial-value problem (1.1.1) by the finite difference equation

(1.1.2a) $$\frac{u_i^{n+1} - u_i^n}{k} = a^2 \frac{u_{i+1}^n - u_i^n + u_{i=1}^n}{h^2} , \quad n > 0$$

with the discrete initial condition

(1.1.2b) $$u_i^0 = f(ih) , \quad n = 0.$$

With the finite difference equation (1.1.2), the continuous problem defined for $t > 0$ and $-\infty < x < \infty$ has been transformed into a discrete problem, defined at discrete points in time $t = nk$, and in space $x = ih$. This finite difference equation can be explicitly solved for u_i^{n+1} in terms of u_{i-1}^n, u_i^n, and u_{i+1}^n,

$$u_i^{n+1} = \lambda u_{i+1}^n + (1-2\lambda)u_i^n + \lambda u_{i-1}^n, \tag{1.1.3}$$

where $\lambda = a^2k/h^2$. The solution is advanced in time from time $t = nk$ to time $t = (n+1)k$ using (1.1.3). To start set $n = 0$, which corresponds to the initial data $u_i^0 = f(ih)$. The solution at time $(n+1)k$ is an average of the solution at three points at the preceeding time level nk.

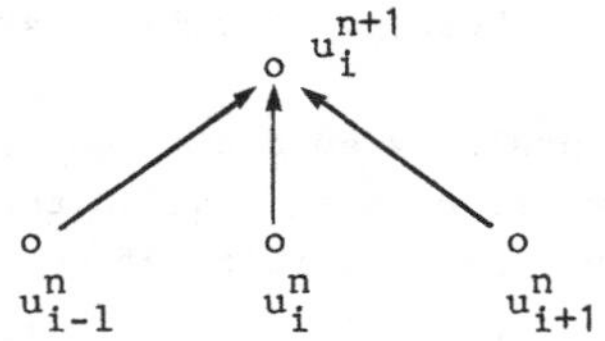

Figure 1.1

One difficulty in dealing with a pure initial value problem is that the discrete form of the real line, that is $x = ih$, $i = 0, \pm 1, \pm 2, \ldots$, is infinite. This countably infinite set must be reduced to a finite set which effectively adds boundaries that physically are not present. This does not present a problem in this case as a necessary condition for the solution to (1.1.1) to exist is that $v(x,t) \to 0$ as $x \to \pm\infty$. At the artificial boundaries impose the condition that $u^n = 0$.

For the solution to (1.1.1) to exist, we must have $v(x,t) \to 0$ as $x \to \pm\infty$. Thus we can impose conditions at the artificial boundaries that $u^n = 0$.

As a numerical example, consider an initial condition given by a smooth function $f(x)$ having support on an interval, centered at a grid point, with length $2h$. One such function $f(x)$ is depicted in Figure 1.2.

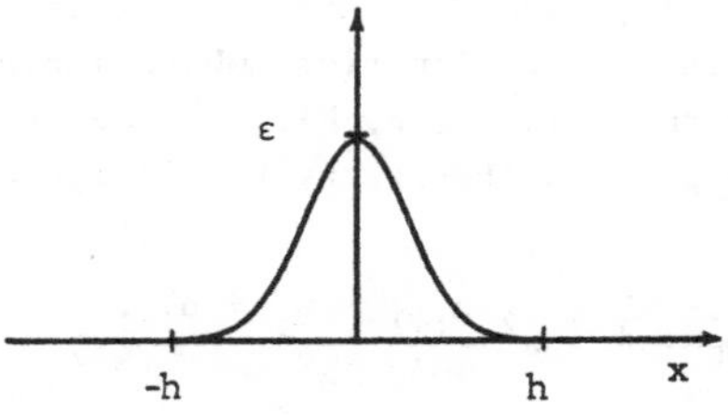

Figure 1.2.

The discrete initial conditions become

$$u_i^0 = \begin{cases} \varepsilon > 0, & i = i' \\ 0 \;, & i \neq i' \;. \end{cases}$$

Replace the real line with the interval $[-1,1]$ and impose boundary conditions $v(\pm 1,t) = 0$. Figure 1.3a depicts the exact solution (represented by a solid line) and the approximate solution (represented by a dashed line) with $h = 0.1$ and $\lambda = 0.25$ at times $t = 0.02$, 0.04, and 0.06. The general structure of the solution is represented by the approximate solution. Figure 1.3b depicts the results with $\lambda = 2$. In this case the approximate solution oscillates with the amplitude of the oscillation growing with the number of time steps.

The partial differential equation has been approximated in the sense that $v(x,t)$ almost satisfies (1.1.2); but, however, does the solution of (1.1.2) approximate the solution of (1.1.1)? This example demonstrates that answer depends on λ.

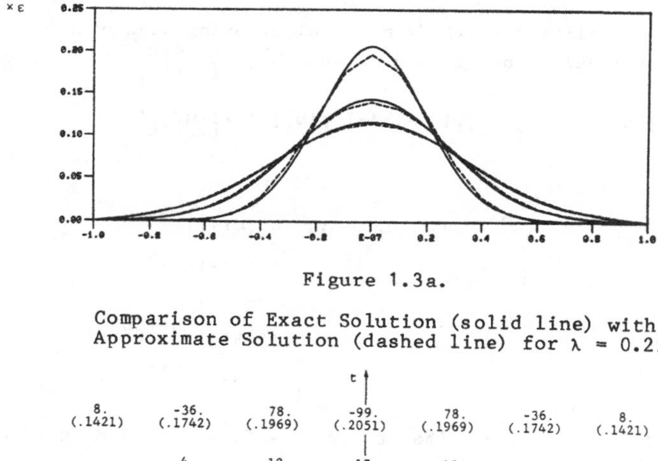

Figure 1.3a.

Comparison of Exact Solution (solid line) with Approximate Solution (dashed line) for $\lambda = 0.25$.

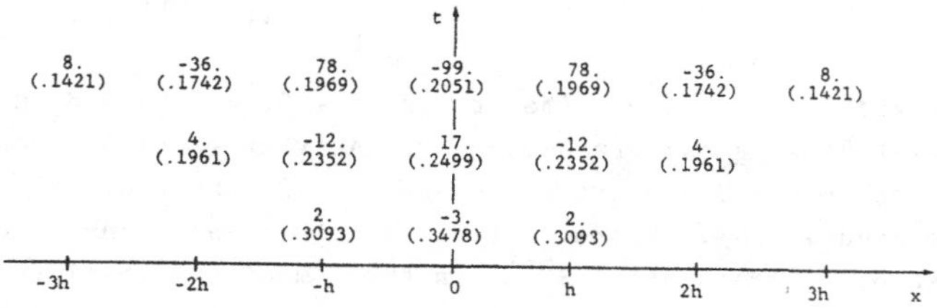

Figure 1.3b.

Comparison of (Exact Solution) with Approximate Solution for $\lambda = 2$.

For $\lambda > 1/2$, $(1-2\lambda) < 0$ and so $(1-2\lambda)\varepsilon < 0$ while $\lambda\varepsilon > 0$. Thus u_i^n is zero or alternates in sign with i for fixed n (typical values for $n = 1$ and 2 are depicted in Figure 1.4.). From which, using (1.1.3)

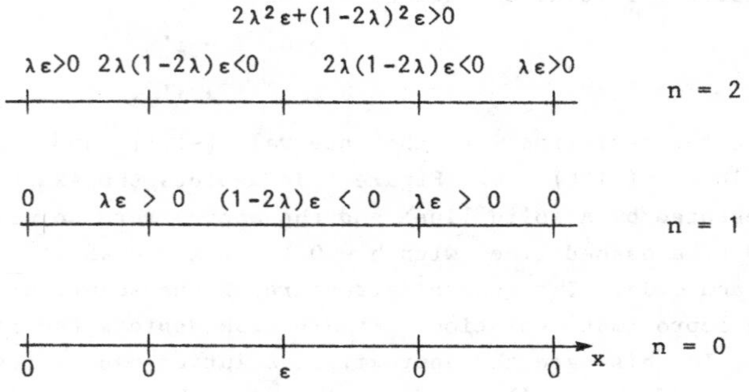

Figure 1.4

$$|u_i^{n+1}| = |\lambda u_{i+1}^n + (1-2\lambda)u_i^n + \lambda u_{i-1}^n|$$

$$= \lambda|u_{i+1}^n| + (2\lambda-1)|u_i^n| + \lambda|u_{i-1}^n| ,$$

where the last equality follows from the alternating sign property of u_i^n. Sum over all values of i and define $S^n = \sum_i |u_i^n|$

$$\sum_i |u_i^{n+1}| = \lambda\sum_i |u_{i+1}^n| + (2\lambda-1)\sum_i |u_i^n| + \lambda\sum_i |u_{i-1}^n|$$

or

$$\begin{aligned} S^{n+1} = \lambda S^n + (2\lambda-1)S^n + \lambda S^n &= (4\lambda-1)S^n \\ &= (4\lambda-1)^2 S^{n-1} \\ &\vdots \\ &= (4\lambda-1)^{n+1} S^0 \\ &= (4\lambda-1)^{n+1}\varepsilon \quad . \end{aligned}$$

Thus for fixed λ and fixed time t as $h \to 0$, $k \to 0$ and $n \to +\infty$, so that $(4\lambda-1)^{n+1}\varepsilon$ grows without bound. At time steps corresponding to $n = 1$ there are 3 nonzero terms and $n = 2$ there are 5 nonzero terms (see Figure 1.4). For a general n, there are $2n+1$ nonzero terms. Hence, at step $n+1$, S^{n+1} is the sum of the absolute value of $2n+3$ terms for which there must be some value of i such that

$$|u_i^{n+1}| > \frac{(4\lambda-1)^{n+1}\varepsilon}{2n+3},$$

which grows without bound as $h \to 0$. This leads to disaster.

If $f(x)$ is uniformly bounded, and $\lambda < 1/2$, then u_i^{n+1} is uniformly bounded. To see this, define $E^n = \max_i |u_i^n|$. Using (1.1.3),

$$|u_i^{n+1}| = |\lambda u_{i+1}^n + (1-2\lambda)u_i^n + \lambda u_{i-1}^n|$$
$$\leq \lambda|u_{i+1}^n| + (1-2\lambda)|u_i^n| + \lambda|u_{i-1}^n|,$$

since, if $\lambda \leq 1/2$, then $1 - 2\lambda \geq 0$. But then

$$\max_i|u_i^{n+1}| \leq \lambda \max_i|u_{i+1}^n| + (1-2\lambda)|u_i^n| + \lambda \max_i|u_{i-1}^n|$$

or

$$E^{n+1} \leq \lambda E^n + (1-2\lambda)E^n + \lambda E^n = E^n.$$

Thus $E^{n+1} \leq E^n \leq E^{n-1} \leq \ldots \leq E^0 = \max_i|f(ih)| \leq M$, which means $\max_i|u_i^{n+1}| \leq M$ or $|u_i^{n+1}| \leq M$.

This shows that if $\lambda \leq 1/2$ the approximate solution is uniformly bounded; but, however, does the solution u_i^n converge to the exact solution $v(ih,t)$ for some fixed time? To answer this make use of the Taylor formula with remainder. Expand $v(x,t)$ in a Taylor series in time about the point (ih,nk) and evaluate it at $t = (n+1)k$

$$v_i^{n+1} \equiv v(ih,(n+1)k) = v_i^n + k(\partial_t v)_i^n + \frac{k^2}{2}\partial_t^2 v(ih,t^*),$$

where $nk < t^* < (n+1)k$. Solving for $(\partial_t v)_i^n$

$$\partial_t v = \frac{v_i^{n+1} - v_i^n}{k} - \tfrac{1}{2}k\partial_t^2 v(ih,t^*). \tag{1.1.5}$$

Similarly, expand $v(x,t)$ in a Taylor series in space about the point (ih,nk) and evaluate at $x = (i-1)h$ and $x = (i+1)h$

$$v_{i-1}^n \equiv v((i-1)h,nk) = v_i^n - h(\partial_x v)_i^n + \frac{h^2}{2}(\partial_x^2 v)_i^n - \frac{h^3}{6}(\partial_x^3 v)_i^n + \frac{h^4}{24}\partial_x^4 v(x_1^*,nk),$$

$$v_{i+1}^n \equiv v((i+1)h,nk) = v_i^n + h(\partial_x v)_i^n + \frac{h^2}{2}(\partial_x^2 v)_i^n + \frac{h^3}{6}(\partial_x^3 v)_i^n, + \frac{h^4}{24}\partial_x^4 v(x_2^*,nk)$$

where $(i-1)h < x_1^* < ih$ and $ih < x_2^* < (i+1)h$, respectively. Adding these two series and solving for $(\partial_x^2 v)_i^n$

$$\partial_x^2 v = \frac{v_{i+1}^n - 2v_i^n + v_{i-1}^n}{h^2} - \frac{h^2}{24}(\partial_x^4 v(x_1^*,nk) + \partial_x^4 v(x_2^*,nk)). \tag{1.1.6}$$

Combining (1.1.5) and (1.1.6)

$$\partial_t v - a^2\partial_x^2 v = \frac{v_i^{n+1} - v_i^n}{k} - a^2\left(\frac{v_{i+1}^n - 2v_i^n + v_{i-1}^n}{h^2}\right) - \tau_i^n$$

or upon using (1.1.1a)

$$(1.1.7)\qquad \frac{v_i^{n+1} - v_i^n}{k} - a^2\left(\frac{v_{i+1}^n - 2v_{i-1}^n v_{i-1}^n}{h^2}\right) = \tau_i^n,$$

where $\tau_i^n = \frac{1}{2}k\partial_t^2 v(ih,t^*) - \frac{1}{24}a^2(h^2(\partial_x^4 v(x_i^*,nk) + \partial_x^4 v(x_2^*,nk))$ is called the <u>local truncation error</u> at (ih,nk). This suggests the following definition of consistency: A finite difference method is called <u>consistent</u> with a differential equation if the solution of the differential equation fails to satisfy the finite difference method by an arbitrarily small amount. This is not a precise definition because it does not tell what this small amount depends on. It will be made precise below.

Equation (1.1.7) may be written in a form that is more compatible with the finite difference method (1.1.3).

$$(1.1.8)\qquad v_i^{n+1} = \lambda v_{i+1}^n + (1-2\lambda)v + \lambda v_{i-1}^\nu + k\tau_i^n.$$

Define the error to be $e_i^n = v_i^n - u_i^n$, then by subtracting (1.1.3) from (1.1.8)

$$e_i^{n+1} = \lambda e_{i+1}^n + (1-2\lambda)e_i^n + \lambda e_{i-1}^n + k\tau_i^n$$

which implies

$$\begin{aligned}|e_i^{n+1}| &= |\lambda e_{i+1}^n + (1-2\lambda)e_i^n + \lambda e_{i-1}^n + k\tau_i^n| \\ &< \lambda|e_{i+1}^n| + (1-2\ )|e_i^n| + \lambda|e_{n-1}^n| + k|\tau_i^n|,\end{aligned}$$

since $\lambda < 1/2$. Let $\varepsilon^n = \max_i|e_i^n|$, the maximum error at time nk, and $\tau^n = \max_i|\tau_i^n|$, the maximum local truncation error at time nk, then

$$\begin{aligned}\max_i|e_i^{n+1}| < \lambda \max_i|e_{i+1}^n| + (1-2\lambda)\max_i|e_i^n| &+ \lambda \max_i|e_{i-1}^n| \\ &+ k \max_i|\tau_i^n|\end{aligned}$$

or

$$\varepsilon^{n+1} < \lambda\varepsilon^n + (1-2\lambda)\varepsilon^n + \lambda\varepsilon^n + k\tau^n = \varepsilon^n + k\tau^n.$$

Applying this inequality repeatedly gives

$$\begin{aligned}\varepsilon^{n+1} &< \varepsilon^n + k\tau^n < \varepsilon^{n-1} + k\tau^{n-1} + k\tau^n \\ &< \varepsilon^{n-2} + k\tau^{n-2} + k\tau^{n-1} + k\tau^n < \dots < \varepsilon + k\sum_{j=0}^{n}\tau^j.\end{aligned}$$

Thus

$$(1.1.9)\qquad \varepsilon^{n+1} < \varepsilon^0 + k\tau$$

where $\tau = \sum_{j=0}^{n}\tau^j$ is the sum of the maximum local truncation errors at times jk, $j = 0,1,\dots,n$. Ignoring the effects of round-off, $\varepsilon^0 = \max_i|v_i^0 - u_i^0| = 0$, as this is just the initial data. With this

$$(1.1.10)\qquad \varepsilon^{n+1} < k\tau.$$

For fixed $\lambda < 1/2$, as $h \to 0$, $k \to 0$ and so by the definition of τ_i^n, τ^n, and τ each approaches 0 as $h \to 0$ and $k \to 0$. Convergence is achieved since by (1.1.10) $\varepsilon^{n+1} \to 0$ as $h \to 0$.

Introduce the "big oh" notation. α is of order β, denoted by $\alpha = O(\beta)$, if there is a positive number K such that $|\alpha| \leq K|\beta|$. The local truncation error τ_i^n and τ in (1.1.10) is $O(k) + O(h^2)$. One thing must now be mentioned that will be given greater attention later. If the truncation error τ_i^n is written as

$$\tau_i^n = O(k) + O(h^2)$$

this can be very misleading because hidden in the $O(k)$ term is $\partial_t^2 v(ih,t^*)$ and in the $O(h^2)$ term is $\partial_x^4 v(x_1^*,nk) + \partial_x^4 v(x_2^*,nk)$. In general, it is not known how large these derivatives are. So even if k and h are small so that $O(k) + O(h^2)$ would seem small, one or both of the derivatives may be very large resulting in a much larger truncation error.

I.2. Basic Difference Operators

Consider a function defined on a discrete set of points $x = ih$, $f_i = f(ih)$. Such a function is called a grid function. Assume that $f_i = 0$ for $|i| > I$, where I is some large natural number. Let f denote

$$f = (\ldots,f_{-2},f_{-1},f_0,f_1,f_2,\ldots) = \{f_i\}.$$

Consider two grid functions $f = \{f_i\}$ and $g = \{g_i\}$. Define the discrete inner product of f and g, (f,g) by

(1.2.1) $$(f,g) = \sum_i f_i g_i h$$

where h is a scaling factor. The Euclidean or ℓ_2-norm is defined to be

(1.2.2) $$\|f\|_2 = \left(\sum_i f_i^2 h\right)^{1/2} = \sqrt{(f,f)}\ .$$

This scaling factor h is important. For without it, as the grid is refined (that is, as h is reduced), the sum over i will involve a greater number of terms and the ℓ_2-norm of f would grow without bound.

Define operators acting on the grid function $f = \{f_i\}$ as follows:

the forward shift operator S_+

(1.2.3) $$(S_+f)_i = f_{i+1},$$

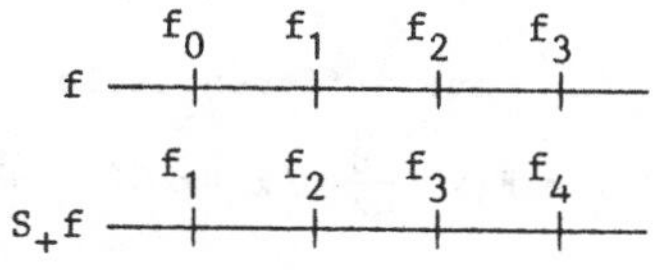

Figure 1.7

the <u>backward shift operator</u> S_-

(1.2.4) $$(S_-f)_i = f_{i-1},$$

the <u>identity operator</u> I

(1.2.5) $$(If)_i = f_i,$$

the <u>forward difference operator</u> D_+

(1.2.6) $$(D_+f)_i = \frac{f_{i+1} - f_i}{h},$$

the <u>backward difference operator</u> D_-

(1.2.7) $$(D_-f)_i = \frac{f_i - f_{i-1}}{h},$$

and the <u>centered difference operator</u> D_0

(1.2.8) $$(D_0f)_i = \frac{f_{i+1}-f_{i-1}}{2h}.$$

Observe that

$$D_+ = (S_+ - I)/h,$$

$$D_- = (I - S_-)/h,$$

and

$$D_0 = \frac{1}{2h}(S_+ - S_-) = \frac{1}{2}(D_+ + D_-).$$

Often for the sake of convenience $(D_+f)_i$ is written as D_+f and so on. Some further properties of these operators are listed here:

(1.2.9) $$S_+S_- = S_-S_+ = I,$$

which implies that $S_- = (S_+)^{-1}$,

(1.2.10) $$S_-D_+ = D_-,$$

and similarly

(1.2.11) $$S_+D_- = D_+.$$

Property (1.2.10) follows from

$$S_-D_+f = S_-(D_+f) = S_-\left(\frac{f_{i+1} - f_i}{h}\right) = \frac{S_-f_{i+1} - s_-f_i}{h} = \frac{f_i - f_{i-1}}{h} = D_-f.$$

Certain analogues of properties of functions defined on the real line shall be derived.

1) Suppose $F(x)$ and $G(x)$ are differentiable,

$$\frac{d}{dx}(F(x)G(x)) = F(x)\frac{d}{dx}G(x) + G(x)\frac{d}{dx}F(x),$$

which is the product rule.

To obtain the discrete analogue of the product rule, let $f = \{f_i\}$ and $g = \{g_i\}$ and define $fg = \{f_i g_i\}$.

$$(D_+fg)_i = \frac{f_{i+1}g_{i+1} - f_i g_i}{h} = \frac{f_{i+1}g_{i+1} - f_i g_{i+1} + f_i g_{i+1} - f_i g_i}{h}$$

$$= \frac{f_{i+1}g_{i+1} - f_i g_{i+1}}{h} + \frac{f_i g_{i+1} - f_i g_i}{h}$$

$$= g_{i+1}(D_+f)_i + f_i(D_+g)_i$$

$$= (S_+g)_i(D_+f)_i + f_i(D_+g)_i$$

$$= S_+(gD_-f)_i + f_i(D_+g)_i ,$$

hence,

(1.2.12) $$D_+fg = S_+(gD_-f) + fD_+g .$$

2) Suppose $F(x)$ and $G(x)$ are differentiable. The discrete analogue of integration by parts

$$\int_a^b F(x) \frac{d}{dx} G(x)dx = F(x)G(x)\Big|_a^b - \int_a^b G(x) \frac{d}{dx} F(x)dx$$

shall be derived. To this end, consider the sum

$$\sum_{i=0}^{N-1} (D_+fg)_i h = \frac{f_1 g_1 - f_0 g_0}{h} h + \frac{f_2 g_2 - f_1 g_1}{h} h + \cdots$$

$$+ \frac{f_N g_N - f_{N-1}g_{N-1}}{h} h$$

$$= f_N g_N - f_0 g_0$$

due to cancellation in pairs. However, since $D_+fg = S_+(gD_-f) + fD_+g$, by summing

$$f_N g_N - f_0 g_0 = \sum_{i=0}^{N-1} S_+(gD_-f)_i h + \sum_{i=0}^{N-1} (fD_+g)_i h$$

$$= \sum_{i=1}^{N} (gD_-f)_i h + \sum_{i=0}^{N-1} (fD_+g)_i h .$$

Assume that f_i (and/or g_i) $= 0$ for $i < 0$ and $i > N$, then

$$0 = \sum_i (gD_-f)_i h + \sum_i (fD_+g)_i h$$

or by the definition of the discrete inner product,

$$0 = (g,D_-f) + (f,D_+g)$$

or

$$(1.2.13) \qquad (g, D_- f) = -(D_+ g, f) \ .$$

3) Suppose $F(x)$ is continuously differentiable on the open interval $(0,1)$ and $F(0) = 0$, then Sobolev's inequality is

$$\| F(x) \|_2 \leq \| \frac{d}{dx} F(x) \|_2 \ ,$$

where $\| G(x) \|_2 = (\int_0^1 G^2(x) dx)^{1/2}$ in this case. To derive the discrete analogue of the Sobolev inequality, assume that $f_0 = 0$ and $i \leq N$ where N is a positive integer with $Nh = 1$. Write f_i in the form (due to cancellation by pairs and $f_0 = 0$)

$$f_i = \frac{f_i - f_{i-1}}{h} h + \frac{f_{i-1} - f_{i-2}}{h} h + \ldots + \frac{f_2 - f_1}{h} h + \frac{f_1 - f_0}{h} h$$

$$= \sum_{j=0}^{i-1} (D_+ f)_j h$$

and, hence,

$$|f_i| = \left| \sum_{j=0}^{i-1} (D_+ f)_j h \right| \leq \sum_{j=0}^{i-1} |(D_+ f)_j| h \leq \sum_{j=0}^{N-1} |(D_+ f)_j| h$$

Applying Schwartz's inequality, $|(x,y)| \leq \|x\|_2 \|y\|_2$, with $x = |D_+ f|$ and $y = 1$,

$$|f_i| \leq \left(\sum_{j=0}^{N-1} (D_+ f)_i^2 \, h \right)^{1/2} \left(\sum_{j=0}^{N-1} 1 \, h \right)^{1/2} .$$

Since $\sum_{j=0}^{N-1} h = Nh = 1$,

$$|f_i| \leq \left(\sum_{j=0}^{N-1} (D_+ f)_j^2 \, h \right)^{1/2} .$$

Squaring both sides, multiplying by h $(= 1/N)$, and summing from 0 to $N - 1$,

$$\sum_{i=0}^{N-1} |f_i|^2 h \leq h \sum_{i=0}^{N-1} (\sum_{j=0}^{N-1} (D_+ f)_j^2 h)$$

$$= hN (\sum_{j=0}^{N-1} (D_+ f)_j^2 h)$$

$$= \sum_{j=0}^{N-1} (D_+ f)_j^2 h \ .$$

So if $f_i = 0$ for $i \leq 0$ and $i \leq N$,

$$\sum_i f_i^2 h \leq \sum_i (D_+ f)_i^2 h$$

or

$$(\sum_i f_i^2 h)^{1/2} \leq (\sum_i (D_+f)_i^2 h)^{1/2}.$$

This gives

(1.2.14) $$\|f\|_2 \leq \|D_+f\|_2,$$

and similarly

(1.2.15) $$\|f\|_2 \leq \|D_-f\|_2.$$

Using the triangle inequality

$$\|D_+f\|_2 = \left\|\frac{S_+f + If}{h}\right\|_2 = \frac{1}{h}\|S_+f + If\|_2 \leq \frac{1}{h}(\|S_+f\|_2 + \|f\|_2).$$

Since $\|S_+f\|_2 = (\sum_i (S_+f)_i^2 h)^{1/2} = (\sum_i f_{i+1}^2 h)^{1/2} = (\sum_i f_i^2 h)^{1/2} = \|f\|_2,$

(1.2.16) $$\|D_+f\|_2 \leq \frac{2}{h}\|f\|_2$$

and similarly,

$$\|D_+f\|_2 \leq \frac{2}{h}\|f\|_2 .$$

Observe that

$$(D_-D_+f) = D_-(\frac{f_{i+1} - f_i}{h}) = \frac{D_-f_{i+1} - D_-f_i}{h}$$

$$= \frac{\frac{f_{i+1} - f_i}{h} - \frac{f_i - f_{i-1}}{h}}{h} = \frac{f_{i+1} - 2f_i + f_{i-1}}{h^2}$$

and similarly

$$(D_+D_-f)_i = \frac{f_{i+1} - 2f_i + f_{i-1}}{h^2} .$$

Return now to the heat equation considered in Section I.1 with boundary conditions

$$\partial_t v = a^2 \partial_x^2 v \quad t > 0 , \quad 0 \leq x \leq 1$$

$$v(x,0) = f(x)$$

$$v(0,t) = 0 \quad \text{and} \quad v(1,t) = 0.$$

Divide the unit interval into N points with $h = \frac{1}{N}$. Let u_i^n denote the approximate solution as described in the first section. Consider the method (1.1.2)

$$u_i^{n+1} = u_i^n + \frac{a^2k}{h^2}(u_{i+1}^n - 2u_i^n - 2u_i^n + u_{i+1}^n)$$

or

(1.2.17) $$u_i^{n+1} = u_i^n + a^2kD_+D_-^n u_i$$

with $u_0 = 0$ and $u_N = 0$ (the boundary conditions). The question of stability arose in the first section, and it was seen that the method was stable, (that is, in an intuitive sense the solution is bounded) for $\lambda = a^2k/h^2 \le 1/2$. How was this condition arrived at? The inequalities derived in this section will be used to answer the question: Is $\|u^n\|_2$ bounded and, if so, for what values of k and h? The method used is called the energy method which will be discussed in detail in later chapters. Take the discrete inner product of (1.2.17), with itself

$$\begin{aligned}(u^{n+1},u^{n+1}) &= (u^n + a^2kD_+D_-u^n,\ u^n + a^2kD_+D_-u^n) \\ &= (u^n,u^n) + (u^n,a^2kD_+D_-u^n) + (a^2kD_+D_-u^n,u^n) \\ &= (u^n,u^n) + 2a^2k(u^n,D_+D_-u^n) + a^4k\ (D_+D_-u^n,D_+D_-u^n)\end{aligned}$$

or

$$\|u^{n+1}\|_2^2 = \|u^n\|_2^2 + 2a^2k(u^n,D_+D_-u^n) + a^4k^2\|D_+D_-u^n\|_2^2 .$$

Applying discrete integration by parts (1.2.13) where $g = D_-u^n$ and $f = u^n$,

$$\begin{aligned}\|u^{n+1}\|_2^2 &= \|u^n\|_2^2 - 2a^2k(D_-u^n,D_-u^n) + a^4k^2\|D_+D_-u^n\|_2^2 \\ &= \|u^n\|_2^2 - 2a^2k\|D_-u^n\|_2^2 + a^4k^2\|D_+D_-u^n\|_2^2 .\end{aligned}$$

Using inequality (1.2.16) with $f = D_-u^n$,

$$\|D_+D_-u^n\|_2 \le \frac{2}{h}\|D_-u^n\|_2$$

and by applying inequality (1.2.16) again, we have

(1.2.18) $$\|D_+D_-u^n\|_2^2 \le \frac{4}{h^2}\|u^n\|_2^2 .$$

Using $\|D_+D_-u^n\|_2^2 \le \frac{4}{h^2}\|D_-u^n\|_2^2$,

$$\begin{aligned}\|u^{n+1}\|_2^2 &\le \|u^n\|_2^2 - 2a^2k\|D_-u^n\|_2^2 + \frac{4a^4k^2}{h^2}\|D_-u^n\|_2^2 \\ &= \|u^n\|_2^2 + \left(\frac{4a^4k^2}{h^2} - 2a^2k\right)\|D_-u^n\|_2^2 ,\end{aligned}$$

so that, if $\frac{4a^4k^2}{h^2} - 2k \le 0$ or $\frac{a^2k}{h^2} \le \frac{1}{2}$, then the solution of (1.2.17) is bounded, that is, $\|u^{n+1}\|_2 \le \|u^n\|_2$.

I.3. Lax's Theorem

We are now in a position to make precise the idea of stability. Consider an initial value problem

(1.3.1) $$\partial_t v = P(\partial_x)v\ ,\quad t > 0.$$

and

(1.3.2) $$v(x,0) = f(x)\ ,$$

where $P(\partial_x)$ is a polynomial in ∂_x. For example, if $P(z) = z^2 + 5z + 1$, then $P(\partial_x) = \partial_x^2 + 5\partial_x + 1$ and (1.3.1) becomes $\partial_t v = \partial_x^2 v + 5\partial_x v + v$.

<u>Definition</u>. The initial value problem (1.3.1)-(1.3.2) is said to be well-<u>posed</u> in a norm $\|\cdot\|$ if a solution exists, is unique, and depends continuously on the initial data, that is, there exist constants C and α such that

$$\|v(x,t)\| \leq Ce^{\alpha t}\|f(t)\|.$$

Let $u^n = \{u_i^n\}$ denote the grid function at time $t = nh$. Let $u_i^0 = f(ih)$ denote the discrete initial conditions. The initial value problem (1.3.1)-(1.3.2) shall be approximated by a finite difference method of the form

$$u^{n+1} = Qu^n \ , \ n \geq 0 \tag{1.3.3}$$

where Q is a polynomial in the forward and backward shift operators S_+ and S_-, that is,

$$Q = Q(S_+,S_-).$$

This is the form of a typical finite difference method. For example, the finite difference method (1.1.3) is

$$\begin{aligned} u^{n+1} &= \lambda u_{i+1}^n + (1 - 2\lambda)u_i^n + \lambda u_{i-1}^n \\ &= \lambda S_+ u_i^n + (1 - 2\lambda)u_i^n + \lambda S_- u_i^n \\ &= (\lambda S_+ + (1 - 2\lambda)I + \lambda S_-)u_i^n \ . \end{aligned}$$

Let $Q(x,y) = \lambda x + \lambda y + (1 - 2\lambda)$ then $Q(S_+,S_-) = \lambda S_+ + \lambda S_- + (1 - 2\ )I$ and (1.1.3) becomes

$$u^{n+1} = Qu^n$$

where

$$Q(S_+,S_-) = \lambda S_+ + (1 - 2\lambda)I + \lambda S_-.$$

<u>Definition</u>. The finite difference method (1.3.3) is called <u>stable</u> if there exist constants K and β and some norm $\|\cdot\|$ such that,

$$\|u^n\| \leq Ke^n\|u^0\| = Ke^{\beta t}\|u^0\|$$

where $t = nk$, and K and β are independent of h and k.

Observe that the definition of stability is similar to the definition of well-posedness for the continuous case. This definition of stability allows growth with time but <u>no</u> growth with the number of time steps.

<u>Definition</u>. A finite difference method is called <u>unconditionally stable</u> if it is stable for any time step k and grid spacing h.

The definition of consistency given in Section I.1 will now be made precise.

Definition. A finite difference method (1.3.3) is consistent up to time T in a norm $\|\cdot\|$ with equation (1.3.1) if the actual solution v to the initial value problem (1.3.1)-(1.3.2) "almost" satisfies the finite difference method (1.3.3), that is,

$$v^{n+1} = Qv^n + k\tau^n, \tag{1.3.5}$$

where $\|\tau^n\| < \tau(h)$, $nk < T$ and $\tau(h) \to 0$ as $k \to 0$. Here it is assumed that h is defined in terms of k and goes to 0 with k. v^n, denotes $v(ih,nk)$, the exact solution evaluated at the grid point (ih,nk). τ^n is the local truncation error at time nk.

Definition. The finite difference method (1.3.3) is accurate of order (p,q) if

$$\|v^{n+1} - Qv^n\| = k(O(h^p) + O(k^q)). \tag{1.3.6}$$

In this case $O(h^p) + O(k^q)$ is called the local truncation error.

The truncation error associated with the finite difference method

$$u^{n+1} = u^n + a^2kD_+D_-u^n$$

is $\tau^n = O(h^2) + O(k)$ (see Section I.1), so the method is accurate of order (2.1).

Definition. The finite difference method (1.3.3) is convergent in a norm $\|\cdot\|$, if

$$\|v(x,t) - u\|_i^n \to 0 \quad \text{as} \quad h,k \to 0$$

It is convergent of order (p,q) in a norm $\|\cdot\|$, if $\|v(x,t) - u_i^n\| = O(h^p) + O(k^q)$.

All of these ideas are connected together in the following theorem due to Lax (sometimes called the Lax equivalence theorem).

Theorem 1.3 (Lax). If a finite difference method is linear, stable, and accurate of order (p,q), then it is convergent of order (p,q).

Before proving this theorem, there are a few comments that should be made.

Convergent of order (p,q) gives the rate of convergence as h and $k \to 0$. So, by Lax's theorem, the more accurate the method the faster the rate of convergence.

It is very important that the finite difference method be linear. This corresponds to a linear partial differential equation (1.3.1). For nonlinear equations, things become much more complicated as we shall see in later chapters.

Proof. Using the finite differences method (1.3.3)

$$u^n = Qu^{n-1} = Q(Qu^{n-2}) \equiv Q^2u^{n-2} = Q^2(Qu^{n-3}) \equiv Q^3u^{n-3}$$
$$= \dots = Q^{n-1}u^1 = Q^{n-1}(Qu^0) \equiv Q^nu^0 \ ,$$

where by Q^j means the finite difference operator Q applied j-times. Using this result ($u^n = Q^n u^0$) and the definition of stability

$$\|u^n\| = \|Q^n u^0\| \le Ke^{\beta t}\|u^0\|$$

or

$$\frac{\|Q^n u^0\|}{\|u^0\|} \le Ke^{\beta t} .$$

Furthermore, by definition of $\|Q^n\|$,

$$\|Q^n\| = \max_{\|u^0\| \ne 0} \frac{\|Q^n u^0\|}{\|u^0\|} \le Ke^{\beta t}.$$

Let $v(x,t)$ denote the solution of the initial-value problem (1.3.1)-(1.3.2); then by the definition of accuracy of order (p,q), $v^n = Qv^{n-1} + k(O(h^p) + O(k^q))$. Define the error at the n-th time step by $w^n = u^n - v^n$, where $w^0 = u^0 - v^0 = 0$, ignoring the effects of round-off error. Then by repeated application of (1.3.3) and the definition of accuracy of order (p,q),

$$\begin{aligned} w^n &= Qw^{n-1} + k(O(h^p) + O(k^q)) \\ &= Q^2w^{n-2} + Q(k(O(h^p) + O(k^q))) + k(O(h^p) + O(k^q)) \\ &= \ldots \\ &= Q^n w^0 + k \sum_{j=0}^{n-1} Q^j(O(h^p) + O(k^q)) \\ &= k \sum_{j=0}^{n-1} Q^j(O(h^p) + O(k^q)). \end{aligned}$$

This gives

$$\begin{aligned} \|w^n\| &\le k \sum_{j=0}^{n-1} \|Q^j\|(O(h^p) + O(k^q)) \\ &\le nkKe^{\beta t}(O(h^p) + O(k^q)) \\ &= tKe^{\beta t}(O(h^p) + O(k^q)) \\ &\le Ke^{(\beta+1)t}(O(h^p) + O(k^q)) \\ &= O(h^p) + O(k^q) . \end{aligned}$$

Thus, the method is convergent of order (p,q).

I.4. The Fourier Method

We shall begin with a brief review of some basic facts about Fourier series and the complex Fourier series. Consider real-valued, integrable functions $f(x)$ that are 2π-periodic, that is, $f(x + 2\pi) = f(x)$ for every x.

Define

$$a_\xi = \frac{1}{\pi} \int_0^{2\pi} f(x) \cos \xi x \, dx, \quad a_0 = \frac{1}{2\pi} \int_0^{2\pi} f(x) \, dx$$

and

$$b_\xi = \frac{1}{\pi} \int_0^{2\pi} f(x) \sin \xi x \, dx.$$

If $S_n(x)$ denotes the partial sum

$$S_n(x) = a_0 + \sum_{\xi=1}^{n} (a_\xi \cos \xi x + b_\xi \sin \xi x)$$

the Fourier series of $f(x)$ is defined by

$$\lim_{n\to\infty} S_n(x) = a_0 + \sum_{\xi=1}^{\infty} (a_\xi \cos (\xi x) + b_\xi \sin (\xi x)).$$

The complex form is much more convenient for our purposes, so consider the expansion of $f(x)$ in terms of a complex function $e^{i\xi x}$, where $i = \sqrt{-1}$ and ξ is any integer. The set of functions $\{\frac{1}{\sqrt{2\pi}} e^{i\xi x} : \xi$ is an integer$\}$ is orthonormal on $[0,2\pi)$, that is

$$\frac{1}{2\pi} \int_0^{2\pi} e^{i\xi_1 x} \overline{e^{i\xi_2 x}} \, dx = \frac{1}{2\pi} \int_0^{2\pi} e^{i(\xi_1-\xi_2)x} \, dx$$

$$= \begin{cases} 0 \ , & \xi_1 \neq \xi_2 \\ 1 \ , & \xi_1 = \xi_2 \ , \end{cases}$$

where the complex conjugate of $e^{i\xi x}$, denoted by $\overline{e^{i\xi x}}$, is $e^{-i\xi x}$. The complex Fourier series for $f(x)$ is defined to be

(1.4.1) $$f(x) = \frac{1}{\sqrt{2\pi}} \sum_{\xi \varepsilon Z} \hat{f}(\xi) e^{i\xi x}$$

where Z denotes the set of all integers. The coefficient $\hat{f}(\xi)$ is obtained by multiplying both sides of (1.4.1) by $e^{i\xi x}$ and integrating with respect to x between 0 and 2π. Since these exponential functions are orthonormal

(1.4.2) $$\hat{f}(x) = \frac{1}{\sqrt{2\pi}} \int_0^{2\pi} f(x) e^{i\xi x} \, dx$$

$\hat{f}(\xi)$ is called the Fourier transform of $f(x)$.

Some properties are:

1) If $f \varepsilon C^{p+1}$ (that is, f has $p+1$ continuous derivatives), then by applying integration by parts to (1.4.2)

$$\hat{f}(\xi) = O(\frac{1}{|\xi|^{p+1}}) \ ,$$

that is, the smoothness of f causes $\hat{f}(\xi)$ to die out quickly for large $|\xi|$.

2) By orthonormality, we have Parseval's relation

$$\int_0^{2\pi} |f(x)|^2 \, dx = \sum_{\xi \sim Z} |f(\xi)|^2 \ .$$

The reason for the interest in these periodic exponential functions $e^{-i\xi x}$, where ξ is an integer, is that they will be seen to be "eigenfunctions", for the forward shift operator S_+ and, therefore, of all finite difference operators made up of linear combinations of powers of S_+.

The discrete Fourier transform will be used. In this approach the roles of f and $\hat{f}$, and $[0,2\pi)$ and Z in (1.4.1) and (1.4.2) are interchanged. Consider the grid function $u = \{u_j\}$, define the discrete Fourier transform of u,

(1.4.4) $$\hat{u}(\xi) = \sum_j u_j e^{ij\xi} , \quad 0 \leq \xi < 2\pi .$$

Thus, $\hat{u}$ is the function whose Fourier coefficients are the u_j's.

If the forward shift operator S_+ is applied to u, the grid function $S_+u = \{u_{j+1}\}$ is obtained. Taking the discrete Fourier transform of S_+u yields

$$\begin{aligned}\widehat{(S_+u)}(\xi) = \sum_j u_{j+1} e^{ij\xi} &= \sum_j u_j e^{i(j-1)\xi} \\ &= \sum_j u_j e^{ij}\ e^{-i\xi} \\ &= e^{-i\xi} \sum_j u_j e^{ij\xi} \\ &= e^{-i\xi}\ \hat{u}(\xi) .\end{aligned}$$

$e^{-i\xi}$ is called the symbol of S_+. Thus the operation of S_+ in the discrete Fourier transform spaces corresponds to multiplying by the function $e^{-i\xi}$. This leads to the heuristic diagram depicted in (Figure 1.8).

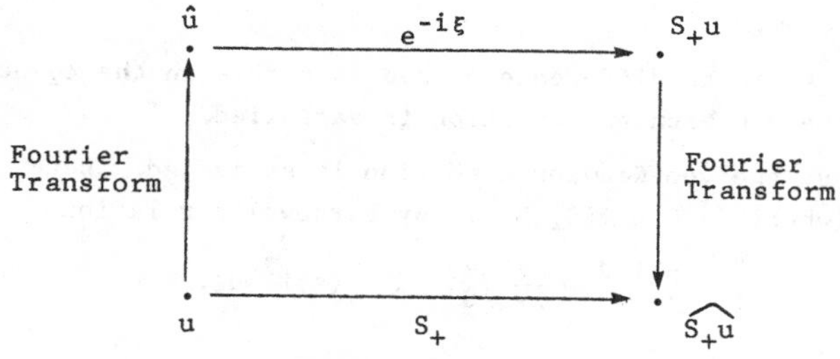

Figure 1.8

Similarly,

$$\widehat{(S_-u)}(\xi) = e^{i\xi}\ \hat{u}(\xi)$$

so $e^{i\xi}$ is the symbol of S_-. From this it follows that if $u^{n+1} = Q(S_+,S_-)u^n$, then

(1.4.5) $$u^{n+1} = Q(e^{-i\xi}, e^{i\xi})\, u^n$$

that is, applying the finite difference operator $Q(S_+,S_-)$ to u^n and taking the discrete Fourier transform is the same as multiplying the discrete Fourier transform u^n by $Q(e^{-i\xi},e^{i\xi})$. The symbol* of the finite difference method (1.3.3) is $\rho(\xi) = Q(e^{-i\xi},e^{i\xi})$.

As an example, consider the finite difference method (1.1.3)

$$\begin{aligned} u_i^{n+1} &= \lambda u_{i+1}^n + (1-2\lambda)u_i^n + \lambda u_{i-1}^n \\ &= (\lambda S_+ + (1-2\lambda)I + \lambda S_-)u_i^n \\ &\equiv Q(S_+,S_-)u_i^n . \end{aligned}$$

The symbol $\rho(\xi)$ is obtained by taking the discrete Fourier transform of both sides

$$\hat{u}^{n+1}(\xi) = \rho(\xi)\hat{u}^n(\xi)$$

where

$$\begin{aligned} \rho(\xi) = Q(e^{-i\xi},e^{i\xi}) &= \lambda e^{-i\xi} + (1-2\lambda) + \lambda e^{i\xi} \\ &= 1-2\lambda + 2\lambda\left(\frac{e^{i\xi}+e^{-i\xi}}{2}\right) \\ &= 1-2\lambda(1 - \cos\xi), \end{aligned}$$

for $0 \leq \xi < 2\pi$.

The symbol $\rho(\xi)$ is said to satisfy the von Neumann condition if there exists a constant $C > 0$ (independent of k, h, n, and ξ) such that

(1.4.6) $$|\rho(\xi)| \leq 1 + Ck$$

for $0 \leq \xi < 2\pi$ and where k denotes the time step.

The importance of the von Neumann condition can be seen from the following theorem.

Theorem 1.4. A finite difference method is stable in the ℓ_2-norm if and only if the von Neumann condition is satisfied.

Proof. Suppose the von Neumann condition is satisfied. Let $u^{n+1} = Qu^n$, (where , $Q = Q(S_+,S_-)$. By Parseval's relation

$$\sum_j (u_j^{n+1})^2 = \frac{1}{2\pi}\int_0^{2\pi} |\hat{u}^{n+1}(\xi)|^2\, d\xi.$$

and upon multiplying by h

$$\begin{aligned} \|u^{n+1}\|_2^2 \equiv \sum_j (u_j^{n+1})^2 h &= \frac{h}{2\pi}\int_0^{2\pi} |\hat{u}^{n+1}(\xi)|^2\, d\xi, \\ &= \frac{h}{2\pi}\int_0^{2\pi} |\rho(\xi)|^2 |\hat{u}^n(\xi)|^2\, d\xi . \end{aligned}$$

---

* $\rho(\xi)$ is also called the amplification factor.

Since the von Neumann condition $(|\rho(\xi)| \le 1 + Ck$ for some $C > 0)$ is satisfied

$$\|u^{n+1}\|_2^2 \le \frac{h}{2\pi}(1 + Ck)^2 \int_0^{2\pi} |\hat{u}^n(\xi)|^2 \, d\xi$$
$$= (1 + Ck)^2 \, h \sum_j (u_j^n)^2 \, .$$

By Parseval's relation again,

$$\|u^{n+1}\|_2^2 \le (1 + Ck)^2 \sum_j (u_j^n)^2 \, h = (1 + Ck)^2 \|u^n\|_2^2$$

or $\|u^{n+1}\|_2 \le (1 + Ck)\|u^n\|_2$. Since $1 + Ck \le e^{Ck}$,

$$\|u^{n+1}\|_2 \le e^{Ck}\|u^n\|_2$$
$$\le e^{Ck}(e^{Ck}\|u^{n-1}\|_2) = e^{C2k}\|u^{n-1}\|_2$$
$$\le \ldots \le e^{C(n+1)k}\|u^0\|_2 = e^{Ct}\|u^0\|_2 ,$$

where $t = (n + 1)k$. So by the definition of stability with $K = 1$ and $\beta = C$ in (1.3.4), the finite difference method is stable.

Conversely, we shall show that if the von Neumann condition is not satisfied, then the finite difference method is not stable. To this end, suppose for each $C > 0$ there exists a number ξ_C, $0 \le \xi_C < 2\pi$, such that $|\rho(\xi_C)| > 1 + Ck$. The symbol $\rho(\xi)$ is a continuous function of ξ so there exists an interval I_C containing ξ_C such that $|\rho(\xi)| > 1 + Ck$ for every value of ξ in that interval (see Figure 1.9).

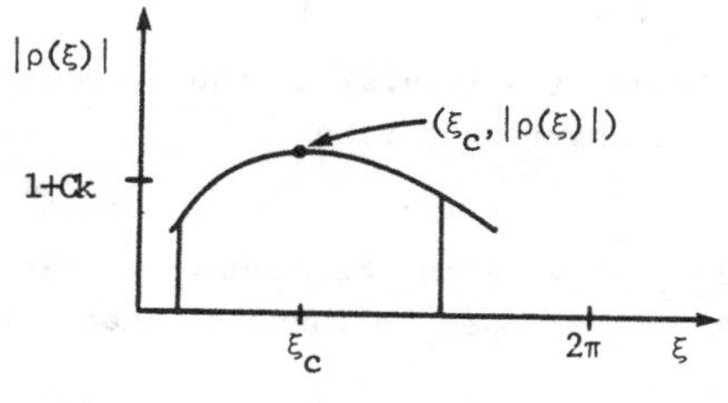

Figure 1.9

Consider the initial data u_j^0 so that the discrete Fourier transform of u_j^0, $\hat{u}^0(\xi)$ is 0 outside the interval I_C. Thus, for all $0 \le \xi < 2\pi$ with $\hat{u}^0(\xi) \ne 0$, $|\rho(\xi)| > 1 + Ck$. From this, it follows that

$$\hat{u}^n(\xi) = \rho(\xi)\hat{u}^{n-1}(\xi) = \rho^2(\xi)\hat{u}^{n-2}(\xi)$$
$$= \ldots = \rho^n(\xi)\hat{u}^n(\xi) > (1 + Ck)^n\hat{u}^0(\xi) \, .$$

Since $|\rho(\xi)| > 1 + Ck$

$$\|u^n\|_2 > (1 + Ck)^n \|u^0\|_2,$$

where as in the first part of the theorem Perseval's relation has been used twice. $u^n = Q^n u^0$ so that the inequality may be written as

$$\|Q^n u^0\|_2 > (1 + Ck)^n \|u^0\|_2$$

or

$$\frac{\|Q^n u^0\|_2}{\|u^0\|_2} > (1 + Ck)^n .$$

Using the definition of the matrix norm
$\|Q^n\|_2 = \max_{\|u\|_2 \neq 0} \frac{\|Q^n u\|_2}{\|u\|_2} > \frac{\|Q^n u^0\|_2}{\|u^0\|_2}$, so that $\|Q^n\|_2 > (1 + Ck)^n$. Thus $\|Q^n\|_2$ is unbounded for all constants $C > 0$, which implies the finite difference method is unstable.

To see how the von Neumann condition can be used to analyze the stability of a finite difference method, consider (1.1.3) and its symbol

$$\rho(\xi) = 1 - 2\lambda(1 - \cos \xi), \quad 0 \leq \xi < 2\pi.$$

For $0 \leq \xi < 2\pi$, $0 \leq 1 - \cos \xi \leq 2$, so that

$$-2 \leq -(1 - \cos \xi) \leq 0 ,$$

$$1 - 4\lambda \leq 1 - 2\lambda\,(1 - \cos \xi) \leq 1 ,$$

or

$$1 - 4\lambda \leq \rho(\xi) \leq 1 .$$

In practice, the inequality (1.4.6) in the von Neumann condition will be replaced by the stronger condition

$$(1.4.7) \qquad |\rho(\xi)| \leq 1.$$

One side of the inequality (1.4.7) in the example is satisfied, $\rho(\xi) \leq 1$. Restrictions on λ must be found so that $-1 \leq \rho(\xi)$. To this end, require

$$-1 \leq 1 - 4\lambda \leq \rho(\xi)$$

so that the von Neumann condition is satisfied if $\lambda \leq 1/2$.

I.5. References

1. Lax, P. D. and R. D. Richtmyer, Survey of the stability of Linear Finite Difference Equations, Comm. Pure Appl. Math., 9, 267 (1956).
2. Richtmyer, R. D. and K. W. Morton, Difference Methods for Initial-Value Problems, 2nd Ed., J. Wiley (Interscience), New York (1967).

II. PARABOLIC EQUATIONS

II.1. Introduction.

The problem of the flow of heat along a rod whose temperature depends only on the x-coordinate and time t leads to the parabolic partial differential equation

$$c\rho \partial_t v = \partial_x (k \partial_x v) + \phi_1(x,t) .$$

In this equation, $v(x,t)$ denotes the temperature, k denotes the thermal conductivity of the rod, ρ denotes the density, and c denotes the specific heat. The function $\phi_1(x,t)$ is a source term, that is, it is the strength of heat sources located in the rod. In general, the coefficients ρ, c, and k depend on x and t. Furthermore, k and c may depend on the temperature v which makes the equation nonlinear.

Consider the case where ρ, c, and k are constant. This leads to the prototype equation

(2.1.1) $$\partial_t v = a^2 \partial_x^2 v + \phi(x,t)$$

where

$$a^2 = \frac{k}{c\rho} \quad \text{and} \quad \phi(x\phi t) = \frac{1}{c\rho}\phi_1(x,t) .$$

If there are no sources of heat, then $\phi(x,t) = 0$ and equation (2.1.1) reduces to the homogeneous equation

(2.1.2) $$\partial_t v = a^2 \partial_x^2 v .$$

Suppose we consider a thin rod, whose lateral surface is not insulated. In this case, the heat loss through the surface is proportional to the difference in temperature between the rod and the ambient temperature, (which, without a loss of generality, we shall take to be 0.) This gives rise to the following parabolic partial differential equation

$$c\rho \partial_t v = \partial_x (k \partial_x v) + \phi_1(x,t) - \frac{hp}{\sigma} v ,$$

where h is the coefficient of external thermal conductivity, p is the perimeter of a cross section of the rod normal to the x-axis, and σ is the area of such a cross section.

Thus in the case where ρ, c, and k are constants and there are no heat sources, that is, $\phi_1(x,t) = 0$, the equation takes the form

$$\partial_t v = a^2 \partial_x^2 v - bv , \tag{2.1.3}$$

where a^2 is as before and

$$b = \frac{hp}{c\rho\sigma} .$$

Introduction of the change of variable

$$v' = e^{-bt} v$$

into equation (2.1.3) yields equation (2.1.2) in v'. In this case v' is the concentration and a^2 is the coefficient of diffusion.

Parabolic partial differential equations also arise in other branches of physics and engineering such as, the diffusion of smoke through the air, the diffusion of neutrons in a nuclear reactor, the fluid flow through a porous media, and boundary layer flow over a flat plate.

Suppose a viscous fluid moves in the y-direction over a flat plate placed along the y-axis. If the x-axis is taken to be normal to the y-axis, the velocity v will vary with x as a result of the viscosity of the fluid and the presence of a boundary (flat plate). If we assume that the velocity v depends on x and t only and neglecting the presure gradients, we obtain the equation

$$\partial_t v = \nu \partial_x^2 v$$

where ν is the coefficient of kinematic viscosity.

Consider the pure initial-value problem, that is, the case where the rod is infinitely thin and infinitely long. This gives rise to the equation (2.1.2)

$$\partial_t v = a^2 \partial_x^2 v , \quad -\infty < x < +\infty , \quad t > 0$$

with initial condition (initial temperature distribution)

$$v(x,0) = f(x) , \quad -\infty < x < +\infty .$$

The <u>fundamental kernal</u> for the operator $\partial_t - a^2 \partial_x^2$ is

$$K(x,t) = \frac{1}{2a\sqrt{\pi t}} e^{-x^2/4a^2 t} , \quad -\infty < x < +\infty , \quad t > 0 .$$

If the initial condition $f(x)$ is continuous and bounded for $-\infty < x < +\infty$, then a unique solution to the initial-value problem is given by

$$v(x,t) = \int_{-\infty}^{+\infty} f(y)K(x-y,t)\,dy$$

$$(2.1.4) \qquad = \frac{1}{2a\sqrt{\pi t}} \int_{-\infty}^{+\infty} f(y)e^{-(x-y)^2/4a^2t}\,dy$$

We shall now formulate the statement of the initial boundary value problem for equation (2.1.1), for the case of heat flow. Other physical problems giving rise to an equation of the form of (2.1.1) generally lead to the same boundary conditions.

Consider equation (2.1.1)

$$\partial_t v = a^2 \partial_x^2 v + \phi(x,t) ,$$

with the initial temperature distribution

$$v(x,0) = f(x) ,$$

and with the boundary conditions of the following types:

Case a) At the end of the rod, the temperatures are specified as functions of time t, that is,

$$v(0,t) = \alpha_0(t) \qquad \text{at} \quad x = 0$$

$$v(\ell,t) = \beta_0(t) \qquad \text{at} \quad x = \ell .$$

Case b) The flow of heat through the ends of the rod is specified as functions of time t. Since the flow of heat at each end is proportional to $\partial_x v$

$$\partial_x v = \alpha_1(t) \qquad \text{at} \quad x = 0$$

$$\partial_x v = \beta_1(t) \qquad \text{at} \quad x = \ell .$$

Case c) There is an exchange of heat with the medium, surrounding the rod, through the ends of the rod, that is,

$$\partial_x v = h(v - \alpha_3 t)) \qquad \text{at} \quad x = 0$$

$$\partial_x v = -h(v) - \beta_3(t)) \qquad \text{at} \quad x = \ell ,$$

where $\alpha_3(t)$ and $\beta_3(t)$ are the temperatures of the surrounding medium at $x = 0$ and $x = \ell$, respectively, as functions of time.

Mixtures of these types of boundary conditions also arise. The boundary conditions are said to be <u>homogeneous</u> if α_0, α_1, α_3, β_0, β_1, and β_3 are zero.

If the initial condition $f(x)$ is continuous on the closed interval $0 \le x \le \ell$ and the source term $\phi(x,t)$ is continuous on $0 \le x \le \ell$ and $t \ge 0$ and bounded, then a unique solution exists to the initial boundary value problem (2.1.1) with homogeneous boundary conditions.

Inhomogeneous boundary conditions can be reduced to homogeneous boundary conditions by the substitution

$$v'(x,t) = v(x,t) + z(x,t) \ ,$$

where z is an arbitrary twice differentiable function which satisfies the inhomogeneous boundary condition and v is the solution of the initial-boundary value problem with homogeneous boundary conditions.

For example, let v denote the solution of

$$\begin{aligned} &\partial_t v = a^2 \partial_x^2 v \ , && 0 < x < \ell \ , \quad t > 0 \\ &v(x,0) = f(x) \ , && 0 \leq x \leq \ell \ , \\ &v(0,t) = v(\ell,t) = 0 \ , \quad t \geq 0 \ . \end{aligned}$$

Suppose the inhomogeneous boundary conditions given in Case a) $v(0,t) = \alpha_0(t)$ and $v(\ell,t) = \beta_0(t)$. Define $z(x,t)$ by

$$z(x,t) = \alpha_0(t) + \frac{x}{\ell} (\beta_0(t) - \alpha_0(t)) \ .$$

Then $v'(x,t) = v(x,t) + z(x,t)$ satisfies

$$\begin{aligned} &\partial_t v' = a^2 \partial_x^2 v' \ , && 0 < x < \ell \ , \quad t > 0 \\ &v'(x,0) = f(x) + z(x,0) \ , && 0 \leq x \leq \ell \ , \\ &v'(0,t) = \alpha_0(t) \ , && t \geq 0 \ . \\ &v'(\ell,t) = \beta_0(t) \ , && t \geq 0 \ . \end{aligned}$$

A property of parabolic equations which is used to prove uniqueness is the maximum principle.

Maximum Principle. Let $v(x,t)$ be continuous in the closed rectangle $a \leq x \leq b$, $T_1 \leq t \leq T_2$ and satisfy

$$\partial_t v = a^2 \partial_x^2 v$$

throughout the interior of the rectangle $a < x < b$, $T_1 < t < T_2$. Then the maximum of $v(x,t)$ is assumed at least at one point on the vertical sides ($x = a$ or $x = b$) or on the bottom ($t = T_1$) of the rectangle.

II.2. Finite Difference Methods for a Single Parabolic Equation in One Space Dimension.

Finite difference methods can be divided into two classes. One class is explicit methods, one which involves only one grid point at the advanced time level $(n + 1)k$. The other class is implicit method, one which involves more than one grid point at the advanced time level $(n + 1)k$.

The finite difference method (1.1.3) given in Chapter I is an example of an explicit finite difference method

$$u_i^{n+1} = \lambda u_{i+1}^n + (1 - 2\lambda)u_i^n + \lambda u_{i-1}^n$$

since the only grid point at the advanced time level $(n + 1)k$ is $x = ih$.

The method has advantages in that it is easy to program and the number of operations (multiplications) per grid point is low. However, there are also disadvantages which, for many problems, outweigh the advantages, the stability requirement that $\lambda = a^2k/h^2 < \frac{1}{2}$ places restrictions on the time step. Also the method is of relatively lower order accuracy, that is, $O(k) + O(h^2)$.

This can be improved. Suppose instead of considering (1.1.3),

$$\frac{u_i^{n+1} - u_i^n}{k} = D_+D_-u_i^n ,$$

we consider

(2.2.1) $$\frac{u_i^{n+1} - u_i^n}{k} = D_+D_-u_i^{n+1} ,$$

This may rewritten in the form

(2.2.2) $$(I - kD_+D_-)u_i^{n+1} = u_i^n$$

or by expanding $I - kD_+D_-$ out

(2.2.2') $$-\lambda u_{i+1}^{n+1} + (1 + 2\lambda)u_i^{n+1} - \lambda u_{i-1}^{n+1} = u_i^n$$

where $\lambda = a^2k/h^2$.

Since there are three grid points $(i - 1)h$, ih , and $(i + 1)h$ involved at the advanced time step $(n + 1)k$, (as seen in (2.2.2')), this method is an implicit method.

To determine the accuracy of the method (2.2.2) expand $v(x,t)$ in a Taylor series in time about the point $(ih,(n+1)k)$, evaluate it at $t = nk$, and solve for $(\partial_t v)_i^{n+1}$

$$\frac{v_i^{n+1} - v_i^n}{k} = (\partial_t v)_i^{n+1} + O(k) .$$

Similarly, expand $v(x,t)$ in a Taylor series in space about the point $(ih,(n+1)k)$, evaluating it at the two values $x = (i - 1)h$ and $x = (i + 1)h$, adding these two series, and solving for $(\partial_x^2 v)_i^{n+1}$

$$\frac{v_{i+1}^{n+1} - 2v_i^{n+1} + v_{i-1}^{n+1}}{h^2} = (\partial_x^2 v)_i^{n+1} + O(h^2) .$$

Combining and using the Equation (1.1.1), $\partial_t v = a^2 \partial_x^2 v$ gives

$$\frac{v_i^{n+1} - v_i^n}{k} - a^2 \left(\frac{v_{i+1}^{n+1} - 2v_i^{n+1} + v_{i-1}^{n+1}}{h^2}\right) = O(k) + O(h^2)$$

Thus the method is accurate of $O(k) + O(h^2)$, which is the same as for the explicit method (1.1.3).

Consider the initial-boundary value problem

(2.2.3a) $\quad \partial_t v = a^2 \partial_x^2 v , \quad 0 < x < 1 , \quad t > 0$

(2.2.3b) $\quad v(x,0) = f(x) , \quad 0 < x < 1$

(2.2.3c) $\quad v(0,t) = g_0(t) , \quad t > 0$

(2.2.3d) $\quad v(1,t) = g_1(t) , \quad t > 0$,

where the compatibility condition between the initial condition and the boundary condition gives rise to

$$f(0) = v(0,0) = g_0(0) ,$$
$$f(1) = v(1,0) = g_1(0) .$$

Let $h = 1/N$ denote the grid spacing. The initial condition becomes

$$u_i^0 = f(ih)$$

and the boundary conditions become

$$u_0^n = g_0(nk) \quad \text{and} \quad u_N^n = g_1(nk) ,$$

where $Nh = 1$.

Writing out (2.2.2') for $i = 1$ to $N - 1$ gives rise to a system of $N - 1$ equations in $N + 1$ unknowns

$j = 1 \quad -\lambda u_0^{n+1} + (1+2\lambda)u_1^{n+1} - \lambda u_2^{n+1} = u_1^n$

$j = 2 \quad -\lambda u_1^{n+1} + (1+2\lambda)u_2^{n+1} - \lambda u_3^{n+1} = u_2^n$

$\vdots$

$j = i \quad -\lambda u_{i-1}^{n+1} + (1+2\lambda)u_i^{n+1} - \lambda u_{i+1}^{n+1} = u_i^n$

$\vdots$

$j = N-1 \quad -\lambda u_{N-2}^{n+1} + (1+2\lambda)u_{N-1}^{n+1} - \lambda u_N^{n+1} = u_{N-1}^n$

In implicit methods some form of boundary conditions are needed to close the system of equations. In this system for $j = 1$, $-\lambda u_0^{n+1}$ is known and for $j = N - 1$, $-\lambda u_N^{n+1}$ is known so these terms are brought over to the other side of the equal sign. This results in a system of $N - 1$ equations in $N - 1$ unkonwns, which may be written in matrix form

$$\underline{A}\,\underline{u}^{n+1} = \underline{b}^n \tag{2.2.4}$$

where $\underline{A}$ is an $(N - 1) \times (N - 1)$ tridiagonal matrix

$$\underline{A} = \begin{pmatrix} 1+2\lambda & -\lambda & & & \\ -\lambda & 1+2\lambda & -\lambda & & \\ & \ddots & \ddots & \ddots & \\ & & -\lambda & 1+2\lambda & -\lambda \\ & & & -\lambda & 1+2\lambda \end{pmatrix}$$

and $\underline{u}^{n+1}$, $\underline{b}^n$ are $(N - 1) \times 1$ vectors

$$\underline{u}^{n+1} = \begin{pmatrix} u_1^{n+1} \\ u_2^{n+1} \\ \vdots \\ u_i^{n+1} \\ \vdots \\ u_{N-2}^{n+1} \\ u_{N-1}^{n+1} \end{pmatrix} \quad \text{and} \quad \underline{b}^n = \begin{pmatrix} u_1^n + \lambda u_0^{n+1} \\ u_2^n \\ \vdots \\ u_i^n \\ \vdots \\ u_{N-2}^n \\ u_{N-1}^n + \lambda u_N^{n+1} \end{pmatrix} .$$

In order to solve equation (2.2.4), Gaussian elimination is used. Consider the general tridiagonal system

$$\begin{pmatrix} d_1 & c_1 & & & \\ a_1 & d_2 & c_2 & & \\ & \ddots & \ddots & \ddots & \\ & & a_{N-3} & d_{N-2} & c_{N-2} \\ & & & a_{N-2} & d_{N-1} \end{pmatrix} \begin{pmatrix} u_1 \\ u_2 \\ \vdots \\ u_{N-2} \\ u_{N-1} \end{pmatrix} = \begin{pmatrix} b_1 \\ b_2 \\ \vdots \\ b_{N-2} \\ b_{N-1} \end{pmatrix} .$$

A tridiagonal matrix is diagonally dominant if, for each i, $1 \le i \le n-1$

$$|d_i| > |c_i| + |a_{i-1}| ,$$

that is, the sum of the absolute value of the elements of the i-th row, excluding the diagonal element, is less than the absolute value of the diagonal element for each row.

Diagonal dominance is sufficient for using Gaussian elimination without pivoting. In system (2.2.4) $a_i = -\lambda$, $d_i = 1 + 2\lambda$, and $c_i = -\lambda$ for $1 \le i \le N-1$, so that $|c_i| + |a_{i-1}| = |-\lambda| + |-\lambda| = 2\lambda$ for $\lambda > 0$. Since

$$|d_i| = 1 + 2\lambda > |c_1| + |a_{i-1}| = 2\lambda ,$$

the tridiagonal matrix $\underline{A}$ in (2.2.4) is diagonally dominant.

The forward elimination step consists of removing the subdiagonal, that is, the elements a_i. The algorithm for this step is given by

$$d_i' = \begin{cases} d_i , & i = 1 \\ d_i - \dfrac{c_{i-1}a_{i-1}}{d_{i-1}'} , & 2 \le i \le N-1 \end{cases}$$

$$b_i' = \begin{cases} b_i , & i = 1 \\ b_i - \dfrac{b_{i-1}a_{i-1}}{d_{i-1}'} , & 2 \le i \le N-1 \end{cases} ,$$

the elements c_i are unchanged. This gives

$$\begin{pmatrix} d_1' & c_1' & & \\ & \ddots & \ddots & \\ & & d_{N-2}' & c_{N-2}' \\ & & & d_{N-1}' \end{pmatrix} \begin{pmatrix} u_1 \\ \vdots \\ u_{N-2} \\ u_{N-1} \end{pmatrix} = \begin{pmatrix} b_1' \\ \vdots \\ b_{N-2} \\ b_{N-1} \end{pmatrix} .$$

The back substitution step consists of solving for u_i for $i = N-1, N-2, \ldots, 1$. The algorithm for this step is

$$u_{N-1} = b'_{N-1}/d'_{N-1} ,$$

$$u_i = \frac{b'_i - c_i u_{i+1}}{d'_i} , \qquad i = N-2, \ldots, 1.$$

How much work is involved in solving (2.2.3) for u^{n+1} ? The number of multiplications (and additions) involved in the forward elimination is $4(N - 2)$ (and $2(N - 2)$) , where the number of unknowns is $N - 1$. The number of multiplication (and additions) involved in the back substitution is $2(N - 2) + 1$ (and $N - 2$). The construction of $\underline{b}$ involves 2 (and 2) multiplications (and additions), one for each of the boundary conditions (if nonzero). Thus the total number of operations, multiplications (and additions) required to solve (2.2.2) is $6(N - 2) + 3$ (and $3(N - 2) + 2$) , or approximately $6N$ (and $3N$) per time step.

How much work is involved in solving (1.1.3) for u^{n+1} ? For efficiency, (1.1.3) may be written in the form

$$u_i^{n+1} = \lambda(u_{i+1}^n + u_{i-1}^n) + (1 - 2\lambda)u_i^n .$$

So we see there are 2 multiplications (and additions) for each grid point $i = 1, \ldots, N-1$. This gives a total number of $2(N - 1)$ (and $2(N - 1)$) multiplications (and additions) or approximately $2N$ (and $2N$) per time step. Thus the implicit method (2.2.2) requires approximately 3 times as many operations as the explicit method (1.1.3).

It is seen that there is more work involved in programming the method and the number of operations is increased. So a natural question arises. Does one gain anything by using this method? To answer this question, consider the stability of the method, using the Fourier method. Taking the discrete Fourier transform of (2.2.2) (or (2.2.2')),

$$\widehat{(I - kD_+D_-)u}^{n+1}(\xi) = \hat{u}^n(\xi) ,$$

we have

$$(-\lambda e^{-i\xi} + (1 + 2\lambda) - \lambda e^{i\xi})\hat{u}^{n+1}(\xi) = \hat{u}^n(\xi) .$$

or

$$(1 + 2\lambda(1 - \cos \xi))\hat{u}^{n+1}(\xi) = \hat{u}^n(\xi) .$$

This may be written in the form

$$\hat{u}^{n+1}(\xi) = \frac{1}{1 + 2\lambda(1 - \cos \xi)} \hat{u}^n(\xi) ,$$

where the symbol $\rho(\xi)$ of (2.2.2) is

$$\rho(\xi) = \frac{1}{1 + 2\lambda(1 - \cos \xi)} .$$

Since $0 \leq 1 - \cos \xi \leq 2$ and $\lambda > 0$,

$$1 \leq 1 + 2\lambda(1 - \cos \xi) \leq 1 + 4\lambda .$$

By taking the reciprocal

$$\frac{1}{1 + 4\lambda} \leq \rho(\xi) \leq 1 .$$

Hence, $|\rho(\xi)| \leq 1$ and the von Neumann condition is satisfied for all $\lambda > 0$. Thus the method is unconditionally stable. Thus no restriction is placed on the time step k due to stability requirements. That is not to say that k can be chosen arbitrarily large. For the order of the method is $O(k) + O(h^2)$ will impose an upper bound on the time step k. Typically, k is chosen so that $k = O(h^2)$. If k is chosen larger than this, the term $O(k)$ will dominate the error. On the other hand, if k is chosen smaller than this, extra work will be required (more time steps) to reach a given time without any gain in accuracy since, in this case, the term $O(h^2)$ will dominate the error. Thus $k = O(h^2)$ is a reasonable choice, for example, $k = ch^2$ where c is some constant.

Compare this with the explicit finite difference method (1.1.3). We see that for $\lambda \leq \frac{1}{2}$, if we choose $\lambda = \frac{1}{2}$, we obtain $k = h^2/2a^2$. Thus the implicit method (2.2.2) allows a larger time step. If one chooses a time step for the implicit method 3 (or more) times that of the explicit method $k = h^2/2a^2$, that is, $k_{\text{implicit}} = 3h^2/2a^2$, then the accuracy is about the same and the number of operations required by each method to reach a fixed time, that is, an integer multiple of k_{implicit} is about the same.

Can one achieve a higher degree of accuracy? Consider, more generally, the finite difference method

(2.2.5) $$\frac{u_i^{n+1} - u_i^n}{k} = a^2 D_+ D_- ((1 - \alpha)u_i^n + \alpha u_i^{n+1}) .$$

with $0 \leq \alpha \leq 1$. For $\alpha = 0$, the method is explicit and reduces to (1.1.3). For $0 < \alpha \leq 1$ the method is implicit. For $\alpha = 1$, the method reduces to (2.2.2).

The finite difference method (2.2.6) can be written in the form

(2.2.6) $$(I - a^2 k\alpha D_+ D_-)u_i^{n+1} = (I + a^2 k(1 - \alpha)D_+ D_-)u_i^n .$$

The Fourier method will be used to analyze the stability of the method. First, note that

(2.2.7) $$\widehat{(D_+D_-u)}(\xi) = \frac{1}{h^2}(S_+ - 2I + S_-)\hat{u}(\xi) = \frac{1}{h^2}(2\cos\xi - 2)\hat{u}(\xi) = -\frac{4}{h^2}\sin^2(\tfrac{\xi}{2})\hat{u}(\xi) .$$

Thus the symbol of D_+D_- is $-\frac{4}{h^2}\sin^2(\frac{\xi}{2})$.

Take the discrete Fourier transform of (2.2.6),

$$\widehat{(I - a^2k\alpha D_+D_-)u^{n+1}}(\xi) = \widehat{(I + a^2k(1-\alpha)D_+D_-)u^n}(\xi) ,$$

which gives, by using the symbol of D_+D_- ,

$$(1 + 4\alpha\lambda \sin^2(\tfrac{\xi}{2}))\hat{u}^{n+1}(\xi) = (1 - 4(1-\alpha)\lambda \sin^2(\tfrac{\xi}{2}))\hat{u}^n(\xi) ,$$

where $\lambda = a^2k/h^2$. This may be written as

$$\hat{u}^{n+1}(\xi) = \frac{1 - 4(1-\alpha)\lambda \sin^2(\frac{\xi}{2})}{1 + 4\alpha\lambda \sin^2(\frac{\xi}{2})}\hat{u}^n(\xi) .$$

We see that the symbol of (2.2.6) is

$$\rho(\xi) = \frac{1 - 4(1-\alpha)\lambda \sin^2(\frac{\xi}{2})}{1 + 4\alpha\lambda \sin^2(\frac{\xi}{2})} .$$

Since $\lambda > 0$, $w = 4a^2\lambda \sin^2(\frac{\xi}{2}) > 0$. By substitution

$$\rho(\xi) = \frac{1 - (1-\alpha)w}{1 + \alpha w} .$$

In order to satisfy the von Neumann condition, require that $|\rho(\xi)| < 1$ or $-1 < \rho(\xi) < 1$. This gives

$$-1 < \frac{1 - (1-\alpha)w}{1 + \alpha w} < 1$$

or

$$2 + 2\alpha w > w > 0 .$$

By the definition of w , $w > 0$ is automatically satisfied. Next consider

$$2 + 2\alpha w > w$$

or

$$(1 - 2\alpha)w < 2 .$$

This last inequality is satisfied independent of λ provided that

$1 - 2\alpha \leqslant 0$ or $\alpha \geqslant 1/2$. However, $0 \leqslant \alpha \leqslant 1$, so the method (2.2.6) is unconditionally stable for $1/2 \leqslant \alpha \leqslant 1$.

For $0 \leqslant \alpha < \frac{1}{2}$, $1 - 2\alpha > 0$. If

(2.2.8)
$$(1 - 2\alpha)4\lambda < 2$$

then

$$(1 - 2\alpha)w = (1 - 2\alpha)4\lambda \sin^2(\tfrac{\xi}{2}) \leqslant (1 - 2\alpha)4\lambda < 2 ,$$

since $\sin^2(\frac{\xi}{2}) \leqslant 1$.

Solving the inequality (2.2.8), we obtain the condition

$$\lambda < \frac{1}{2 - 4\alpha} , \qquad 0 \leqslant \alpha < \tfrac{1}{2} .$$

For $\alpha = 0$, $\lambda < \frac{1}{2}$ which is the stability condition for (1.1.3). As $\alpha \to \frac{1}{2}$, $2 - 4\alpha \to 0$ and $\frac{1}{2 - 4\alpha} \to +\infty$ so , as $\alpha \to \frac{1}{2}$, the method has an increasingly weaker stability condition.

For $0 < \alpha \leqslant 1$, the solution of (2.2.6) involves solving a system of equations. Why consider this method? The answer lies in the truncation error of the method. Expanding $v(x,t)$ in a Taylor series in both time and space about the point $(ih, (n + \frac{1}{2})k)$, we obtain the following approximations

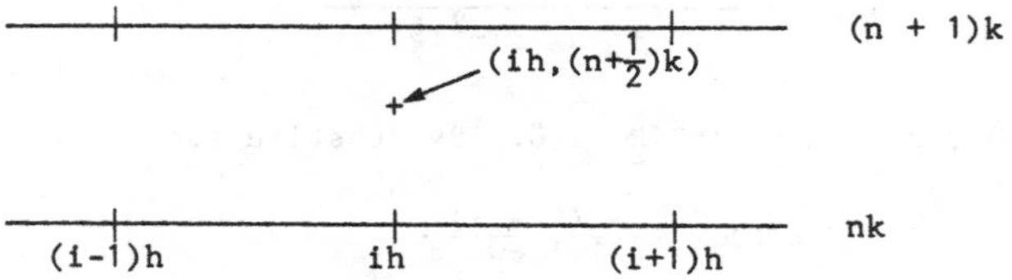

Figure 2.1

$$\frac{v_{i-1}^{n+1} + 2v_i^{n+1} + v_{i+1}^{n+1}}{h^2} = (\partial_x^2 v)_i^{n+\frac{1}{2}} + \frac{k}{2}(\partial_x^2 \partial_t v)_i^{n+\frac{1}{2}} + O(k^4) + O(h^2) ,$$

$$\frac{v_{i-1}^n - 2v_i^n + v_{i+1}^n}{h^2} = (\partial_x^2 v)_i^{n+\frac{1}{2}} - \frac{k}{2}(\partial_x^2 \partial_t v)_i^{n+\frac{1}{2}} + O(k^4) + O(h^2) ,$$

and

$$\frac{v_i^{n+1} - v_i^n}{k} = (\partial_t v)_i^{n+\frac{1}{2}} + \frac{k^2}{24}(\partial_t^3 v)_i^{n+\frac{1}{2}} + O(k^3) + O(h^4) .$$

Combining these three results

$$\frac{v_i^{n+1} - v_i^n}{k} - a^2\{(1-\alpha)(\frac{v_{i+1}^n - 2v_i^n + v_{i-1}^n}{h^2}) + \alpha(\frac{v_{i+1}^{n+1} - 2v_i^{n+1} + v_{i-1}^{n+1}}{h^2})\}$$

$$= (\partial_t v)_i^{n+\frac{1}{2}} + \frac{k^2}{24}(\partial_t^3 v)_i^{n+\frac{1}{2}} + O(k^3) + O(h^4)$$

$$- a^2(1-\alpha)((\partial_x^2 v)_i^{n+\frac{1}{2}} - \frac{k}{2}(\partial_x^2\partial_t v)_i^{n+\frac{1}{2}})$$

$$- a^2\alpha((\partial_x^2 v)_i^{n+\frac{1}{2}} + \frac{k}{2}(\partial_x^2\partial_t v)_i^{n+\frac{1}{2}}) + O(k^4) + O(h^2)$$

$$= (\partial_t v)_i^{n+\frac{1}{2}} - a^2(\partial_x^2 v)_i^{n+\frac{1}{2}} + \frac{k^2}{24}(\partial_t^3 v)_i^{n+\frac{1}{2}}$$

$$+ a^2(1-2\alpha)\frac{k}{2}(\partial_x^2\partial_t v)_i^{n+\frac{1}{2}} + O(k^3) + O(h^2) .$$

Using the differential equation $\partial_t v = a^2\partial_x^2 v$ evaluated at $(ih, (n+\frac{1}{2})k)$

$$\frac{v_i^{n+1} - v_i^n}{k} - a^2\{(1-\alpha)(\frac{v_{i+1}^n - 2v_i^n + v_{i-1}^n}{h^2}) + \alpha(\frac{v_{i+1}^{n+1} - 2v_i^{n+1} + v_{i-1}^{n+1}}{h^2})\}$$

$$= \tau_i^{n+\frac{1}{2}} ,$$

the location trucation error at $(ih, (n+\frac{1}{2})k)$, where

$$\tau_i^{n+\frac{1}{2}} = a^2(1-2\alpha)\frac{k}{2}(\partial_x^2\partial_t v)_i^{n+\frac{1}{2}} + O(k^3) + O(h^2) . \tag{2.2.9}$$

Thus

$$\tau_i^{n+\frac{1}{2}} = \begin{cases} O(k) + O(h^2) , & \alpha \neq \frac{1}{2} \\ O(k^2) + O(h^2) , & \alpha = \frac{1}{2} \end{cases} .$$

This special case when $\alpha = \frac{1}{2}$ which is accurate of order (2,2), that is, $O(k^2) + O(h^2)$ is called the Crank-Nicolson method

$$u_i^{n+1} = u_i^n + \frac{k}{2} D_+D_-(u_i^n + u_i^{n+1}) . \tag{2.2.10}$$

It remains to describe how the method (2.2.6) is implemented. Consider the initial-boundary value problem (2.2.3). This gives rise to the system of $N - 1$ equations in $N - 1$ unknowns u_i^{n+1} , $1 \leq i \leq N-1$, written in matrix form

$$\underline{A}\ \underline{u}^{n+1} = \underline{b}^n$$

where $\underline{A}$ is an $(N - 1) \times (N - 1)$ tridiagonal matrix

$$\underline{A} = \begin{pmatrix} 1+2\alpha\lambda & & & & \\ -\alpha\lambda & -1+2\alpha\lambda & -\alpha\lambda & & \\ & \ddots & \ddots & \ddots & \\ & & -\alpha\lambda & 1+2\alpha\lambda & -\alpha\lambda \\ & & & -\alpha\lambda & 1+2\alpha\lambda \end{pmatrix}$$

and $\underline{u}^{n+1}$, $\underline{b}^n$ are $(N - 1) \times 1$ vectors

$$\underline{u}^{n+1} = \begin{pmatrix} u_1^{n+1} \\ \cdot \\ \cdot \\ \cdot \\ \cdot \\ \cdot \\ u_{N-1}^{n+1} \end{pmatrix} \quad \text{and}$$

$$\underline{b}^n = \begin{pmatrix} (I + a^2k(1-\alpha)D_+D_-)u_1^n + \alpha\lambda u_0^{n+1} \\ (I + a^2k(1-\alpha)D_+D_-)u_2^n \\ \cdot \\ \cdot \\ \cdot \\ \cdot \\ \cdot \\ (I + a^2k(1-\alpha)D_+D_-)u_{N-2}^n \\ (I + a^2k(1-\alpha)D_+D_-)u_{N-1}^n + \alpha\lambda u_N^{n+1} \end{pmatrix}$$

$$= \begin{pmatrix} (1-\alpha)\lambda u_0^n + (1 - 2(1-\alpha)\lambda)u_1^n + (1-\alpha)\lambda u_2^n + \alpha\lambda u_0^{n+1} \\ (1-\alpha)\lambda u_1^n + (1 - 2(1-\alpha)\lambda)u_2^n + (1-\alpha)\lambda u_3^n \\ \cdot \\ \cdot \\ \cdot \\ \cdot \\ \cdot \\ (1-\alpha)\lambda u_{N-3}^n + (1 - 2(1-\alpha)\lambda)u_{N-2}^n + (1-\alpha)\lambda u_{N-1}^n \\ (1-\alpha)\lambda u_{N-2}^n + (1 - 2(1-\alpha)\lambda)u_{N-1}^n + (1-\alpha)\lambda u_N^n + \alpha\lambda u_N^{n+1} \end{pmatrix} .$$

II.3. Finite Difference Methods for a Single Parabolic Equation in Two-Space Dimensions.

Consider the prototype of this two-dimensional parabolic equation

$$(2.3.1)\qquad \partial_t v = \partial_x^2 v + \partial_y^2 v\ ,\ -\infty < x < +\infty,\ -\infty < y < +\infty,\ t > 0$$

with the initial condition

$$(2.3.2)\qquad v(x,y,0) = f(x,y),\ -\infty < x < +\infty,\ -\infty < y < +\infty$$

In order to use the notation of Section I.3, we need to modify the shift operator notation to take the different space-directions into account.

Divide the x-axis into discrete points $x_i = ih_x$, where h_x is the mesh spacing in the x-direction. Similarly, divide the y-axis into discrete points $y_j = jh_y$, where h_y is the mesh spacing in the y-direction. Let $u_{i,j} = u(ih_x, jh_y, nk)$ approximate the exact solution v at the point (jh_x, jh_y, nk).

Denote the forward and backward shift operators in the x-direction by

$$S_{+1}u_{i,j} = u_{i+1,j}\ ,$$
$$S_{-1}u_{i,j} = u_{i-1,j}\ .$$

Similarly, denote the forward and backward shift operators in the y-direction by

$$S_{+2}u_{i,j} = u_{i,j+1}\ ,$$
$$S_{-2}u_{i,j} = u_{i,j-1}\ .$$

From this the obvious extensions of D_0, D_+, and D_- to multiple space dimensions are

$$D_{+1}u_{i,j} = \frac{u_{i+1,j} - u_{i,j}}{h_x}\ ,\quad D_{+2}u_{i,j} = \frac{u_{i,j+1} - u_{i,j}}{h_y}$$

$$D_{-1}u_{i,j} = \frac{u_{i,j} - u_{i-1,j}}{h_x}\ ,\quad D_{-2}u_{i,j} = \frac{u_{i,j} - u_{i,j-1}}{h_y}$$

$$D_{01}u_{i,j} = \frac{u_{i+1,j} - u_{i-1,j}}{2h_x},\quad D_{02}u_{i,j} = \frac{u_{i,j+1} - u_{i,j-1}}{2h_y}$$

which gives rise to

$$D_+D_{-1}u_{i,j} = \frac{1}{h_x^2}(u_{i+1,j} - 2u_{i,j} + u_{i-1,j})$$

and

$$D_+D_{-2}u_{i,j} = \frac{1}{h_y^2}(u_{i,j+1} - 2u_{i,j} + u_{i,j-1}).$$

The two-dimensional discrete Fourier transform is defined by

$$\hat{u}(\xi_1,\xi_2) = \sum_{i,j} u_{i,j}\, e^{\sqrt{-1}i\xi_1}\, e^{\sqrt{-1}j\xi_2},$$

$$0 < \xi_1 < 2\pi,\ 0 < \xi_2 < 2\pi.$$

Thus $\hat{u}$ is the function whose Fourier coefficients are the $u_{i,j}$. As in one-dimension, Parseval's relation is

$$\sum_{i,j} |u_{i,j}|^2 = \frac{1}{(2\pi)^2}\int_0^{2\pi}\int_0^{2\pi} |\hat{u}(\xi_1,\xi_2)|^2\, d\xi_1 d\xi_2$$

The ℓ_2-norm in two-space dimensions must be modified to reflect the grid spacing in the x and y-directions, so for $u = \{u_{i,j}\}$

$$\|u\|_2^2 = \sum_{i,j} u_{ij}^2 h_x h_y$$

Let $u^n = \{u_{i,j}^n\}$ denote the grid function at time nk. Let $u_{i,j}^0 = f(ih_x, jh_y)$ denote the discrete initial conditions. Approximate the initial-value problems (2.3.1)-(2.3.2) by a finite difference method of the form

(2.3.3) $$Q_1u^{n+1} = Q_2u^n,\ n \geq 0$$

where Q_1 and Q_2 are polynomials in the forward and backward shift operators S_{+1}, S_{-1}, S_{+2}, and S_{-2},

$$Q_\ell = Q_\ell(S_{+1}, S_{-1}, S_{+2}, S_{-2})$$

for $\ell = 1,2$. Since $S_- = S_+^{-1}$ (by 1.2.9), we may write Q_1 and Q_2 in the form

(2.3.4a) $$Q_1 = Q_1(S_{+1}, S_{+2}) = \sum_{r,s} b_{r,s}\, S_{+1}^r S_{+2}^s$$

and

(2.3.4b) $$Q_2 = Q_2(S_{+1}, S_{+2}) = \sum_{r,s} a_{r,s}\, S_{+1}^r S_{+2}^s$$

where the $a_{r,s}$ and $b_{r,s}$ are constants. Observe that if $Q_1 \equiv 1$ (or a constant), then (2.3.3) represents an explicit method, otherwise (2.3.3) represents an implicit method.

By inverting Q_1, (2.3.3) may be written in the form

$$u^{n+1} = Qu^n,\ n \geq 0$$

where $Q = Q_1^{-1}Q_2$.

As in Section I.4, the symbol of S_{+1} and S_{+2} is $e^{-i\xi_1}$ and $e^{-i\xi_2}$, respectively. This follows from taking the discrete Fourier transform of $S_{+1}u$ and $S_{+2}u$

$$\widehat{(S_{+1}u)}(\xi_1,\xi_2) = e^{-i\xi_1}\hat{u}(\xi_1,\xi_2)$$
$$\widehat{(S_{+2}u)}(\xi_1,\xi_2) = e^{-i\xi_2}\hat{u}(\xi_1,\xi_2)$$

Taking the discrete Fourier transform of (2.3.3) with (2.3.4),

$$(\widehat{\sum_{r,s} b_{r,s}S_{+1}^{r}S_{+2}^{s}u^{n+1}})(\xi_1,\xi_2) = (\widehat{\sum_{r,s} a_{r,s}S_{+1}^{r}S_{+2}^{s}u^{n}})(\xi_1,\xi_2) ,$$

gives rise to

$$\sum_{r,s} b_{r,s}e^{-ir\xi_1}e^{-is\xi_2}\hat{u}^{n+1}(\xi_1,\xi_2) = \sum_{r,s} a_{r,s}e^{-ir\xi_1}e^{-is\xi_2}\hat{u}^{n}(\xi_1,\xi_2)$$

or by solving for $\hat{u}^{n+1}(\xi_1,\xi_2)$

$$\begin{aligned}\hat{u}^{n+1}(\xi_1,\xi_2) &= (\sum_{r,s} b_{r,s}e^{-ir\xi_1}e^{-is\xi_2})^{-1}(\sum_{r,s} a_{r,s}e^{-ir\xi_1}e^{-is\xi_2})\hat{u}^{n}(\xi_1,\xi_2)\\ &= Q(e^{-i\xi_1},e^{-i\xi_2})\hat{u}^{n}(\xi_1,\xi_2),\end{aligned}$$

where $\rho(\xi_1,\xi_2) = Q(e^{-i\xi_1},e^{-i\xi_2})$ is called the symbol of the finite difference method (2.3.3).

The definitions of stability, consistency, and accuracy are identical to those in Chapter I, except for the corresponding changes for two-dimensions.

The symbol $\rho(\xi_1,\xi_2)$ is said to satisfy the von Neumann condition if there exists a constant $C > 0$ (independent of k,h_x,h_y,n,ξ_1, and ξ_2) such that

$$|\rho(\xi_1,\xi_2)| \leq 1 + Ck$$

for $0 \leq \xi_1 < 2\pi$, $0 \leq \xi_2 < 2\pi$, and k denotes the timestep.

Lax's theorem (Theorem 1.3) in Section I.3 and the Theorem 1.4 in Section I.4 remain valid in two-dimensions.

Consider the explicit finite difference method analogous to (1.1.3)

(2.3.5) $$u^{n+1} = u^n + kD_+D_{-1}u^n + kD_+D_{-2}u^n,$$

where D_+D_{-1} and D_+D_{-2} approximates ∂_x^2 and ∂_y^2, respectively in (2.3.1).

This may be written in the form

$$u^{n+1} = [\frac{k}{h_x^2}(S_{+1} + S_{-1}) + \frac{k}{h_y^2}(S_{+2} + S_{-2}) + (1 - \frac{2k}{h_x^2} - \frac{2k}{h_y^2})I]u^n$$

(2.3.6)

$$= [\frac{k}{h_x^2}(S_{+1} + S_{+1}^{-1}) + \frac{k}{h_y^2}(S_{+2} + S_{+2}^{-1}) + (1 - \frac{2k}{h_x^2} - \frac{2k}{h_y^2})I]u^n .$$

In this case (2.3.4) becomes $Q_1 = 1$ and $Q_2 = \sum_{r=-1}^{1} \sum_{s=-1}^{1} a_{r,s} S_{+1}^r S_{+2}^s$ where

$a_{-1,-1} = 0$	$a_{0,-1} = k/h_y^2$	$a_{1,-1} = 0$
$a_{-1,0} = k/h_x^2$	$a_{0,0} = 1 - 2k/h_x^2 - 2k/h_y^2$	$a_{1,0} = k/h_y^2$
$a_{-1,1} = 0$	$a_{0,1} = k/h_y^2$	$a_{1,1} = 0$.

The symbol of (2.3.5) (or (2.3.6)) is

$$\rho(\xi_1,\xi_2) = [\frac{k}{h_x^2}(e^{-i\xi_1} + e^{i\xi_1}) + \frac{k}{h_y^2}(e^{-i\xi_2} + e^{i\xi_2}) + (1 - \frac{2k}{h_x^2} - \frac{2k}{h_y^2})]$$

$$= \frac{2k}{h_x^2}(\cos \xi_1 - 1) + \frac{2k}{h_y^2}(\cos \xi_2 - 1) + 1.$$

As in the one-dimensional case (Section II.1), for $0 \leq \xi_1, \xi_2 < 2\pi$

$$\frac{-4k}{h_x^2} \leq \frac{2k}{h_x^2}(\cos \xi_1 - 1) \leq 0$$

and

$$\frac{-4k}{h_y^2} \leq \frac{2k}{h_y^2}(\cos \xi_2 - 1) \leq 0 ,$$

so that

(2.3.7) $$1 - 4k(\frac{1}{h_x^2} + \frac{1}{h_y^2}) \leq \rho(\xi_1,\xi_2) \leq 1 .$$

In order to satisfy the von Neumann condition, we require $|\rho(\xi_1,\xi_2)| \leq 1$. From (2.3.7) $\rho(\xi_1,\xi_2) \leq 1$. In order to satisfy $-1 \leq \rho(\xi_1,\xi_2)$, require

$$-1 \leq 1 - 4k(\frac{1}{h_x^2} + \frac{1}{h_y^2})$$

or

(2.3.8) $$\frac{k}{h_x^2} + \frac{k}{h_y^2} \leq \frac{1}{2} .$$

If we consider the special case where the grid spacing in the x and y directions is the same, that is, $h_x = h_y = h$, then the stability condition (2.3.8) reduces to

$$\lambda < \frac{1}{4}$$

where $\lambda = k/h^2$. This places a stronger restriction on the time step k than in the one-dimensional case of the explicit method. This will be a major disadvantage of the method.

The accuracy of the method may be determined directly. From the one-dimensional case, we obtain for multiple space-dimensions

$$\frac{v_{i,j}^{n+1} - v_{i,j}^{n}}{k} = \partial_t v + O(k) ,$$

$$D_+D_{-1}v_{i,j}^n = \partial_x^2 v + O(h_x^2) ,$$

and

$$D_+D_{-2}v_{i,j}^n = \partial_y^2 v + O(h_y^2) .$$

Thus combining these results and using the equation (2.3.3)

$$\frac{v_{i,j}^{n+1} - v_{i,j}^{n}}{k} - D_+D_{-1}v_{i,j}^n - D_+D_{-2}v_{i,j}^n = O(h_x^2) + O(h_y^2) . \tag{2.3.9}$$

And, in the special case where $h_x = h_y = h$, the local truncation error of this explicit method is

$$O(k) + O(h^2),$$

as in the one-dimensional case.

In order to evaluate $u_{i,j}^{n+1}$ in (2.3.5), sweeps are made in one space direction. For example, if one sweeps in the x-direction, j is fixed at 1 for the first sweep. Then (2.3.5) is solved for $u_{i,j}^{n+1}$ for $i = 1,\ldots,N-1$. After this has been completed, j is increased to 2 and (2.3.5) is solved for $u_{i,j}^{n+1}$ for $i = 1,\ldots,N-1$ again. This is continued until the last sweep has been completed corresponding to $j = M-1$. This is represented by Figure 2.2.

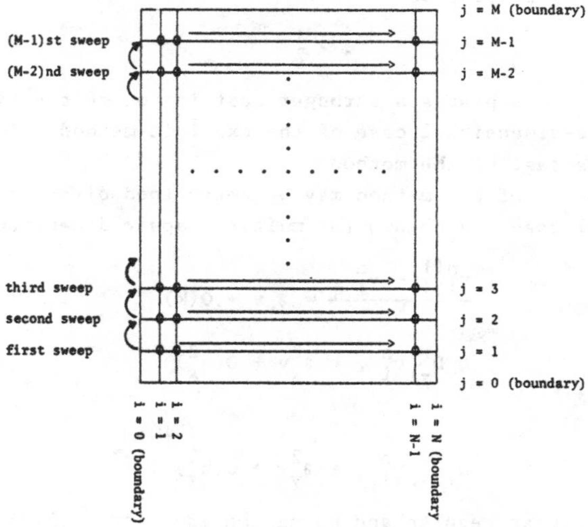

Figure 2.2

As in the case of one space dimension, consider an implicit method in an attempt to improve the stability condition. Consider the implicit method analogous to (2.2.1)

$$\frac{u_{i,j}^{n+1} - u_{i,j}^{n}}{k} = D_+D_{-1}u_{i,j}^{n+1} + D_+D_{-2}u_{i,j}^{n+1}$$

or

$$u_{i,j}^{n+1} = u_{i,j}^{n} + kD_+D_{-1}u_{i,j}^{n+1} + kD_+D_{-2}u_{i,j}^{n+1}.$$

This may be written in the form

(2.3.10) $$(I - kD_+D_{-1} - kD_+D_{-2})u_{i,j}^{n+1} = u_{i,j}^{n}$$

or by expanding $I = kD_+D_{-1} - kD_+D_{-2}$ out

(2.3.10')

$$-\lambda_y u_{i,j-1}^{n+1} - \lambda_x u_{i-1,j}^{n+1} + (1+2\lambda_x + 2\lambda_y)u_{i,j}^{n+1} - \lambda_x u_{i+1,j}^{n+1} - \lambda_y u_{i,j+1}^{n+1} = u_{i,j}^{n},$$

where $\lambda_x = k/h_x^2$ and $\lambda_y = k/h_y^2$. By expanding v in a Taylor series about the point $(ih_x, jh_y, (n+1)k)$, we see that this method is accurate $O(k) + O(h_x^2) + O(h_y^2)$.

To simplify the stability analysis, use (2.2.10) to write

$$(D_+D_{-1}u^n)(\xi_1,\xi_2) = -\frac{4}{h_x^2}\sin^2\left(\frac{\xi_1}{2}\right)\hat{u}^n(\xi_1,\xi_2)$$

and

$$(D_+D_{-2}u^n)(\xi_1,\xi_2) = -\frac{4}{h_y^2}\sin^2(\frac{\xi_2}{2})\hat{u}^n(\xi_1,\xi_2) .$$

Thus the symbol D_+D_{-1} is $\frac{-4}{h_x^2}\sin^2(\frac{\xi_1}{2})$ and, similarly, the symbol D_+D_{-2} is $\frac{-4}{h_y^2}\sin^2(\frac{\xi_2}{2})$.

Taking the discrete Fourier transform in (2.3.10),

$$(I - kD_+D_{-1} - kD_+D_{-2})u^{n+1}(\xi_1,\xi_2) = \hat{u}^n(\xi_1,\xi_2)$$

gives rise to

$$(1 + \frac{4k}{h_x^2}\sin^2(\frac{\xi_1}{2}) + \frac{4k}{h_y^2}\sin^2(\frac{\xi_2}{2})\hat{u}^{n+1}(\xi_1,\xi_2) = \hat{u}^n(\xi_1,\xi_2) ,$$

or

$$\hat{u}^{n+1}(\xi_1,\xi_2) = \rho(\xi_1,\xi_2)\hat{u}^n(\xi_1,\xi_2) ,$$

where

$$\rho(\xi_1,\xi_2) = \frac{1}{(1 + \frac{4k}{h_x^2}\sin^2(\frac{\xi_1}{2}) + \frac{4k}{h_y^2}\sin^2(\frac{\xi_2}{2}))} .$$

Since the denominator is always $\leqslant 1$, we see that $|\rho(\xi_1,\xi_2)| \leqslant 1$ independent of λ_x or λ_y. This the method is unconditionally stable.

Consider the initial-boundary value problem

(2.3.11a) $\quad \partial_t v = \partial_x^2 v + \partial_y^2 v , \quad 0 < x < 1 , \quad 1 < y < 1 , \quad t > 0 ,$

(2.3.11b) $\quad v(x,y,0) = f(x,y) , \quad 0 \leqslant x \leqslant 1 , \quad 0 \leqslant y \leqslant 1 ,$

(2.3.11c) $\quad v(0,y,t) = g_1(y,t) , \quad 0 \leqslant y \leqslant 1 , \quad t \geqslant 0 ,$

(2.3.11d) $\quad v(1,y,t) = g_2(y,t) , \quad 0 \leqslant y \leqslant 1 , \quad t \geqslant 0 ,$

(2.3.11e) $\quad v(x,0,t) = g_3(x,t) , \quad 0 \leqslant x \leqslant 1 , \quad t \geqslant 0 ,$

(2.3.11f) $\quad v(x,1,t) = g_4(x,t) , \quad 0 \leqslant x \leqslant 1 , \quad t \geqslant 0 .$

Let $h_x = 1/N$ and $h_y = 1/M$ denote the grid spacing in the s-direction and y-direction, respectively. The initial condition becomes

$$u_{i,j}^0 = f(ih_x,jh_y) ,$$

and the boundary conditions become

$$u_{0,j}^n = g_1(jh_y,nk) , \quad u_{N,j}^n = g_2(jh_y,nk)$$

$$u_{i,0}^n = g_3(ih_x,nk) , \quad u_{i,M}^n = g_4(ih_x,nk) ,$$

where $Mh_y = Nh_x = 1$. We need to solve a set of equations for the unknowns $u_{i,j}^{n+1}$ in terms of $u_{i,j}^{n}$ for $1 < i < N-1$ and $i < j < M-1$.

In order to solve (2.3.10), sweep in the one space direction. For example, sweep in the x-direction by fixing j as described for the explicit method. The grid points are ordered in the way in which the sweeps were performed for the explicit method.

Let

$$\underline{u}^{n+1} = \left(\begin{array}{c} u_{1,1}^{n+1} \\ u_{2,1}^{n+1} \\ \vdots \\ u_{N-1,1}^{n+1} \\ \hline u_{1,2}^{n+1} \\ u_{2,2}^{n+1} \\ \vdots \\ u_{N-1,2}^{n+1} \\ \hline \vdots \\ \hline u_{1,M-1}^{n+1} \\ u_{2,M-1}^{n+1} \\ \vdots \\ u_{N-1,M-1}^{n+1} \end{array}\right)$$

Let $\underline{bc}_x^{n+1}$ denote the vector that contains the boundary values at $x = 0$ and $x = 1$, and $\underline{bc}_y^{n+1}$ denote the vector that contains the boundary values at $y = 0$ and $y = 1$ at time $(n + 1)k$. These vectors are given by

$$\underline{bc}_x^{n+1} = \left(\begin{array}{c} \lambda_x u_{0,1}^{n+1} \\ 0 \\ \vdots \\ 0 \\ \lambda_x u_{N,1}^{n+1} \\ \hline \lambda_x u_{0,2}^{n+1} \\ 0 \\ \vdots \\ 0 \\ \lambda_x u_{N,2}^{n+1} \\ \hline \vdots \\ \hline \lambda_x u_{0,M-2}^{n+1} \\ 0 \\ \vdots \\ 0 \\ \lambda_x u_{N,M-2}^{n+1} \\ \hline \lambda_x u_{0,M-1}^{n+1} \\ 0 \\ \vdots \\ 0 \\ \lambda_x u_{N,M-1}^{n+1} \end{array}\right) \quad \text{and} \quad \underline{bc}_y^{n+1} = \left(\begin{array}{c} \lambda_y u_{1,0}^{n+1} \\ \lambda_y u_{2,0}^{n+1} \\ \vdots \\ \lambda_y u_{N-1,0}^{n+1} \\ \hline 0 \\ \vdots \\ 0 \\ \hline \vdots \\ \hline 0 \\ \vdots \\ 0 \\ \hline \lambda_y u_{1,M}^{n+1} \\ \lambda_y u_{2,M}^{n+1} \\ \vdots \\ \lambda_y u_{N-1,M}^{n+1} \end{array}\right)$$

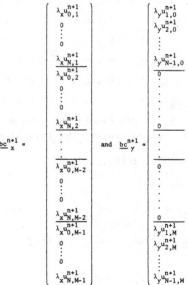

This gives rise to a system of $(N - 1)(M - 1)$ equations in $(N - 1)(M - 1)$ unknowns

$$\underline{A}\,\underline{u}^{n+1} = \underline{u}^{n} + \underline{bc}_x^{n+1} + \underline{bc}_y^{n+1} ,$$

where $\underline{A}$ is the block tridiagonal matrix

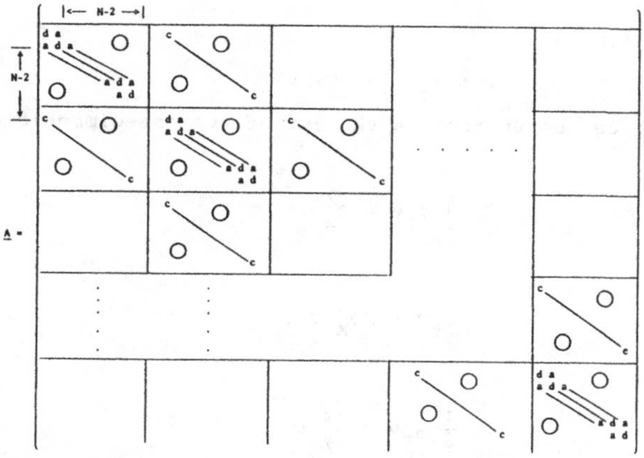

where $a = -\lambda_x$, $d = 1 + 2\lambda_x + 2\lambda_y$, and $c = -\lambda_y$. Each block has dimension $(N - 1) \times (N - 1)$. There are $M - 1$ blocks in each row or column of the block tridiagonal matrix.

This block tridiagonal matrix is much more complicated to solve than the simple tridiagonal matrix. As in the case of the simple tridiagonal matrix, the block tridiagonal matrix is diagonally dominant, so pivoting is not required.

As an illustration example of the set-up of the system, consider the case where $N = M = 4$. The system becomes, where $\lambda = \lambda_x = \lambda_y = k/h^2$,

$$\left(\begin{array}{ccc|ccc|ccc}
1+4\lambda & -\lambda & & -\lambda & & & & & \\
-\lambda & 1+4\lambda & -\lambda & & -\lambda & & & & \\
 & -\lambda & 1+4\lambda & & & -\lambda & & & \\
\hline
-\lambda & & & 1+4\lambda & -\lambda & & -\lambda & & \\
 & -\lambda & & -\lambda & 1+4\lambda & -\lambda & & -\lambda & \\
 & & -\lambda & & -\lambda & 1+4\lambda & & & -\lambda \\
\hline
 & & & -\lambda & & & 1+4\lambda & -\lambda & \\
 & & & & -\lambda & & -\lambda & 1+4\lambda & -\lambda \\
 & & & & & -\lambda & & -\lambda & 1+4\lambda
\end{array}\right)
\left(\begin{array}{c}
u_{11}^{n+1} \\ u_{21}^{n+1} \\ u_{31}^{n+1} \\ \hline u_{12}^{n+1} \\ u_{22}^{n+1} \\ u_{32}^{n+1} \\ \hline u_{13}^{n+1} \\ u_{23}^{n+1} \\ u_{33}^{n+1}
\end{array}\right)
=
\left(\begin{array}{c}
u_{11}^{n} + \lambda u_{10}^{n+1} + \lambda u_{01}^{n+1} \\
u_{21}^{n} + \lambda u_{20}^{n+1} \\
u_{31}^{n} + \lambda u_{30}^{n+1} + \lambda u_{41}^{n+1} \\ \hline
u_{12}^{n} + \lambda u_{02}^{n+1} \\
u_{22}^{n} \\
u_{32}^{n} + \lambda u_{42}^{n+1} \\ \hline
u_{13}^{n} + \lambda u_{14}^{n+1} + \lambda u_{03}^{n+1} \\
u_{23}^{n} + \lambda u_{24}^{n+1} \\
u_{33}^{n} + \lambda u_{34}^{n+1} + \lambda u_{43}^{n+1}
\end{array}\right)$$

II.4. Fractional Step Methods.

In this section a compromise between the explicit and implicit finite difference methods considered in Section II.3 is introduced. Consider the initial value problem (2.3.1)-(2.3.2),

$$\partial_t v = \partial_x^2 v + \partial_y^2 v \tag{2.4.1}$$

with initial condition

$$v(x,y,0) = f(x,y) \ . \tag{2.4.2}$$

Equation (2.4.1) can be written as the sum of two one-dimensional equations

$$\tfrac{1}{2}\,\partial_t v = \partial_x^2 v \tag{2.4.3}$$

or

$$\partial_t v = 2\partial_x^2 v \ , \tag{2.4.3'}$$

and

$$\tfrac{1}{2}\,\partial_t v = \partial_y^2 v \tag{2.4.4}$$

or

$$\partial_t v = 2\partial_y^2 v \ . \tag{2.4.4'}$$

In order to advance the solution from nk to $(n+1)k$, it is assumed that equation (2.4.3) holds from nk to $(n+\frac{1}{2})k$ and that equation (2.4.4) holds from $(n+\frac{1}{2})k$ to $(n+1)k$. This gives rise to replacing (2.4.1)-(2.4.2) by two initial-value problems, each one dimensional

$$\begin{cases} \partial_t v' = 2\partial_x^2 v' \ , & nk < t < (n+\frac{1}{2})k \\ v'(x,y,nk) = v''(x,y,nk) & (= f(x,y) \ \text{ if } \ n = 0) \ , \end{cases} \tag{2.4.5}$$

and

$$\begin{cases} \partial_t v'' = 2\partial_y^2 v'' \ , & (n+\frac{1}{2})k < t < (n+1)k \\ v''(x,y,(n+\frac{1}{2})k) = v'(x,y,(n+\frac{1}{2})k) \ . \end{cases} \tag{2.4.6}$$

Consider first an explicit method (of the form of (1.1.3)) for the solutions of (2.4.5) and (2.4.6). Approximate (2.4.5) by

$$\frac{u_{ij}^{n+\frac{1}{2}} - u_{ij}^{n}}{k/2} = 2D_+D_-{}_1 u_{ij}^{n}$$

or

(2.4.7) $$u_{ij}^{n+\frac{1}{2}} = (I + kD_+D_{-1})u_{ij}^n \ .$$

Approximate (2.4.6) by

$$\frac{u_{ij}^{n+1} - u_{ij}^{n+\frac{1}{2}}}{k/2} = 2D_+D_{-2}u_{ij}^{n+\frac{1}{2}}$$

or

(2.4.8) $$u_{ij}^{n+1} = (I + kD_+D_{-2})u_{ij}^{n+\frac{1}{2}} .$$

Such a method is called a <u>franctional step method</u> (see Yaneuko [22]). How is the stability of this method analyzed? First consider the stability of each of the fractional steps (2.4.7) and (2.4.8). Taking the discrete Fourier transform of (2.4.7).

$$\hat{u}^{n+\frac{1}{2}}(\xi_1,\xi_2) = \rho_1(\xi_1)\hat{u}^n(\xi_1,\xi_2) ,$$

where

$$\rho_1(\xi_1) = 1 - 4\lambda_x \sin^2(\tfrac{\xi_1}{2}) \ .$$

As seen in Section 1.4 $|\rho_1(\xi_1)| \le 1$ provided $\lambda_x \le \frac{1}{2}$. This is just the one dimensional explicit finite difference method (1.1.3). Similarly, taking the discrete Fourier transform of (2.4.8),

$$\hat{u}^{n+\frac{1}{2}}(\xi_1,\xi_2) = \rho_2(\xi_2)\hat{u}^{n+\frac{1}{2}}(\xi_1,\xi_2) ,$$

where

$$\rho_2(\xi_2) = 1 - 4\lambda_y \sin^2(\tfrac{\xi_2}{2}) \ .$$

Again, $|\rho_2(\xi_2)| \le 1$ provided $\lambda_y \le \frac{1}{2}$.

The combination of (2.4.7) and (2.4.8), results in a method that advances the solution from nk to $(n + 1)k$

(2.4.9) $$\hat{u}_{ij}^{n+1} = (I + kD_+D_{-2})(I + kD_+D_{-1})u_{ij}^n \ .$$

The stability of this combined method must be considered. Take the discrete Fourier transform of (2.4.9).

$$\begin{aligned}\hat{u}_{ij}^{n+1}(\xi_1,\xi_2) &= (I + kD_+D_{-2})(I + kD_+D_{-1})\hat{u}^n(\xi_1,\xi_2) \\ &= \rho(\xi_1,\xi_2)\hat{u}^n(\xi_1,\xi_2) ,\end{aligned}$$

where

$$\rho(\xi_1,\xi_2) = \rho_1(\xi_1)\rho(\xi_2) ,$$

and

$$|\rho(\xi_1,\xi_2)| \le |\rho_1(\xi_1)||\rho_2(\xi_2)| \le 1$$

provided $\lambda_x \le \frac{1}{2}$ and $\lambda_y \le \frac{1}{2}$. Thus by considering the two-dimensional initial-value problem as two one-dimensional initial-value problems, an improved stability condition has been achieved. Assume $h_x = h_y$ so that $\lambda = \lambda_x = \lambda_y$, and $\lambda \le \frac{1}{2}$ for the fractional step method while $\lambda \le \frac{1}{4}$ for the two-dimensional explicit method (2.3.5).

In general if two symbols $\rho_1(\xi_1,\xi_2)$ and $\rho_2(\xi_1,\xi_2)$ satisfy the von Neumann condition, that is,

$$|\rho_1(\xi_1,\xi_2)| \le 1 + C_1 k ,$$

$$|\rho_2(\xi_1,\xi_2)| \le 1 + C_2 k ,$$

then the symbol $\rho(\xi_1,\xi_2) = \rho_1(\xi_1,\xi_2)\rho_2(\xi_1,\xi_2)$ satisfies the von Neumann condition, for

$$\begin{aligned}|\rho(\xi_1,\xi_2)| &\le |\rho_1(\xi_1,\xi_2)||\rho_2(\xi_1,\xi_2)| \\ &\le (1 + C_1 k)(1 + C_2 k) \\ &= 1 + (C_1 + C_2)k + O(k^2) \\ &= 1 + Ck .\end{aligned}$$

For fractional step methods, the question of stability reduces to the question of stability of the two one-dimensional methods. This is generally much easier to analyze. What about the accuracy of the method? Consider the combined method (2.4.9) expanded out

$$u_{ij}^{n+1} = (I + kD_+D_{-1} + kD_+D_{-2} + k^2 D_+D_{-2}D_+D_{-1})u_{i,j}^n$$

or

$$\frac{u_{ij}^{n+1} - u_{ij}^n}{k} = (D_+D_{-1} + D_+D_{-2} + kD_+D_{-2}D_+D_{-1})u_{i,j}^n .$$

The term $kD_+D_{-2}D_+D_{-1}u_{ij}^n$ is an extra term introduced by using the two fractional steps. Expanding $v(x,y,t)$ in a Taylor series about the point (ih_x, jh_y, nk), gives rise to

$$\frac{v_{ij}^{n+1} - v_{ij}^n}{k} = \partial_t v + O(k) ,$$

$$D_+D_{-1}v_{ij}^n = \partial_x^2 v + O(h_x^2) ,$$

and

$$D_+D_{-2}v_{ij}^n = \partial_y^2 v + O(h_y^2) .$$

What does the extra term $kD_+D_{-2}D_+D_{-1}u^n_{ij}$ approximate? Again, by expanding v in a Taylor series in space about the point (ih_x, jh_x, nk),

$$D_+D_{-2}D_+D_{-1}v^n_{ij} = \partial^2_y\partial^2_x v + O(h^2_x) + O(h^2_y) .$$

Thus the local truncation error becomes

$$\begin{aligned}\frac{v^{n+1}_{ij} - v^n_{ij}}{k} &= (D_+D_{-1} + D_+D_{-2} + kD_+D_{-2}D_+D_{-1})v^n_{ij} \\ &= \partial_t v + O(k) - \partial^2_x v + O(h^2_x) - \partial^2_y v + O(h^2_y) \\ &\quad - k(\partial^2_y\partial^2_x v) + O(h^2_x) + O(h^2_y)) \\ &= O(k) + O(h^2_x) + O(h^2_y) + k\partial^2_y\partial^2_x v \\ &= O(k) + O(h^2_x) + O(h^2_y) .\end{aligned}$$

Thus the term $kD_+D_{-2}D_+D_{-1}u^n_{ij}$ adds $k\partial^2_y\partial^2_x v$ to the truncation error which is harmless since it is absorbed in the $O(k)$ term. It can be seen from the local truncation error that the fractional step method also solves the equation

$$\partial_t v = \partial^2_x v + \partial^2_y v + k\partial^2_y\, \partial^2_x v$$

with accuracy $O(k^2) + O(h^2_x) + O(h^2_y)$.

To implement the method, first solve (2.4.7) to obtain $u^{n+\frac{1}{2}}_{ij}$. For a fixed j, (2.4.7) represents a purely one-dimensional explicit method in the x-direction,

$$u^{n+\frac{1}{2}}_{ij} = \lambda_x u^n_{i+1,j} + (1 - 2\lambda_x)u^n_{ij} + \lambda_x u^n_{1-i,j} .$$

This is called an <u>x-sweep</u>. Thus for each value of j, (2.4.7) is solved for $u^{n+\frac{1}{2}}_{i,j}$.

Next, solve (2.4.8) for $u^{n+1}_{i,j}$. For a fixed i, (2.4.8) represents a purely one-dimensional explicit method in the y-direction,

$$u^{n+1}_{ij} = \lambda_y u^{n+\frac{1}{2}}_{i,j+1} + (1 - 2\lambda_y)u^{n+\frac{1}{2}}_{i,j} + \lambda_y u^{n+\frac{1}{2}}_{i,j-1} .$$

This is called a <u>y-sweep</u>. Thus for each value of i, (2.4.8) is solved for $u^{n+1}_{i,j}$.

If one considers an initial-boundary value problem (2.4.3), intermediate boundary conditions at $x = 0$ and $x = 1$ will be needed in (2.4.7). Since $(n + \frac{1}{2})k$ actually represents a real time the intermediate boundary conditions are

$$u_{0,j}^{n+\frac{1}{2}} = g_1(jh_y,(n + \tfrac{1}{2})k)$$

and

$$u_{N,j}^{n+\frac{1}{2}} = g_2(jh_y,(n + \tfrac{1}{2})k) \ .$$

A natural question arises. Can the stability constraint on λ be improved greatly without increasing the amount of computational labor? Suppose equations (2.4.6) and (2.4.7) are approximated by an implicit finite difference method (of the form of 2.2.1)). Approximate (2.4.6) by

$$\frac{u_{i,j}^{n+\frac{1}{2}} - u_{i,j}^{n}}{k/2} = 2D_+D_{-1}u_{i,j}^{n+\frac{1}{2}}$$

or

$$(I - kD_+D_{-1})u_{i,j}^{n+\frac{1}{2}} = u_{i,j}^{n} \ . \tag{2.4.10}$$

Approximate (2.4.7) by

$$\frac{u_{i,j}^{n+1} - u_{i,j}^{n+\frac{1}{2}}}{k/2} = 2D_+D_{-2}u_{i,j}^{n+1}$$

or

$$(I - kD_+D_{-2})u_{i,j}^{n+1} = u_{i,j}^{n+\frac{1}{2}} \ . \tag{2.4.11}$$

The stability is analyzed in the same way as for the explicit fractional step method. First, consider the stability of each of the fractional steps (2.4.10) and (2.4.11). Take the discrete Fourier transform of (2.4.10).

$$\hat{u}^{n+\frac{1}{2}}(\xi_1,\xi_2) = \rho_1(\xi_1)\hat{u}^n(\xi_1,\xi_2) \ ,$$

where

$$\rho_1(\xi_1) = \frac{1}{1 + 4\lambda_x \sin^2(\frac{\xi_1}{2})} \ .$$

Since $1 + 4\lambda_x \sin^2(\frac{\xi_1}{2}) \geq 1$, for every $\lambda_x > 0$, $|\rho_1(\xi_1)| \leq 1$ for every λ_x. Thus (2.4.10) is unconditionally stable. Observe that this is just the one-domensional implicit method (2.2.1).

Similarly, take the discrete Fourier transform of (2.4.11),

$$\hat{u}^{n+1}(\xi_1,\xi_2) = \rho_2(\xi_2)\hat{u}^{n+\frac{1}{2}}(\xi_1,\xi_2) ,$$

where

$$\rho_2(\xi_2) = \frac{1}{1 + 4\lambda_x \sin^2(\frac{\xi_2}{2})} .$$

As above, $|\rho_2(\xi_2)| \leq 1$ for every $\lambda_y > 0$. Thus (2.4.11) is unconditionally stable.

Combining (2.4.10) and (2.4.11) results in a method that advances the solution from nk to $(n + 1)k$. Rewriting (2.4.10) and (2.4.11),

$$u_{i,j}^{n+\frac{1}{2}} = (I - kD_+D_{-1})^{-1} u_{i,j}^n$$

and

$$u_{i,j}^{n+1} = (I - kD_+D_{-1})^{-1} u_{i,j}^{n+\frac{1}{2}} .$$

Combining these two expressions gives

$$u_{i,j}^{n+1} = (I - kD_+D_{-2})^{-1}(I - kD_+D_{-1})^{-1} u_{i,j}^n$$

$$= [(I - kD_+D_{-1})(I - kD_+D_{-2})]^{-1} u_{i,j}^n$$

or

(2.4.12) $$(I - kD_+D_{-1})(I - kD_+D_{-2})u_{i,j}^{n+1} = u_{i,j}^n .$$

The combined method (2.4.12) is the equation that is to be solved. Take the discrete Fourier transform of (2.4.12),

$$\left(\frac{1}{\rho_1(\xi_1)} \cdot \frac{1}{\rho_2(\xi_2)}\right)\hat{u}^{n+1}(\xi_1,\xi_2) = \hat{u}^n(\xi_1,\xi_2)$$

or

$$\hat{u}^{n+1}(\xi_1,\xi_2) = \rho_1(\xi_1)\rho_2(\xi_2)\hat{u}^n(\xi_1,\xi_2) ,$$

where $\rho_1(\xi_1)$ and $\rho_2(\xi_2)$ are defined above. Thus

$$\hat{u}^{n+1}(\xi_1,\xi_2) = \rho_1(\xi_1,\xi_2)\hat{u}^n(\xi_1,\xi_2) ,$$

where $\rho(\xi_1,\xi_2) = \rho_1(\xi_1)\rho_2(\xi_2)$ and

$$|\rho(\xi_1,\xi_2)| \leq |\rho_1(\xi_1)||\rho_2(\xi_2)| \leq 1$$

for every λ_x and λ_y. Thus the fractional step method (2.4.10)-(2.4.11) or (2.4.12) is conditionally stable.

Write (2.4.12) out

$$(I - kD_+D_{-1} - kD_+D_{-2} + k^2D_+D_{-2}D_+D_{-1})u_{i,j}^{n+1} = u_{i,j}^n \; .$$

As in the case of the explicit fractional step method, by expanding $v(x,y,t)$ in a Taylor series about the point $(ih_x, jh_y, (n+1)k)$, this implicit method is accurate $O(k) + O(h_x^2) + O(h_y^2)$. The term $kD_+D_{-2}D_+D_{-1}$ adds $k\partial_y^2\partial_x^2 v$ to the local truncation error which is absorbed in the $O(k)$ term.

To discuss the implementation of the implicit fractional step method (2.4.10)-(2.4.12), consider the initial-boundary value problem (2.3.13). The first fractional step (2.4.10) represents an approximation to the one-dimensional parabolic equation in the x-direction. Thus, we sweep in the x-direction by fixing j. This will be illustrated for the special case where $N = M = 4$ and $\lambda = \lambda_x = \lambda_y$, where following system of equations is obtained.

$$\left(\begin{array}{ccc|ccc|ccc}
1+2\lambda & -\lambda & & & & & & & \\
-\lambda & 1+2\lambda & -\lambda & & 0 & & & 0 & \\
 & -\lambda & 1+2\lambda & & & & & & \\
\hline
 & & & 1+2\lambda & -\lambda & & & & \\
 & 0 & & -\lambda & 1+2\lambda & -\lambda & & 0 & \\
 & & & & -\lambda & 1+2\lambda & & & \\
\hline
 & & & & & & 1+2\lambda & -\lambda & \\
 & 0 & & & 0 & & -\lambda & 1+2\lambda & -\lambda \\
 & & & & & & & -\lambda & 1+2\lambda
\end{array}\right)
\left(\begin{array}{c}
u_{1,1}^{n+1} \\ u_{1,2}^{n+1} \\ u_{1,3}^{n+1} \\ \hline u_{2,1}^{n+1} \\ u_{2,2}^{n+1} \\ u_{2,3}^{n+1} \\ \hline u_{3,1}^{n+1} \\ u_{3,2}^{n+1} \\ u_{3,3}^{n+1}
\end{array}\right)
=
\left(\begin{array}{c}
u_{1,1}^{n+\frac12} \\ u_{1,2}^{n+\frac12} \\ u_{1,3}^{n+\frac12} \\ \hline u_{2,1}^{n+\frac12} \\ u_{2,2}^{n+\frac12} \\ u_{2,3}^{n+\frac12} \\ \hline u_{3,1}^{n+\frac12} \\ u_{3,2}^{n+\frac12} \\ u_{3,3}^{n+\frac12}
\end{array}\right)
+
\left(\begin{array}{c}
\lambda u_{1,0}^{n+1} \\ 0 \\ \lambda u_{1,4}^{n+1} \\ \hline \lambda u_{2,0}^{n+1} \\ 0 \\ \lambda u_{2,4}^{n+1} \\ \hline \lambda u_{3,0}^{n+1} \\ 0 \\ \lambda u_{3,4}^{n+1}
\end{array}\right)$$

The second vector on the right-hand side represents the boundary conditions.

The matrices are block diagonal matrices. Thus each of the individual blocks can be solved separately, where each block is tridiagonal.

To implement the general method, first solve (2.4.10) to obtain $u_{i,j}^{n\frac{1}{2}}$ for each value of j. This is called the <u>x-sweep</u>. For each j, $1 \le j \le M-1$, solve the tridiagonal system of equations of order $N-1$.

(2.4.13)
$$\begin{pmatrix} 1+2\lambda_x & -\lambda_x & & & \\ -\lambda_x & 1+2\lambda_x & -\lambda_x & & \\ & \ddots & \ddots & \ddots & \\ & & -\lambda_x & 1+2\lambda_x & -\lambda_x \\ & & & -\lambda_x & 1+2\lambda_x \end{pmatrix} \begin{pmatrix} u_{1,j}^{n+\frac{1}{2}} \\ u_{2,j}^{n+\frac{1}{2}} \\ \vdots \\ u_{N-2,j}^{n+\frac{1}{2}} \\ u_{N-1,j}^{n+\frac{1}{2}} \end{pmatrix}$$

$$= \begin{pmatrix} u_{1,j}^{n} + \lambda_x u_{0,j}^{n+\frac{1}{2}} \\ u_{2,j}^{n} \\ \vdots \\ u_{N-2j}^{n} \\ u_{N-1,j}^{n} + \lambda_x u_{N,j}^{n+\frac{1}{2}} \end{pmatrix} .$$

The x-sweep consists of solving $M - 1$ systems of equations.

The second step consists of solving (2.4.11) to obtain $u_{i,j}^{n+1}$ for each value of i. This is called the <u>y-sweep</u>. For each i, $1 < i < N-1$, solve the tridiagonal system of equations of order $N - 1$.

(2.4.14)
$$\begin{pmatrix} 1+2\lambda_y & -\lambda_y & & & \\ -\lambda_y & 1+2\lambda_y & -\lambda_y & & \\ & \ddots & \ddots & \ddots & \\ & & -\lambda_y & 1+2\lambda_y & -\lambda_y \\ & & & -\lambda_y & 1+2\lambda_y \end{pmatrix} \begin{pmatrix} u_{i,1}^{n+1} \\ u_{i,2}^{n+1} \\ \vdots \\ u_{i,M-2}^{n+1} \\ u_{i,M-1}^{n+1} \end{pmatrix}$$

$$= \begin{pmatrix} u_{i,1}^{n+\frac{1}{2}} + \lambda_y u_{i,0}^{n+1} \\ u_{i,2}^{n+\frac{1}{2}} \\ \cdot \\ \cdot \\ \cdot \\ u_{i,M-2}^{n+\frac{1}{2}} \\ u_{i,M-1}^{n+\frac{1}{2}} + \lambda_y u_{i,M}^{n+1} \end{pmatrix} .$$

The y-sweep consists of solving $N - 1$ system of equations.

The primary advantage of this form of the implementation is that it requires far less storage than using the block diagonal matrix. The implicit fractional step method reduces the problem of solving one large $(N - 1) - (M - 1)$ block tridiagonal system of equations to solving a large number, $(N - 1) + (M - 1)$, of small $((N - 1)$ or $(M - 1))$ tridiagonal system of equations.

Consider the three-dimensional parabolic equation

$$\partial_t v = \partial_x^2 v + \partial_y^2 v + \partial_z^2 v , \quad -\infty < x < +\infty , \quad -\infty < y < +\infty , \quad -\infty < z < +\infty , \quad t > 0 \tag{2.4.15}$$

with the initial conditions

$$v(x,y,z,0) = f(x,y,z) , \quad -\infty < x < +\infty , \quad -\infty < y < +\infty , \quad -\infty < z < +\infty . \tag{2.4.16}$$

Use notation analogous to that in Section II.3. Let $u_{ij\ell}^n = u(ih_x, ih_y, \ell h_z, nk)$ approximate the exact solution v at the point $(ih_x, jh_y, \ell h_z, nk)$. The notation for the shift and difference operators will be suitable adapted. A subscript 3 in a shift or difference operators will indicate application of that operator in the z-direction. For example, $S_{+3}u_{i,j,\ell}^n = u_{i,j,\ell+1}^n$.

The definitions of stability, consistency, and accuracy are identical to those in Chapter I except for the corresponding changes for three-dimensions. The three-dimensional discrete Fourier transform and the von Neumann conditions undergo the appropriate changes to three-dimensions as in the two-dimensional case in Section II.3.

Equation (2.4.1) can be written as the sum of three one-dimensional equations

$$\partial_t v = 3\partial_x^2 v , \tag{2.4.17}$$

(2.4.18)
$$\partial_t v = 3\partial_y^2 v ,$$

and

(2.4.19)
$$\partial_t v = 3\partial_z^2 v .$$

In order to advance the solution from time nk to $(n + 1)k$, it is assumed that equation (2.4.17) holds from nk to $(n + \frac{1}{3})k$, that equation (2.4.18) holds from $(n + \frac{1}{3})k$ to $(n + \frac{2}{3})k$, and that equation (2.4.19) holds from $(n + \frac{2}{3})k$ to $(n + 1)k$. This gives rise to replacing (2.4.15)-(2.4.16) with three initial-value problems, each one-dimensional

(2.4.20)
$$\begin{cases} \partial_t v' = 3\partial_x^2 v' & nk < t < (n + \frac{1}{3})k \\ v'(x,y,z,nk) = v'''(x,y,z,nk) & (= f(x,y,z) \quad \text{if} \quad n = 0) , \end{cases}$$

(2.4.21)
$$\begin{cases} \partial_t v'' = 3\partial_y^2 v'' & (n + \frac{1}{3})k < t < (n + \frac{2}{3})k \\ v''(x,y,z,(n + \frac{1}{3})k) = v'(x,y,z,(n + \frac{1}{3})k) , \end{cases}$$

and

(2.4.22)
$$\begin{cases} \partial_t v''' = 3\partial_z^2 v''' & (n + \frac{2}{3})k < t < (n + 1)k \\ v'''(x,y,z,(n + \frac{2}{3})k) = v''(x,y,z,(n + \frac{2}{3})k) . \end{cases}$$

First consider an explicit method for the solution of (2.4.20)-(2.4.22) similar to (2.4.7)-(2.4.8). Approximate (2.4.20)-(2.4.21) by

(2.4.23)
$$u_{ij\ell}^{n+\frac{1}{3}} = (I + kD_+D_{-1})u_{ij\ell}^n ,$$

(2.4.24)
$$u_{ij\ell}^{n+\frac{2}{3}} = (I + kD_+D_{-2})u_{ij\ell}^{n+\frac{1}{3}} ,$$

and

(2.4.25)
$$u_{ij\ell}^{n+1} = (I + kD_+D_{-3})u_{ij\ell}^{n+\frac{2}{3}} ,$$

respectively. The combined method is

(2.4.26)
$$u_{ij\ell}^{n+1} = (I + kD_+D_{-3})(I + kD_+D_{-2})(I + kD_+D_{-1})u_{ij\ell}^n .$$

Take the discrete Fourier transform of (2.4.26) to obtain

$$\hat{u}^{n+1}(\xi_1,\xi_2,\xi_3) = (I + kD_+D_{-3})(I + kD_+D_{-2})(I + kD_+D_{-1})\hat{u}^n(\xi_1,\xi_2,\xi_3)$$
$$= \rho(\xi_1,\xi_2,\xi_3)\hat{u}^n(\xi_1\ \xi_2\ \xi_3) ,$$

where $\rho(\xi_1,\xi_2,\xi_3) = \rho_3(\xi_3)\rho_2(\xi_2)\rho_1(\xi_1)$, $\rho_1(\xi_1)$, $\rho_2(\xi_2)$, and $\rho_3(\xi_3)$ are the symbols of the fractional steps (2.4.23)-(2.4.25), respectively. As in the two-dimensional case

$$\rho_1(\xi_1) = 1 - 4\lambda_x \sin^2(\frac{\xi_1}{2}) ,$$

$$\rho_2(\xi_2) = 1 - 4\lambda_y \sin^2(\frac{\xi_2}{2}) ,$$

and

$$\rho_1(\xi_3) = 1 - 4\lambda_z \sin^2(\frac{\xi_3}{2}) ,$$

so that $|\rho_1(\xi_1)| \leq 1$, $|\rho_2(\xi_2)| \leq 1$, and $|\rho_3(\xi_3)| \leq 1$ and hence $|\rho(\xi_1,\xi_2,\xi_3)| \leq 1$ provided that $\lambda_x \leq \frac{1}{2}$, $\lambda_y \leq \frac{1}{2}$, and $\lambda_t \leq \frac{1}{2}$. Thus the von Neumann condition is satisfied for $\lambda_x \leq \frac{1}{2}$, $\lambda_y \leq \frac{1}{2}$, and $\lambda_z \leq \frac{1}{2}$.

Expanding (2.4.26) out

$$\frac{u_{ij\ell}^{n+1} - u_{ij\ell}^n}{k} = [D_+D_{-1} + D_+D_{-2} + D_+D_{-3} + k(D_+D_{-2}D_+D_{-1} + D_+D_{-3}D_+D_{-1} + D_+D_{-3}D_+D_{-2}) + k^2D_+D_{-3}D_+D_{-1}]u_{ij\ell}^n .$$

The last two terms on the right hand side are introduced by using the three fractional steps. Expanding $v(x,y,z,t)$ in a Taylor series about the point $(ih_x,ih_y,\ell h_z,nk)$. The local truncation error becomes

$$\frac{v_{ij\ell}^{n+1} - v_{ij\ell}^n}{k} - [D_+D_{-1} + D_+D_{-2} + D_+D_{-3} + k(D_+D_{-2}D_+D_{-1} + D_+D_{-3}D_+D_{-1} + D_+D_{-3}D_+D_{-2}) + k^2D_+D_{-3}D_+D_{-2}D_+D_{-1}]v_{ij\ell}^n$$
$$= \partial_t v + O(k) - \partial_x^2 v + O(h_x^2) - \partial_y^2 v + O(h_y^2) - \partial_z^2 v + O(h_z^2)$$
$$- k[\partial_y^2\partial_x^2 v + O(h_x^2) + O(h_y^2) + \partial_z^2\partial_x^2 v + O(h_x^2) + O(h_z^2) + \partial_z^2\partial_y^2 v + O(h_y^2) + O(h_z^2)] + k^2[\partial_z^2\partial_y^2\partial_x^2 v + O(h_x^2) + O(h_y^2) + O(h_z^2)]$$
$$= O(k) + O(h_x^2) + O(h_y^2) + O(h_z^2) .$$

Thus the additional terms introduced by the fractional step method are absorbed in the $O(k)$ term in the local fruncation error.

To implement the method, first solve (2.4.23) to obtain $u_{ij\ell}^{n+\frac{1}{3}}$. For a fixed j and ℓ (2.4.23) represents a purely one-dimensional explicit method in the x-direction

$$u_{ij\ell}^{n+\frac{1}{3}} = \lambda_x u_{i+1,j,\ell}^n + (1 - 2\lambda_x) u_{i,j,\ell}^n + \lambda_x u_{i-1,j,1}^n .$$

This is called an <u>x-sweep</u>, as in the two-dimensional case.

Next solve (2.4.24) for $u_{ij\ell}^{n+\frac{2}{3}}$. For a fixed i and ℓ, (2.4.24) represents a purely one-dimensional explicit method in the y-direction,

$$u_{i,j,\ell}^{n+\frac{2}{3}} = \lambda_y u_{i,j+1,\ell}^{n+\frac{1}{3}} + (1 - 2\lambda_y) u_{i,j,\ell}^{n+\frac{1}{3}} + \lambda_y u_{i,j-1,\ell}^{n+\frac{1}{3}} .$$

This is called a <u>y-sweep</u>, as in the two-dimensional case.

Finally, solve (2.4.25) for $u_{i,j,\ell}^{n+1}$. For a fixed i and j, (2.4.25) represents a purely one-dimensional explicit method in the z-direction,

$$u_{ij\ell}^{n+1} = \lambda_z u_{i,j,\ell+1}^{n+\frac{2}{3}} + (1 - 2\lambda_z) u_{i,j,\ell}^{n+\frac{2}{3}} + \lambda_z u_{i,j,\ell-1}^{n+\frac{2}{3}} .$$

This is called a <u>z-sweep</u>.

As in the case of two dimensions the stability constraints on λ_x, λ_y, and λ_z can be improved by considering an implicit franctional step method for the solution of (2.4.20)-(2.4.22). Approximate (2.4.20)-(2.4.22) by

(2.4.27) $$(I - kD_+D_{-1}) u_{ij\ell}^{n+\frac{1}{3}} = u_{ij\ell}^n ,$$

(2.4.28) $$(I - kD_+D_{-2}) u_{ij\ell}^{n+\frac{2}{3}} = u_{ij\ell}^{n+\frac{1}{3}} ,$$

and

(2.4.29) $$(I - kD_+D_{-3}) u_{ij\ell}^{n+1} = u_{ij\ell}^{n+\frac{2}{3}} ,$$

respectively. The combined method is

(2.4.30) $$(I - kD_+D_{-1})(I - kD_+D_{-2})(I - kD_+D_{-3}) u_{ij\ell}^{n+1} = u_{ij\ell}^n .$$

Take the discrete Fourier transform of (2.4.30).

$$\left(\frac{1}{\rho_1(\xi)} \cdot \frac{1}{\rho_2(\xi)} \cdot \frac{1}{\rho_3(\xi)}\right) \hat{u}^{n+1}(\xi_1,\xi_2,\xi_3) = \hat{u}^n(\xi_1,\xi_2,\xi_3)$$

or

$$\hat{u}^{n+1}(\xi_1,\xi_2,\xi_3) = \rho_1(\xi_1)\rho_2(\xi_2)\rho_3(\xi_3)\hat{u}^n(\xi_1,\xi_2,\xi_3)$$
$$= \rho(\xi_1,\xi_2,\xi_3)\hat{u}^n(\xi_1,\xi_2,\xi_3) ,$$

where $\rho_1(\xi_1)$, $\rho_2(\xi_2)$, and $\rho_3(\xi_3)$ are the symbols of the fractional steps (2.4.27)-(2.4.29). As in the two-dimensional case we see that

$$\rho_1(\xi_1) = \frac{1}{1 + 4\lambda_x \sin^2(\frac{\xi_1}{2})} ,$$

$$\rho_2(\xi_2) = \frac{1}{1 + 4\lambda_y \sin^2(\frac{\xi_2}{2})} ,$$

and

$$\rho_3(\xi_3) = \frac{1}{1 + 4\lambda_z \sin^2(\frac{\xi_3}{2})} ,$$

so that $|\rho_1(\xi_1)| \le 1$, $|\rho_2(\xi_2)| \le 1$, and $|\rho_3(\xi_3)| \le 1$ and hence $|\rho(\xi_1,\xi_2,\xi_3)| \le |\rho_1(\xi_1)||\rho_2(\xi_2)||\rho_3(\xi_3)| \le 1$ for any $\lambda_x > 0$, $\lambda_y > 0$, $\lambda_z > 0$. Thus, the von Neumann condition is satisfied for any $\lambda_x > 0$, $\lambda_y > 0$, $\lambda_z > 0$ and the implicit fractional step method (2.4.27)-(2.4.29) is unconditionally stable.

As in the case of the two-dimensional implicit fractional step method, expand $v(x,y,z,t)$ in a Taylor series about the point $(ih_x,jh_y,\ell h_z,(n+1)k)$ to obtain the local truncation error which is $O(k) + O(h_x^2) + O(h_y^2) + O(h_z^2)$.

To discuss the implementation of the implicit fractional step method (2.4.27)-(2.4.29), consider the initial-boundary value problem

(2.4.31a) $\partial_t v = \partial_x^2 v + \partial_y^2 v + \partial_z^2 v$, $0 < x < 1$, $0 < y < 1$, $0 < z < 1$, $t > 0$,

(2.4.31b) $v(x,y,z,0) = f(x,y,t)$, $0 \le x \le 1$, $0 \le y \le 1$, $0 \le z \le 1$,

(2.4.31c) $v(0,y,z,t) = g_1(y,z,t)$, $0 \le y \le 1$, $0 \le z \le 1$, $t \ge 0$,

(2.4.31d) $v(1,y,z,t) = g_2(y,z,t)$, $0 \le y \le 1$, $0 \le z \le 1$, $t \ge 0$,

(2.4.31e) $v(x,0,z,t) = g_3(x,z,t)$, $0 \le x \le 1$, $0 \le z \le 1$, $t \ge 0$,

(2.4.31f) $v(x,1,z,t) = g_4(x,z,t)$, $0 \le x \le 1$, $0 \le z \le 1$, $t \ge 0$,

(2.4.31g) $v(x,y,0,t) = g_5(x,y,t)$, $0 \le x \le 1$, $0 \le y \le 1$, $t \ge 0$,

(2.4.31h) $v(x,y,1,t) = g_6(x,y,t)$, $0 \le x \le 1$, $0 \le y \le 1$, $t \ge 0$.

Let $h_x = 1/N$, $h_y = 1/M$, and $h_z = 1/R$, where N , M , and R are positive integers. To implement the implicit fractional step method (2.4.27)-(2.4.29), first solve (2.4.27) to obtain $u_{ij\ell}^{n+\frac{1}{3}}$ for each value of j and ℓ. This is called the x-sweep. For each j, $1 \le j \le M-1$, and each ℓ, $1 \le \ell \le R-1$, solve the tridiagonal system of equations of order $N-1$

$$(2.4.32)\quad \begin{pmatrix} 1+2\lambda_x & -\lambda_x & & & \\ -\lambda_x & 1+2\lambda_x & -\lambda_x & & \\ & \ddots & \ddots & \ddots & \\ & & -\lambda_x & 1+2\lambda_x & -\lambda_x \\ & & & -\lambda_x & 1+2\lambda_x \end{pmatrix} \begin{pmatrix} u_{1,j,\ell}^{n+\frac{1}{3}} \\ u_{2,j,\ell}^{n+\frac{1}{3}} \\ \vdots \\ u_{N-2,j,\ell}^{n+\frac{1}{3}} \\ u_{N-1,j,\ell}^{n+\frac{1}{3}} \end{pmatrix}$$

$$= \begin{pmatrix} u_{1,j,\ell}^{n} + \lambda_x u_{0,j,\ell}^{n+\frac{1}{3}} \\ u_{2,j,\ell}^{n} \\ \vdots \\ u_{N-2,j,\ell}^{n} \\ u_{N-1,j,\ell}^{n} + \lambda_x u_{N,j,\ell}^{n+\frac{1}{3}} \end{pmatrix},$$

where $u_{0,j,\ell}^{n+\frac{1}{3}} = g_1(jh_y, \ell h_z, (n+\frac{1}{3})k)$ and $u_{N,j,\ell}^{n+\frac{1}{3}} = g_2(jh_y, \ell h_z, (n+\frac{1}{3})k)$. The x-sweep consists of solving $(M-1) \times (R-1)$ systems of equations.

The second step consists of solving (2.4.28) to obtain $u_{i,j,\ell}^{n+\frac{2}{3}}$ for each i and ℓ. This is called the y-sweep. For each i, $1 \le i \le N-1$, and each ℓ, $1 \le \ell \le R-1$, we solve the tridiagonal system of equations of order $M-1$

$$(2.4.33)\qquad \begin{pmatrix} 1+2\lambda_y & -\lambda_y & & & \\ -\lambda_y & 1+2\lambda_y & -\lambda_y & & \\ & \ddots & \ddots & \ddots & \\ & & -\lambda_y & 1+2\lambda_y & -\lambda_y \\ & & & -\lambda_y & 1+2\lambda_y \end{pmatrix} \begin{pmatrix} u^{n+\frac{2}{3}}_{i,1,\ell} \\ u^{n+\frac{2}{3}}_{i,2,\ell} \\ \vdots \\ u^{n+\frac{2}{3}}_{i,M-2,\ell} \\ u^{n+\frac{2}{3}}_{i,M-1,\ell} \end{pmatrix}$$

$$= \begin{pmatrix} u^{n+\frac{1}{3}}_{i,1,\ell} + \lambda_y u^{n+\frac{2}{3}}_{i,0,\ell} \\ u^{n+\frac{1}{3}}_{i,2,\ell} \\ \vdots \\ u^{n+\frac{1}{3}}_{i,M-2,\ell} \\ u^{n+\frac{1}{3}}_{i,M-1,\ell} + \lambda_y u^{n+\frac{2}{3}}_{i,M,\ell} \end{pmatrix},$$

where $u^{n+\frac{2}{3}}_{i,0,\ell} = g_3(ih_x, \ell h_z, (n+\frac{2}{3})k)$ and $u^{n+\frac{2}{3}}_{i,M,\ell} = g_4(ih_x, \ell h_z, (n+\frac{2}{3})k)$. The y-sweep consists of solving $(N-1) \times (R-1)$ systems of equations.

The third step consists of solving (2.4.29) to obtain $u^{n+1}_{i,j,\ell}$ for each i and j. This is called the <u>z-sweep</u>. For each i, $1 \leq i \leq N-1$, and each j, $1 \leq j \leq M-1$, we solve the tridiagonal system of equation of order $R-1$

$$(2.4.34)\qquad \begin{pmatrix} 1+2\lambda_z & -\lambda_z & & & \\ -\lambda_z & 1+2\lambda_z & -\lambda_z & & \\ & \ddots & \ddots & \ddots & \\ & & -\lambda_z & 1+2\lambda_z & -\lambda_z \\ & & & -\lambda_z & 1+2\lambda_z \end{pmatrix} \begin{pmatrix} u^{n+1}_{i,j,1} \\ u^{n+1}_{i,j,2} \\ \vdots \\ u^{n+1}_{i,j,R-2} \\ u^{n+1}_{i,j,R-1} \end{pmatrix}$$

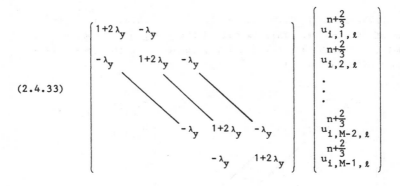

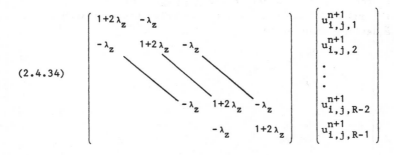

$$= \begin{pmatrix} u_{i,j,1}^{n+\frac{2}{3}} + \lambda_z u_{i,j,0}^{n+1} \\ u_{i,j,2}^{n+\frac{2}{3}} \\ \cdot \\ \cdot \\ \cdot \\ u_{i,j,R-2}^{n+\frac{2}{3}} \\ u_{i,j,R-1}^{n+\frac{2}{3}} + \lambda_z u_{i,j,R}^{n+1} \end{pmatrix},$$

where $u_{i,j,0}^{n+1} = g_5(ih_x,jh_y,(n+1)k)$ and $u_{i,j,R}^{n+1} = g_6(ih_x,jh_y,(n+1)k)$. The z-sweep consists of solving $(N-1) \times (M-1)$ systems of equations.

II.5. <u>Alternating Direction Implicit (A.D.I.) Methods.</u>

Consider the two-dimensional Crank-Nicolson method

$$(I - \frac{k}{2} D_+D_{-1} - \frac{k}{2} D_+D_{-2})u_{i,j}^{n+1} = (I + \frac{k}{2} D_+D_{-1} + \frac{k}{2} D_+D_{-2})u_{i,j}^{n},$$

which is accurate of $O(k^2) + O(h_x^2) + O(h_y^2)$. This method involves solving a block tridiagonal system of equations, as does the implicit method discussed in Section II.3. Can the idea of fractional steps be applied to the Crank-Nicolson method preserving the improved accurary?

The fractional step version of Crank-Nicolson, first proposed by Douglas and Rachford [9], takes the form

$$(I - \frac{k}{2} D_+D_{-1})u_{i,j}^{n+\frac{1}{2}} = (I + \frac{k}{2} D_+D_{-1})u_{i,j}^{n}$$

and

$$(I - \frac{k}{2} D_+D_{-2})u_{i,j}^{n+1} = (I + \frac{k}{2} D_+D_{-2})u_{i,j}^{n+\frac{1}{2}}.$$

By the same arguement used in Section II.4, it can be seen that this method is unconditionally stable.

Solve the first fractional step for $u_{i,j}^{n+\frac{1}{2}}$ to obtain

$$u_{i,j}^{n+\frac{1}{2}} = (I - \frac{k}{2} D_+D_{-1})^{-1}(I + \frac{k}{2} D_+D_{-1})u_{i,j}^{n}.$$

Combining this with the second fractional step

(2.5.1)
$$(I - \frac{k}{2} D_+D_{-2})u_{i,j}^{n+1} = (I + \frac{k}{2} D_+D_{-2})(I - \frac{k}{2} D_+D_{-1})^{-1}(I + \frac{k}{2} D_+D_{-1})u_{i,j}^n .$$

If the computational domain is a reactangle with its sides parallel to the coordinate axes, then the finite difference operators D_+D_{-1} and D_+D_{-2} commute, that is, $D_+D_{-1}D_+D_{-2} = D_+D_{-2}D_+D_{-1}$. Thus, under these conditions, (2.5.1) may be written in the form

$$(I - \frac{k}{2} D_+D_{-2})u_{i,j}^{n+1} = (I - \frac{k}{2} D_+D_{-1})^{-1}(I + \frac{k}{2} D_+D_{-1})(I + \frac{k}{2} D_+D_{-2})u_{i,j}^n ,$$

or

$$(I - \frac{k}{2} D_+D_{-1})(I - \frac{k}{2} D_+D_{-2})u_{i,j}^{n+1} = (I + \frac{k}{2} D_+D_{-1})(I + \frac{k}{2} D_+D_{-2})u_{i,j}^n .$$

or

$$(I + \frac{k}{2} D_+D_{-1} - \frac{k}{2} D_+D_{-2} + \frac{k^2}{4} D_+D_{-1}D_+D_{-2})u_{i,j}^{n+1}$$
$$= (I + \frac{k}{2} D_+D_{-1} + \frac{k}{2} D_+D_{-2} + \frac{k^2}{4} D_+D_{-1}D_+D_{-2})u_{i,j} .$$

As in the other fractional step methods discussed, the term $\frac{k}{4} D_+D_{-1}D_+D_{-2}$ adds $\frac{k}{4} \partial_y^2\partial_x^2 v$ to the local trucation error. However, the other terms in this finite difference method form the two-dimensional Crank-Nicolson method, which is accurate to $O(k^2) + O(h_x^2) + O(h_y^2)$. Since the term $\frac{k}{4} \partial_y^2\partial_x^2 v$ is $O(k)$ it is no longer harmless. The method becomes accurate to $O(k) + O(h_x^2) + O(h_y^2)$. Thus there is no benefit to be gained from using this method.

Can one remove the restriction of first order accuracy in time, that is, $O(k)$? Yes, by carefully allowing for the neglected terms in each step.

Rather than considering two one-dimensional equations (2.4.3) and (2.4.4), we consider the original equation (2.4.1) and solve by a mean of a two-step method which is a refinement of a fractional step method, called an <u>alternating direction implicit method</u>.

An alternating direction method due to Peaceman-Richford [19] and Douglas [4] is a two-step method, which involves an intermediate value $u_{i,j}^{n,\frac{1}{2}}$. The Peaceman-Rachford-Douglas method is

(2.5.2)
$$(I - \frac{k}{2} D_+D_{-1})u_{i,j}^{n,\frac{1}{2}} = (I + \frac{k}{2} D_+D_{-2})u_{i,j}^n$$

and

(2.5.3)
$$(I - \frac{k}{2} D_+D_{-2})u_{i,j}^{n+1} = (I + \frac{k}{2} D_+D_{-1})u_{i,j}^{n,\frac{1}{2}}$$

To analyze the stability of this method, take the discrete

Fourier transform of (2.5.2) obtaining

$$(1 + 2\lambda_x \sin^2(\tfrac{\xi_1}{2}))\hat{u}^{n,\frac{1}{2}}(\xi_1,\xi_2) = (1 - 2\lambda_y \sin^2(\tfrac{\xi_2}{2}))\hat{u}^n(\xi_1,\xi_2) .$$

This gives rise to

$$\hat{u}^{n,\frac{1}{2}}(\xi_1,\xi_2) = \rho(\xi_1,\xi_2)\hat{u}^n(\xi_1,\xi_2) ,$$

where the symbol of (2.5.2) is

$$\rho_1(\xi_1,\xi_2) = \frac{1 - 2\lambda_y \sin^2(\frac{\xi_2}{2})}{1 + 2\lambda_x \sin^2(\frac{\xi_1}{2})} . \tag{2.5.4}$$

Clearly, $\rho_1(\xi_1,\xi_2) \leq 1$, since $1 - 2\lambda_y \sin^2(\frac{\xi_2}{2}) \leq 1$ and $1 \leq 1 + 2\lambda_x \sin^2(\frac{\xi_1}{2}) \leq 1 + 2\lambda_x$ or $(1 + 2\lambda_y \sin^2(\frac{\xi_1}{2}))^{-1} \leq 1$. In order that $-1 \leq \rho_1(\xi_1,\xi_2)$, consider the worst possible case in (2.5.4) which is when the numerator in (2.5.4) is the largest $(\xi_2 = \pi)$ and the denominator in (2.5.4) the smallest $(\xi_1 = 0)$. This gives

$$-1 \leq 1 - 2\lambda_y$$

or

$$\lambda_y \leq 1 .$$

This stability requirement is due to the $\frac{k}{2} D_+D_{-2}$ term in (2.5.2). Since the right-hand side of (2.5.2) involves an explicit finite difference method to represent the ∂_y^2 term, it should not be surprising that there is a restriction on λ_y. For the one-dimensional explicit finite difference method (in y), $\lambda_y \leq \frac{1}{2}$. However, $\frac{k}{2}$ multiplies D_+D_{-2} , not k , so that $(\frac{k}{2})/h_y^2 \leq \frac{1}{2}$ or $\lambda_y \leq 1$.

Similarly, by taking the discrete Fourier transform of (2.5.3),

$$(1 + 2\lambda_y \sin^2(\tfrac{\xi_2}{2}))\hat{u}^{n+1}(\xi_1,\xi_2) = (1 - 2\lambda_x \sin^2(\tfrac{\xi_1}{2}))\hat{u}^{n,\frac{1}{2}}(\xi_1,\xi_2) ,$$

which gives rise to

$$\hat{u}^{n+1}(\xi_1,\xi_2) = \rho_2(\xi_1,\xi_2)\hat{u}^{n,\frac{1}{2}}(\xi_1,\xi_2) ,$$

where the symbol of (2.5.3) is

$$\rho_2(\xi_1,\xi_2) = \frac{1 - 2\lambda_x \sin^2(\frac{\xi_1}{2})}{1 + 2\lambda_y \sin^2(\frac{\xi_2}{2})} . \tag{2.5.5}$$

$|\rho_2(\xi_1,\xi_2)| \le 1$ provided that $\lambda_x \le 1$. This is due to the $\frac{k}{2} D_+D_{-1}$ term in the right-hand side of (2.5.3).

However, if the two steps (2.5.2) and (2.5.3) are combined,

$$\begin{aligned}\hat{u}^{n+1}(\xi_1,\xi_2) &= \rho_2(\xi_1,\xi_2)\hat{u}^{n,\frac{1}{2}}(\xi_1,\xi_2) \\ &= \rho_2(\xi_1,\xi_2)\rho_1(\xi_1,\xi_2)\hat{u}^n(\xi_1,\xi_2) .\end{aligned}$$

Thus the symbol $\rho(\xi_1,\xi_2)$ of the combined method is

$$\begin{aligned}\rho(\xi_1,\xi_2) &= \rho_1(\xi_1,\xi_2)\rho_2(\xi_1,\xi_2) \\ &= \frac{1 - 2\lambda_y \sin^2(\frac{\xi_2}{2})}{1 + 2\lambda_x \sin^2(\frac{\xi_2}{2})} \cdot \frac{1 - 2\lambda_x \sin^2(\frac{\xi_1}{2})}{1 + 2\lambda_y \sin^2(\frac{\xi_2}{2})} \\ &= \frac{1 - 2\lambda_x \sin^2(\frac{\xi_1}{2})}{1 + 2\lambda_x \sin^2(\frac{\xi_1}{2})} \cdot \frac{1 - 2\lambda_y \sin^2(\frac{\xi_2}{2})}{1 + 2\lambda_y \sin^2(\frac{\xi_2}{2})} .\end{aligned}$$

By regrouping, each quotient is a function of only one argument, the first is a function of ξ_1 and the second a function of ξ_2. Set $a = 2\lambda_x \sin^2(\frac{\xi_1}{2})$, so that the first quotient becomes $(1 - a)/(1 + a)$ which, for every a, satisfies

$$\left|\frac{1 - a}{1 + a}\right| \le 1 .$$

Thus

$$\left|\frac{1 - 2\lambda_x \sin^2(\frac{\xi_1}{2})}{1 + 2\lambda_x \sin^2(\frac{\xi_1}{2})}\right| \le 1$$

for every λ_x and, similarly,

$$\left|\frac{1 - 2\lambda_y \sin^2(\frac{\xi_2}{2})}{1 + 2\lambda_y \sin^2(\frac{\xi_2}{2})}\right| \le 1$$

for every λ_y. Then

$$|\rho(\xi_1,\xi_2)| \le 1$$

for all λ_x and λ_y and the combined method is unconditionally stable.

If one chooses $\lambda_y > 1$, then the first step (2.5.2) may become unstable and $\|u^{n,\frac{1}{2}}\|_2$ may become large compared to $\|u^n\|_2$ and $\|u^{n+1}\|_2$. However, this is corrected when the second step (2.5.3) is applied.

Write (2.5.2) and (2.5.3) in the form

$$\frac{u_{i,j}^{n,\frac{1}{2}} - u_{i,j}^n}{k} = \frac{1}{2}\,(D_+D_{-1}u_{i,j}^{n,\frac{1}{2}} + D_+D_{-2}u_{i,j}^n) \tag{2.5.6}$$

and

$$\frac{u_{i,j}^{n+1} - u_{i,j}^{n,\frac{1}{2}}}{k} = \frac{1}{2}\,(D_+D_{-2}u_{i,j}^{n+1} + D_+D_{-1}u_{i,j}^{n,\frac{1}{2}}) . \tag{2.5.7}$$

Adding (2.5.6) and (2.5.7)

$$\frac{u_{i,j}^{n+1} - u_{i,j}^n}{k} = (D_+D_{-1}u_{i,j}^{n,\frac{1}{2}} + \frac{1}{2}\,D_+D_{-2}(u_{i,j}^{n+1} + u_{i,j}^n) . \tag{2.5.8}$$

By subtracting (2.5.7) from (2.5.6)

$$\frac{2u_{i,j}^{n,\frac{1}{2}} - u_{i,j}^n - u_{i,j}^{n+1}}{k} = \frac{1}{2}\,D_+D_{-2}(u_{i,j}^n - u_{i,j}^{n+1}) ,$$

and solving for $u_{i,j}^{n,\frac{1}{2}}$,

$$u_{i,j}^{n,\frac{1}{2}} = \frac{1}{2}\,(u_{i,j}^{n+1} + u_{i,j}^n) - \frac{k^2}{4}\,D_+D_{-2}\left(\frac{u_{i,j}^{n+1} - u_{i,j}^n}{k}\right) . \tag{2.5.9}$$

By substitution of (2.5.9) into (2.5.8) yields

$$\begin{aligned}\frac{u_{i,j}^{n+1} - u_{i,j}^n}{k} &= (D_+D_{-1} + D_+D_{-2})\left(\frac{u_{i,j}^n + u_{i,j}^{n+1}}{2}\right)\\ &\quad - \frac{k^2}{4}\,D_+D_{-1}D_+D_{-2}\left(\frac{u_{i,j}^{n+1} - u_{i,j}^n}{k}\right) .\end{aligned} \tag{2.5.10}$$

The left-hand side and the first term on the right-hand side represent the two dimensional Crank-Nicolson method which is accurate to $O(k^2) + O(h_x^2) + O(h_y^2)$. The extra term is approximately $\frac{k^2}{4}\,\partial_x^2\partial_y^2\partial_t v = O(k^2)$. Thus the extra term is harmless as it is absorbed in the $O(k^2)$ term from the Crank-Nicolson method. Observe that the extra term is only harmless as long as the solution $v(x,y,t)$ is sufficiently smooth.

The Peaceman-Rachford-Douglas method is implemented in the same way as the implicit fractional step method (2.4.7)-(2.4.8). Each step (2.5.2) and (2.5.3) involves the solution of a tridiagonal system.

To consider the initial-boundary value problem (2.3.13) boundary conditions for the intermediate value $u_{i,j}^{n,\frac{1}{2}}$ are required. In the original formulation by Peaceman and Rachford, the boundary condition used for the fractional step method was used, for example, at $x = 0$

$$u_{0,j}^{n,\frac{1}{2}} = g_1(jh_y, (n + \tfrac{1}{2})k) .$$

This is not an exact value, for $u_{i,j}^{n,\frac{1}{2}}$ does not represent the approximate solution at time $(n + \frac{1}{2})k$. This boundary condition approximation is not $O(k^2)$ and, thus, the method (2.5.2) and (2.5.3) is not $O(k^2) + O(h_x^2) + O(h_y^2)$.

How can this be corrected? The following procedure is due to Mitchell and Fairweather [18]. The representation of $u_{i,j}^{n,\frac{1}{2}}$ in (2.5.9) can be rewritten as

$$\text{(2.5.9')} \qquad u_{i,j}^{n,\frac{1}{2}} = \frac{1}{2} (I + \frac{k}{2} D_+D_{-2})u_{i,j}^{n} + \frac{1}{2} (I - \frac{k}{2} D_+D_{-2})u_{i,j}^{n+1} .$$

Also, (2.5.10) may be written in the form

$$\text{(2.5.10')} \qquad (I - \frac{k}{2} D_+D_{-1})(I - \frac{k}{2} D_+D_{-2})u_{i,j}^{n+1} = (I + \frac{k}{2} D_+D_{-1})(I + \frac{k}{2} D_+D_{-2})u_{i,j}^{n} ,$$

which is accurate to $O(k^2) + O(h_x^2) + O(h_y^2)$. This was obtained by substituting (2.5.9) (or (2.5.9')) in (2.5.2).

Consider (2.5.2) at $i = 1$, which yields

$$-\frac{\lambda_x}{2} u_{0,j}^{n,\frac{1}{2}} + (1 + \lambda_x)u_{1,j}^{n,\frac{1}{2}} - \frac{\lambda_x}{2} u_{2,j}^{n,\frac{1}{2}} = (I + \frac{k}{2} D_+D_{-2})u_{1,j}^{n}$$

or

$$\text{(2.5.11)} \qquad (1 + \lambda_x)u_{1,j}^{n,\frac{1}{2}} - \frac{\lambda_x}{2} u_{2,j}^{n,\frac{1}{2}} = (I + \frac{k}{2} D_+D_{-2})u_{1,j}^{n} + \frac{\lambda_x}{2} u_{0,j}^{n,\frac{1}{2}} .$$

Substituting (2.5.9') into (2.5.11)

$$(1 + \lambda_x)[\frac{1}{2} (I + \frac{k}{2} D_+D_{-2})u_{1,j}^{n} + \frac{1}{2} (I - \frac{k}{2} D_+D_{-2})u_{1,j}^{n+1}] - \frac{\lambda_x}{2} [\frac{1}{2} (I + \frac{k}{2} D_+D_{-2})u_{2,j}^{n} + \frac{1}{2} (I - \frac{k}{2} D_+D_{-2})u_{2,j}^{n+1}]$$

$$= (I + \frac{k}{2} D_+D_{-2})u_{1,j}^{n} + \frac{\lambda_x}{2} u_{0,j}^{n,\frac{1}{2}} .$$

By adding $-\frac{\lambda_x}{2}[\frac{1}{2}(I+\frac{k}{2}D_+D_{-2})u_{0,j}^n + \frac{1}{2}(I-\frac{k}{2}D_+D_{-2})u_{0,j}^{n+1}]$ to both sides

$$(I-\frac{k}{2}D_+D_{-1})[\frac{1}{2}(I+\frac{k}{2}D_+D_{-2})u_{1,j}^n + \frac{1}{2}(I-\frac{k}{2}D_+D_{-2})u_{1,j}^{n+1}]$$
$$= (I+\frac{k}{2}D_+D_{-1})u_{1,j}^n + \frac{\lambda_x}{2}[u_{0,j}^{n,\frac{1}{2}} - \frac{1}{2}(I+\frac{k}{2}D_+D_{-2})u_{0,j}^n$$
$$- \frac{1}{2}(I-\frac{k}{2}D_+D_{-2})u_{0,j}^{n+1}]$$

or

$$(I-\frac{k}{2}D_+D_{-1})(I-\frac{k}{2}D_+D_{-2})u_{1,j}^{n+1} \tag{2.5.12}$$
$$= (I+\frac{k}{2}D_+D_{-1})(I+\frac{k}{2}D_+D_{-2})u_{1,j}^n$$
$$+ \frac{\lambda_x}{2}[u_{0,j}^{n,\frac{1}{2}} - \frac{1}{2}(I+\frac{k}{2}D_+D_{-2})u_{0,j}^n - \frac{1}{2}(I-\frac{k}{2}D_+D_{-2})u_{0,j}^{n+1}] .$$

In order to maintain the accuracy of (2.5.10') at $i = 1$, (2.5.12) must reduce to (2.5.10'), that is, the second term on the right-hand side of (2.5.12) must vanish. This gives the boundary condition for the intermediate value at $i = 0$ (or $x = 0$)

$$u_{0,j}^{n,\frac{1}{2}} = \frac{1}{2}(I+\frac{k}{2}D_+D_{-2})g_1(jh_y,nk) \tag{2.5.13}$$
$$+ \frac{1}{2}(I+\frac{k}{2}D_+D_{-2})g_1(jh_y,(n+1)k) .$$

Similarly, at $i = N-1$, the equation

$$(I-\frac{k}{2}D_+D_{-1})(I-\frac{k}{2}D_+D_{-2})u_{N-1,j}^{n+1}$$
$$= (I+\frac{k}{2}D_+D_{-1})(I+\frac{k}{2}D_+D_{-2})u_{N-1,j}^n$$
$$+ \frac{\lambda_x}{2}[u_{N,j}^{n,\frac{1}{2}} - \frac{1}{2}(I+\frac{k}{2}D_+D_{-2})u_{N,j}^n - \frac{1}{2}(I-\frac{k}{2}D_+D_{-2})u_{N,j}^{n+1}] ,$$

which gives rise to the boundary condition for the intermediate value at $i = N$ (or $x = 1$)

$$u_{N,j}^{n,\frac{1}{2}} = \frac{1}{2}(I+\frac{k}{2}D_+D_{-2})g_2(jh_y,nk) \tag{2.5.14}$$
$$+ \frac{1}{2}(I-\frac{k}{2}D_+D_{-2})g_2(jh_y,(n+1)k) .$$

Observe that if the boundary conditions are time-independent, then (2.5.12) and (2.5.13) reduce to

$$u_{0,j}^{n,\frac{1}{2}} = g_1(jh_y)$$

and

$$u_{N,j}^{n,\frac{1}{2}} = g_2(jh_y) .$$

The Peaceman-Rachford-Douglas method does not directly extend to three dimensions. However, Douglas [8] and Brian [1] independently developed a three-dimensional ADI method which reduces to the Peaceman-Rachford-Douglas method in two-dimensions and the Crank-Nicolson method in one dimension. This method is a three-step method, which in advancing the solution from time nk to $(n + 1)k$ involves two intermediate values $u_{ij\ell}^{n,\frac{1}{3}}$ and $u_{ij\ell}^{n,\frac{2}{3}}$.

The Douglas-Brian method is

$$\text{(2.5.15)} \qquad \frac{u_{ij\ell}^{n,\frac{1}{3}} - u_{ij\ell}^{n}}{k} = \frac{1}{2} D_+D_{-1}(u_{ij\ell}^{n,\frac{1}{3}} + u_{ij\ell}^{n}) + D_+D_{-2}u_{ij\ell}^{n} + D_+D_{-3}u_{ij\ell}^{n} ,$$

$$\text{(2.5.16)} \qquad \frac{u_{ij\ell}^{n,\frac{2}{3}} - u_{ij\ell}^{n}}{k} = \frac{1}{2} D_+D_{-1}(u_{ij\ell}^{n,\frac{1}{3}} + u_{ij\ell}^{n}) + \frac{1}{2} D_+D_{-2}(u_{ij\ell}^{n,\frac{2}{3}} + u_{ij\ell}^{n}) + D_+D_{-3}u_{ij\ell}^{n} ,$$

and

$$\text{(2.5.17)} \qquad \frac{u_{ij\ell}^{n+1} - u_{ij\ell}^{n}}{k} = \frac{1}{2} D_+D_{-1}(u_{ij\ell}^{n,\frac{1}{3}} + u_{ij\ell}^{n}) + \frac{1}{2} D_+D_{-2}(u_{ij\ell}^{n,\frac{2}{3}} + u_{ij\ell}^{n}) + \frac{1}{2} D_+D_{-3}(u_{ij\ell}^{n+1} + u_{ij\ell}^{n}) .$$

This may be written in a more practical form

$$\text{(2.5.18)} \qquad (I - \frac{k}{2} D_+D_{-1})u_{1j\ell}^{n,\frac{1}{3}} = (I + \frac{k}{2} (D_+D_{-1} + 2\, D_+D_{-2} + 2\, D_+D_{-3}))u_{ij\ell}^{n} ,$$

$$\text{(2.5.19)} \qquad (I - \frac{k}{2} D_+D_{-2})u_{1j\ell}^{n,\frac{2}{3}} = u_{ij\ell}^{n,\frac{1}{3}} - \frac{k}{2} D_+D_{-2}\, u_{ij\ell}^{n} ,$$

and

$$\text{(2.5.20)} \qquad (I - \frac{k}{2} D_+D_{-3})u_{1j\ell}^{n+1} = u_{ij\ell}^{n,\frac{2}{3}} - \frac{k}{2} D_+D_{-3}\, u_{ij\ell}^{n} .$$

Equation (2.5.18) is obtained by regrouping equation (2.5.15). Equation (2.5.19) is obtained by subtracting (2.5.15) from (2.5.16) and regrouping. Finally, equation (2.5.20) is obtained by subtracting (2.5.16) from (2.5.17).

The combined method becomes

$$(2.5.21) \quad (I - \frac{k}{2} D_+D_{-1})(I - \frac{k}{2} D_+D_{-2})(I - \frac{k}{2} D_+D_{-3})u_{ij\ell}^{n+1}$$
$$= [(I + \frac{k}{2} D_+D_{-1} + D_+D_{-2} + D_+D_{-3})$$
$$+ \frac{k^2}{4} (D_+D_{-1}D_+D_{-2} + D_+D_{-1}D_+D_{-3} + D_+D_{-2}D_+D_{-3})$$
$$- \frac{k^3}{8} (D_+D_{-1}D_+D_{-2}D_+D_{-3}]u_{ij\ell}^n .$$

Take the discrete Fourier transform of (2.5.21).

$$\hat{u}^{n+1}(\xi_1,\xi_2,\xi_3) = \rho(\xi_1,\xi_2,\xi_3)\hat{u}^n(\xi_1,\xi_2,\xi_3) ,$$

where

$$\rho(\xi_1,\xi_2,\xi_3) = \frac{1 - X_1 - X_2 - X_3 + X_1X_2 + X_1X_3 + X_2X_3 + X_1X_2X_3}{(1 + X_1)(1 + X_2)(1 + X_3)}$$
$$= \frac{1 - X_1 - X_2 - X_3 + X_1X_2 + X_1X_3 + X_2X_3 + X_1X_2X_3}{1 + X_1 + X_2 + X_3 + X_1X_2 + X_1X_3 + X_2X_3 + X_1X_2X_3} ,$$

$$X_1 = 2\lambda_x \sin^2(\frac{\xi_1}{2}) ,$$

$$X_2 = 2\lambda_y \sin^2(\frac{\xi_1}{2}) ,$$

and

$$X_3 = 2\lambda_z \sin^2(\frac{\xi_1}{2}) .$$

Since $X_1, X_2, X_3 > 0$ for $\lambda_x, \lambda_y, \lambda_z > 0$, $|\rho(\xi_1,\xi_2,\xi_3)| < 1$ and hence the von Neumann condition is satisfied. This shows that the Douglas-Brian method is unconditionally stable.

It can be shown that this method is $O(k^2) + O(h_x^2) + O(h_y^2) + O(h_z^2)$.

As with the Peaceman-Rachford-Douglas method, $u_{ij\ell}^{n,\frac{1}{3}}$ and $u_{ij\ell}^{n,\frac{2}{3}}$ are only intermediate values used to obtained $u_{ij\ell}^{n+1}$. These values do not represent approximate solutions at times $(n + \frac{1}{3})k$ and $(n + \frac{2}{3})k$ as with fractional step methods. Boundary condition must be provided for these intermediate values.

Consider the initial-boundary value problem (2.4.30). From (2.5.18), intermediate boundary conditions for $u_{ij\ell}^{n,\frac{1}{3}}$ are needed at $x = 0$ and $x = 1$, that is, values for $u_{0,j\ell}^{n,\frac{1}{3}}$ and $u_{N,j,\ell}^{n,\frac{1}{3}}$, respectively. Similarly, from (2.5.19), intermediate boundary conditions are needed for $u_{ij\ell}^{n,\frac{2}{3}}$ at $y = 0$ and $y = 1$, that is, values for $u_{i,0,\ell}^{n,\frac{2}{3}}$ and $u_{i,M,\ell}^{n,\frac{2}{3}}$, respectively. By preserving the second order accuracy in time and space, following the procedure of Mitchell and Fairweather for the Peaceman-Rachford-Douglas method, the following intermediate boundary conditions are; at $i = 0$ (or $x = 0$)

$$u_{0,j\ell}^{n,\frac{1}{3}} = g_1(jh_y,\ell h_z,nk) + (I + \frac{k}{2} D_+D_{-2})(I - \frac{k}{2} D_+D_{-3}) (g_1(jh_y,\ell h_z,(n+1)k) - g_1(jh_y,\ell h_z,nk)) , \tag{2.5.22}$$

at $i = N$ (or $x = 1$)

$$u_{N,j,\ell}^{n,\frac{1}{3}} = g_2(jh_y,\ell h_z,nk) + (I + \frac{k}{2} D_+D_{-2})(I - \frac{k}{2} D_+D_{-3}) (g_2(jh_y,\ell h_z,(n+1)k) - g_2(jh_y,\ell h_z,nk)) , \tag{2.5.23}$$

at $j = 0$ (or $y = 0$)

$$u_{i,0,\ell}^{n,\frac{2}{3}} = g_3(ih_x,\ell h_z,nk) + (I - \frac{k}{2} D_+D_{-3}) (g_3(ih_x,\ell h_z,(n+1)k) - g_3(ih_y,\ell h_z,nk)) , \tag{2.5.24}$$

and at $j = M$ (or $y = 1$)

$$u_{i,M,\ell}^{n,\frac{2}{3}} = g_4(ih_x,\ell h_z,nk) + (I - \frac{k}{2} D_+D_{-3}) (g_4(ih_y,\ell h_z,(n+1)k) - g_4(ih_y,\ell h_z,nk)) . \tag{2.5.25}$$

The implementation of this method is similar to the Peaceman-Rachford-Douglas method and will not be further explained here.

II.6. Effect of Lower Order Terms.

In this section, we demonstrate by an example of how stability is affected by lower order terms.

Consider an equation of the form

$$\partial_t v = a\partial_x^2 v + b\partial_x v + cv , \tag{2.6.1}$$

where a, b, and c are constants with $a > 0$.

In considering the earlier prototype equation

$$\partial_t v = \partial_x^2 v ,$$

D_+D_- was used to approximate ∂_x^2 to $O(h^2)$. This will be used to approximate ∂_x^2 in (2.6.1). What should be used to approximate $\partial_x v$? We could use the analogous approximation to $\partial_t v$, that is, D_+. However,

$$D_+v_i^n = \partial_x v + O(h) .$$

This error will dominate the error $O(h^2)$ from D_+D_- reducing the overall accuracy to $O(h)$. So consider a second order approximation to $\partial_x v$.

Expanding $v(x,t)$ in a Taylor series in space about the point (ih,nk) in x, gives rise to

$$D_0 v_i^n \equiv \frac{v_{i+1}^n - v_{i-1}^n}{2h} = (v_x)_i^n + O(h^2) .$$

Combining these terms, an explicit finite difference method that approximates equation (2.6.1) is

(2.6.2) $$u_i^{n+1} = u_i^n + akD_+D_-u_i^n + bkD_0u_i^n + cku_i^n .$$

This method is accurate to $O(k) + O(h^2)$, where the undifferentiated term cku_i^n does not introduce a truncation error since it is not being approximated.

To anaylyze the stability of (2.6.2), take the discrete Fourier transform yielding

(2.6.3) $$\begin{aligned} \hat{u}^{n+1}(\xi) &= (\widehat{(1 + ck)I + akD_+D_- + bkD_0)u}^n(\xi) . \\ &= \rho(\xi)\hat{u}^n(\xi). \end{aligned}$$

Since

(2.6.4) $$\begin{aligned} D_0 u &= \frac{(S_+ - S_-)}{2h} u \\ &= \frac{1}{2h}(e^{-i\xi} - e^{i\xi})\hat{u}(\xi) \\ &= (\frac{-i}{h}\sin\xi)\hat{u}(\xi) , \end{aligned}$$

the symbol of D_0 is $\frac{-i}{h}\sin\xi$. Substituting this and the symbol of D_+D_- (given in (2.2.12)) into (2.6.3), the symbol $\rho(\xi)$ is given by

$$\rho(\xi) = 1 + ck - 4\lambda\, a\sin^2(\tfrac{\xi}{2}) - ib\lambda h\sin\xi .$$

Where $h = \sqrt{k/\lambda}$,

$$\begin{aligned} |\rho(\xi)|^2 &= \rho(\xi)\overline{\rho(\xi)} \\ &= (1 + ck - 4\lambda a\sin^2(\tfrac{\xi}{2}))^2 + (b\lambda h\sin\xi)^2 \\ &= (1 + ck - 4\lambda a\sin^2(\tfrac{\xi}{2}))^2 + (b\sqrt{\tfrac{k}{\lambda}}\,\lambda\sin\xi)^2 , \\ &= (1 - 8\lambda a\sin^2(\tfrac{\xi}{2}) + (16\lambda^2a^2\sin^4(\tfrac{\xi}{2})) \\ &\quad + k(2c + b^2\lambda\sin^2\xi - 8\lambda ac\sin^2(\tfrac{\xi}{2})) + k^2c^2 . \end{aligned}$$

Let $f_1(\xi) = 1 - 8\lambda a\sin^2(\frac{\xi}{2}) + 16\lambda^2 a^2\sin^4(\frac{\xi}{2})$, $f_2(\xi) = 2c + b^2\lambda\sin^2\xi - 8\lambda ac\sin^2(\frac{\xi}{2})$, and $f_3(\xi) = c^2$, then by substitution

$$|\rho(\xi)| = (f_1(\xi) + kf_2(\xi) + k^2 f_3(\xi))^{1/2} .$$

Require $|f_1(\xi)| \leq 1$, $|f_2(\xi)| \leq M_2$, and $|f_3(\xi)| \leq M_3$, so that

$$|\rho(\xi)| \leq (1 + M_2 k + M_3 k^2)^{1/2} = 1 + O(k)^* .$$

Thus if $|f_1(\xi)| \leq 1$ and $f_2(\xi)$ and $f_3(\xi)$ are bounded, the von Neuman condition is satisfied and the explicit finite difference method (2.6.2) is stable. Since λ is fixed, the condition that $f_2(\xi)$ and $f_3(\xi)$ are bounded is satisfied if the constants a , b , and c are bounded. Since $f_1(\xi) \geq 0$, it remains to determine under what conditions $f_1(\xi) \leq 1$, or

$$(1 - 4\lambda a\sin^2(\tfrac{\xi}{2}))^2 \leq 1 , \quad 0 \leq \xi < 2\pi$$

or

$$\frac{1}{2a} \geq \lambda\sin^2(\tfrac{\xi}{2}) \geq 0 .$$

The worst case occurs when $\xi = \pi$, so that $\sin^2(\frac{\xi}{2}) = 1$ and

$$\lambda \leq \frac{1}{2a} . \qquad (2.6.5.)$$

Thus the finite difference method (2.6.2) is stable if (2.6.5) is satisfied. This says that the higher derivative $\partial_x^2 v$ governs the stability.

An implicit finite difference method that approximate equation (2.6.1) based on the explicit method (2.6.2) is

$$u_i^{n+1} = u_i^n + kaD_+D_-u_i^{n+1} + kbD_0u_i^{n+1} + kcu_i^{n+1} . \qquad (2.6.6)$$

This method is also accurate $O(k) + O(h^2)$. It will be shown that this implicit method is unconditionally stable, using the energy method.

Take the discrete inner product of (2.6.6) with u^{n+1}.

$$(u^{n+1},u^{n+1}) = (u^{n+1},u^n) + (u^{n+1},kaD_+D_-u^{n+1}) + (u^{n+1},kbD_0u^{n+1}) + (u^{n+1},kcu^{n+1})$$

---

* Using Taylor series for $(1 + z)^{1/2}$ about the point $z = 0$

$$(1 + z)^{1/2} = 1 + 1/2z + O(z^2) .$$

Let $z = M_2k + M_3k^2$. Upon substitution

$$(1 + M_2k + M_3k^2) = 1 + M_2k + M_3k^2 + O(k^2) = 1 + O(k) .$$

or

$$\|u^{n+1}\|_2^2 = (u^{n+1},u^n) + ka(u^{n+1},D_+D_-u^{n+1})$$
$$+ kb(u^{n+1},D_0u^{n+1}) + kc\|u^{n+1}\|_2^2 .$$

Applying the discrete integration by parts (1.2.13) to the second term on the right-hand side

$$(u^{n+1},D_+D_-u^{n+1}) = -(D_-u^{n+1},D_-u^{n+1}) = -\|D_-u^{n+1}\|_2^2 ,$$

and upon substitution

$$\|u^{n+1}\|_2^2 = (u^{n+1},u^n) - ka\|D_-u^{n+1}\|$$
$$+ kb(u^{n+1},D_0u^{n+1}) + kc\|u^{n+1}\|_2^2 ,$$

or

$$\|u^{n+1}\|_2^2 + ka\|D_-u^{n+1}\|_2^2 = (u^{n+1},u^n) + kb(u^{n+1},D_0u^{n+1})$$
$$+ kc\|u^{n+1}\|_2^2 .$$

Using the triangle inequality and the Schwarz inequality

(2.6.7) $$\|u^{n+1}\|_2^2 + ka\|D_-u^{n+1}\|_2^2$$
$$< \|u^{n+1}\|_2\|u^n\|_2 + kb|(u^{n+1},D_0u^{n+1})| + kc\|u^{n+1}\|_2^2$$
$$< \|u^{n+1}\|_2\|u^n\|_2 + kb\|u^{n+1}\|_2\|D_0u^{n+1}\|_2 + kc\|u^{n+1}\|_2^2 .$$

The goal is to remove the $ka\|D_-u^{n+1}\|_2^2$ term from the left-hand side of (2.6.7). To this end using the inequality

$$|AB| < \frac{A^2 + B^2}{2}$$

with $A = b\sqrt{\frac{k}{2a}}\,\|u^{n+1}\|_2$ and $B = \sqrt{2ka}\,\|D_0u^{n+1}\|_2$,

(2.6.8) $$kb\|u^{n+1}\|_2\,\|D_0u^{n+1}\|_2 < \frac{kb^2}{4a}\,\|u^{n+1}\|_2^2 + ka\|D_0u^{n+1}\|_2^2 .$$

Since $D_0 = \frac{1}{2}(D_+ + D_-)$,

$$\|D_0u^{n+1}\|_2 = \|\tfrac{1}{2}(D_+ + D_-)u^{n+1}\|_2 < \tfrac{1}{2}(\|D_+u^{n+1}\|_2 + \|D_-u^{n+1}\|_2)$$
$$= \|D_-u^{n+1}\|_2 \text{ (or } \|D_+u^{n+1}\|_2) ,$$

by substitution into (2.6.8),

$$kb\|u^{n+1}\|_2\,\|D_0u^{n+1}\|_2 < \frac{kb^2}{4a}\,\|u^{n+1}\|_2^2 + ka\|D_-u^{n+1}\|_2^2 ,$$

and by substitution into (2.6.7),

$$\|u^{n+1}\|_2^2 + ka\|D_-u^{n+1}\|_2^2 < \|u^{n+1}\|_2 \|u^n\|_2 + \frac{kb^2}{4a}\|u^{n+1}\|_2^2$$
$$+ ka\|D_-u^{n+1}\|_2^2 + kc\|u^{n+1}\|_2^2 .$$

After cancellation and regrouping

$$(1 - (\frac{4ac + b^2}{4a})k \|u^{n+1}\|_2 < \|u^n\|_2 ,$$

or

$$\|u^{n+1}\|_2 < \frac{1}{1 - (\frac{4ac + b^2}{4a})k} \|u^n\|_2 = (1 + O(k)) \|u^n\|_2^* .$$

Thus the method (2.6.6) is unconditionally stable.

It should be mentioned that even though the constant b does not play a role in the stability of the explicit and implicit methods, it can have a dramatic effect on the solution. Suppose that $|b| >> a$ and $c = 0$. In this case $b\partial_x v$ is the dominant term in equation (2.6.1). Consider the approximate solution of $x = ih$, u_i^{n+1}. Since $D_0u_i = \frac{1}{2h}(u_{i+1} - u_{i-1})$, that is, D_0u_i depends on the values of u at the two adjacent points and does not depend on u_i, the value of u_i^n from the approximations to the spatial derivatives, will have little effect in determining u_i^{n+1}. This can result in the values of u at a given time oscillating about the exact solution.

To see this, consider equation (2.6.1) with a, b positive and $c = 0$. Approximate (2.6.1) by the explicit finite difference method (2.6.2). Suppose the initial condition is given by

$$u_i^0 = \begin{cases} 0, & i \neq j \\ \varepsilon, & i = j \end{cases} ,$$

where ε is some positive number and jh is some point on the grid. Consider the first time level corresponding to $n = 1$. By substitution of the initial condition into (2.6.2), we obtain at $i = j - 1$, j, and $j + 1$

$$u_{j-1}^1 = (a + \frac{bh}{2})\lambda\varepsilon ,$$

$$u_j^1 = (1 + ck - 2a\lambda)\varepsilon ,$$

---

* Using Taylor series for $(1 - z)^{-1}$ about the point $z = 0$

$$(1 - z)^{-1} = 1 + z + O(z^2) .$$

Let $z = (\frac{4ac + b^2}{4a})k$, then upon substitution

$$\frac{1}{1 - (\frac{4ac + b^2}{4a})k} = 1 + (\frac{4ac + b^2}{4a})k + O(k^2) = 1 + O(k) .$$

$$u_{j+1}^1 = (a - \frac{bh}{2})\lambda\varepsilon .$$

The exact solution is > 0 for all time. The stability condition (2.6.5) $\lambda < \frac{1}{2a}$ guarantees that $u_j^1 > 0$. However, if

$$a - \frac{bh}{2} < 0$$

or

$$\frac{bh}{2} > a ,$$

then $u_{j+1}^1 < 0$ and $u_j^1 > 0$. This is the beginning of the oscillation. We see that if the grid spacing is chosen so that $h < 2a/b$, then u_{j+1}^1 would be positive. This would be a very expensive solution to the problem of oscillations.

This type of oscillation, as well as other remedies, will be discussed in Chapter III.

II.7. Systems of Equations.

Consider the system of differential equations

$$\underline{v}_t = P(\partial_x)\underline{v} , \quad t > 0 \tag{2.7.1}$$

where $\underline{v}$ denotes a vector with p-components of the form $(v_1,v_2,\ldots,v_p)^T$ and P is a polynomial in ∂_x whose coefficients are matrices, that is,

$$P(z) = \sum_{j=0}^{\ell} \underline{B}_j \ z^j$$

or upon replacing z with ∂_x

$$P(\partial_x) = \sum_{j=0}^{\ell} \underline{B}_j(\partial_x)^j$$

where the $\underline{B}_j$'s are matrices of order $p \times p$.

For example, let $\ell = 2$ (the highest derivative is second order) and $p = 3$ (the number of equations) with $\underline{B}_0 = \underline{B}_1 = \underline{0}$ (3×3 matrices with all elements 0) and

$$\underline{B}_2 = \begin{pmatrix} 1 & 0 & 0 \\ 0 & 1 & 0 \\ 0 & 0 & 1 \end{pmatrix} = \underline{I} .$$

Then $P(z) = \underline{B}_2 \ z^2 = \begin{pmatrix} z^2 & 0 & 0 \\ 0 & z^2 & 0 \\ 0 & 0 & z^2 \end{pmatrix}$ and $P(\partial_x) = \underline{B}_2 \ \partial_x^2 = \begin{pmatrix} \partial_x^2 & 0 & 0 \\ 0 & \partial_x^2 & 0 \\ 0 & 0 & \partial_x^2 \end{pmatrix}$.

Equation (2.7.1) becomes

$$\partial_t \underline{v} = \partial_x^2\ \underline{v}$$

with $\underline{v} = (v_1, v_2, v_3)$ or

$$\begin{pmatrix} \partial_t v_1 \\ \partial_t v_2 \\ \partial_t v_3 \end{pmatrix} = \begin{pmatrix} \partial_x^2 & 0 & 0 \\ 0 & \partial_x^2 & 0 \\ 0 & 0 & \partial_x^2 \end{pmatrix} \begin{pmatrix} v_1 \\ v_2 \\ v_3 \end{pmatrix} = \begin{pmatrix} \partial_x^2 v_1 \\ \partial_x^2 v_2 \\ \partial_x^2 v_3 \end{pmatrix} .$$

Suppose $\ell = 2$ and $p = 2$ with

$$\underline{B}_0 = \underline{0}\ ,\quad \underline{B}_1 = \begin{pmatrix} 2 & 0 \\ 0 & 0 \end{pmatrix},\ \text{and}\ \underline{B}_2 = \begin{pmatrix} 1 & 1 \\ 0 & 1 \end{pmatrix}.$$

Then $P(z) = \underline{B}_2 z^2 + \underline{B}_1 z = \begin{pmatrix} z^2 & z^2 \\ 0 & z^2 \end{pmatrix} + \begin{pmatrix} 2z & 0 \\ 0 & 0 \end{pmatrix} = \begin{pmatrix} z^2+2z & z^2 \\ 0 & z^2 \end{pmatrix}$ and

$$P(\partial_x) = \underline{B}_2 \partial_x^2 + \underline{B}_1 \partial_x = \begin{pmatrix} \partial_x^2 + 2\partial_x & \partial_x^2 \\ 0 & \partial_x^2 \end{pmatrix} .$$

Equation (2.7.1) becomes

$$\partial_t \underline{v} = (\underline{B}_2\ \partial_x^2 + \underline{B}_1\ \partial_x)\underline{v}$$

with $\underline{v} = (v_1, v_2)$ or

$$\begin{pmatrix} \partial_t v_1 \\ \partial_t v_2 \end{pmatrix} = \begin{pmatrix} \partial_x^2 + 2\partial_x & \partial_x^2 \\ 0 & \partial_x^2 \end{pmatrix} \begin{pmatrix} v_1 \\ v_2 \end{pmatrix} = \begin{pmatrix} \partial_x^2 v_1 + \partial_x^2 v_2 + 2\partial_x v_1 \\ \partial_x^2 v_2 \end{pmatrix} .$$

Consider the initial-value problem given by equation (2.7.1) with the initial condition

(2.7.2) $$\underline{v}(x,0) = \underline{f}(x) \equiv (f_1(x), \ldots, f_p(x))^T\ ,\quad -\infty < x < =a\ .$$

Let $\underline{u} = (u_1, \ldots, u_p)^T$, where each component u_j is a grid function. Let $\underline{u}_i^0 = \underline{f}(ih)$ denote the discrete initial conditions. Approximate the initial-value problem (2.7.1)-(2.7.2) by a finite difference method of the form

(2.7.3) $$\underline{Q}_1 u^{n+1} = \underline{Q}_2 \underline{u}^n\ ,\quad u > 0$$

or

(2.7.3') $$\underline{u}^{n+1} = \underline{Q}_1^{-1} \underline{Q}_2 \underline{u}^n \equiv \underline{Q}\ \underline{u}^n\ ,\quad n > 0$$

where $\underline{Q}_1$ and $\underline{Q}_2$ are $p \times p$ matrices of the form

(2.7.4a) $$\underline{Q}_1 = \sum_{j=-m_3}^{m_4} \underline{Q}_{1j}\ S_+^j\ ,$$

and

(2.7.4b)
$$\underline{Q}_2 = \sum_{j=-m_1}^{m_2} \underline{Q}_{2j}\, S_+^j ,$$

with the $\underline{Q}_{1j}$'s and $\underline{Q}_{2j}$'s $p \times p$ constant matrices. As in the scalar case (where $p = 1$), if $\underline{Q}_1 = \alpha \underline{I}$ where α is a constant $\neq 0$, then (2.7.3) is an explicit method, otherwise, it is an implicit method.

As in the scalar case, define the discrete Fourier transform of a grid function $\underline{u}$ as

(2.7.5)
$$\hat{\underline{u}}(\xi) = \sum_j \underline{u}_j e^{ij\xi} = \sum_j \begin{pmatrix} u_1 \\ \vdots \\ u_p \end{pmatrix}_j e^{ij\xi} .$$

Since Parseval's relation holds for each component equation in (2.7.5),

$$\sum_j (u_\ell)_j^2 = \frac{1}{2\pi} \int_0^{2\pi} |u_\ell(\xi)|^2 d\xi \qquad \ell = 1,\ldots,p ,$$

$$\|\underline{u}\| \equiv \frac{1}{2\pi} \int_0^{2\pi} \left(\sum_{\ell=1}^{p} |u_\ell|^2\right) d\xi_\ell \qquad \text{(by definition of } \|\underline{u}\|_2^2)$$

$$= \sum_{\ell=1}^{p} \left(\frac{1}{2\pi} \int_0^{2\pi} |u_\ell|^2 d\xi\right) = \sum_{\ell=1}^{p} \left(\sum_j (u_\ell)_j^2\right)$$

$$= \sum_j \left(\sum_{\ell=1}^{p} (u_\ell)_j^2\right) = \sum_j \|u_j\|^2 ,$$

where $\|\cdot\|$ denotes the vector norm, that is,

$$\|\underline{v}\| = (v_1^2 + \ldots + v_p^2)^{1/2} .$$

Again, as in Chapter I, we obtain

$$\widehat{(S_+\underline{u})}(\xi) = e^{-i\xi}\hat{\underline{u}}(\xi)$$

and

$$\widehat{(S_-\underline{u})}(\xi) = e^{i\xi}\hat{\underline{u}}(\xi) ,$$

since this holds for each component of $\underline{u}$. Thus by applying the discrete Fourier transform to (2.7.3), using (2.7.4),

$$\sum_{j=-m_3}^{m_4} \underline{Q}_{1j}\, e^{-ij\xi}\, \hat{\underline{u}}^{n+1}(\xi) = \sum_{j=-m_1}^{m_2} \underline{Q}_{2j}\, e^{ij\xi}\, \hat{\underline{u}}^{n}(\xi)$$

or

$$(2.7.5)\qquad \underline{\hat{u}}^{n+1}(\xi) = \left(\sum_{j=-m_3}^{m_4} \underline{Q}_{1j}\, e^{-ij\xi}\right)^{-1} \sum_{j=-m_1}^{m_2} \underline{Q}_{2j}\, e^{-ij\xi}\ \underline{\hat{u}}^{n}(\xi)\ .$$

As in the scalar case, define the symbol or amplification matrix of the finite difference method (2.7.3)

$$(2.7.6)\qquad \underline{G}(\xi) = \left(\sum_{j=-m_3}^{m_4} \underline{Q}_{1j}\, e^{-ij\xi}\right)^{-1} \sum_{j=-m_1}^{m_2} \underline{Q}_{2j}\, e^{-ij\xi}\ ,$$

a $p \times p$ matrix. Thus by substitution of (2.7.6) into (2.7.5)

$$\underline{\hat{u}}^{n+1} = \underline{G}(\xi)\underline{\hat{u}}^{n}\ ,$$

and by repeated application of (2.7.6) and (2.7.5),

$$\underline{\hat{u}}^{n+1} = \underline{G}^{n+1}(\xi)\underline{\hat{u}}^{0}\ ,$$

where $\underline{u}^0$ is the discrete initial condition. This leads to the following theorem.

Theorem 2.1. A necessary and sufficient condition for stability in the ℓ_2-norm is that there exist constants K and α independent of h and k such that

$$\|\underline{G}^{n}(\xi)\| \le Ke^{\alpha t}\ .$$

The proof is analogous to that of Chapter I and will be omitted.

The definitions of accuracy, consistency, and convergence are as defined in Chapter I.

Definition. The finite difference method (2.7.3) is said to satisfy the von Neumann condition if there exists a constant C independent of h, k, n, and ξ such that

$$\sigma(\underline{G}(\xi)) \le 1 + Ck$$

where k denotes the time step and $\sigma(G(\xi))$ denotes the spectral radius* of the amplification matrix (2.7.6).

This leads to the following theorem.

Theorem 2.2. The von Neumann condition is necessary for stability in the ℓ_2 norm.

The proof is identical to the corresponding theorem in Chapter I using the fact that

$$\sigma(\underline{G}(\xi)) \le \|\underline{G}(\xi)\|\ .$$

* The spectral radius of a $p \times p$ matrix $\underline{G}(\xi)$, $\sigma(\underline{G}(\xi))$, is

$$\sigma(\underline{G}(\xi)) = \max_{1\le i\le p} |\lambda_i(\xi)|\ ,$$

where $\lambda_i(\xi)$, $i = 1,\ldots,p$ are the eigenvalues of $\underline{G}(\xi)$.

However, the von Neumann condition is not sufficient for stability as it is in the scalar case. The cause of the difficulty is that $\underline{G}(\xi)$ is a matrix. For, in the scalar case, $|\rho^n(\xi)| = |\rho(\xi)|^n$ and, in the vector case, $\|\underline{G}^n(\xi)\| \leq \|\underline{G}(\xi)\|^n$. The inequality is in the wrong direction to show that the von Neumann condition is sufficient for stability.

Proposition. A sufficient condition for stability is that

$$\|\underline{G}(\xi)\| \leq 1 + Ck$$

where C is a constant independent of h, k, n, and ξ, and k is the time step.

Using the inequality

$$\|\underline{G}^n(\xi)\| \leq \|\underline{G}(\xi)\|^n$$

$$\leq (1 + Ck)^n \leq e^{Cnk} = e^{Ct} .$$

Hence, by Theorem 2.1, stability follows.

The following counter example due to Kreiss demonstrates that the von Neumann condition is not sufficient for stability in the vector case. Consider the differential equation

$$\partial_t \underline{v} = 0$$

where $\underline{v} = \begin{pmatrix} v_1 \\ v_2 \end{pmatrix}$ approximated by

$$\underline{u}_i^{n+1} = \underline{u}_i^n + h^2 \begin{pmatrix} 0 & 1 \\ 0 & 0 \end{pmatrix} D_+D_- \underline{u}_i^n$$

$$= \begin{pmatrix} u_{1i}^n + h^2 D_+D_- u_{2i}^n \\ u_{2i}^n \end{pmatrix} .$$

This is consistent with the differential equation. The term $h^2D_+D_-u_{2i}^n$ is small. Applying the discrete Fourier transform to the difference method, we obtain

$$\hat{\underline{u}}^{n+1}(\xi) = \underline{G}(\xi)\hat{\underline{u}}^n(\xi)$$

where

$$\underline{G}(\xi) = \begin{bmatrix} 1 & h^2D_+D_- \\ 0 & 1 \end{bmatrix} = \begin{bmatrix} 1 & -4\sin^2(\frac{\xi}{2}) \\ 0 & 1 \end{bmatrix} .$$

The eigenvalues of $\underline{G}(\xi)$ are both 1, thus $\sigma(\underline{G}(\xi)) = 1$ so the von Neumann condition is satisfied. However,

$$\underline{G}^n(\xi) = \begin{bmatrix} 1 & -4\sin^2(\frac{\xi}{2}) \\ 0 & 1 \end{bmatrix}^n = \begin{bmatrix} 1 & -4n\sin^2(\frac{\xi}{2}) \\ 0 & 1 \end{bmatrix} ,$$

which is not uniformly bounded, that is, grows with the number of time step. Hence, the finite difference method is unstable.

It turns out that just a little extra is needed in addition to the von Neumann condition for stability. There are many theorems which give sufficient conditions for stability, such as the Kreiss matrix theorem (see Appendix A). However, many of these results are difficult to use in practice. The following theorem gives practical conditions, which are sufficient for stability.

Theorem 2.3. Consider the finite difference method (2.7.3) whose amplification matrix $\underline{G}(\xi)$ satisfies the von Neumann condition. Then any one of the following conditions is sufficient for stability:

i) $\underline{G}(\xi)$ (or $\underline{Q}_1$ and $\underline{Q}_2$ in (2.7.3)) is symmetric,

ii) $\underline{G}(\xi)$ (or $\underline{Q}_1$ and $\underline{Q}_2$ in (2.7.3)) is similar* to a symmetric matrix, that is $\tilde{\underline{G}}(\xi) = \underline{S}(\xi)\underline{G}(\xi)\underline{S}^{-1}(\xi)$, where $\tilde{\underline{G}}(\xi)$ is symmetric and $\|\underline{S}(\xi)\| \le C_1$ and $\|\underline{S}^{-1}(\xi)\| \le C_2$,

iii) $\underline{G}(\xi)\ \underline{G}^*(\xi) \le \underline{I}$, where $\underline{G}^*(\xi)$ is the complex conjugate of $G(\xi)$,

iv) all elements of $\underline{G}(\xi)$ are bounded and all but one of the eigenvalues of $G(\xi)$ lie in a circle inside the unit circle, that is, all but one eigenvalue λ_i are such that $|\lambda_i| \le r < 1$ (where r is the radius of the circle inside the unit circle).

Proof. i) Since $\underline{G}(\xi)$ is symmetric, there exists a matrix $\underline{U}(\xi)$ such that

$$\underline{G}(\xi) = \underline{U}(\xi)\underline{D}(\xi)\underline{U}^T(\xi) ,$$

where

$$\underline{D}(\xi) = \mathrm{diag}(\lambda_1(\xi),\dots,\lambda_p(\xi))$$

with $\lambda_1,\dots,\lambda_p$ the eigenvalues of $\underline{G}(\xi)$ and $\underline{U}(\xi)\underline{U}^T(\xi) = \underline{I}$ (the identity matrix). For any vector $\underline{x}$,

$$\begin{aligned}
\|\underline{G}(\xi)\underline{x}\|^2 &= \underline{x}^T\underline{G}^T(\xi)\underline{G}(\xi)\underline{x} \\
&= \underline{x}^T(\underline{U}(\xi)\underline{D}(\xi)\underline{U}^T(\xi))^T\underline{U}(\xi)\underline{D}(\xi)\underline{U}^T(\xi)\underline{x} \\
&= \underline{x}^T\underline{U}(\xi)\underline{D}(\xi)\underline{U}^T(\xi)\underline{U}(\xi)\underline{D}(\xi)\underline{U}^T(\xi)\underline{x} \\
&= \underline{x}^T\underline{U}(\xi)\underline{D}^2(\xi)\underline{U}^T(\xi)\underline{x} \\
&= (\underline{U}^T(\xi)\underline{x})^T\underline{D}^2(\xi)(\underline{U}^T(\xi)x) \\
&= \underline{y}^T\underline{D}^2(\xi)\underline{y} ,
\end{aligned}$$

where $\underline{y} = \underline{U}^T(\xi)x$. This gives, where $\underline{y} = (y_1,\dots,y_p)^T$,

* A matrix $\underline{A}$ is similar to a matrix $\underline{B}$ if there exists a nonsingular matrix $\underline{S}$ such that $\underline{B} = \underline{S}\ \underline{A}\ \underline{S}^{-1}$. The matrix $\underline{S}\ \underline{A}\ \underline{S}^{-1}$ is a similarity transformation of $\underline{A}$.

$$\|\underline{G}(\xi)\underline{x}\|^2 = \sum_{\ell=1}^{p} \lambda_\ell^2 \, y_\ell^2$$

$$< \sigma(\underline{G}(\xi))^2 \sum_{\ell=1}^{p} y_\ell^2$$

$$= \sigma(\underline{G}(\xi))^2 (\underline{U}^T(\xi)\underline{x})^T (\underline{U}^T(\xi)\underline{x})$$

$$= \sigma(\underline{G}(\xi))^2 \underline{x}^T \underline{x}$$

$$= \sigma(\underline{G}(\xi))^2 \|\underline{x}\|_2^2 \, .$$

By the von Neumann condition,

$$\|\underline{G}(\xi)\underline{x}\|^2 < (1 + Ck)^2 \|\underline{x}\|^2$$

or

$$\frac{\|\underline{G}(\xi)\underline{x}\|}{\|\underline{x}\|} < (1 + Ck) \, .$$

Since this inequality holds for any vector $\underline{x}$,

$$\|\underline{G}(\xi)\| = \max_{\|\underline{x}\| \neq 0} \frac{\|\underline{G}(\xi)\underline{x}\|}{\|\underline{x}\|} < 1 + Ck \, ,$$

from which we conclude, using Theorem 2.2 that the finite difference method is stable.

An alternate way to show this result is to use the fact that if $\underline{G}(\xi)$ is symmetric, then $\sigma(\underline{G}(\xi)) = \|\underline{G}(\xi)\|$. So if the von Neumann condition is satisfied,

$$\|\underline{G}(\xi)\| = \sigma(\underline{G}(\xi)) < 1 + Ck$$

from which we conclude, using the proposition in this section, that the finite difference method is stable.

Observe, that if $\underline{Q}$ is symmetric, then $\underline{G}(\xi)$ is symmetric. Thus, if $\underline{Q}$ is symmetric, then the finite difference method is stable.

ii) Suppose $\underline{G}(\xi)$ is similar to $\underline{\tilde{G}}(\xi)$, where $\underline{\tilde{G}}(\xi)$ is symmetric. Then

$$\underline{G}(\xi) = \underline{S}^{-1}(\xi)\underline{\tilde{G}}(\xi)\underline{S}(\xi) \, .$$

This leads to

$$\underline{G}^n(\xi) = \underline{S}^{-1}(\xi)\underline{\tilde{G}}^n(\xi)\underline{S}(\xi) \, ,$$

from which

$$\|\underline{G}^n(\xi)\| < \|\underline{S}^{-1}(\xi)\| \|\underline{\tilde{G}}^n(\xi)\| \|\underline{S}(\xi)\|$$

$$< C_1 C_2 \|\underline{\tilde{G}}^n(\xi)\|$$

$$< C_1 C_2 \|\underline{G}(\xi)\|^n$$

$$= C_1 C_2 ((\underline{G}(\xi)))^n \, ,$$

since $\tilde{\underline{G}}(\xi)$ is symmetric, $\sigma(\tilde{\underline{G}}(\xi)) = \|\tilde{\underline{G}}(\xi)\|$. Furthermore, since $\underline{G}(\xi)$ is similar to $\tilde{\underline{G}}(\xi)$,

$$\sigma(\underline{G}(\xi)) = \sigma(\tilde{\underline{G}}(\xi)) \ ,$$

which gives, using the von Neumann condition,

$$\begin{aligned}\|\underline{G}^n(\xi)\| &\le C_1C_2(\sigma(\tilde{\underline{G}}(\xi)))^n \\ &= C_1C_2(\sigma(\underline{G}(\xi)))^n \\ &\le C_1C_2(1 + Ck)^n \\ &\le C_1C_2e^{Ct} \ .\end{aligned}$$

We conclude, from Theorem 2.1, that the finite difference method is stable.

iii) Let $\underline{x}$ be any vector, since $\underline{G}(\xi)\underline{G}^*(\xi) \le \underline{I}$, we have

$$\|\underline{G}(\xi)\underline{x}\|^2 = \underline{x}\ \underline{G}(\xi)\underline{G}^*(\xi)\underline{x}^* \le \underline{x}\ \underline{x}^* = \|\underline{x}\|^2$$

This gives $\|\underline{G}(\xi)\underline{x}\| \le \|\underline{x}\|$ or

$$\frac{\|\underline{G}(\xi)\underline{x}\|}{\|\underline{x}\|} \le 1 \ .$$

Since this inequality holds for every vector $\underline{x}$, we have

$$\|\underline{G}(\xi)\| = \max_{\|\underline{x}\|\neq 0} \frac{\|\underline{G}(\xi)\underline{x}\|}{\|\underline{x}\|} \le 1 \ .$$

Thus, by Theorem 2.2 the finite difference method is stable.

The proof of part iv) is given in Appendix A.

It should be commented that the von Neumann condition controls the size of the largest eigenvalue of $\underline{G}(\xi)$. However, part iv) of the theorem states how large the remaining eigenvalues may be to ensure stability.

We end this section with a perturbation theorem due to Strang.

Theorem 2.4. Consider the stable finite difference method

$$\underline{u}^{n+1} = \underline{Q}\ \underline{u}^n \tag{2.7.7}$$

and a bounded matrix $\underline{B}$, then the operator $\underline{Q} + k\underline{B}$ is stable.

Proof. The main difficulty in the proof of this theorem is that the matrices $\underline{Q}$ and $\underline{B}$ do not necessarily commute, that is, $\underline{Q}\ \underline{B} \neq \underline{B}\ \underline{Q}$.

Let C_1 and C_2 be bounds as above, such that $\|\underline{Q}^n\| \le C_1$ and $\|B\| \le C_2$. Using the binomial theorem

$$(\underline{Q} + k\underline{B})^n = \underline{Q}^n + n \text{ terms of the form } \begin{pmatrix} k\ \underline{B}\ \underline{Q} \ldots \underline{Q} \\ k\ \underline{Q}\ \underline{B}\ \underline{Q} \ldots \underline{Q} \\ k\ \underline{Q}\ \underline{Q}\ \underline{B}\ \underline{Q} \ldots \underline{Q} \end{pmatrix}$$
etc.

$$+ \binom{n}{2} \text{ terms with } 2\underline{B}\text{'s and } n - 2\ \underline{Q}\text{'s} + \ldots + k^n\underline{B}^n.$$

If an expression contains one $\underline{B}$, it divides the $\underline{Q}$'s into, at most, two groups. And in general, if an expression contains m $\underline{B}$'s, that is, $(\underline{Q}\,\underline{Q} \ldots \underline{Q}\,\underline{B}\,\underline{Q} \ldots \underline{B}\,\underline{Q} \ldots \underline{B}\,\underline{Q})$, they divide the $\underline{Q}$'s into, at most $(m+1)$ groups. This gives rise to

$$\|\underline{Q}\,\underline{Q} \ldots \underline{Q}\,\underline{B}\,\underline{Q} \ldots \underline{B}\,\underline{Q} \ldots \underline{B}\,\underline{Q})\| \leq \|\underline{Q} \ldots \underline{Q}\|\ \|\underline{B}\|\ \|\underline{Q} \ldots \underline{Q}\| \ldots \|\underline{B}\|\ \|\underline{Q}\|$$
$$\leq C_1^{m+1} C_2^m ,$$

by regrouping the terms on the right. Hence,

$$\begin{aligned}\|(\underline{Q} + k\underline{B})^n\| &\leq \|\underline{Q}^n + k\,(n \text{ terms with } 1\ \underline{B} \text{ and } n-1\ \underline{Q}\text{'s}) \\ &\quad + k^2(\tbinom{n}{2} \text{ terms with } 2\ \underline{B}\text{'s and } n-2\ \underline{Q}\text{'s}) + \ldots\| \\ &\leq C_1 + knC_1^2C_2 + k^2\tbinom{n}{2}C_1^3C_2^2 + \ldots + k^m\tbinom{n}{m}C^{m+1}C_2^m + \ldots \\ &= C_1(1 + knC_1C_2 + k^2\tbinom{n}{2}C_1^2C_2^2 + \ldots + k^m\tbinom{n}{m}C_1^mC_2^m + \ldots) \\ &= C_1(1 + kC_1C_2)^n \\ &\leq C_1 e^{C_1C_2nk} = C_1 e^{C_1C_2t} ,\end{aligned}$$

and, by Theorem 2.1, the operator $\underline{Q} + k\underline{B}$ is stable. This theorem states that a stable finite difference method remains stable after a small perturbation. In particular, a stable finite difference method remains stable after the addition of bounded undifferentiated terms.

II.8. Weak Stability.

In Section II.7 an example was given showing that the von Neumann condition was not sufficient for stability. How bad are such examples?

A typical matrix which violates the von Neumann condition is

$$\underline{G}^n = \begin{pmatrix} 2 & 0 \\ 0 & 2 \end{pmatrix}^n = \begin{pmatrix} 2^n & 0 \\ 0 & 2^n \end{pmatrix}. \tag{2.8.1}$$

$\|G^n\|$ can grow very rapidly and errors can grow quite dramatically. While the matrix

$$\underline{G}^n = \begin{pmatrix} 1 & 1 \\ 0 & 1 \end{pmatrix}^n = \begin{pmatrix} 1 & n \\ 0 & 1 \end{pmatrix}. \tag{2.8.2}$$

satisfies the von Neumann condition. However $\|\underline{G}^n\|$ is not uniformly bounded so that a scheme whose amplification matrix is given by $\underline{G}$ is not stable. The propagation of errors is not really too bad. There is a qualitative difference between the two cases (2.8.1) and (2.8.2).

This idea leads to a new type of stability called weak stability.

<u>Definition</u>. A finite difference method (2.7.3) is called <u>weakly stable</u> if

$$\|\underline{u}^n\| \leq C(t)h^{-(p-1)}\ \|\underline{u}^0\|$$

or

$$\|\underline{Q}^n\| \le C(t)h^{-(p-1)}$$

where p is the order of $\underline{Q}$. (If $k = O(h)$ then h may be replaced by k.)

The matrix given by (2.8.2) will satisfy this definition of weak stability while the matrix given by (2.8.1) will not.

This leads to the following theorem which is a direct consequence of the definition of weak stability.

Theorem 2.5. A necessary and sufficient condition for weak stability is

$$\|\underline{G}^n\| \le C(t)h^{-p} \tag{2.8.3}$$

for fixed p.

Theorem 2.6. Consider the finite difference method (2.7.3) which is weakly stable and accurate of order of $q > p$ (where p is from definition of weakly stable) and such that the solution $v(x,t)$ of the partial differential equation is smooth enough, then the finite difference method is convergent of order $q - p$.

Proof. Using the definition of accurate of order q

$$\underline{v}^n = \underline{Q}\,\underline{v}^{n-1} + kQ(h^q) .$$

Define $\underline{w}^n = \underline{u}^n - \underline{v}^n$, then

$$\begin{aligned}\underline{w}^n &= \underline{Q}\,\underline{w}^{n-1} + kO(h^q)\\ &= \underline{Q}^2\,\underline{w}^{n-2} + k\underline{Q}(O(h^q)) + kO(h^q)\\ &\;\;\vdots\\ &= \sum_{j=1}^{n} k\underline{Q}^j(O(h^q)) .\end{aligned}$$

Thus, by using the definition of weak stability,

$$\begin{aligned}\|\underline{w}^n\| &= \Big\|\sum_{j=1}^{n} k\underline{Q}^j(O(h^q))\Big\|\\ &\le \sum_{j=1}^{n} k\|\underline{Q}^j\|(O(h^q))\\ &\le \sum_{j=1}^{n} kC(t)h^{-p}O(h^q)\\ &\le nkC(t)h^{-p}O(h^q)\end{aligned}$$

or

$$\|\underline{w}^n\| \le \tilde{C}(t)O(h^{q-p}) .$$

If $q - p > 0$ then $\|\underline{w}^n\| \to 0$ as $h \to 0$.

We now come to the central theorem of this section which is due to Kreiss [16].

Theorem 2.7. Consider the finite difference method (2.7.3), where the amplication matrix $\underline{G}(\xi)$ given by (2.7.6) is uniformly bounded. The von Neumann condition is sufficient for weak stability.

Proof. There exists a unitary matrix† $\underline{U}(\xi)$ such that $\underline{U}(\xi)\underline{G}(\xi)\underline{U}^*(\xi)$ is an upper triangular matrix. This may be written in the form

$$\underline{U}(\xi)\underline{G}(\xi)\underline{U}^*(\xi) = \underline{D}(\xi) + \underline{S}(\xi) ,$$

where $\underline{D}(\xi)$ is a diagonal matrix and $\underline{S}(\xi)$ is an upper triangular matrix with zeros along the main diagonal. It follows that $\underline{S}(\xi)$ is nilpotent of order p (where p is the order of $\underline{Q}$ in (2.7.3) and $\underline{G}(\xi)$ in (2.7.6), that is, $\underline{S}^{p-1} \neq \underline{0}$ and $\underline{S}^{\ell} = \underline{0}$ for $\ell > p$. Write

$$\underline{G}(\xi) = \underline{U}^*(\xi)(\underline{D}(\xi) + \underline{S}(\xi))\underline{U}(\xi) ,$$

or

$$\underline{G}^n(\xi) = \underline{U}^*(\xi)(\underline{D}(\xi) + \underline{S}(\xi))^n\underline{U}(\xi) .$$

Since $\underline{U}(\xi)$ is unitary $\|\underline{U}(\xi)\| = \|\underline{U}^*(\xi)\| = 1$, from which it follows that

$$\begin{aligned}\|\underline{G}^n(\xi)\| &= \|\underline{U}^*(\xi)(\underline{D}(\xi) + \underline{S}(\xi))^n\underline{U}(\xi)\| , \\ &= \|(\underline{D}(\xi) + \underline{S}(\xi))^n\| .\end{aligned}$$

Using the binomial theorem to expand $(\underline{D}(\xi) + \underline{S}(\xi))^n$

$$\|\underline{G}^n(\xi)\| < \sum_{j=0}^{n} \binom{n}{j} \|(\underline{D}(\xi)\|^{n-j}\|\underline{S}(\xi))\|^j .$$

However, $\underline{S}^j = \underline{0}$ for $j > p$ so that

$$\|\underline{G}^n(\xi)\| < \sum_{j=0}^{p-1} \binom{n}{j} \|(\underline{D}(\xi)\|^{n-j}\|\underline{S}(\xi))\|^j . \tag{2.8.4}$$

Since $\underline{G}(\xi)$ is uniformly bounded, $\underline{S}(\xi)$ is uniformly bounded with bounded M and $\|\underline{S}(\xi)\|^j < M^j$. The elements of $\underline{D}(\xi)$ are the eigenvalues of $\underline{G}(\xi)$ which by the von Neumann condition $\|D(\xi)\| < 1 + ck$, where c is a constant independent of k , h , ξ , and n. Thus $\|D(\xi)\|^{n-j} < (1 + ck)^{n-j} < e^{cnk} = e^{ct}$. By substitution into (2.8.4)

$$\begin{aligned}\|\underline{G}^n(\xi)\| &< \sum_{j=0}^{p-1} \binom{n}{j} e^{ct}M^i \\ &= C(t) \sum_{j=0}^{p-1} \binom{n}{j}\end{aligned}$$

---

† A matrix $\underline{A}$ is unitary if $\underline{A}\,\underline{A}^* = \underline{A}^*\underline{A} = \underline{I}$, where $\underline{A}^*$ denotes the adjoint of $\underline{A}$. The adjoint of $\underline{A} = (a_{ij})$ is the complex conjugate transpose of $\underline{A}$, that is, $\underline{A}^* = (\overline{a_{ji}})$.

$$\leq n^{p-1}C(t)$$
$$= k^{-(p-1)}\tilde{C}(t) ,$$

where $n = t/k$ and $\tilde{C}(t) = t^{p-1}C(t)$. If $k = O(h)$ then $\|\underline{G}^n(\xi)\|$ satisfies (2.8.3) from which weak stability follows.

Under the above hypothesis the von Neumann condition is also necessary for weak stability (see Kreiss [15]).

The Strang perturbation theorem (Theorem 2.4) states that a stable finite difference method remains stable after a bounded perturbation. This result does not hold for weak stability, that is, weak stability is not invariant under bounded perturbation. To see this, we consider a counter example due to Kreiss.

Consider the differential equation

$$\partial_t \underline{v} = 0 ,$$

with $\underline{v} = (v_1, v_2)$, approximated by

$$\underline{u}^{n+1} = (\underline{I} + k^2 \underline{S}\, D_+)\underline{u}^n ,$$

where $\underline{S} = \begin{pmatrix} 0 & 1 \\ 0 & 0 \end{pmatrix}$. The amplification matrix is

$$\underline{G}(\xi) = \begin{pmatrix} 1 & (e^{-i\xi} - 1)^2 \\ 0 & 0 \end{pmatrix}.$$

Hence $\sigma(\underline{G}(\xi)) = 1$ so that the von Neumann condition is satisfied and by Theorem 2.7 the finite difference method is weakly stable.

Now consider the same finite difference method with a perturbation

$$\underline{u}^{n+1} = (\underline{I} + k^2 \underline{S}\, D_+)\underline{u}^n + \begin{pmatrix} 0 & 0 \\ k & 0 \end{pmatrix} \underline{u}^n .$$

In this case

$$\underline{G}(\xi) = \begin{pmatrix} 1 & (e^{-i\xi} - 1)^2 \\ k & 0 \end{pmatrix}$$

and $\sigma(\underline{G}(\xi)) = 1 + O(\sqrt{k})$ so that the von Neumann condition is violated and hence the method is not weakly stable. To see this more directly observe that

$$\|\underline{G}^n(\xi)\| \leq (1 + c(\sqrt{k})^n \leq e^{n\sqrt{k}} = e^{\sqrt{n}\sqrt{t}}$$

which grows exponentially with n.

We must look at each weakly stable finite difference method on an individual basis.

II.9. Multilevel Methods.

A multilevel method is a finite difference method that involves more than two time levels, that is, the advancement of the solution to time $(n+1)k$ depends on more than one previous time level. Multi-level methods are typically used to improve accuracy or stability.

An explicit finite difference method introduced by Richardson [20] to solve the prototype equation

$$\partial_t v = \partial_x^2 v$$

is

$$(2.9.1) \qquad \frac{u_i^{n+1} - u_i^{n-1}}{2k} = D_+D_-u_i^n .$$

Expanding $v(x,t)$ in a Taylor series about the point (ih,nk)

$$\frac{v_i^{n+1} - v_i^{n-1}}{2k} = (\partial_t v)_i^n + O(k^2) ,$$

and

$$D_+D_-v_i^n = (\partial_x^2 v)_i^n + O(h^2) .$$

Substitution into (2.9.1) gives the local truncation error

$$\frac{v_i^{n+1} - v_i^{n-1}}{2k} - D_+D_-v_i^n = (\partial_t v)_i^n - (\partial_x^2 v)_i^n + O(k^2) + O(h^2) = O(k^2) + O(h^2).$$

Thus Richardson's method is accurate to $O(k^2) + O(h^2)$. How does one analyze the stability of Richardson's method? In order to use the Fourier transform approach, two time levels are needed in order to define the symbol. Richardson's method involves three time levels $(n+1)k$, nk, and $(n-1)k$. The symbol of this method cannot be defined in the usual manner.

Write this as a system of two equations. Define $\underline{u}_i^n = (u_{1i}^n, u_{2i}^n)^T$. Richardson's method (2.9.1) can then be written in the form

$$(2.9.2a) \qquad u_{1i}^{n+1} = 2kD_+D_-u_{1i}^n + u_{2i}^n$$

$$(2.9.2b) \qquad u_{2i}^{n+1} = u_{1i}^n .$$

The actual solution is given by the first component of $\underline{u}_i^n$, namely, u_{1i}^n. The second component given in (2.9.2b) is an intermediate result so that the method can be written in the form $\underline{u}^{n+1} = \underline{Q}\,\underline{u}^n$, involving only two time levels.

By (2.9.2b), $u_{2i}^n = u_{1i}^{n-1}$, upon substitution into (2.9.2a), (2.9.1) is recovered in the form

$$u_{1i}^{n+1} = 2kD_+D_-u_{1i}^n + u_{1i}^{n-1} .$$

System (2.9.2a)-(2.9.2b) can be written as a system of the form

$$(2.9.3)\qquad \underline{u}_i^{n+1} = \begin{pmatrix} 2kD_+D_- & 1 \\ 1 & 0 \end{pmatrix} \underline{u}_i^n .$$

Take the discrete Fourier transform of (2.9.3),

$$\hat{u}^{n+1}(\xi) = \underline{G}(\xi)\underline{\hat{u}}^n(\xi) ,$$

where

$$\underline{G}(\xi) = \begin{pmatrix} -8\lambda \sin^2(\frac{\xi}{2}) & 1 \\ 1 & 0 \end{pmatrix},$$

which has eigenvalues

$$\mu_\pm = -4\lambda \sin^2(\tfrac{\xi}{2}) \pm (1 + 16\lambda^2 \sin^4(\tfrac{\xi}{2}))^{1/2} , \quad j = 1,2 .$$

One eigenvalue

$$\mu_- = -4\lambda \sin^2(\tfrac{\xi}{2}) - (1 + 16\lambda^2 \sin^4(\tfrac{\xi}{2}))^{1/2}$$

is < -1 for any λ, since μ_- is in the form $-a - \sqrt{1 + a^2} < -1$, where $a = 4\lambda \sin^2(\frac{\xi}{2})$. Thus Richardson's method is unconditionally unstable. This is an example demonstrating the need for stability analysis.

Writing (2.9.1) out,

$$(2.9.4)\qquad u_i^{n+1} = 2\lambda u_{i+1}^n - 4\lambda u_i^n + 2\lambda u_{i-1}^n + u_i^{n-1} .$$

Dufort and Frankel [11] suggested replacing the u_i^n term in (2.9.4) with an average over the two adjacent time levels, $\frac{1}{2}(u_i^{n+1} + u_i^{n-1})$. This gives

$$(1 + 2\lambda)u_i^{n+1} = 2\lambda(u_{i+1}^n + u_{i-1}^n) + (1 + 2\lambda)u_i^{n-1}$$

or

$$(2.9.5)\qquad u_i^{n+1} = \frac{2\lambda}{1 + 2\lambda}(u_{i+1}^n + u_{i-1}^n) + \frac{1 - 2\lambda}{1 + 2\lambda} u_i^{n-1} .$$

To analyze the stability of the Dufort-Frankel method (2.9.5), write as a system analogous to Richardson's method,

$$u_{1i}^{n+1} = \frac{2\lambda}{1 + 2\lambda}(u_{1i+1}^n + u_{1i-1}^n) + \frac{1 - 2\lambda}{1 + 2\lambda} u_{2i}^n .$$

$$u_{2i}^{n+1} = u_{1i}^n ,$$

which may be written in the form

$$(2.9.6)\qquad \underline{u}_i^{n+1} = \begin{pmatrix} \frac{2\lambda}{1 + 2\lambda}(S_+ + S_-) & \frac{1 - 2\lambda}{1 + 2\lambda} \\ 1 & 0 \end{pmatrix} \underline{u}_i^n .$$

Take the discrete Fourier transform of (2.9.6),

$$\hat{\underline{u}}^{n+1}(\xi) = \underline{G}(\xi)\,\hat{\underline{u}}^{n}(\xi)\ ,$$

where

$$\underline{G}(\xi) = \begin{pmatrix} \frac{4\lambda}{1+2\lambda}\cos\xi & \frac{1-2\lambda}{1+2\lambda} \\ 1 & 0 \end{pmatrix}\ ,$$

which has eigenvalues

$$\mu_{\pm} = \frac{2\lambda\cos\xi \pm \sqrt{1-4\lambda^2\sin^2\xi}}{1+2\lambda}\ .$$

If $1 - 4\lambda^2\sin^2\xi < 0$, then $\mu_{\pm}$ are complex and

$$\mu_{\pm} = \frac{2\lambda\cos\xi}{1+2\lambda} \pm i\,\frac{\sqrt{4\lambda^2\sin^2\xi - 1}}{1+2\lambda}\ ,$$

so that

$$|\mu_{\pm}| = \sqrt{\left|\frac{2\lambda-1}{2\lambda+1}\right|} < 1$$

for any λ. If $1 - 4\lambda^2\sin^2\xi > 0$, then $1 - 4\lambda^2\sin^2\xi \le 1$ and

$$\mu_{\pm} \le \frac{2\lambda|\cos\xi| \pm \sqrt{1-4\lambda^2\sin^2\xi}}{1+2\lambda} \le \frac{2\lambda|\cos\xi| \pm 1}{2\lambda+1} \le 1\ ,$$

for any λ. Also for μ_- ,

$$-1 < -\frac{1}{1+2\lambda} < \frac{2\lambda\cos\xi - 1}{1+2\lambda} \le \mu_- < \frac{2\lambda\cos\xi}{1+2\lambda} \le \frac{2\lambda}{1+2\lambda} < 1$$

or $|\mu_-| < 1$, for any λ. Thus the von Neumann condition is satisfied and by Theorem 2.3(iv) in Section II.7, we see that the Dufort-Frankel method is unconditionally stable.

Two things remain to be determined: the accuracy of the method and whether the method is consistant.

Expand $v(x,t)$ in a Taylor series about the (ih,nk) and substituting into (2.9.5) to obtain

$$(1+\frac{2k}{h^2})v_i^{n+1} - \frac{2k}{h^2}(v_{i+1}^n + v_{i-1}^n) - (1-\frac{2k}{h^2})v_i^{n-1}$$

$$= (1+\frac{2k}{h^2})(v_i^n + k(\partial_t v)_i^n + \frac{k^3}{2}(\partial_t^2 v)_i^n + O(k^3)) - \frac{2k}{h^2}(2v_i^n + h^2(\partial_x^2 v)_i^n$$

$$+ \frac{h^4}{12}(\partial_x^4 v)_i^n + O(h^5) - (1 - \frac{2k}{2})(v_i^n - k(\partial_t v)_i^n + \frac{k^2}{2}(\partial_t^2 v)_i^n + O(k^3))$$

$$= k(\frac{2k^2}{h^2}(\partial_t^2 v)_i^n - \frac{h^2}{6}(\partial_x^4 v)_i^n + O(k^2) + O(\frac{k^3}{h^2}) + O(h^3))\ .$$

The leading term in the local truncation error is

$$\tau = \frac{2k^2}{h^2}(\partial_t^2 v)_i^n - \frac{h^2}{6}(\partial_x^4 v)_i^n . \tag{2.9.7}$$

By the definition of consistency, τ must approach 0 as $h,k \to 0$. From (2.9.7), we see that the Dufort-Frankel method is consistent only if the ratio $\frac{k}{h} \to 0$ as $k,h \to 0$.

If $\lambda = \frac{k}{h^2}$ is constant, then $\frac{k}{h} = \lambda h$ which approaches 0 as $h \to 0$. Suppose $\frac{k}{h} = c$, a constant then by (2.9.7),

$$\tau \to 2c^2 \partial_t^2 v \tag{2.9.8}$$

as $h,k \to 0$. Thus the Dufort-Frankel method is <u>not</u> consistent with the equation

$$\partial_t v = \partial_x^2 v$$

if $\frac{k}{h}$ = constant. However, we see from (2.9.8) that the Dufort-Frankel method is consistent with the equation

$$\partial_t + c^2 \partial_t^2 v = \partial_x^2 v ,$$

if $\frac{k}{h}$ = constant.

The Dufort-Frankel method is explicit and unconditionally stable. However, it requires a time level of values u_i^1, in addition to the initial data u_i^0. Thus some other method must be used to generate u_i^1 at the time k.

One disadvantage of multilevel methods is that it can be rather messy to change the time step within the course of a calculation. This is of particular importance for equations with variable coefficients or nonlinear equations. This will be discussed in Section II.10 and II.11. In single level methods, one just changes the time step and proceeds with the calculation. This is not the case with multilevel methods. Suppose just before beginning the computation of u_i^{n+1}, the time step is to be changed. One will have to go back and recompute u_i^n based on the new time step using a single level method with u_i^{n-1} as the initial data or by some form of interpolation. Then one can proceed to compute u_i^{n+1} using the Dufort-Frankel method.

II.10. <u>Boundary Conditions Involving Derivatives</u>.

In considering the stability of a finite difference method using the discrete Fourier transform, it was assumed that the problem was defined on $-\infty < x < +\infty$, that is, it was a pure initial-value problem. When boundary conditions are introduced, the Fourier method is no longer valid.

Dirichlet Boundary conditions. The initial boundary value problem in Section II.2 (where the solution is prescribed at the boundaries)

$$v(0,t) = g_0(t)$$
$$v(1,t) = g_1(t)$$

presents no problem to stability. This will be discussed in detail in Chapter V.

Consider the initial-boundary value problem

$$\partial_t v = \partial_x^2 v \ , \quad 0 < x < 1 \ , \quad t > 0$$

$$v(x,0) = f(x) \ , \quad 0 < x < 1 \ ,$$

with Neumann boundary conditions, that is, derivative boundary conditions

$$\partial_x v(0,t) = g_0(t) \ , \quad t > 0$$
$$\partial_x v(1,t) = g_1(t) \ , \quad t > 0 \ .$$

In solving the initial-boundary value problem using a finite difference method, the derivative boundary condition must be approximated. With this, two things must be taken into consideration when selecting an approximation to the boundary conditions. First, the approximation used when combined with the finite difference method must be result in a stable method. Second, the approximation of the derivative boundary condition should be as accurate as the finite difference. Suppose the boundary condition approximation is of a lower order of accuracy. The solution from the boundary condition approximation will propagate into the interior. This will lower the order of accuracy of the finite difference method to that of the boundary condition approximation.

To demonstrate how stability in the presence of boundaries is analyzed, we shall use the Crank-Nicolson method (2.2.15) which is accurate to $O(k^2) + O(h^2)$. Consider the initial-Neumann boundary value problem just described. The approximation to $\partial_x v$ in the boundary condition should be accurate to $O(h^2)$. To this end, use the finite difference operator D_0 , so the Neumann boundary conditions become

(2.10.1)
$$D_0 u_0^n = g_0(nk)$$

and

(2.10.2)
$$D_0 u_N^n = g_1(nk) \ .$$

By writing out (2.10.1) and (2.10.2),

$$D_0 u_0^n = \frac{u_1^n - u_{-1}^n}{2h} = g_0(nk) \ ,$$

which involves a "false" point corresponding to $i = -1$, that is, u_{-1}^n which lies outside of the domain, and

$$D_0 u_N^n = \frac{u_{N+1}^n - u_{N-1}^n}{2h} = g_1(nk) ,$$

which involves a "false" point corresponding to $i = N + 1$, that is, u_{N+1}^n which lies outside the domain.

In the case of Dirichlet boundary conditions, the solution at the boundaries $x = 0$ $(i = 0)$ u_0^{n+1} and $x = 1$ $(i = N)$ u_N^{n+1} is given by the boundary conditions $g_0((n + 1)k)$ and $g_1((n + 1)k)$, respectively. Thus the finite difference method is solved at the interior grid points, that is, $u_1^{n+1}, \ldots, u_{N-1}^{n+1}$. In the case of Neumann boundary conditions, the solution at the boundaries u_0^{n+1} and u_N^{n+1} must also be obtained. Thus the finite difference method is solved at the points $u_0^{n+1}, u_1^{n+1}, \ldots, u_{N-1}^{n+1}, u_N^{n+1}$. In this case, the approximation to the Neumann boundary condition is used to close the system of equations.

How are the "false" grid points eliminated? Consider the boundary at $x = 0$. Solve the boundary condition (2.10.1) for u_{-1}^{n+1}, yielding

$$u_{-1}^{n+1} = u_1^{n+1} - 2hg_0((n + 1)k) , \tag{2.10.3}$$

and similarly

$$u_{-1}^n = u_1^n - 2hg_0(nk) . \tag{2.10.4}$$

Writing out the Crank-Nicolson method (2.2.15) for $i = 0$,

$$\begin{aligned} -\frac{\lambda}{2} u_{-1}^{n+1} + (1 + \lambda)u_0^{n+1} - \frac{\lambda}{2} u_1^{n+1} \\ = \frac{\lambda}{2} u_{-1}^n + (1 - \lambda)u_0^n + \frac{\lambda}{2} u_1^n . \end{aligned} \tag{2.10.5}$$

Substitution of (2.10.3) and (2.10.4) into (2.10.5), gives

$$\begin{aligned} -\frac{\lambda}{2} u_1^{n+1} + \lambda hg_0((n + 1)k) + (1 + \lambda)u_0^{n+1} - \frac{\lambda}{2} u_1^{n+1} \\ = \frac{\lambda}{2} u_1^n + \lambda hg_0(nk) + (1 - \lambda)u_0^n + \frac{\lambda}{2} u_1^n , \end{aligned}$$

or by regrouping,

$$\begin{aligned} (1 + \lambda)u_0^{n+1} - \lambda u_1^{n+1} \\ = (1 + \lambda)u_0^n + \lambda u_1^n - \lambda h(g_0((n + 1)k) + g_0(nk)) . \end{aligned} \tag{2.10.6}$$

Similarly, by solving the boundary condition (2.10.2) at $x = 1$, for u_{N+1}^n and u_{N+1}^{n+1} and by substitution into the Crank-Nicolson method for $i = N$

(2.10.7)
$$-\frac{\lambda}{2} u_{N-1}^{n+1} + (1 + \lambda)u_N^{n+1}$$
$$= \lambda u_{N-1}^n + (1 - \lambda)u_N^n + \lambda h(g_1((n + 1)k) + g_1(nk)) .$$

Define $\underline{u}^n = (u_0^n, u_1^n, \ldots, u_N^n)^T$, using (2.10.6) and (2.10.7) the following system is obtained

(2.10.8)
$$\underline{A}\ \underline{u}^{n+1} = \underline{B}\ \underline{u}^n + \underline{bc} ,$$

where $\underline{A}$ and $\underline{B}$ are tridiagonal matrices of order $(N + 1) \times (N + 1)$

$$\underline{A} = \begin{pmatrix} 1+\lambda & -\lambda & & & \\ -\frac{\lambda}{2} & 1+\lambda & -\frac{\lambda}{2} & & \\ & \ddots & \ddots & \ddots & \\ & & -\frac{\lambda}{2} & 1+\lambda & -\frac{\lambda}{2} \\ & & & -\lambda & 1+\lambda \end{pmatrix} ,$$

$$\underline{B} = \begin{pmatrix} 1-\lambda & \lambda & & & \\ \frac{\lambda}{2} & 1-\lambda & \frac{\lambda}{2} & & \\ & \ddots & \ddots & \ddots & \\ & & \frac{\lambda}{2} & 1-\lambda & \frac{\lambda}{2} \\ & & & \lambda & 1-\lambda \end{pmatrix} ,$$

and

$$\underline{bc} = \begin{pmatrix} -\lambda h(g_0((n+1)k) + g_0(nk)) \\ \cdot \\ \cdot \\ \cdot \\ \cdot \\ \cdot \\ \lambda h(g_1((n+1)k) + g_1(nk)) \end{pmatrix} .$$

The vector $\underline{bc}$ contains the boundary conditions at $i = 0$ and $i = N$. Observe that $\underline{A}$ and $\underline{B}$ are not symmetric. This is due to the approximation of the derivative boundary condition.

Write

$$\underline{A} = \underline{I} - \frac{\lambda}{2}\underline{Q} \quad \text{and} \quad \underline{B} = \underline{I} + \frac{\lambda}{2}\underline{Q} ,$$

where $\underline{I}$ denotes the identity matrix and $\underline{Q}$ is a non symmetric tridagonal matrix given by

$$\underline{Q} = \begin{bmatrix} -2 & 2 & & & \\ 1 & -2 & 1 & & \\ & \ddots & \ddots & \ddots & \\ & & 1 & -2 & 1 \\ & & & 2 & -2 \end{bmatrix} .$$

Write (2.10.8) in the form

(2.10.9) $$\underline{u}^{n+1} = \underline{A}^{-1}\underline{B}\ \underline{u}^n + \underline{A}^{-1}\ \underline{bc}\ .$$

To analyze the (2.10.12) consider

$$\underline{u}^{n+1} = \underline{A}^{-1}\underline{B}\ \underline{u}^n\ .$$

Take the discrete Fourier transform,

$$\hat{\underline{u}}^{n+1}(\xi) = \underline{G}(\xi)\hat{\underline{u}}^n(\xi)\ ,$$

where the amplication matrix $\underline{G}(\xi) = \underline{A}^{-1}\underline{B}$. Introduce a matrix $\underline{S}$ of order $(N + 1) \times (N + 1)$ given by

$$\underline{S} = \begin{bmatrix} \sqrt{2} & & & \\ & 1 & & \\ & & \ddots & \\ & & & 1 & \\ & & & & \sqrt{2} \end{bmatrix} ,$$

then $\underline{S}^{-1} = \text{diag}(\frac{1}{\sqrt{2}},1,\ldots,1,\frac{1}{\sqrt{2}})$. Define $\tilde{\underline{Q}}$ by the similarity transformation

$$\tilde{\underline{Q}} = \underline{S}^{-1}\underline{Q}\ \underline{S} = \begin{bmatrix} -2 & \sqrt{2} & & & \\ \sqrt{2} & -2 & \sqrt{2} & & \\ & \ddots & \ddots & \ddots & \\ & & \sqrt{2} & -2 & \sqrt{2} \\ & & & \sqrt{2} & -2 \end{bmatrix} .$$

$\tilde{\underline{Q}}$ is symmetric. Thus $\underline{Q}$ is similar to a symmetric matrix. Define the matrix $\underline{A}^{-1}\underline{B}$ by the similarity transformation

$$\begin{aligned} \underline{A}^{-1}\underline{B} &= \underline{S}^{-1}\underline{A}^{-1}\underline{B}\ \underline{S} \\ &= \underline{S}^{-1}(\underline{I} - \tfrac{\lambda}{2}\ \underline{Q})^{-1}(\underline{I} + \tfrac{\lambda}{2}\ \underline{Q})\underline{S} \\ &= \underline{S}^{-1}(\underline{I} - \tfrac{\lambda}{2}\ \underline{Q})^{-1}\underline{S}\ \underline{S}^{-1}(\underline{I} + \tfrac{\lambda}{2}\ \underline{Q})\underline{S} \\ &= (\underline{S}^{-1}(\underline{I} - \tfrac{\lambda}{2}\ \underline{Q})\underline{S})^{-1}\underline{S}^{-1}(\underline{I} + \tfrac{\lambda}{2}\ \underline{Q})\underline{S} \end{aligned}$$

$$= (\underline{I} - \frac{\lambda}{2}\tilde{\underline{Q}})^{-1}(\underline{I} + \frac{\lambda}{2}\tilde{\underline{Q}}) .$$

Since $\tilde{\underline{Q}}$ and $\underline{I}$ are symmetric, $(\underline{I} - \frac{\lambda}{2}\tilde{\underline{Q}})^{-1}$ and $(\underline{I} + \frac{\lambda}{2}\tilde{\underline{Q}})$ are symmetric and, hence, $\underline{A}^{-1}\underline{B}$ is symmetric. Thus $\underline{A}^{-1}\underline{B}$ (and hence $\underline{G}(\xi)$) is similar to a symmetric matrix.

We must verify the von Neumann condition, that is, $\sigma(\underline{G}(\xi)) \equiv \sigma(\underline{A}^{-1}\underline{B}) \leq 1 + O(k)$. To this end, consider the eigenvalues of $\underline{A}^{-1}\underline{B}$, devoted by μ_j, $j = 0,1,\ldots,N$, where

$$\mu_j = (1 - \frac{\lambda}{2}\varepsilon_j)^{-1}(1 + \frac{\lambda}{2}\varepsilon_j) = \frac{1 + \frac{\lambda}{2}\varepsilon_j}{1 - \frac{\lambda}{2}\varepsilon_j} . \tag{2.10.10}$$

and ε_j are the eigenvalues of $\underline{Q}$. The requirement that $|\mu_j| \leq 1$ is equivalent to the requirement that $\varepsilon_j \leq 0$. This can be verified directly. To determine ε_j, consider the equation

$$\det(\underline{Q} - \varepsilon\underline{I}) = 0 ,$$

Expanding in minors by the elements of the first and last rows this determinant may be written as

$$((\varepsilon + 2)^2 - 4) \det (\underline{Q}' - \varepsilon\underline{I}) = 0 ,$$

where $\underline{Q}'$ is the symmetric tridiagonal matrix of order $(N - 1) \times (N - 1)$ obtained by removing the first row and column and the last row and column of $\underline{Q}$, that is,

$$Q' = \begin{pmatrix} -2 & 1 & & & \\ 1 & -2 & 1 & & \\ & \ddots & \ddots & \ddots & \\ & & 1 & -2 & 1 \\ & & & 1 & -2 \end{pmatrix} .$$

For a tridiagonal matrix of order $M \times M$ of the form

$$\begin{pmatrix} d & c & & & \\ a & d & c & & \\ & \ddots & \ddots & \ddots & \\ & & a & d & c \\ & & & a & d \end{pmatrix}$$

where a, d, and c are real with $ac > 0$; the eigenvalues are given by

$$\rho_j = d + 2\sqrt{ac}\cos\left(\frac{j\pi}{M + 1}\right) , \quad j = 1,\ldots,M .$$

In the case considered here, $d = -2$ and $a = c = 1$, so for

$j = 1,\dots,N-1$

(2.10.11) $$\varepsilon_j = -2 + 2\cos(\tfrac{j\pi}{N}) = -4\cos^2(\tfrac{j\pi}{2N}) .$$

Solving $(\varepsilon + 2)^2 - 4 = 0$ gives $\varepsilon = 0$ or -4, which can be combined with (2.10.10) to yield the complete set of eigenvalues of $\underline{Q}$

$$\varepsilon_j = -4\cos^2(\tfrac{j\pi}{2N})$$

for $j = 0,1,\dots,N$, where $\varepsilon_0 = -4$ and $\varepsilon_N = 0$. Clearly $\varepsilon_j \leq 0$. Hence, the Crank-Nicolson method with the boundary approximations (2.10.1) and (2.10.2) is unconditionally stable Theorem 2.3.ii). $|\mu_j| < 1$ for $j = 0,1,\dots,N-1$ and $\mu_j = 1$ (corresponding to $\varepsilon_N = 0$).

II.11. Parabolic Equations with Variable Coefficients.

Consider the partial differential equation of the form

(2.11.1) $$\partial_t v = a(x,t)\partial_x^2 v + b(x,t)\partial_x v + c(x,t)v , \quad -\infty < x < +\infty , \quad 0 \leq t \leq T ,$$

with initial condition

(2.11.2) $$v(x,0) = f(x) , \quad -\infty < x < +\infty .$$

Suppose this is approximated by the explicit finite difference method

$$u^{n+1} = \sum_{j=-m_1}^{m_2} (d_j(x,t)S_+^j)u^n , \quad n \geq 0 .$$

How does one discuss stability? The symbol $\rho(x,t,\xi)$ can be defined by

$$\rho(x,t,\xi) = \sum_j d_j(x,t)e^{-ij\xi} .$$

It is not true in general, however, that

$$\hat{u}^{n+1} = \rho(x,t,\xi)\hat{u}^n$$

because in using the discrete Fourier transform technique developed in Chapter I, the following fact was used

$$\widehat{aS_+u} = ae^{-i\xi}\hat{u}$$

which depends on a being a constant or a function of t only. For if a is a function of t alone

$$a(t)\widehat{S_+u} = \sum_j a(t)u_{j+1}e^{ij\xi} = \sum_j a(t)u_j e^{i(j-1)\xi}$$

$$= \sum_j a(t)u_j e^{ij\xi}e^{-i\xi} = a(t)e^{-i\xi} \sum_j u_j e^{ij} = a(t)e^{-i\xi}\hat{u} .$$

In the case where a depends on x , however, this does not extend over.

An obvious approach is to consider the frozen coefficient case in which one picks a value $\bar{x}$ and considers the difference method

$$u^{n+1} = Au^n$$

where $A = \sum_j d_j(\bar{x},t)S_+^j$ is a constant coefficient (in space) difference operator. For this problem (which is different from the original problem), perform a Fourier analysis and consider the question of stability using the techniques developed in Chapters I and II. There is, in general, no relationship between the frozen coefficient problem and the original one. To see this, consider the following example due to Kreiss and Richtmyer

$$\partial_t v = i\,\partial_x(\sin x\,\partial_x v)$$

$$= i \sin x\,\partial_x^2 v + i \cos x\,\partial_x v .$$

The frozen coefficient case at $\bar{x} = 0$ gives the equation

$$\partial_t v = i\,\partial_x v .$$

This is <u>not</u> a well-posed problem. However, the original problem is well-posed. The source of the difficulty is that for an ill-posed problem, one cannot construct a stable difference method. This is due to the fact that stability is the discrete analogue of well-posedness for the continuous case.

John [15] considered the pure initial-value problem (2.11.1)-(2.11.2) approximated by

$$u_i^{n+1} = \sum_{j=-m_1}^{m_2} ((d_j)_i^n S_+^j)u^n , \quad n \geq 0 , \tag{2.11.3}$$

where $(d_j)_i^n = d_j(ih,nk)$.

As a first step, we shall determine the restrictions on $(d_j)_i^n$ in order that the finite difference method (2.11.3) is consistent with the equation (2.11.1). Expanding $v_{i+j}^n \equiv v((i + j)h,nk)$ in a Taylor series in space about the point (ih,nk)

$$v_{i+j}^n = v_i^n + jh\partial_x v + \frac{1}{2}(jh)^2\partial_x^2 v + O(h^3) ,$$

where the derivatives $\partial_x v$ and $\partial_x^2 v$ are evaluated at (ih,nk) , for $j = -m_1,\ldots,m_2$. Substituting into the right-hand side of the difference method (2.11.3)

$$(2.11.4)\qquad \sum_{j=-m_1}^{m_2} (d_j)_i^n \, S_+^j \, v_i^n$$

$$= \sum_{j=-m_1}^{m_2} (d_j)_i^n \left(v + jh\partial_x v + \frac{1}{2}(jh)^2 \partial_x^2 v + O(h^3)\right)$$

$$= \sum_{j=-m_1}^{m_2} (d_j)_i^n v + h \sum_{j=-m_1}^{m_2} j(d_j)_i^n \partial_x v$$

$$+ \frac{1}{2} h^2 \sum_{j=-m_1}^{m_2} j^2 (d_j)_i^n \partial_x^2 v + O(h^3) \; .$$

Expand v_i^{n+1} in a Taylor series in time about (ih,nk),

$$v_i^{n+1} = v_i^n + k\partial_t v + O(k^2) \; ,$$

which, by substitution of equation (2.11.3), gives

$$(2.11.5)\qquad v_i^{n+1} = v_i^n + kc_i^n v_i^n + kb_i^n \partial_x v + ka_i^n \partial_x^2 v + O(k^2) \; ,$$

where $a_i^n = a(ih,nk)$, etc. By equation (2.11.5), we see that (2.11.4) and (2.11.5) are equal,

$$(2.11.6)\qquad v_i^{n+1} - \sum_j (d_j)_i^n S_+^j v_i^n = (1 + kc_i^n)v_i^n + kb_i^n \partial_x v + ka_i^n \partial_x^2 v -$$

$$\sum_j (d_j)_i^n v + h \sum_j j(d_j)_i^n \partial_x v + \frac{1}{2} h^2 \sum_j j^2 (d_j)_i^n \partial_x^2 v$$

$$+ O(k^2) + O(h^3)$$

$$= [1 + kc_i^n - \sum_j (d_j)_i^n]v + [kb_i^n - h \sum_j j(d_j)_i^n]\partial_x v$$

$$+ [ka_i^n - \frac{1}{2} h^2 \sum_j j^2 (d_j)_i^n]\partial_x^2 v + O(k^2) + O(h^3) \; .$$

By consistency (as defined in Chapter II)

$$\frac{v_i^{n+1} - \sum_j (d_j)_i^n v_i^n}{k} \to 0$$

as $h \to 0$. By dividing (2.11.6) by k , this results three conditions for consistency, as $h \to 0$

$$[1 + kc_i^n - \sum_j (d_j)_i^n] \to 0$$

or

(2.11.7) $$\frac{1}{k}\left(\sum_{j=-m_1}^{m_2}(d_j)_i^n - 1\right) \to c_i^n ,$$

$$[kb_i^n - h \sum_j j(d_j)_i^n] \to 0$$

or

(2.11.8) $$\frac{h}{k}\sum_{j=-m_1}^{m_2} j(d_j)_i^n \to b_i^n ,$$

and

$$[ka_i^n - \frac{1}{2} h^2 \sum_j j^2(d_j)_i^n] \to 0$$

or

(2.11.9) $$\frac{1}{2}\frac{h^2}{k}\sum_{j=-m_1}^{m_2} j^2(d_j)_i^n \to a_i^n .$$

Assume that $a(x,t) \geq a_* > 0$, then by (2.11.9), there is a positive constant λ such that

$$\frac{h^2}{k} \to \frac{1}{\lambda}$$

as $h \to 0$. Thus $\lambda = k/h^2$ or k is a function of h , $k = \lambda h^2$. Expand the coefficients $(d_j)_i^n$ in powers of h ,

(2.11.10) $$(d_j)_i^n = (d_j^0)_i^n + h(d_j^1)_i^n + \frac{1}{2} h^2(d_j^2(h))_i^n ,$$

where we assume that $(d_j^0)_i^n$, $(d_j^1)_i^n$, and $(d_j^2(h))_i^n$ are uniformly bounded for $-\infty < ih < +\infty$ and $0 \leq nk \leq T$ and

$$(d_j^2(h))_i^n \to (d_j^2(0))_i^n \quad \text{as} \quad h \to 0 .$$

As $h \to 0$, write (2.11.7) in the form

$$\sum_j (d_j)_i^n \to 1 + kc_i^n = 1 + \lambda h^2 c_i^n .$$

By substitution of (2.11.10),

$$\sum_j [(d_j^0)_i^n + h(d_j^1)_i^n + \frac{1}{2} h^2(d_j^2(0))_i^n] \to 1 + \lambda h^2 c_i^n$$

or

$$\sum_j (d_j^0)_i^n + h \sum_j (d_j^1)_i^n + \frac{1}{2} h^2 \sum_j (d_j^2(0))_i^n \to 1 + \lambda h^2 c_i^n .$$

Equating coefficients of like powers of h

(2.11.11) $$\sum_{j=-m_1}^{m_2} (d_j^0)_i^n \to 1 , \quad \sum_{j=-m_1}^{m_2} (d_j^1)_i^n \to 0 , \text{ and}$$

$$\sum_{j=-m_1}^{m_2} (d_j^2(0))_i^n \to 2\lambda c_i^n , \quad \text{as} \quad h \to 0 .$$

As $h \to 0$, write (2.11.8) in the form

$$\sum_j j(d_j)_i^n + \frac{k}{n} b_i^n = \lambda h b_i^n ,$$

which by substitution of (2.11.10)

$$\sum_j j[(d_j^0)_i^n + h(d_j^1)_i^n + \frac{1}{2} h^2(d_j^2(0))_i^n] \to \lambda h b_i^n$$

or

$$\sum_j j(d_j^0)_i^n + h \sum_j j(d_j^1)_i^n + \frac{1}{2} h^2 \sum_j j(d_j^2(0))_i^n] = \lambda h b_i^n .$$

Equating like powers of h for h^0 and h^1

(2.11.12) $$\sum_j j(d_j^0)_i^n \to 0 \quad \text{and} \quad \sum_j j(d_j^1)_i^n \to \lambda b_i^n .$$

Finally, as $h \to 0$ write (2.11.9) in the form

$$\sum_j j^2(d_j)_i^n \to 2\lambda a_i^n ,$$

and by substitution of (2.11.10),

$$\sum_j j^2[(d_j^0)_i^n + h(d_j^1)_i^n + \frac{1}{2} h^2(d_j^2(0))] \to 2\lambda a_i^n$$

or

$$\sum_j j^2(d_j^0)_i^n + h \sum_j j^2(d_j^1)_i^n + \frac{1}{2} h^2 \sum_j j^2(d_j^2(0))_i^n \to 2\lambda a_i^n .$$

Equating the coefficients of the h^0 term,

(2.11.13) $$\sum_{j=-m_1}^{m_2} j^2(d_j^0)_i^n \to 2\lambda a_i^n .$$

Thus the consistency requirement given by (2.11.7)-(2.11.9) can be written as six component relations given by (2.11.11)-(2.11.13).

John [15] proved the following stability theorem.

<u>Theorem 2.8</u>. The finite difference method (2.11.3) is stable under the following conditions:

i) $(d_j)_i^n$ has the form (2.11.10),

ii) $(d_j^0)_i^n \geq d > 0$ for some constant d, $j = -m_1,\ldots,m_2$,

iii) $(d_j^1)_i^n$ and $(d_j^2)_i^n$ are uniformly bounded, and

iv) $\sum_{j=-m_1}^{m_2} (d_j^0)_i^n \to 1$ and $\sum_{j=-m_1}^{m_2} (d_j^1)_i^n \to 0$ as $h \to 0$.

Proof. Since $(d_j)_i^n$ can be written in the form (2.11.10) and by iii) $(d_j^1)_i^n$ and $(d_j^2)_i^n$ are uniformly bounded, for h sufficiently small

$$h(d_j^1)_i^n + \frac{1}{2} h^2(d_j^2)_i^n > -d .$$

So that by (ii)

$$(d_j)_i^n = (d_j^0)_i^n + h(d_j^1)_i^n + \frac{1}{2} h^2(d_j^2)_i^n > d - d = 0$$

or

$$(d_j)_i^n > 0$$

for sufficiently small h. This gives

$$\sum_{j=-m_1}^{m_2} |(d_j)_i^n| = \sum_{j=-m_1}^{m_2} (d_j)_i^n = \sum_{j=-m_1}^{m_2} [(d_j^0)_i^n + h(d_j^1)_i^n + \frac{1}{2} h^2(d_j^2(h))_i^n]$$

$$= 1 + \frac{1}{2} h^2 \sum_{j=-m_1}^{m_2} (d_j^0(0))_i^n \leq 1 + h^2 D$$

for sufficiently small h, where $D = \text{l.u.b.}\ \frac{1}{2} \sum_{j=-m_1}^{m_2} (d_j^0(0))_i^n$ which exists by (iii). This gives, by (2.11.3),

$$\|u^n\|_2 \leq (1 + Dh^2)\ \|u^{n-1}\|_2$$
$$\leq (1 + Dh^2)^2 \|u^{n-2}\|_2$$
$$\leq \ldots \leq (1 + Dh^2)^n \|u^0\|_2$$
$$\leq e^{nDh^2}\ \|u^0\|_2 = e^{Dt/\lambda}\ \|u^0\|_2 ,$$

for h sufficiently small. Thus, by the definition of stability, the finite difference method (2.11.3) is stable.

A finite difference method (2.11.3) is said to be of <u>positive type</u> if $(d_j)_i^n > 0$ for h sufficiently small, for $j = -m_1,\ldots,m_2$. John's theorem states that finite difference methods of positive type are stable.

As an example, suppose equation (2.11.3) is approximated by

(2.11.14) $$\begin{aligned} u_i^{n+1} &= u_i^n + ka_i^n D_+ D_- u_i^n + kb_i^n D_+ u_i^n + kc_i^n u_i^n , \quad u > 0 \\ &= \lambda a_i^n u_{i-1}^n + (1 + 2\lambda a_i^n - \lambda h b_i^n + \lambda h^2 c_i^n) u_i^n \\ &\qquad + (\lambda a_i^n + \lambda h b_i^n) u_{i+1}^n \\ &= \sum_{j=-1}^{1} (d_j)_i^n S_+^j u_i^n \end{aligned}$$

where

$$(d_{-1})_i^n = \lambda a_i^n$$

$$(d_0)_i^n = 1 - 2\lambda a_i^n - \lambda h b_i^n + \lambda h^2 c_i^n$$

and

$$(d_i)_i^n = \lambda a_1^n + \lambda h b_i^n .$$

To determine whether this finite difference method is stable, the conditions of John's theorem must be verified. $(d_{-1})_i^n$, $(d_0)_i^n$, and $(d_1)_i^n$ are in the form of (2.11.10). For example,

$$\begin{aligned} (d_0)_i^n &= (1 - 2\lambda a_i^n) + h(-\lambda b_i^n) + \frac{1}{2} h^2 (2\lambda c_i^n) \\ &= (d_0^0)_i^n + h(d_0^1)_i^n + \frac{1}{2} h^2 (d_0^2)_i^n . \end{aligned}$$

Thus condition i) is satisfied. If the functions $a(x,t)$, $b(x,t)$, and $c(x,t)$ are uniformly bounded, then condition iii) is satisfied. Since $(d_{-1}^0)_i^n = \lambda a_i^n$, $(d_0^0)_i^n = 1 - 2\lambda a_i^n$, and $(d_1^0)_i^n = \lambda a_i^n$,

$$\sum_{j=-1}^{1} (d_j^0)_i^n = \lambda a_i^n + 1 - 2\lambda a_i^n + \lambda a_i^n = 1 .$$

Also, $(d_{-1}^1)_i^n = 0$, $(d_0^1)_i^n = -\lambda b_i^n$, and $(d_1^1)_i^n = \lambda b_i^n$ so that

$$\sum_{j=-1}^{1} (d_j^1)_i^n = 0 - \lambda b_i^n + \lambda b_i^n = 0 .$$

Hence condition (iv) is satisfied. Assume that $a(x,t) \geq a_* > 0$ so $(d_{-1}^0)_i^n$ and $(d_1^0)_i^n \geq a_* > 0$. Furthermore, $(d_0^0)_i^n = 1 - 2\lambda a_i^n$ and if $\lambda \leq \lambda^* < \frac{1}{2a_*}$, then $(d_0^0)_i^n \geq d > 0$ and condition (ii) is satisfied. So the explicit method (2.11.14) is of positive type and stable provided that $a(x,t)$, $b(x,t)$, and $c(x,t)$ are bounded, $a(x,t) \geq a_* > 0$ and

(2.11.15) $$\lambda \leq \lambda^* < \frac{1}{2a(x,t)} .$$

In practice one usually uses a slightly weaker form of (2.11.15)

$$\lambda < \frac{1}{2a(x,t)} .$$

In order to determine the accuracy of the method, expand the solution in a Taylor series about the point (ih,nk). The coefficients $a(x,t)$, $b(x,t)$, $c(x,t)$ in the finite difference method (2.11.14) are evaluated at (ih,nk) so their Taylor series are just a_i^n, b_i^n, and c_i^n respectively. Upon substitution into (2.11.14)

$$\frac{v_i^{n+1} - v_i^n}{k} - a_i^n D_+ D_- v_i^n - b_i^n D_+ v_i^n - c_i^n v$$

$$= (\partial_t v)_i^n - a_i^n (\partial_x^2 v)_i^n - b_i^n (\partial_x v)_i^n - c_i^n v_i^n + O(k) + O(h^2) + O(h)$$

$$= O(k) + O(h) ,$$

where equation (2.11.1) has been used to obtain

$$(\partial_t v)_i^n - a_i^n (\partial_x^2 v)_i^n - b_i^n (\partial_x v)_i^n - c_i^n v_i^n = 0 .$$

So the truncation error $\tau = O(k) + O(h)$. This also shows that (2.11.14) is consistent. However, consistency may be determined by (2.11.7)-(2.11.9) directly

$$\frac{1}{k} \sum_{j=-1}^{1} (d_j)_i^n = \frac{1}{k} (\lambda a_i^n + 1 - 2\lambda a_i^n - \lambda h b_i^n + \lambda h^2 c_i^n + \lambda a_i^n + \lambda h b_i^n)$$

$$= \frac{1}{k} + \lambda \frac{h^2}{k} c_i^n = \frac{1}{k} + c_i^n .$$

Hence, condition (2.11.9) is satisfied

$$\frac{h}{k} \sum_{j=-1}^{1} j (d_j)_i^n = \frac{h}{k} (-1 \cdot \lambda a_i^n + 0 \cdot (1 - 2\lambda a_i^n - \lambda h b_i^n + \lambda h^2 c_i^n)$$

$$+ 1 \cdot (\lambda a_i^n + \lambda h b_i^n))$$

$$= \lambda \frac{h^2}{k} b_i^n = b_i^n ,$$

and condition (2.11.8) is satisfied

$$\frac{1}{2} \frac{h^2}{k} \sum_{j=-1}^{1} j^2 (d_j)_i^n = \frac{1}{2} \frac{h^2}{k} [(-1)^2 (\lambda a_i^n) + 0 \cdot (1 - 2\lambda a_i^n - \lambda h b_i^n + \lambda h^2 c_i^n)$$

$$+ (1)^2 (\lambda a_i^n + \lambda h b_i^n)]$$

$$= \frac{1}{2} \frac{h^2}{k} (2\lambda a_i^n + \lambda h b_i^n) = a_i^n + \frac{h}{2} b_i^n$$

$\to a_i^n$ as $h \to 0$. Hence, condition (2.11.9) is satisfied.

Let

(2.11.16) $u_i^0 = f(ih) \qquad -\infty < ih < +\infty$.

Suppose that v is the solution to the pure initial-value problem (2.11.1)-(2.11.2) where v has two continuous derivatives in x and one continuous derivative in t. Furthermore, suppose u_i^n is the solution to (2.11.3) and satisfies the initial condition (2.11.16). Under these conditions, John proved that if the finite difference method (2.11.3) is consistent and stable, then it is covergent; that is, then u_i^n converges uniformly to v as $h \to 0$.

In Forsythe and Wasow [14] this theorem is extended to the initial-boundary value problem (2.11.1)-(2.11.2), along with the boundary conditions

$$v(0,t) = g_0(t) \ , \qquad 0 < t < T$$

and

$$v(1,t) = g_1(t) \ , \qquad 0 < t < T \ ,$$

where g_0 and g_1 are bounded.

It must be pointed out that a finite difference method (2.11.3) does not have to be of positive type to be stable. There are explicit methods that are stable and not of positive type. However, they will not be discussed here.

As in the case of the parabolic equation with constant coefficients, explicit finite difference methods have restrictions of small λ which leads to taking a very large number of time steps to complete the solution of the problem. Again, we shall see that implicit methods will alleviate this difficulty.

Douglas [6] considered a special case of (2.11.1)

(2.11.17) $$r(x,t)\,\partial_t v = \partial_x(p(x,t)\,\partial_x v) - q(x,t)v \ , \quad 0 < x < 1 \ , \ 0 < t < T \ ,$$

where $p(x,t)$ and $r(x,t)$ are positive functions, with initial condition

(2.11.18) $$v(x,0) = f(x) \ , \qquad 0 < x < 1$$

and boundary conditions

(2.11.19) $$v(0,t) = g_0(t) \ , \qquad 0 < t < T \ ,$$

and

(2.11.20) $$v(1,t) = g_1(t) \ , \qquad 0 < t < T \ .$$

Define

(2.11.21) $$w(x,t) = p(x,t)\ \partial_x v \ ,$$

then (2.11.17) can be rewritten as

(2.11.22) $$r(x,t)\partial_t v = \partial_x w - q(x,t)v .$$

Expand $v(x,t)$ in a Taylor series about the point $(ih,(n+1)k)$

(2.11.23) $$\frac{v_i^{n+1} - v_i^n}{k} = (\partial_t v)_i^{n+1} + O(k)$$

and

(2.11.24) $$\frac{v_{i+\frac{1}{2}}^{n+1} - v_{i-\frac{1}{2}}^{n+1}}{h} = (\partial_x v)_i^{n+1} + O(h^2) .$$

Observe that (2.11.24) is a centered difference. Use (2.11.23) and (2.11.24) to approximate ∂_t and ∂_x in (2.11.22), respectively,

(2.11.25) $$r_i^{n+1} \left(\frac{u_i^{n+1} - u_i^n}{k}\right) = \frac{w_{i+\frac{1}{2}}^{n+1} - w_{i-\frac{1}{2}}^{n+1}}{h} - q_i^{n+1} u_i^{n+1} ,$$

where $r_i^{n+1} = r(ih,(n+1)k)$ and $q_i^{n+1} = q(ih,(n+1)k)$.

Rewrite (2.11.21) as

$$\partial_x v = \frac{w}{p} .$$

Integrating over the interval $[(i-1)h,ih]$

$$\int_{(i-1)h}^{ih} (\partial_x v)dx \equiv v_i - v_{i-1} = \int_{(i-1)h}^{ih} \left(\frac{w}{p}\right) dx .$$

Approximate the integral on the right-hand side by the midpoint rule,

$$v_i - v_{i-1} = \frac{w_{i-\frac{1}{2}}}{P_{i-\frac{1}{2}}} h + O(h^2)$$

or upon solving for $w_{i-\frac{1}{2}}$

(2.11.26) $$w_{i-\frac{1}{2}} = P_{i-\frac{1}{2}} \left(\frac{v_i - v_{i-1}}{h}\right) + O(h^2) .$$

Similarly,

(2.11.27) $$w_{i+\frac{1}{2}} = P_{i+\frac{1}{2}} \left(\frac{v_{i+1} - v_i}{h}\right) + O(h^2) .$$

Substituting the approximations (2.11.26) and (2.11.27) to $w_{i-\frac{1}{2}}$ and $w_{i+\frac{1}{2}}$ into (2.11.25)

(2.11.28) $$r_i^{n+1}\left(\frac{u_i^{n+1}-u_i^n}{k}\right) = \frac{p_{i+\frac{1}{2}}^{n+1}\left(\frac{u_{i+1}^{n+1}-u_i^{n+1}}{h}\right) - p_{i-\frac{1}{2}}^{n+1}\left(\frac{u_i^{n+1}-u_{i-1}^{n+1}}{h}\right)}{h} - q_i^{n+1}u_i^{n+1} ,$$
$$= D_-(p_{i+\frac{1}{2}}^{n+1} D_+u_i^{n+1}) - q_i^{n+1}u_i^{n+1} .$$

Since (2.11.26) and (2.11.27) are accurate to order $O(h^2)$, this fully implicit method is accurate to order $O(k) + O(h^2)$. This can also be directly verified.

Douglas [6] analyzed the stability of the implicit method (2.11.28) by considering the eigenvalues of the difference operator

$$D_-(p_{i+1}D_+u_i) - q_iu_i .$$

The details of this argument are rather long and will be omitted. The results only will be summarized here. Make the following assumptions on the coefficients p, q, and r and the time step k; there exists constants p_*, r^*, r_*, and q^* such that

(2.11.29a) $$p(x,t) \geq p_* > 0 ,$$

(2.11.29b) $$r^* \geq r(x,t) \geq r_* > 0 ,$$

(2.11.29c) $$|q(x,t)| < q^* < +\infty ,$$

(2.11.29d) $$\frac{r_*}{k} - q^* > 0 .$$

Douglas [6] prove the following theorem.

<u>Theorem 2.9</u>. If the relations (2.11.29) are satisfied, then the implicit finite difference method (2.11.28) is stable.

Observe that there is a restriction on the time step k by (2.11.29d). However, this is dependent only upon the coefficients of (2.11.17) and not the finite difference method. Furthermore, there is no relationship between k and h, so the method is unconditionally stable.

If the coefficients p, q, and r are twice boundedly differentiable on the $0 \leq x \leq 1$, $0 \leq t \leq T$ and $\partial_x^4 v$ and $\partial_t^2 v$ are bounded in this same region, then u_i^n converges to v in the ℓ_2-norm.

It remains to describe how the method is implemented. Write (2.11.28) in the form

(2.11.30) $$r_i^{n+1}u_i^{n+1} - kD_-(p_{i+\frac{1}{2}}^{n+1} D_+u_i^{n+1}) + kq_i^{n+1}u_i^{n+1} = r_i^{n+1}u_i^n .$$

For $i = 1,\ldots,N-1$, we have for $\lambda = k/h^2$

(2.11.31) $$-\lambda p^{n+1}_{i-\frac{1}{2}} u^{n+1}_{i-1} + (r^{n+1}_i + \lambda(p^{n+1}_{i+\frac{1}{2}} + p^{n+1}_{i-\frac{1}{2}}) + kq^{n+1}_i)u^{n+1}_i - \lambda p^{n+1}_{i+\frac{1}{2}} u^{n+1}_{i+1} = r^{n+1}_i u^n_i .$$

At $i = 1$, impose the boundary condition (2.11.17) at $i = 0$ given by $u^{n+1}_0 = g_0((n + 1)k)$,

(2.11.32) $$(r^{n+1}_1 + \lambda(p^{n+1}_{\frac{3}{2}} + p^{n+1}_{\frac{1}{2}}) + kq^{n+1}_1)u^{n+1}_1 - \lambda p^{n+1}_{\frac{3}{2}} u^{n+1}_2 = r^{n+1}_1 u^n_1 + \lambda p^{n+1}_{\frac{1}{2}} g_0((n + 1)k) .$$

Similarly at $i = N - 1$, impose the boundary condition (2.11.20) at $i = N$ given by $u^{n+1}_N = g_1((n + 1)k)$,

(2.11.33) $$-\lambda p^{n+1}_{N-\frac{3}{2}} u^{n+1}_{N-2} + (r^{n+1}_{N-1} + \lambda(p^{n+1}_{N-\frac{1}{2}} + p^{n+1}_{N-\frac{3}{2}}) + kq^{n+1}_{N-1})u^{n+1}_{N-1} = r^{n+1}_{N-1}u^n_{N-1} + \lambda p^{n+1}_{N-\frac{1}{2}} g_1((n + 1)k) .$$

Combining (2.11.31)-(2.11.33), this may be written as a tridiagonal system of equations of the form

$$\underline{A}\ \underline{u}^{n+1} = \underline{\tilde{u}}^n + \underline{bc}$$

where $\underline{u}^{n+1} = (u^{n+1}_1,\ldots,u^{n+1}_{N-1})^T$, $\underline{\tilde{u}}^n = (r^{n+1}_1 u^n_1,\ldots,r^{n+1}_{N-1}u^n_{N-1})^T$, $\underline{bc} = (\lambda p^{n+1}_{\frac{1}{2}} g_0((n + 1)k) , 0,\ldots, 0, \lambda p^{n+1}_{N-\frac{1}{2}} g_1((n + 1)k))^T$, and $\underline{A}$ is an $(N - 1) \times (N - 1)$ tridiagonal matrix of the form

$$\begin{pmatrix} d_1 & c_1 & & & \\ a_2 & d_2 & c_2 & & \\ & \ddots & \ddots & \ddots & \\ & & a_{N-2} & d_{N-2} & c_{M-2} \\ & & & a_{N-1} & d_{N-1} \end{pmatrix} .$$

The elements of $\underline{A}$ are given by

$$d_i = r^{n+1}_i + \lambda(p^{n+1}_{i+\frac{1}{2}} + p^{n+1}_{i-\frac{1}{2}}) + kq^{n+1}_i \quad ; \quad i = 1,\ldots,N-1 ,$$

$$a_i = -\lambda p^{n+1}_{i-\frac{1}{2}} , \quad i = 2,\ldots,N-1 ,$$

and

$$c_i = -\lambda p^{n+1}_{i+\frac{1}{2}} , \quad i = 1,\dots,N-2 .$$

As in the case of constant coefficients, the accuracy of this implicit method can be improved by using the Crank-Nicolson method. One way to obtain $O(k^2)$ accuracy (in time) is to approximate ∂_t by a finite difference operator over two-time levels. For example, in Richardson's method, ∂_t was approximated by

$$\frac{v_i^{n+1} - v_i^{n-1}}{2k} = (\partial_t v)_i^n + O(k^2) .$$

However, this method has been shown to be unconditionally unstable. In this case, the finite difference operator was centered about the point (ih,nk) , resulting in the cancellation of the $O(k^2)$ in the Taylor series expansion of v_i^{n+1} and v_i^{n-1}.

This idea can still be used to gain $O(k^2)$ accuracy. Consider a finite difference operator centered about the point $(ih,(n+\frac{1}{2})k)$. Expand v_i^{n+1} and v_i^n in a Taylor series about this point. Due to the cancellation of the $O(k^2)$ terms,

$$\frac{v_i^{n+1} - v_i^n}{k} = (\partial_t v)_i^{n+\frac{1}{2}} + O(k^2) .$$

This is the basis of the Crank-Nicolson method. Furthermore, by averaging $D_+D_-u_i^n$ and $D_+D_-u_i^{n+1}$, an $O(h^2)$ approximation is obtained to ∂_x^2 without requiring the evaluation of u at the point $(ih,(n+\frac{1}{2})k)$. For by expanding in a Taylor series about this point

$$D_+D_- \left(\frac{v_i^{n+1} + v_i^n}{2}\right) = (\partial_x^2 v)_i^{n+\frac{1}{2}} + O(h^2) .$$

Douglas [6] applied this approach to the initial-boundary value problem (2.11.17)-(2.11.20). The coefficients $r(x,t)$, $p(x,t)$ and $q(x,t)$ are evaluated at the point $(ih,(n+\frac{1}{2})k)$. With this in mind, (2.11.28) is rewritten in the spirit of the Crank-Nicolson method

(2.11.34)
$$r_i^{n+\frac{1}{2}} \left(\frac{u_i^{n+1} - u_i^n}{k}\right)$$
$$= D_- \left(p^{n+\frac{1}{2}}_{i+\frac{1}{2}} D_+ \left(\frac{u_i^{n+1} + u_i^n}{2}\right)\right) - q_i^{n+\frac{1}{2}} \left(\frac{u_i^{n+1} + u_i^n}{2}\right) .$$

Expanding the exact solution $v(x,t)$ and the coefficients in a Taylor series about the point $(ih,(n+\frac{1}{2})k)$

$$(2.11.35)\qquad \left(p_{i+\frac12}^{n+\frac12}\left(\frac{v_{i+1}^{n+1}-v_i^{n+1}}{h}\right) - q_{i-\frac12}^{n+\frac12}\left(\frac{v_i^{n+1}-v_{i-1}^{n+1}}{h}\right)\right)/2h$$

$$= \frac12 p_i^{n+\frac12}(\partial_x^2 v)_i^{n+\frac12} + \frac12(\partial_x p)_i^{n+\frac12}(\partial_x v)_i^{n+\frac12} + \frac{k}{4}(\partial_x p)_i^{n+\frac12}(\partial_t\partial_x v)_i^{n+\frac12}$$

$$+ \frac{hk}{8}(\partial_t\partial_x^2 v)_i^{n+\frac12} + O(k^2) + O(h^2) ,$$

and

$$(2.11.36)\qquad \left(p_{i+\frac12}^{n+\frac12}\left(\frac{v_{i+1}^{n+1}-v_i^{n+1}}{h}\right) - q_{i-\frac12}^{n+\frac12}\left(\frac{v_i^{n+1}-v_{i-1}^{n+1}}{h}\right)\right)/2h$$

$$= \frac12 p_i^{n+\frac12}(\partial_x^2 v)_i^{n+\frac12} + \frac12(\partial_x p)_i^{n+\frac12}(\partial_x v)_i^{n+\frac12} - \frac{k}{4}(\partial_x p)_i^{n+\frac12}(\partial_t\partial_x v)_i^{n+\frac12}$$

$$- \frac{hk}{8}(\partial_t\partial_x^2 v)_i^{n+\frac12} + O(k^2) + O(h^2) ,$$

Combining (2.11.35) and (2.11.36)

$$(2.11.37)\qquad D_-\left(p_{i+\frac12}^{n+\frac12} D_+\left(\frac{v_i^{n+1}+v_i^n}{2}\right)\right) = p_i^{n+\frac12}(\partial_x^2 v)_i^{n+\frac12} + (\partial_x p)_i^{n+\frac12}(\partial_x v)_i^{n+\frac12}$$

$$+ O(h^2) + O(k^2) .$$

Also,

$$\left(\frac{v_i^{n+1}+v_i^n}{2}\right) = v_i^{n+\frac12} + O(k^2) ,$$

so upon substitution in (2.11.34)

$$r_i^{n+\frac12}\left(\frac{v_i^{n+1}+v_i^n}{2}\right) - D_-\left(p_{i+\frac12}^{n+\frac12} D_+\left(\frac{v_i^{n+1}+v_i^n}{2}\right)\right) + q_i^{n+\frac12}\left(\frac{v_i^{n+1}+v_i^n}{2}\right)$$

$$= r_i^{n+\frac12}(\partial_t v)_i^{n+\frac12} - p_i^{n+\frac12}(\partial_x^2 v)_i^{n+\frac12} - (\partial_x p)_i^{n+\frac12}(\partial_x v)_i^{n+\frac12} + q_i^{n+\frac12} v_i^{n+\frac12}$$

$$+ O(h^2) + O(k^2)$$

$$= O(k^2) + O(h^2) ,$$

using equation (2.11.17). Thus the method is $O(k^2) + O(h^2)$ and is consistent.

Douglas [6] proved the following theorem.

Theorem 2.10. If the relations (2.11.29) are satisfied, then the Crank-Nicolson type method (2.11.34) is unconditionally stable.

If the cofficients p, q, and r are twice boundedly

differentiable on $0 < x < 1$, $0 < t < T$ and ∂_x^4 and ∂_t^3 are bounded in this same region, then u_i^n converges to v in the ℓ_2-norm. Notice that the Crank-Nicolson method requires one extra derivative in time than the fully implicit method for convergence.

We shall now describe the implementation. Write equation (2.11.34) as

$$r_i^{n+\frac{1}{2}} u_i^{n+1} - \frac{k}{2} D_- (p_{i+\frac{1}{2}}^{n+\frac{1}{2}} D_+ u_i^{n+1}) + \frac{k}{2} q_i^{n+\frac{1}{2}} u_i^{n+1}$$

$$= r_i^{n+\frac{1}{2}} u_i^n + \frac{k}{2} D_- (p_{i+\frac{1}{2}}^{n+\frac{1}{2}} D_+ u_i^n) - \frac{k}{2} q_i^{n+\frac{1}{2}} u_i^n .$$

For $i = 1,\ldots,N-1$ where $\lambda = k/h^2$ this may be expanded out as

(2.11.37)
$$- \frac{\lambda}{2} p_{i-\frac{1}{2}}^{n+\frac{1}{2}} u_{i-1}^{n+1} + (r_i^{n+\frac{1}{2}} + \frac{\lambda}{2}(p_{i+\frac{1}{2}}^{n+\frac{1}{2}} + p_{i-\frac{1}{2}}^{n+\frac{1}{2}}) + \frac{k}{2} q_i^{n+\frac{1}{2}})u_i^{n+1} - \frac{\lambda}{2} p_{i+\frac{1}{2}}^{n+\frac{1}{2}} u_{i+1}^{n+1}$$

$$= \frac{\lambda}{2} p_{i-\frac{1}{2}}^{n+\frac{1}{2}} u_{i-1}^n + (r_i^{n+\frac{1}{2}} - \frac{\lambda}{2}(p_{i+\frac{1}{2}}^{n+\frac{1}{2}} + p_{i-\frac{1}{2}}^{n+\frac{1}{2}}) + \frac{k}{2} q_i^{n+\frac{1}{2}})u_i^n + \frac{\lambda}{2} p_{i+\frac{1}{2}}^{n+\frac{1}{2}} u_{i+1}^n .$$

At $i = 1$, impose the boundary condition (2.11.12) at $i = 0$ given $u_0^{n+1} = g_0((n+1)k)$ and $u_0^n = g_0(nk)$. Thus, at $i = 1$, (2.11.37) reduces to

(2.11.38)
$$(r_1^{n+\frac{1}{2}} + \frac{\lambda}{2}(p_{\frac{1}{2}}^{n+\frac{1}{2}} + p_{\frac{3}{2}}^{n+\frac{1}{2}}) + \frac{k}{2} q_1^{n+\frac{1}{2}})u_1^{n+1} - \frac{\lambda}{2} p_{\frac{3}{2}}^{n+\frac{1}{2}} u_2^{n+1}$$

$$= (r_1^{n+\frac{1}{2}} - \frac{\lambda}{2}(p_{\frac{1}{2}}^{n+\frac{1}{2}} + p_{\frac{3}{2}}^{n+\frac{1}{2}}) + \frac{k}{2} q_1^{n+\frac{1}{2}})u_1^n + \frac{\lambda}{2} p_{\frac{3}{2}}^{n+\frac{1}{2}} u_2^n$$

$$+ \frac{\lambda}{2} p_{\frac{1}{2}}^{n+\frac{1}{2}} (g_0(nk) + g_0((n+1)k)) .$$

Similarly, at $i = N - 1$, impose the boundary condition (2.11.20) at $i = N$ given by $u_N^{n+1} = g_1((n + 1)k)$ and $u_N^n = g_1(nk)$. Thus, at $i = N - 1$, (2.11.37) reduces to

$$(2.11.39) \qquad -\frac{\lambda}{2} p_{N-\frac{3}{2}}^{n+\frac{1}{2}} u_{N-2}^{n+1} + (r_{N-1}^{n+\frac{1}{2}} + \frac{\lambda}{2}(p_{N-\frac{3}{2}}^{n+\frac{1}{2}} + p_{N-\frac{1}{2}}^{n+\frac{1}{2}}) + \frac{k}{2} q_{N-1}^{n+\frac{1}{2}})u_{N-1}^{n+1}$$

$$= \frac{\lambda}{2} p_{N-\frac{3}{2}}^{n+\frac{1}{2}} u_{N-2}^{n} + (r_{N-1}^{n+\frac{1}{2}} - \frac{\lambda}{2}(p_{N-\frac{3}{2}}^{n+\frac{1}{2}} + p_{N-\frac{1}{2}}^{n+\frac{1}{2}}) - \frac{k}{2} q_{N-1}^{n+\frac{1}{2}})u_{N-1}^{n}$$

$$+ \frac{\lambda}{2} p_{N-\frac{1}{2}}^{n+\frac{1}{2}} (g_1(nk) + g_1((n+1)k)) .$$

Combining (2.11.37)-(2.11.39), this may be written as a tridiagonal system of equations of this form

$$\underline{A}\ \underline{u}^{n+1} = \underline{\tilde{u}}^n + \underline{bc}$$

where $\underline{u}^{n+1} = (u_1^{n+1},\dots,u_{N-1}^{n+1})^T$, $\underline{bc} = (\frac{\lambda}{2} p_{\frac{1}{2}}^{n+\frac{1}{2}} (g_0(nk) + g_0((n+1)k)), 0, \dots, 0, \frac{\lambda}{2} p_{N-\frac{1}{2}}^{n+\frac{1}{2}} (g_1(nk) + g_1((n+1)k)))^T$, $\underline{\tilde{u}}^n$ contains the right-hand side of (2.11.37)-(2.11.39) excluding the boundary condition terms at $i = 1$ and $i = N - 1$. $\underline{A}$ is an $(N - 1) \times (N - 1)$ tridiagonal matrix of the form

$$\begin{pmatrix} d_1 & c_1 & & & \\ a_2 & d_2 & c_2 & & \\ & \ddots & \ddots & \ddots & \\ & & a_{N-2} & d_{N-2} & c_{N-2} \\ & & & a_{N-1} & d_{N-1} \end{pmatrix} ,$$

where the elements are given by

$$d_i = r_i^{n+\frac{1}{2}} + \frac{\lambda}{2} (p_{i-\frac{1}{2}}^{n+\frac{1}{2}} + p_{i+\frac{1}{2}}^{n+\frac{1}{2}}) + \frac{k}{2} q_i^{n+\frac{1}{2}} , \quad i = 1,\dots,N-1$$

$$a_i = -\frac{\lambda}{2} p_{i-\frac{1}{2}}^{n+\frac{1}{2}} , \quad i = 2,\dots,N-1 ,$$

and

$$c_i = -\frac{\lambda}{2} p_{i+\frac{1}{2}}^{n+\frac{1}{2}} , \quad i = 2,\dots,N-2 .$$

II.12. An Example of a Nonlinear Parabolic Equation.

Consider a prototype equation known as Burger's equation

$$(2.12.1) \qquad \partial_t v = \nu\partial_x^2 v + v\partial_x v \ , \quad 0 < x < 1 < \ , \quad 0 < t < T \ ,$$

where ν denotes a positive constant with initial condition

$$v(x,0) = f(x) \ , \quad 0 < x < 1 \ .$$

and boundary conditions

$$v(0,t) = v(1,t) = 0 \ , \quad 0 < t < T \ .$$

Approximate Burger's equation (2.12.1) by the implicit finite difference method

$$(2.12.2) \qquad \frac{u_i^{n+1} - u_i^n}{k} = \nu D_+D_-u_i^{n+1} + u_i^n D_0 u_i^{n+1}$$

with initial condition

$$u_i^0 = f(ih) \qquad 0 < i < N$$

and boundary conditions

$$u_0^n = u_N^n = 0 \ , \qquad 0 < nk < T \ .$$

The notion of stability for the linear case has not been extended to the nonlinear case and the standard convergence proofs do not hold. Consistency shall be shown in the usual manner. It will be shown that the solution u_i^n exists for $0 < nk < T$ that $\|u^n - v^n\| \to 0$ as $h \to 0$, that is, convergence.

We shall begin by considering four basic results.

Lemma 1. $\|u\|_{max} < \frac{1}{\sqrt{h}} \|u\|_2$.

Let $\|u\|_{max} = \max_i |u_i| \equiv |u_I|$, then

$$\|u\|_{max} < \frac{1}{\sqrt{h}} \{\|u\|_2\ h\}^{1/2} = \frac{1}{\sqrt{h}}\{|u_I|^2 h\}^{1/2}$$
$$< \frac{1}{\sqrt{h}} \{\sum_i |u_i|^2 h\}^{1/2} = \frac{1}{\sqrt{h}} \|u\|_2 \ .$$

Lemma 2. $|(u,v)| < \|u\|_2\|v\|_2 < \frac{\|u\|_2^2 + \|v\|_2^2}{2}$.

The Schwarz inequality gives $|(u,v)| < \|u\|_2\|v\|_2$. The inequality follows from the fact that

$$|ab| < \frac{a^2 + b^2}{2}.$$

where $a = \|u\|_2$ and $b = \|v\|_2$.

Lemma 3. Consider three grid functions u , v , and w. Let $vw = \{v_i w_i\}$ then

$$|(u,vw)| \le \|v\|_{max}(\|u\|_2 \|w\|_2) \le \|v\|_{max} \frac{\|u\|_2^2 + \|w\|_2^2}{2} .$$

By definition, $(u,vw) = \sum_i u_i(v_i w_i)h$. This gives rise to

$$\begin{aligned}|(u,vw)| &\le \sum_i |u_i v_i w_i| h = \sum_i |u_i \| v_i \| w_i| h \\ &\le \max_i |v_i| \sum_i |u_i||w_i| h \\ &= \|v\|_{max}(|u|,|w|).\end{aligned}$$

Using $\||u|\|_2 = \{\sum_i |u_i|^2 h\}^{1/2} = \{\sum_i u_i^2 h\}^{1/2} = \|u\|_2$ and the Schwarz inequality

$$\begin{aligned}|(u,vw)| &\le \|v\|_{max}(|u|,|w|) \\ &\le \|v\|_{max} \||u|\|_2 \||w|\|_2 \\ &= \|v\|_{max} \|u\|_2 \|w\|_2 .\end{aligned}$$

The results follow by using Lemma 2

$$|u,vw| \le \|v\|_{max}(\frac{\|u\|_2^2 + \|w\|_2^2}{2}) .$$

Lemma 4. $\|D_0 u\|_2 \le \|D_+ u\|_2 = \|D_- u\|_2$.

Observe that $D_0 = \frac{1}{2}(D_+ + D_-)$, thus

$$\begin{aligned}\|D_0 u\|_2 &= \|\tfrac{1}{2}(D_+ + D_-)u\|_2 \\ &\le \tfrac{1}{2}(\|D_+ u\|_2 + \|D_- u\|_2) \\ &= \tfrac{1}{2}(\|D_+ u\|_2 + \|D_+ u\|_2) \\ &= \|D_+ u\|_2 = \|D_- u\|_2 .\end{aligned}$$

We shall first show that (2.12.2) is consistent. By expanding the exact solution $v(x,t)$ in a Taylor series about the point $(ih,(n+1)k)$ and substitution into the finite difference method (2.12.2)

$$\begin{aligned}&\frac{v_i^{n+1} - v_i^n}{k} - \nu D_+ D_- v_i^{n+1} - v_i^n D_0 v_i^{n+1} \\ &= (\partial_t v)_i^{n+1} + O(k) - \nu(\partial_x^2 v)_i^{n+1} + O(h^2) - (v_i^{n+1} - k(\partial_t v)_i^{n+1} + O(k^2)) \\ &\qquad \cdot ((\partial_x v)_i^{n+1} + O(h^2)) \\ &= (\partial_t v)_i^{n+1} - \nu(\partial_x^2 v)_i^{n+1} - v_i^{n+1}(\partial_x v)_i^{n+1} + O(k) + O(h^2) ,\end{aligned}$$

which using Burger's equation (2.12.1) evaluated at $(ih,(n+1)k)$ gives

$$\text{(2.12.3)} \qquad \frac{v_i^{n+1} - v_i^n}{k} - \nu D_+D_-v_i^{n+1} - v_i^n D_0 v_i^{n+1} = O(k) + O(h^2) .$$

Thus the method is consistent and accurate of $O(k) + O(h^2)$.

The energy method will be used to show that the solution u^n exists for $0 \le nk \le T$. Take inner product of (2.12.2) with u^{n+1}

$$(u^{n+1},u^{n+1}) = (u^{n+1},u^n) + k\nu(u^{n+1},D_+D_-u^{n+1}) + k(u^{n+1},u^nD_0u^n) .$$

By discrete integration by parts

$$(u^{n+1},D_+D_-u^{n+1}) = -(D_-u^{n+1},D_-u^{n+1}) = -\|D_-u^{n+1}\|_2^2 ,$$

which gives by substitution

$$\begin{aligned}\|u^{n+1}\|_2^2 + k\nu\|D_-u^{n+1}\|_2^2 &= (u^{n+1},u^n) + k(u^{n+1},u^nD_0u^{n+1}) \\ &\le |(u^{n+1},u^n)| + k|(u^{n+1},u^nD_0u^{n+1})| \\ &\le \|u^{n+1}\|_2\|u^n\|_2 + k|(u^{n+1},u^nD_0u^{n+1})| .\end{aligned}$$

Applying Lemmas 3 and 4 to the second term on the right

$$\begin{aligned}\|u^{n+1}\|_2^2 + k\nu\|D_-u^{n+1}\|_2^2 &\le \|u^{n+1}\|_2\|u^n\|_2 + k\|u^n\|_{max}\|u^{n+1}\|_2\|D_0u^{n+1}\|_2 \\ &\le \|u^{n+1}\|_2\|u^n\|_2 + k\|u^n\|_{max}\|u^{n+1}\|_2\|D_-u^{n+1}\|_2\end{aligned}$$

Using the inequality $|ab| \le \frac{a^2+b^2}{2}$, written in the form

$$|ab| = |(\sqrt{2\nu}a)(\frac{b}{\sqrt{2\nu}})| \le \nu a^2 + \frac{b^2}{4\nu} ,$$

where $a = \|D_-u^{n+1}\|_2$ and $b = \|u^n\|_{max}\|u^{n+1}\|_2$

$$\begin{aligned}\|u^{n+1}\|_2^2 + k\nu\|D_-u^{n+1}\|_2^2 &\le \|u^{n+1}\|_2\|u^n\|_2 + k\nu\|D_-u^{n+1}\|_2^2 \\ &\quad + \frac{k\|u^n\|_{max}^2}{4\nu}\|u^{n+1}\|_2^2\end{aligned}$$

or

$$(1 - \frac{k\|u^n\|_{max}^2}{4\nu})\|u^{n+1}\|_2^2 \le \|u^{n+1}\|_2\|u^n\|_2 .$$

This finally gives

$$\text{(2.12.4)} \qquad (1 - \frac{k\|u^n\|_{max}^2}{4\nu})\|u^{n+1}\|_2^2 \le \|u^n\|_2 .$$

which gives rise to the following lemma.

Lemma 5. If $\frac{k\|u^n\|^2_{max}}{4\nu} < 1$, then u^{n+1} exists.

Using (2.12.4), if $\frac{k\|u^n\|^2_{max}}{4\nu} < 1$, then $\|u^n\|_2 = 0$ implies that $\|u^{n+1}\|_2 = 0$. So that the null space of the finite difference operator is the zero vector only and by the Fredholm alternative (see Epstein [12]), there exists a unique solution u^{n+1}.

Observe that the restriction on the timestep k (and hence on the mesh spacing h where we assume that $k = O(h^2)$, the optimal choice based on the accuracy of (2.12.2) $O(k) + O(h^2)$) becomes severe as ν becomes smaller.

We shall now prove a sort of stability condition. If the solution does not grow, then the error does not grow. This is summarized in the following lemma which gives a bound on growth.

Lemma 6. If $\|u^n\|_{max} \leq K$ for some constant K and $\frac{kK^2}{4\nu} < 1$, then

$$\|e^{n+1}\|_2 \leq (1 + C_1k)\ \|e^n\|_2 + kO(h^2) ,$$

where $e^n = v^n - u^n$ and C_1 is a constant depending on ν and K only.

Assume that v has one bounded derivative with respect to x and

$$M = \max_{\substack{0\leq x\leq 1\\ 0\leq t\leq T}} |\partial_x v| .$$

By the mean value theorem, for some ξ_i with $(i-1)h < \xi_i < (i+1)h$

$$\begin{aligned} v^{n+1}_{i+1} - v^{n+1}_{i-1} &= \partial_x v(\xi_i,(n+1)k)((i+1)h - (i-1)h) \\ &= \partial_x v(\xi_i,(n+1)h)2h , \end{aligned}$$

which gives rise to

$$D_0 v^{n+1}_i = \partial_x v(\xi_i,(n+1)k)$$

so that

$$|D_0 v^{n+1}_i| = |\partial_x v(\xi_i,(n+1)k)|$$

or

$$\text{(2.12.5)}\qquad \|D_0 v^{n+1}\|_{max} = \max_i |D_0 v^{n+1}_i| = \max_i |\partial_x v(\xi_i,(n+1)k)| \leq M .$$

Subtracting (2.12.2) from (2.12.3)

$$\begin{aligned} e^{n+1} &= e^n + k\nu D_+D_- e^{n+1} + kv^n D_0 v^{n+1} - ku^n D_0 u^{n+1} + kO(h^2) \\ &= e^n + k\nu D_+D_- e^{n+1} + kv^n D_0 v^{n+1} - ku^n D_0 u^{n+1} \\ &\qquad + kv^n D_0 v^{n+1} - ku^n D_0 u^{n+1} + kO(h^2) \end{aligned}$$

or

(2.12.6) $$e^{n+1} = e^n + k\nu D_+D_-e^{n+1} + ke^nD_0v^{n+1} + ku^nD_0e^{n+1} + kO(h^2) .$$

Taking the inner product of (2.12.6) with e^{n+1},

$$(e^{n+1},e^{n+1}) = (e^{n+1},e^n) + k\nu(e^{n+1},D_+D_-e^{n+1}) + k(e^{n+1},e^nD_0v^{n+1}) + k(e^{n+1},u^nD_0e^{n+1}) + kO(h^2) .$$

Proceed as in the first part. Using discrete integration by parts

$$\|e^{n+1}\|_2^2 + k\nu\|D_-e^{n+1}\|_2^2 = (e^{n+1},e^n) + k(e^{n+1},e^nD_0v^{n+1}) + k(e^{n+1},u^nD_0e^{n+1}) + kO(h^2)$$
$$\le \|e^{n+1}\|_2\|e^n\|_2 + k|(e^{n+1},e^nD_0v^{n+1})| \; k|(e^{n+1},u^nD_0e^{n+1})| + kO(h^2) .$$

Applying Lemma 3 to the second and third terms on the right

$$|(e^{n+1},e^nD_0(v^{n+1})| \le \|D_0v^{n+1}\|_{max} \|e^{n+1}\|_2 \|e^n\|_2 \le M\|e^{n+1}\|_2 \|e^n\|_2 ,$$

by (2.12.5), and

$$|(e^{n+1},u^nD_0(e^{n+1})| \le \|u^n\|_{max} \|e^{n+1}\|_2 \|D_0e^n\|_2 \le K\|e^{n+1}\|_2 \|D_0e^{n+1}\|_2 \le K\|e^{n+1}\|_2 \|D_-e^{n+1}\|_2 ,$$

by Lemma 4. This gives by substitution

$$\|e^{n+1}\|_2^2 + k\nu\|D_-e^{n+1}\|_2^2 \le \|e^{n+1}\|_2 \|e^n\|_2 + kM\|e^{n+1}\|_2 \|e^n\|_2 + kK\|e^{n+1}\|_2 \|D_-e^{n+1}\|_2 + kO(h^2) .$$

Again, using the inequality $|ab| \le \frac{a^2 + b^2}{2}$,

$$K\|e^{n+1}\|_2\|D_-e^{n+1}\|_2 \le \nu\|D_-e^{n+1}\|_2^2 + \frac{K^2}{4\nu} \|e^{n+1}\|_2^2$$

and

$$\|e^{n+1}\|_2^2 \le \|e^{n+1}\|_2\|e^n\|_2 + KM\|e^{n+1}\|_2\|e^n\|_2 + \frac{kK^2}{4\nu} \|e^{n+1}\|_2^2 + kO(h^2)$$

or

$$(1 - \frac{kK^2}{4\nu})\|e^{n+1}\|^2 \le (1 + Mk)\|e^n\|_2 + kO(h^2) .$$

This may be written as

$$\|e^{n+1}\|_2 \le \frac{1}{1-\frac{kK^2}{4\nu}}(1+Mk)\|e^n\|_2 + kO(h^2)$$

$$= (1 + \frac{K^2}{4\nu}k + O(h^2))^*(1+Mk)\|e^n\|_2 + kO(h^2)$$

$$= (1 + C_1 k)\|e^n\|_2 + kO(h^2) \ ,$$

where $C_1 = \frac{K^2}{4\nu} + M$.

The final stage is to prove convergence. Let

$$V = \max_{\substack{0 \le x \le 1 \\ 0 \le t \le T}} |v|$$

and k_0 be defined by

$$\frac{k_0(2V)^2}{4\nu} = 1 \ .$$

Thus k_0 is the upper bound on the timestep, that is, $0 < k < k_0$. Define

$$T_0(k) = k \max\{n : \|u^n\|_{max} \le 2V\} \ , \quad 0 < k < k_0 \ ,$$

that is, the time interval over which the estimate holds. For $0 \le nk \le T_0(k)$,

$$\frac{k\|u^n\|^2_{max}}{4\nu} \le \frac{k(2V)^2}{4\nu} < \frac{k_0(2V)^2}{4\nu} = 1$$

and the condition of Lemma 5 is satisfied. Hence, u^n exists for $0 \le nk \le T_0(k)$.

At time $t = 0$, $n = 0$ and $u^0 = v^0$ (so that $e^0 = 0$). Thus $\|u^0\|_{max} = \|v^0\|_{max} \le V < 2V$ and the hypothesis of Lemma 5 is satisfied. Hence u^1 exists, and we can at least start. For $nk \le T_0(k)$, u^n is uniquely defined and by Lemma 6

$$\|e^n\|_2 \le (1 + C_1 k)\|e^{n-1}\|_2 + C_2 kh^2 \ ,$$

for some constant C_2. Let $C = \max(C_1, C_2)$, then

$$\|e^n\|_2 \le (1 + Ck)\|e^{n-1}\|_2 + Ckh^2 \ .$$

---

*This is obtained by expanding $\frac{1}{1-z}$ in a Taylor series about $z = 0$ retaining only two terms

$$\frac{1}{1-z} = 1 + z + O(z^2) \ ,$$

and setting $z = \frac{kK^2}{4\nu}$.

Applying this estimate repeatedly

$$\begin{aligned}\|e^n\|_2 &\le (1 + Ck)[(1 + Ck)\|e^{n-1}\|_2 + Ckh^2\\ &= (1 + Ck)^2\|e^{n-2}\|_2 + (1 + (1 + Ck))Ckh^2\\ \le \dots &\le (1 + Ck)^n\|e^0\|_2 + (1 + (1 + Ck) + \dots + (1 + Ck)^{n-1})Ckh^2\\ &= (\sum_{j=0}^{n-1} (1 + Ck)^{n-1})Ckh^2 \le e^{Cnk}h^2 = e^{Ct}h^2\end{aligned}$$

or

$$\|e^n\|_2 \le e^{Cnk}h^2 = O(h^2)$$

or by Lemma 1 of this section,

$$\|e^n\|_{max} \le e^{Cnk}h^{3/2} = O(h^{3/2}) .$$

Thus we have convergence in both the ℓ_2-norm and the maximum norm for $nk \le T_0(k)$, provided that $\partial_x^4 v$ and $\partial_t^2 v$ are bounded. Since $\|u^n\|_{max} - \|v^n\|_{max} \le \|v^n - u^n\|_{max} = \|e^n\|_{max}$,

$$\|u^n\|_{max} - \|v^n\|_{max} \le e^{Cnk}h^{3/2}$$

or

$$\|u^n\|_{max} \le \|v^n\|_{max} + e^{Cnk}h^{3/2} \le V + e^{Cnk}h^{3/2} < 2V$$

for h sufficiently small. From this it follows that for T_1 where

$$e^{CT_1}h^{3/2} = V ,$$

$T_0(k) \ge T_1$. Clearly, by taking h sufficiently small, T_1 can be chosen as large as we like. Thus for h sufficiently small, $T_1 > T$, where T is defined by (2.12.1) and u^n is defined on the entire interval $0 \le t \le T$ with estimates

$$\|e^n\|_2 = O(h^2)$$

and

$$\|e^n\|_{max} = O(h^{3/2}) .$$

II.13. Nonlinear Parabolic Equations.

Consider the quasi-linear* parabolic equation

$$r(x,t,w)\partial_t w = \partial_x(p(x,t)p_1(w)\partial_x w) - q(x,t,w) .$$

* A quasi-linear equation is an equation whose coefficients depend on x , t , and the solution but not on any derivative of the solution.

Introducing the change of variable $v = \int^{w} p_1(z)dz$, this equation reduces to

(2.13.1) $$r(x,t,v)\partial_t v = \partial_x(p(x,t)\partial_x v) - q(x,t,v) \; .$$

Consider this equation on $0 < x < 1$, $0 < t < T$ along with the initial condition

(2.13.2) $$v(x,0) = f(x) \; , \quad 0 < x < 1 \; ,$$

and the boundary conditions

(2.13.3) $$v(0,t) = g_0(t) \; , \quad 0 < t < T \; ,$$

and

(2.13.4) $$v(1,t) = g_1(t) \; , \quad 0 < t < T \; .$$

As in the case of variable coefficients (Section II.11) consider

(2.13.5a) $$r_i^{n+1,n}\left(\frac{u_i^{n+1} - u_i^n}{k}\right) = D_-(p_{i+\frac{1}{2}}^{n+1} D_+ u_i^{n+1}) - q_i^{n+1,n} \; ,$$

where

$$r_i^{n+1,n} = r(ih,(n+1)k,u_i^n)$$

and

$$q_i^{n+1,n} = q(ih,(n+1)h,u_i^n) \; .$$

The initial condition becomes

(2.13.5b) $$u_i^0 = f(ih) \; , \quad 0 < ih < 1$$

and the boundary conditions become

(2.13.5c) $$u_0^n = g_0(nk) \; , \quad 0 < nk < T \; ,$$

and

(2.13.5d) $$u_N^n = g_1(nk) \; , \quad 0 < nk < T \; .$$

Expanding $v(x,t)$ in a Taylor series about the point $(ih,(n+1)k)$

$$D_-(p_{i+\frac{1}{2}}^{n+1} D_+ v_i^{n+1}) = (\partial_x(p\,\partial_x v))_i^{n+1} + O(h^2) \; ,$$

$$r(ih,(n+1)k,v_i^n) = r(ih,(n+1)k,v_i^{n+1}) - (v_i^{n+1} - v_i^n)\,\frac{\partial\bar{r}}{\partial v} \; ,$$

where $\frac{\partial\bar{r}}{\partial v} = \frac{\partial r}{\partial v}(ih,(n+1)h,v^*)$ with $v_i^n < v^* < v_i^{n+1}$. By substitution of $v_i^{n+1} - v_i^n = -k(\partial_t v)_i^{n+1} + O(k^2)$,

$$r(ih,(n+1)k,v_i^n) = r(ih,(n+1)k,v_i^{n+1}) - (k(\partial_t v)_i^{n+1} + O(k^2))\overline{\frac{\partial r}{\partial v}} .$$

Similarly,

$$q(ih,(n+1)k,v_i^n) = q(ih,(n+1)k,v_i^{n+1}) - (k(\partial_t v)_i^{n+1} + O(k^2))\overline{\frac{\partial q}{\partial v}} .$$

where $\overline{\frac{\partial q}{\partial v}} = \frac{\partial q}{\partial v}(ih,(n+1)k,v^{**})$ with $v_i^n < v^{**} < v_i^{n+1}$.

Combining these four terms

(2.13.6) $$r(ih,(n+1)k,v_i^n)\left(\frac{v_i^{n+1} - v_i^n}{k}\right) - D_-(p_{i+\frac{1}{2}}^{n+1} D_+v_i^{n+1}) + q(ih,(n+1)k,v_i^n)$$

$$= (r(ih,(n+1)k,v_i^{n+1}) - (k(\partial_t v)_i^{n+1} + O(k^2))\overline{\frac{\partial r}{\partial v}})((\partial_t v)_i^{n+1} + O(k^2))$$

$$- (\partial_x(p\partial_x v))_i^{n+1} + O(h^2) + (q(ih,(n+1)k,v_i^{n+1})$$

$$- (k(\partial_t v)_i^{n+1} + O(k^2))\overline{\frac{\partial q}{\partial v}})$$

$$= r(ih,(n+1)k,v_i^{n+1})(\partial_t v)_i^{n+1} - (\partial_x(p\partial_x v))_i^{n+1} + q(ih,(n+1)k,v_i^{n+1})$$

$$+ O(k) + O(h^2) .$$

However, by Equation (2.13.1), evaluated at the point $(ih,(n+1)k)$,

$$r(ih,(n+1)k,v_i^n)\left(\frac{v_i^{n+1} - v_i^n}{k}\right) - D_-(p_{i+\frac{1}{2}}^{n+1} D_+v_i^{n+1}) + q(ih,(n+1)k,v_i^n)$$

$$= O(k) + O(h^2) .$$

Thus, the method (2.13.5) is consistent and accurate of $O(k) + O(h^2)$.

Douglas [6] proved the following convergence theorem.

<u>Theorem 2.11</u>. Suppose that p has three bounded derivatives with respect to x and that q and r have a bounded derivative with respect to v. Further, suppose that v has four bounded derivatives with respect to x and two bounded derivatives with respect to t. Then the solution u_i^n of (2.13.5) converges in the ℓ_2-norm to the solution of (2.13.1)-(2.13.4).

Observe that the requirements for the boundedness of the various derivatives of p , q , r and the solution v are due to the truncation errors of the Taylor series.

It remains to describe how the method is implemented. Since the coefficients r and q are evaluated at u_i^n (the previous time level), the resulting algebraic equations at each timestep are linear. Write (2.13.5a) in the form (similar to (2.11.30))

(2.13.7) $$r_i^{n+1,n}u_i^{n+1} - kD_-(p_{i+\frac{1}{2}}^{n+1} D_+u_i^{n+1}) = r_i^{n+1,n}u_i^n - kq_i^{n+1,n} .$$

For $i = 1,\ldots,N-1$, and $\lambda = k/h^2$

$$(2.13.8)\qquad -\lambda p^{n+1}_{i-\frac{1}{2}}u^{n+1}_{i-1} + (r^{n+1,n}_{i} + \lambda(p^{n-1}_{i-\frac{1}{2}} + p^{n+1}_{i+\frac{1}{2}}))u^{n+1}_{i} - \lambda p^{n+1}_{i+\frac{1}{2}}u^{n+1}_{i+1}$$
$$= r^{n+1,n}_{i}u^{n}_{i} - kq^{n+1,n}_{i} .$$

At $i = 1$ impose the boundary condition (2.13.3) at $i = 0$ given by (2.13.5c). Thus, at $i = 1$, (2.13.8) reduces to

$$(2.13.9)\qquad (r^{n+1,n}_{i} + \lambda(p^{n+1}_{\frac{1}{2}} + p^{n+1}_{\frac{3}{2}}))u^{n+1}_{1} - \lambda p^{n+1}_{\frac{3}{2}}u^{n+1}_{2}$$
$$= r^{n+1,n}_{1}u^{n}_{1} - kq^{n+1,n}_{1} + \lambda p^{n+1}_{\frac{1}{2}}\, g_0((n+1)k) .$$

Similarly, at $i = N - 1$, impose the boundary condition (2.13.4) at $i = N$ given by (2.13.5d). Thus, at $i = N - 1$, (2.13.8) reduces to

$$(2.13.10)\qquad -\lambda p^{n+1}_{N-\frac{3}{2}}u^{n+1}_{N-2} + (r^{n+1,n}_{N-1} + \lambda(p^{n+1}_{N-\frac{3}{2}} + p^{n+1}_{N-\frac{1}{2}}))u^{n+1}_{N-1}$$
$$= r^{n+1,n}_{N-1}u^{n}_{N-1} - kq^{n+1,n}_{N-1} + \lambda p^{n+1}_{N-\frac{1}{2}}g_1((n+1)k) .$$

Combining (2.13.8)-(2.13.10), this may be written as a tridiagonal system of equations of the form

$$\underline{Au}^{n+1} = \tilde{\underline{u}}^{n} + \mathrm{bc}$$

where $\underline{u}^{n+1} = (u^{n+1}_{1},\ldots,u^{n+1}_{N-1})^T$, $\tilde{\underline{u}}^{n} = (r^{n+1,n}_{1}u^{n}_{1} - kq^{n+1,n}_{1},\ldots,$ $r^{n+1,n}_{N-1}u^{n}_{N-1} - kq^{n+1,n}_{N-1})^T$, $\underline{bc} = (\lambda p^{n+1}_{\frac{1}{2}}g_0((n+1)k),0,\ldots,$ $0,\lambda p^{n+1}_{N-\frac{1}{2}}g_1((n+1)k))^T$, and $\underline{A}$ is an $(N - 1) \times (N - 1)$ tridiagonal matrix of the form

$$\begin{pmatrix} d_1 & c_1 & & & \\ a_2 & d_2 & c_2 & & \\ & \ddots & \ddots & \ddots & \\ & & a_{N-2} & d_{N-2} & c_{N-2} \\ & & & a_{N-1} & d_{N-1} \end{pmatrix} .$$

The elements of $\underline{A}$ are given by

$$d_i = r^{n+1,n}_{i} + \lambda(p^{n+1}_{i-\frac{1}{2}} + p^{n+1}_{i+\frac{1}{2}}) , \quad i = 1,\ldots,N-1 ,$$

$$a_i = -\lambda p^{n+1}_{i-\frac{1}{2}} , \qquad i = 2,\ldots,N-1 ,$$

and

$$c_i = -\lambda p^{n+1}_{i+\frac{1}{2}} , \qquad i = 1,\ldots,N-2 .$$

As in the case of constant and variable coefficients, the accuracy of the implicit method (2.13.5a) can be improved by using a Crank-Nicolson-type method. The Crank-Nicolson method, as described in Section II.2, consists of evaluating the coefficients at the points $(ih,(n+1/2)k)$ and averaging the values of v at nk and $(n+1)k$ in the evaluation of v and its derivatives. This gives rise to

$$(2.13.11) \qquad r_i^{n+\frac{1}{2},n+\frac{1}{2}} \left(\frac{u_i^{n+1} - u_i^n}{k}\right) = D_-\left(p^{n+\frac{1}{2}}_{i+\frac{1}{2}} D_+\left(\frac{u_i^{n+1} + u_i^n}{2}\right)\right) - q_i^{n+\frac{1}{2},n+\frac{1}{2}}$$

where

$$r_i^{n+\frac{1}{2},n+\frac{1}{2}} = r(ih,(n + \tfrac{1}{2})k,u_i^{n+\frac{1}{2}})$$

and

$$q^{n+\frac{1}{2},n+\frac{1}{2}} = q(ih,(n + \tfrac{1}{2})k,u_i^{n+\frac{1}{2}}) .$$

However, in order to avoid evaluating $u_i^{n+\frac{1}{2}}$, expand r and q in a Taylor series about the point $(ih,(n+\frac{1}{2})k,v_i^{n+\frac{1}{2}})$. In the case of r,

$$r(ih,(n+\tfrac{1}{2})k,v) = r(ih,(n + \tfrac{1}{2})k,v_i^{n+\frac{1}{2}}) + \overline{\frac{\partial r}{\partial v}} (v - v_i^{n+\frac{1}{2}})$$

where $\overline{\frac{\partial r}{\partial v}} = \frac{\partial r}{\partial v} (ih,(n + \frac{1}{2})k,v^*)$, $v_i^{n+\frac{1}{2}} < v^* < v$. Evaluating this Taylor series at $v = v_i^{n+1}$ and $v = v_i^n$, respectively,

$$r(ih,(n+\tfrac{1}{2})k,v_i^{n+1}) = r(ih,(n + \tfrac{1}{2})k,v_i^{n+\frac{1}{2}}) + (v_i^{n+1} - v_i^{n+\frac{1}{2}}) \overline{\frac{\partial r}{\partial v}}^1$$

where $\overline{\frac{\partial r}{\partial v}}^1 = \frac{\partial r}{\partial v} (ih,(n + \frac{1}{2})k,v^*)$, with $v_i^{n+\frac{1}{2}} < v^* < v_i^{n+1}$, and

$$r(ih,(n+\tfrac{1}{2})k,v_i^n) = r(ih,(n + \tfrac{1}{2})k,v_i^{n+\frac{1}{2}}) - (v_i^{n+\frac{1}{2}} - v_i^n) \overline{\frac{\partial r}{\partial v}}^2 ,$$

where $\overline{\frac{\partial r}{\partial v}}^2 = \frac{\partial r}{\partial v} (ih,(n + \frac{1}{2})k,v^{**})$, with $v_i^n < v^{**} < v_i^{n+\frac{1}{2}}$.

Averaging these two results

(2.13.12)
$$\tfrac{1}{2}r(ih,(n+\tfrac{1}{2})k,v_i^{n+1}) + r(ih,(n+\tfrac{1}{2})k,v_i^n)$$
$$= r(ih,(n+\tfrac{1}{2})k,v_i^{n+\frac{1}{2}}) + (v_i^{n+1} - 2v_i^{n+\frac{1}{2}} + v_i^n)\,\overline{\frac{\partial r}{\partial v}}$$

where $\overline{\frac{\partial r}{\partial v}} = \frac{\partial r}{\partial v}\,(ih,(n+\tfrac{1}{2})h,v^{***})$ with $v_i^n < v^{***} < v_i^{n+1}$. Expanding v_i^{n+1} and v_i^n in a Taylor series about the point $(ih,(n+\frac{1}{2})k)$

$$v_i^{n+1} - 2v_i^{n+\frac{1}{2}} + v_i^n = \frac{k^2}{8}\,(\partial_t^2 v)_i^{n+\frac{1}{2}} + O(k^3)$$

which, by substitution in (2.13.12), gives

(2.13.13)
$$\tfrac{1}{2}(r(ih,(n+\tfrac{1}{2})k,v_i^{n+1}) + r(ih,(n+\tfrac{1}{2})k,v_i^n))$$
$$= r(ih,(n+\tfrac{1}{2})k,v_i^{n+\frac{1}{2}}) + (\frac{k^2}{8}\,(\partial_t^2 v)_i^{n+\frac{1}{2}} + O(k^3))\,\overline{\frac{\partial r}{\partial v}}$$
$$= r(ih,(n+\tfrac{1}{2})k,v_i^{n+\frac{1}{2}}) + O(k^2)\ .$$

Similarly, we have for q

(2.13.14)
$$\tfrac{1}{2}(q(ih,(n+\tfrac{1}{2})k,v_i^{n+1}) + q(ih,(n+\tfrac{1}{2})k,v_i^n))$$
$$= q(ih,(n+\tfrac{1}{2})k,v_i^{n+\frac{1}{2}}) + (\frac{k^2}{8}\,(\partial_t^2 v)_i^{n+\frac{1}{2}} + O(k^3))\,\overline{\frac{\partial q}{\partial v}}$$
$$= q(ih,(n+\tfrac{1}{2})k,v_i^{n+\frac{1}{2}}) + O(k^2)\ .$$

Use these averages (2.13.13) and (2.13.14) in (2.13.11) to obtain

(2.13.15)
$$\tfrac{1}{2}(r_i^{n+\frac{1}{2},n+1} + r_i^{n+\frac{1}{2},n})(\frac{u_i^{n+1} - u_i^n}{k})$$
$$= D_-(p_{i+\frac{1}{2}}^{n+\frac{1}{2}}\,D_+(\frac{u_i^{n+1} + u_i^n}{2})) - \tfrac{1}{2}(q_i^{n+\frac{1}{2},n+1} + q_i^{n+\frac{1}{2},n})\ .$$

From (2.13.7)

(2.13.16)
$$D_-(p_{i+\frac{1}{2}}^{n+\frac{1}{2}}\,D_+(\frac{v_i^{n+1} + v_i^n}{2})) = (\partial_x(p\partial_x v))_i^{n+\frac{1}{2}} + O(k^2) + O(h^2)\ .$$

Combining (2.13.13)-(2.13.16) along with

$$\frac{v_i^{n+1} + v_i^n}{k} = (\partial_t v)_i^{n+\frac{1}{2}} + O(k^2) \ ,$$

gives

$$\frac{1}{2}(r(ih,(n+\tfrac{1}{2})k,v_i^{n+1}) + r(ih,(n+\tfrac{1}{2})k,v_i^n))(\frac{v_i^{n+1} - v_i^n}{k})$$

$$- D_-(p_{i+\frac{1}{2}}^{n+\frac{1}{2}} D_+(\frac{v_i^{n+1} + v_i^n}{2})) + \frac{1}{2}(q(ih,(n+\tfrac{1}{2})k,v_i^{n+1}) + q(ih,(n+\tfrac{1}{2})k,v_i^n))$$

$$= (r(ih,(n+\tfrac{1}{2})k,v_i^{n+\frac{1}{2}}) + O(k^2)) \cdot ((v_t)_i^{n+\frac{1}{2}} + O(k^2)) - ((\partial_x(p\partial_x v))_i^{n+\frac{1}{2}}$$

$$+ O(k^2) + O(h^2) + q(ih,(n+\tfrac{1}{2})k,v_i^{n+\frac{1}{2}}) + O(k^2)$$

$$= (r(ih,(n+\tfrac{1}{2})k,v_i^{n+\frac{1}{2}})(v_t)_i^{n+\frac{1}{2}} - (\partial_x(p\partial_x v))_i^{n+\frac{1}{2}} + q(ih,(n+\tfrac{1}{2})k,v_i^{n+\frac{1}{2}})$$

$$+ O(k^2) + O(h^2) \ .$$

However, by Equation (2.13.1) evaluated at the point $(ih,(n+\frac{1}{2})k)$

$$\frac{1}{2}(r(ih,(n+\tfrac{1}{2})k,v_i^{n+1}) + r(ih,(n+\tfrac{1}{2})k,v_i^n))(\frac{v_i^{n+1} - v_i^n}{k})$$

$$- D_-(p_{i+\frac{1}{2}}^{n+\frac{1}{2}} D_+(\frac{v_i^{n+1} + v_i^n}{2})) + \frac{1}{2}(q(ih,(n+\tfrac{1}{2})k,v_i^{n+1}) + q(ih,(n+\tfrac{1}{2})k,v_i^n))$$

$$= O(k^2) + O(h^2) \ .$$

Thus, the method (2.13.15) is consistent and accurate of $O(k^2) + O(h^2)$.

Douglas [6] proved the following convergence theorem.

<u>Theorem 2.12</u>. Suppose that p has three bounded derivatives with respect to x and that q and r have a bounded derivative with respect to v. Further, suppose that v has four bounded derivatives with respect to x and t. Assume that k/h = constant. Then the solution of (2.13.17) converges in the ℓ_2-norm to the solution of (2.13.1)-(2.13.4).

Observe that in order to have convergence, a restriction on k and h must be imposed

$$k/h = \text{constant} \ .$$

This was not true for the Crank-Nicolson method for constant and variable coefficients.

Also, we see that the resulting set of Equation (2.13.15) is nonlinear and must be solved by an iterative process. Let $u_i^{n+1,0} = u_i^n$ and suppose the iterate $u_i^{n+1,m-1}$ is known, then (2.13.15) may be written in the form

$$\frac{1}{2}\left(r_i^{n+\frac{1}{2},n+1,m-1} + r_i^{n+\frac{1}{2},n}\right)\left(\frac{u_i^{n+1,m} - u_i^n}{k}\right) \tag{2.13.17}$$

$$= D_-\left(p_{i+\frac{1}{2}}^{n+\frac{1}{2}} D_+\left(\frac{u_i^{n+1,m} + u_i^n}{2}\right)\right) - \frac{1}{2}\left(q_i^{n+\frac{1}{2},n+1,m-1} + q_i^{n+\frac{1}{2},n}\right) ,$$

where $\quad r_i^{n+\frac{1}{2},n+1,m-1} = r(ih,(n+\frac{1}{2})k,u^{n+1,m-1})$

and $\quad q_i^{n+\frac{1}{2},n+1,m-1} = q(ih,(n+\frac{1}{2})k,u^{n+1,m-1})$.

This results in a set of linear equations since the coefficients are evaluated at the previous iterate $u_i^{n+1,m-1}$. The iteration is continued until

$$\max_i \left| \frac{u_i^{n+1,m} - u_i^{n+1,m-1}}{u_i^{n+1,m-1}} \right| < \varepsilon$$

for some tolerance ε. This method of iteration is known as functional iteration (or the method of successive approximations). When the sequence of iterates $\{u_i^{n+1,m}\}$ converges, convergence is linear, that is $u_i^{n+1,m} - u_i^{n+1} = O(u_i^{n+1,m-1} - v_i^{n+1})$ as $m \to \infty$. It is a difficult task to prove convergence of an iteration method (see Lees [17]). For $i = 1,\ldots,N-1$ with $\lambda = k/h^2$

$$- \frac{\lambda}{2} p_{i-\frac{1}{2}}^{n+\frac{1}{2}} u_{i-1}^{n+1,m} + \left(\frac{1}{2}\left(r_i^{n+\frac{1}{2},n+1,m-1} + r_i^{n+\frac{1}{2},n}\right)\right. \tag{2.13.18}$$

$$\left. + \frac{\lambda}{2}\left(p_{i-\frac{1}{2}}^{n+\frac{1}{2}} \; p_{i+\frac{1}{2}}^{n+\frac{1}{2}}\right)\right)u_{i-1}^{n+1,m} - \frac{\lambda}{2} p_{i-\frac{1}{2}}^{n+\frac{1}{2}} u_{i-1}^{n+1,m}$$

$$= \frac{\lambda}{2} p_{i-\frac{1}{2}}^{n+\frac{1}{2}} u_{i-1}^n + \left(\frac{1}{2}\left(r_i^{n+\frac{1}{2},n+1,m-1} + r_i^{n+\frac{1}{2},n}\right)\right.$$

$$\left. - \frac{\lambda}{2}\left(p_{i-\frac{1}{2}}^{n+\frac{1}{2}} + p_{i-\frac{1}{2}}^{n+\frac{1}{2}}\right)\right) + \frac{\lambda}{2} p_{i-\frac{1}{2}}^{n+\frac{1}{2}} u_{i+1}^n - \frac{k}{2}\left(q_i^{n+\frac{1}{2},n+1,m-1} + q_i^{n+\frac{1}{2},n}\right).$$

At $i = 1$ impose the boundary condition (2.13.3) at $i = 0$ given by (2.13.5c). Thus at $i = 1$, (2.13.18) reduces to

$$(2.13.19)\quad \left(\tfrac{1}{2}\left(r_i^{n+\frac{1}{2},n+1,m-1} + r_i^{n+\frac{1}{2},n}\right) + \tfrac{\lambda}{2}\left(p_{\frac{1}{2}}^{n+\frac{1}{2}} + p_{\frac{3}{2}}^{n+\frac{1}{2}}\right)\right)u_1^{n+1,m}$$

$$- \tfrac{\lambda}{2}\, p_{\frac{3}{2}}^{n+\frac{1}{2}}\, u_2^{n+1,m} = \left(\tfrac{1}{2}\left(r_i^{n+\frac{1}{2},n+1,m-1} + r_i^{n+\frac{1}{2},n}\right)\right)$$

$$- \tfrac{\lambda}{2}\left(p_{\frac{1}{2}}^{n+\frac{1}{2}} + p_{\frac{3}{2}}^{n+\frac{1}{2}}\right)u_1^n + \tfrac{\lambda}{2}\, p_{\frac{3}{2}}^{n+\frac{1}{2}} u_2^n - \left(\tfrac{k}{2}\left(q_1^{n+\frac{1}{2},n+1,m-1} + q_i^{n+\frac{1}{2},n}\right)\right)$$

$$+ \tfrac{\lambda}{2}\, p_{\frac{1}{2}}^{n+\frac{1}{2}}\left(g_0((n+1)k) + g_0(nk)\right) .$$

Similarly, at $i = N - 1$, impose the boundary condition (2.13.4) at $i = N$ given by (2.13.5d). Thus at $i = N - 1$, (2.13.18) reduces to

$$(2.13.20)\quad - \tfrac{\lambda}{2}\, p_{N-\frac{3}{2}}^{n+\frac{1}{2}}\, u_{N-2}^{n+1,m} + \left(\tfrac{1}{2}\left(r_{N-1}^{n+\frac{1}{2},n+1,m-1} + r_{N-1}^{n+\frac{1}{2},n}\right)\right)$$

$$+ \tfrac{\lambda}{2}\left(p_{N-\frac{3}{2}}^{n+\frac{1}{2}}\ p_{N-\frac{1}{2}}^{n+\frac{1}{2}}\right)u_{N-1}^{n+1,m} = \tfrac{\lambda}{2}\, p_{N-\frac{3}{2}}^{n+\frac{1}{2}}\, u_{N-2}^n$$

$$+ \left(\tfrac{1}{2}\left(r_{N-1}^{n+\frac{1}{2},n+1,m-1} + r_{N-1}^{n+\frac{1}{2},n}\right)\right) - \tfrac{\lambda}{2}\left(p_{N-\frac{3}{2}}^{n+\frac{1}{2}} + p_{N-\frac{1}{2}}^{n+\frac{1}{2}}\right)u_{N-1}^n$$

$$- \tfrac{k}{2}\left(q_{N-1}^{n+\frac{1}{2},n+1,m-1} + q_{N-1}^{n+\frac{1}{2},n}\right) + \tfrac{\lambda}{2}\, p_{i-\frac{1}{2}}^{n+\frac{1}{2}}\left(g_1((n+1)k) = g_1(nk)\right) .$$

Combining (2.13.18)-(2.13.20), this may be written as a tridiagonal system of equations of the form

$$\underline{A}^{m-1}\underline{u}^{n+1,m} = \underline{\tilde{u}}^{n,m-1} + \underline{bc}$$

where $\underline{u}^{n+1,m} = (u_1^{n+1,m},\ldots,u_{N-1}^{n+1,m})^T$,

$$\underline{bc} = \left(\tfrac{\lambda}{2}\, p_{\frac{1}{2}}^{n+\frac{1}{2}}(g_0(nk) + g_0((n+1)k)),0,\ldots,0,\tfrac{\lambda}{2}\, p_{N-\frac{1}{2}}^{n+\frac{1}{2}}(g_1(nk) + g_1((n+1)k))\right)^T,$$

and $\underline{\tilde{u}}^{n,m-1}$ contains the right-hand side of (2.13.18)-(2.13.20) excluding the boundary condition terms at $i = 1$ and $i = N - 1$. $\underline{A}^{m-1}$ is an $(N - 1) \times (N - 1)$ tridiagonal matrix of the form

$$\begin{bmatrix} d_1^{m-1} & c_1 & & & \\ a_2 & d_2^{m-1} & c_2 & & \\ & \ddots & \ddots & \ddots & \\ & & a_{N-2} & d_{N-2}^{m-1} & c_{N-2} \\ & & & a_{N-1} & d_{N-1}^{m-1} \end{bmatrix} ,$$

where the elements are given by

$$d_i^{m-1} = \tfrac{1}{2}(r_i^{n+\frac{1}{2},n+1,m-1} + r_i^{n+\frac{1}{2},n}) + \tfrac{\lambda}{2}(p_{i-\frac{1}{2}}^{n+\frac{1}{2}} + p_{i+\frac{1}{2}}^{n+\frac{1}{2}}) , \quad i = 1,\ldots,N-1 ,$$

$$a_i = -\tfrac{\lambda}{2} p_{i-\frac{1}{2}}^{n+\frac{1}{2}} , \quad i = 2,\ldots,N-1 , \quad \text{and}$$

$$c_i = -\tfrac{\lambda}{2} p_{i+\frac{1}{2}}^{n+\frac{1}{2}} , \quad i = 1,\ldots,N-2.$$

The nonlinear equation (2.13.15) results from replacing $r_i^{n+\frac{1}{2},n+\frac{1}{2}}$ and $q_i^{n+\frac{1}{2},n+\frac{1}{2}}$ in (2.13.11) with $\frac{1}{2}(r_i^{n+\frac{1}{2},n+\frac{1}{2}} + r_i^{n+\frac{1}{2},n})$ and $\frac{1}{2}(q_i^{n+\frac{1}{2},n+\frac{1}{2}} + q_i^{n+\frac{1}{2},n})$, respectively, using (2.13.13) and (2.13.14).

Suppose, instead, $u_i^{n+\frac{1}{2}}$ is approximated using u_i^n. To this end, expand $v(x,t)$ in a Taylor series about the point (ih,nk) and evaluate at the point $(ih,(n+\frac{1}{2})k)$

$$v_i^{n+\frac{1}{2}} = v_i^n + \tfrac{k}{2}(\partial_t v)_i^n + O(k^2) .$$

By substitution of the equation (2.13.1) written in the form

$$\partial_t v = \frac{1}{r(x,t,v)} [\partial_x(p(x,t)\partial_x v) - q(x,t,v)]$$

evaluated at (ih,nk) gives

$$\text{(2.13.21)} \quad v_i^{n+\frac{1}{2}} = v_i^n + \frac{k}{2r(ih,nk,v_i^n)} [(\partial_x(p\partial_x v))_i^n - q(ih,nk,v_i^n)] + O(k^2)$$

$$= v_i^n + \frac{k}{2r(ih,nk,v_i^n)} [D_-(p_{i+\frac{1}{2}}^n D_+ v_i^n) - q(ih,nk,v_i^n)] + O(k^2) + O(kh^2).$$

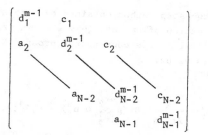

This gives rise to the two-step method obtained by combining (2.13.11) and (2.13.21). The evaluation of p at $(n + \frac{1}{2})k$ in (2.13.21) was used by Douglas [6] to simplify the convergence proof. Nothing is lost in practice by using this value.

Define

(2.13.22) $$\hat{u}_i^n = u_i^n + \frac{k}{2r_i^{n,n}} [D_-(p_{i+\frac{1}{2}}^{n+\frac{1}{2}} D_+u_i^n) - q_i^{n,n}] \ .$$

This is the first step which is an approximation to $v_i^{n+\frac{1}{2}}$ accurate to $O(k^2) + O(kh^2)$. The second step consists of replacing $u_i^{n+\frac{1}{2}}$ in (2.13.11) with u_i^n , that is,

(2.13.23) $$\hat{r}_i^{n,n} \left(\frac{u_i^{n+1} - u_i^n}{k}\right) = D_-(p_{i+\frac{1}{2}}^{n+\frac{1}{2}} D_+(\frac{u_i^{n+1} + u_i^n}{2})) - \hat{q}_i^{n,n} \ ,$$

where

$$\hat{r}_i^{n,n} = r(ih,nk,\hat{u}_i^n)$$

and

$$\hat{q}_i^{n,n} = q(ih,nk,\hat{u}_i^n) \ .$$

Rewrite (2.13.21) in the form

(2.13.24) $$v_i^n + \frac{k}{2r(ih,nk,v_i^n)} [D_-(p_{i+\frac{1}{2}}^{n+\frac{1}{2}} D_+v_i^n) - q(ih,nk,v_i^n)]$$

$$= v_i^{n+\frac{1}{2}} + O(k^2) + O(h^2) \ .$$

Expanding r in a Taylor series about the point $(ih,(n+\frac{1}{2})k,v_i^{n+\frac{1}{2}})$

$$r(ih,(n+\tfrac{1}{2})k,v) = r(ih,(n+\tfrac{1}{2})k,v_i^{n+\frac{1}{2}}) + (v - v_i^{n+\frac{1}{2}})\overline{\frac{\partial r}{\partial v}} \ ,$$

where $\overline{\frac{\partial r}{\partial v}} = \frac{\partial r}{\partial v}(ih,(n+\frac{1}{2})k,v^*)$ with $v_i^{n+\frac{1}{2}} < v^* < v$. Replacing v with the left-hand side of (2.13.24)

$$(2.13.25)\quad r(ih,(n+\tfrac{1}{2})k,v_i^n + \frac{k}{2r(ih,nk,v_i^n)}[D_-(p_{i+\frac{1}{2}}^{n+\frac{1}{2}} D_+v_i^n) - q(ih,nk,v_i^n)])$$

$$= r(ih,(n+\tfrac{1}{2})k,v_i^{n+\frac{1}{2}}) + (v_i^{n+\frac{1}{2}} + O(k^2) + O(h^2) - v_i^{n+\frac{1}{2}})\overline{\frac{\partial r}{\partial v}}$$

$$= r(ih,(n+\tfrac{1}{2})k,v_i^{n+\frac{1}{2}}) + O(k^2) + O(h^2).$$

A similar result is obtained for q,

$$(2.13.26)\quad q(ih,(n+\tfrac{1}{2})k,v_i^n + \frac{k}{2r(ih,nk,v_i^n)}[D_-(p_{i+\frac{1}{2}}^{n+\frac{1}{2}} D_+v_i^n) - q(ih,nk,v_i^n)])$$

$$= q(ih,(n+\tfrac{1}{2})k,v_i^{n+\frac{1}{2}}) + O(k^2) + O(h^2).$$

Combining (2.13.15) and (2.13.16) with (2.13.25) and (2.13.26) and substituting into (2.13.11), we obtain

$$(2.13.27)\quad r(ih,(n+\tfrac{1}{2})k,v_i^n + \frac{k}{2r(ih,nk,v_i^n)}[D_-(p_{i+\frac{1}{2}}^{n+\frac{1}{2}} D_+v_i^n) - q(ih,nk,v_i^n)])$$

$$(\frac{v_i^{n+1} - v_i^n}{k}) - D_-(p_{i+\frac{1}{2}}^{n+\frac{1}{2}} D_+(\frac{v_i^{n+1} + v_i^n}{2}))$$

$$+ q(ih,(n+\tfrac{1}{2})k,v_i^n + \frac{k}{2r(ih,nk,v_i^n)}[D_-(p_{i+\frac{1}{2}}^{n+\frac{1}{2}} D_+v_i^n) - q(ih,nk,v_i^n)])$$

$$= r(ih,(n+\tfrac{1}{2})k,v_i^{n+\frac{1}{2}}) + O(k^2) + O(h^2))((\partial_t v)_i^{n+\frac{1}{2}} + O(k^2))$$

$$- (\partial_x(p\partial_x v))_i^{n+\frac{1}{2}} + O(k^2) + O(h^2) + q(ih,(n+\tfrac{1}{2})k,v_i^{n+\frac{1}{2}})$$

$$+ O(k^2) + O(h^2)$$

$$= r(ih,(n+\tfrac{1}{2})k,v_i^{n+\frac{1}{2}})(\partial_t v)_i^{n+\frac{1}{2}} - (\partial_x(p\partial_x v))_i^{n+\frac{1}{2}} - q(ih,(n+\tfrac{1}{2})k,v_i^{n+\frac{1}{2}})$$

$$+ O(k^2) + O(h^2).$$

However, by Equation (2.13.1) evaluated at the point $(ih,(n+\frac{1}{2})k)$, we see that (2.13.27) is $O(k^2) + O(h^2)$. Thus, the method (2.13.22)-(2.13.23) is consistent and accurate to $O(k^2) + O(h^2)$.

Let

$$p^* = \max_i p(ih,nk)$$

and

$$r_* = \min\ r(ih,nk,v_i^n)\ .$$

Define

(2.13.28)
$$a_i = -\frac{1}{\hat{r}_i^{n,n}}\ [\overline{\frac{\partial q}{\partial v}} + \overline{\frac{\partial r}{\partial v}}\ \overline{\frac{\partial v}{\partial t}}]\ ,$$

where each of the barred derivatives is evaluated at some point in the intervals $(i-1)k,(i+1)k)$, $(nk,(n+1)k)$. Douglas [6] proved the following convergence theorem.

Theorem 2.13. Suppose that p has three bounded derivatives with respect to x and that q and r have a bounded derivative with respect to v. Further, suppose that v has four bounded derivatives with respect to x and t. Assume that k/h = constant. Then the solution of (2.13.24)-(2.13.25) and (2.13.5b)-(2.13.5d) converges in the ℓ_2-norm to the solution of (2.13.1)-(2.13.4) provided one of the three conditions is satisfied:

1) if a_i defined by (2.13.28) is $\geqslant 0$ and $a^* = \max_i a_i$, then

(2.13.29)
$$\frac{k}{h} < (\frac{r_*}{a^*p^*})^{1/2}\ ;$$

2) if a_i defined by (2.13.28) is $\leqslant 0$ and $a_* = \max_i (-a_i)$, then

(2.13.30)
$$\frac{k}{h} < (\frac{p_*}{a_*r^*})^{1/2}\ ;$$

3) if the sign of a_i is indeterminate, then both (2.13.29) and (2.13.30) hold.

Observe by conditions (2.13.29) and (2.13.30) that this modified Crank-Nicolson type method does not converge for any size k/h.

To implement this, write out (2.13.23) for $i = 1,\ldots,N-1$ with $\lambda = k/h^2$

(2.13.31)
$$-\frac{\lambda}{2} p_{i-\frac{1}{2}}^{n+\frac{1}{2}} u_{i-1}^{n+1} + (\hat{r}_i^{n,n} + \frac{\lambda}{2}(p_{i-\frac{1}{2}}^{n+\frac{1}{2}} + p_{i+\frac{1}{2}}^{n+\frac{1}{2}}))u_i^{n+1} - \frac{\lambda}{2} p_{i+\frac{1}{2}}^{n+\frac{1}{2}} u_{i-1}^{n+1}$$

$$= \frac{\lambda}{2} p_{i-\frac{1}{2}}^{n+\frac{1}{2}} u_{i-1}^n + (\hat{r}_i^{n,n} + \frac{\lambda}{2}(p_{i-\frac{1}{2}}^{n+\frac{1}{2}} + p_{i+\frac{1}{2}}^{n+\frac{1}{2}}))u_i^n + \frac{\lambda}{2} p_{i+\frac{1}{2}}^{n+\frac{1}{2}} u_{i-1}^n - k\hat{q}_i^{n,n}.$$

At $i = 1$ impose the boundary condition (2.13.3) at $i = 0$ given by (2.13.5c). Thus at $i = 1$, (2.13.31) reduces to

$$(2.13.32)\qquad (\hat{r}_i^{n,n} + \tfrac{\lambda}{2}(p_{\frac{1}{2}}^{n+\frac{1}{2}} + p_{\frac{3}{2}}^{n+\frac{1}{2}}))u_1^{n+1} - \tfrac{\lambda}{2}\, p_{\frac{3}{2}}^{n+\frac{1}{2}}\, u_2^{n+1}$$

$$= (\hat{r}_i^{n,n} - \tfrac{\lambda}{2}(p_{\frac{1}{2}}^{n+\frac{1}{2}} + p_{\frac{3}{2}}^{n+\frac{1}{2}}))u_1^{n} + \tfrac{\lambda}{2}\, p_{\frac{3}{2}}^{n+\frac{1}{2}}\, u_2^{n} - k\hat{q}_1^{n,n}$$

$$+ \tfrac{\lambda}{2}\, p_{\frac{1}{2}}^{n+\frac{1}{2}}\, (g_0(nk) + g_0((n+1)k)) \ .$$

Similarly, at $i = N - 1$ impose the boundary condition (2.13.4) at $i = N$ given by (2.13.5d). Thus, at $i = N - 1$, (2.13.31) reduces to

$$(2.13.33)\qquad -\tfrac{\lambda}{2}\, p_{N-\frac{3}{2}}^{n+\frac{1}{2}}\, u_{N-2}^{n+1} + (\hat{r}_{N-1}^{n,n} + \tfrac{\lambda}{2}(p_{N-\frac{3}{2}}^{n+\frac{1}{2}} + p_{N-\frac{1}{2}}^{n+\frac{1}{2}}))u_{N-1}^{n+1}$$

$$= \tfrac{\lambda}{2}\, p_{N-\frac{3}{2}}^{n+\frac{1}{2}}\, u_{N-2}^{n} + (\hat{r}_{N-1}^{n,n} + \tfrac{\lambda}{2}(p_{N-\frac{3}{2}}^{n+\frac{1}{2}} + p_{N-\frac{1}{2}}^{n+\frac{1}{2}}))u_{N-1}^{n} - k\hat{q}_{N-1}^{n,n}$$

$$+ \tfrac{\lambda}{2}\, p_{N-\frac{1}{2}}^{n+\frac{1}{2}}\, (g_1(nk) + g_1((n+1)k)) \ .$$

Combining (2.13.31)-(2.13.33), this may be written as a tridiagonal system of equations of the form

$$(2.13.34)\qquad \underline{A}\underline{u}^{n+1} = \underline{\tilde{u}}^{n} + \underline{bc}$$

where $\underline{u}^{n+1} = (u_i^{n+1})^T$, $\underline{u}^{n} = (\frac{\lambda}{2}\, p_{i-\frac{1}{2}}^{n+\frac{1}{2}}\, u_{i-1}^{n} + (\hat{r}_i^{n,n} - \frac{\lambda}{2}(p_{i+\frac{1}{2}}^{n+\frac{1}{2}} + p_{i+\frac{1}{2}}^{n+\frac{1}{2}}\, u_{i+1}^{n} - k\hat{q}_i^{n,n})^T$, $i = 1,\ldots,N-1$,

$$\underline{bc} = (\tfrac{\lambda}{2}\, p_{\frac{1}{2}}^{n+\frac{1}{2}}(g_0(nk) + g_0((n+1)k)),0,\ldots,0,\tfrac{\lambda}{2}\, p_{N-\frac{1}{2}}^{n+\frac{1}{2}}(g_1(nk) + g_1((n+1)k)))^T,$$

and $\underline{A}$ is an $(N - 1) \times (N - 1)$ tridiagonal matrix of the form

$$\begin{pmatrix} d_1 & c_1 & & & \\ a_2 & d_2 & c_2 & & \\ & \ddots & \ddots & \ddots & \\ & & a_{N-2} & d_{N-2} & c_{N-2} \\ & & & a_{N-1} & d_{N-1} \end{pmatrix},$$

where the elements are given by

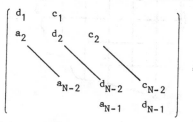

$$d_i = \hat{r}_i^{n,n} + \frac{\lambda}{2}(p_{i-\frac{1}{2}}^{n+\frac{1}{2}} + p_{i+\frac{1}{2}}^{n+\frac{1}{2}}) \ , \quad i = 1,\ldots,N-1 \ ,$$

$$c_i = -\frac{\lambda}{2} p_{i+\frac{1}{2}}^{n+\frac{1}{2}} \ , \quad i = 1,\ldots,N-2 \ ,$$

and

$$a_i = -\frac{\lambda}{2} p_{i-\frac{1}{2}}^{n+\frac{1}{2}} \ , \quad i = 2,\ldots,N-1 \ .$$

To advance the solution from time nk to time $(n + 1)k$, first evaluate (2.13.22) so that $\hat{r}_i^{n,n}$ and $\hat{q}_i^{n,n}$ can be evalueted. The second step consists of solving the tridiagonal system (2.13.34).

A major disadvange of the modified Crank-Nicolson method (2.13.22)-(2.13.23) is that the restriction on the timestep given by (2.13.29) or (2.13.30) requires knowing a bound on the derivatives $\frac{\partial q}{\partial v}$, $\frac{\partial r}{\partial v}$, and $\frac{\partial v}{\partial t}$. This may not be possible in practice. In this situation, this method should be used with caution.

II.14. Irregular Boundaries

A boundary of a two or three dimensional region that does not lie on a rectangular grid is said to be irregular (as depicted in Figure 2.3). Near these boundaries divided differences used to approximate derivatives must be modified.

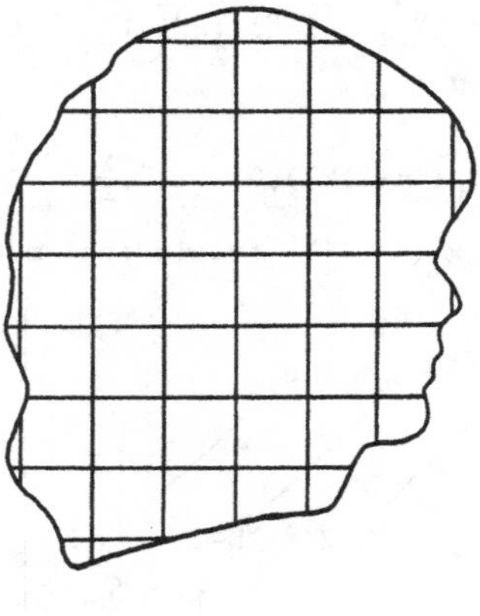

Figure 2.3. Irregular Boundary

Consider a typical point A with coordinates (x_A, y_A) near a curved boundary, as depicted in Figure 2.3.

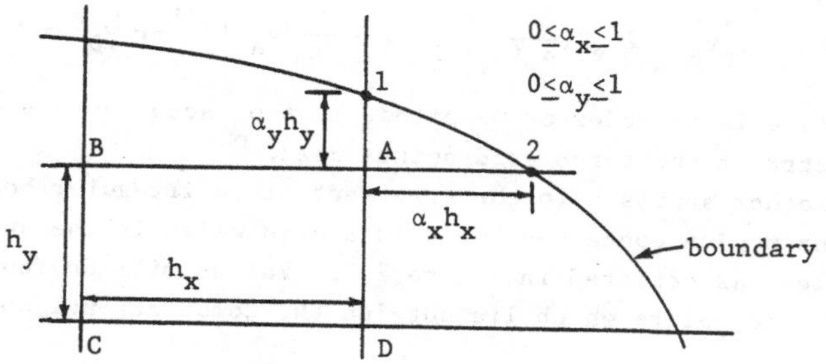

Figure 2.4

Expand $v(x,y,t)$ in a Taylor series in space about the point (x_A, y_A)

$$v(x,y,t) = v_A + \partial_x v_A (x-x_A) + \partial_y v_A (y-y_A) + \frac{1}{2} \partial_x^2 v_A (x-x_A)^2 + \partial_{xy}^2 v_A (x-x_A)(y-y_A) + \frac{1}{2} \partial_y^2 v_A (y-y_A)^2 + \ldots ,$$

where $v_A \equiv v(x_A, y_A, t)$. Evaluating the Taylor series at the point B with coordinate (x_B, y_B) and the boundary point 2 yields

(2.14.1) $$v_B = v_A - h_x \partial_x v_A + \frac{h_x^2}{2} \partial_x^2 v_A + O(h_x^3)$$

and

(2.14.2) $$v_2 = v_A + \alpha_x h_x \partial_x v_A + \frac{\alpha_x^2 h_x^2}{2} \partial_x^2 v_A + O(h^3) ,$$

where v_2 denotes the value of v at the boundary point 2.

In order to construct an approximation to ∂_x accurate to $O(h_x^2)$, more than two points are required because the points B and 2 are not the same distance from the point A and hence the $O(h_x)$ terms do not cancel. Combining (2.14.1) and (2.14.2), in the x-direction

(2.14.3) $$\partial_x v_A = h_x^{-1} \left(\frac{1}{\alpha_x(\alpha_x+1)} v_2 - \frac{1-\alpha_x}{\alpha_x} v_A - \frac{\alpha_x}{\alpha_x+1} v_B\right) + O(h_x^2) .$$

Similarly, in the y-direction

(2.14.4) $$\partial_y v_A = h_y^{-1} \left(\frac{1}{\alpha_y(\alpha_y+1)} v_1 - \frac{1-\alpha_y}{\alpha_y} v_A - \frac{\alpha_y}{\alpha_y+1} v_D\right) + O(h_y^2) .$$

where v_1 denotes v evaluated at the boundary point 1.

We also obtain the approximations to ∂_x^2 and ∂_y^2,

$$(2.14.5) \qquad \partial_x^2 v_A = 2h_x^{-2}\left(\frac{1}{\alpha_x(\alpha_x+1)} v_2 - \frac{1}{\alpha_x} v_A + \frac{1}{\alpha_x+1} v_B\right) + O(h_x) .$$

and

$$(2.14.6) \qquad \partial_y^2 v_A = 2h_y^{-2}\left(\frac{1}{\alpha_y(\alpha_y+1)} v_1 - \frac{1}{\alpha_y} v_A + \frac{1}{\alpha_y+1} v_D\right) + O(h_y) .$$

There is an order of magnitude loss of accuracy due to the lack of symmetry of the three grid points used.

Another approach in the treatment of an irregular boundary is to imbed the region contained by it in a grid which is the union of rectangles, as depicted in Figure 2.5. Values will be found for the "false" grid points which lie outside the computational domain.

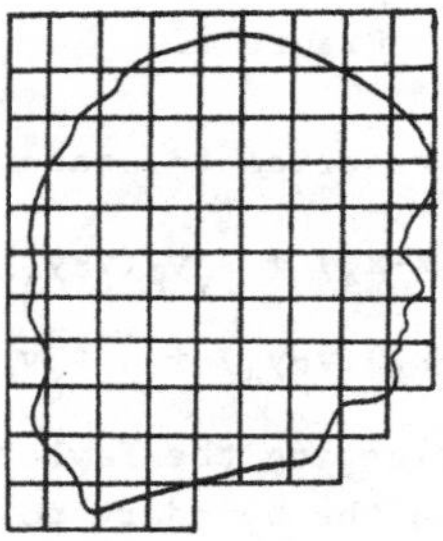

Figure 2.5. Irregular Boundary Imbedded in Rectangular Grid

If Figure 2.5 represents a neighborhood of a portion of this irregular boundary, then by linear interpolation

$$v_2 = \alpha_x v_A + (1-\alpha_x) v_D + O(h_x) .$$

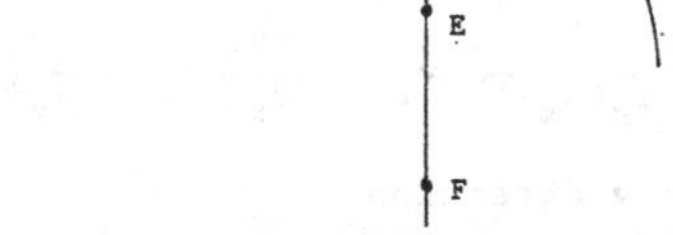

Figure 2.6

Suppose that the boundary conditions are Dirichlet type, so that v is specified on the boundary and v_2 is known, then we may solve for the false point D which lies outside the domain

(2.14.7) $$v_D = \frac{v_2 - \alpha_x v_A}{1 - \alpha_x} + O(h_x) .$$

Similarly, for the false point G which lies outside the domain,

(2.14.8) $$v_G = \frac{v_2 - \alpha_y v_A}{1 - \alpha_y} + O(h_y) .$$

Using quadratic interpolation, for the basic points D and G

(2.14.9) $$v_D = \frac{2}{(1-\alpha_x)(2-\alpha_x)} v_2 - \frac{2\alpha_x}{1-\alpha_x} v_A + \frac{\alpha_x}{2-\alpha_x} v_B + O(h_x^2)$$

and

(2.14.10) $$v_G = \frac{2}{(1-\alpha_y)(2-\alpha_y)} v_1 - \frac{2\alpha_y}{1-\alpha_y} v_A + \frac{\alpha_y}{2-\alpha_y} v_E + O(h_y^2) ,$$

respectively.

We now consider the case where the boundary conditions are of Neumann type so that the normal derivative of v, $\partial_n v$, is specified on the boundary. Approximate the boundary Γ by a piecewise linear curve Γ_h, as depicted in Figure 2.7.

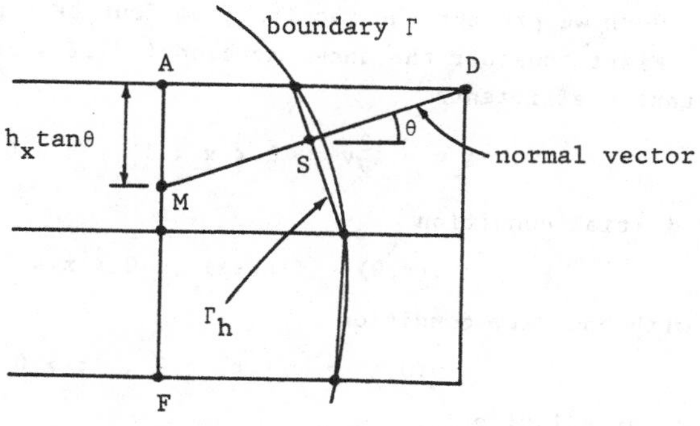

Figure 2.7. Irregular Boundary and Neumann Boundary Conditions

Let θ denote the angle between the positive x-axis and the outward pointing normal vector at the point S on Γ_h. Approximate the normal derivative at S, $\partial_n v_S$, by the divided difference along the normal segment $\overline{DM}$

(2.14.11) $$\partial_n v_S = \frac{v_D - v_M}{|\overline{DB}|} + O(h_x) ,$$

where $|\overline{DB}| = h_x \sec$. Again, this is not $O(h_x^2)$ since the points M and D are not symmetric about the point S. Since $\partial_n v_S$ is specified, we may solve for v_D , the value of v at the false point D ,

(2.14.12) $$v_D = v_M + |\overline{DB}|\partial_n v_S + O(h_x) .$$

The value of v at the point M is obtained by linear interpolation along the segment $\overline{AB}$,

$$v_M = \frac{|\overline{AM}|v_A + |\overline{MB}|v_B}{h_y} ,$$

where $|\overline{AM}| = h_x \tan\theta$ and $|\overline{MB}| = h_y - h_x \tan\theta$. Substitution into (2.14.12) gives

$$v_D = \frac{|\overline{AM}|v_A + |\overline{MB}|v_B}{h_y} + |\overline{DB}|\partial_n v_S + O(h_x) .$$

For other methods of treating irregular boundaries, see Forsythe and Wasow [14].

II.15. Numerical Examples.

Here we present the results from four test problems.

First consider the one-dimensional diffusion equation with constant coefficients

$$\partial_t v = \partial_x^2 v , \quad 0 < x < 1 , \quad t > 0 ,$$

with initial condition

$$v(x,0) = \sin(\pi x) , \quad 0 < x < 1$$

and with boundary condition

$$v(0,t) = v(1,t) = 0 , \quad t > 0 .$$

The exact solution is

$$v(x,t) = e^{-\pi^2 t} \sin(\pi x) .$$

The grid spacing is taken to be $h = 0.1$. Figure 2.8 depicts the results for time $T = 0.1$ where the approximate solutions are represented by dashed lines and the exact solutions are represented by solid lines. In Table 2.1 the numerical methods, values of λ , and the corresponding ℓ_2-errors are presented.

For the Dufort-Frankel method (2.9.8)(see Figure 2.8d). The first time level of values needed to start the method was computed from the exact solution. In this way the error is the error of the Duford-Frankel method alone; and not a combination of the start up error and error from the method.

Figure	Method	λ	$\|e\|_2$
2.8a	Explicit (1.1.3)	0.5	0.4358×10^{-2}
2.8b	Implicit (2.2.1)	1.0	0.1437×10^{-1}
2.8c	Crank-Nicolson (2.2.10)	1.0	0.1933×10^{-2}
2.8d	Dufort-Frankel (2.9.5)	1.0	0.2303×10^{-1}
2.8e	Crank-Nicolson (2.10.7)	1.0	0.3368×10^{-2}

Table 2.1. Numerical Methods and ℓ_2-Errors for One-Dimensional Diffusion Equation with Constant Coefficients.

The Crank-Nicolson method (2.10.7) involves derivative boundary conditions. The exact solution with Dirichet boundary conditions is used to generate Neumann boundary conditions. This gives rise to the initial-boundary-value problem

$$\partial_t v = \partial_x^2 v\ , \quad 0 < x < 1\ , \ t > v$$

with initial condition

$$v(x,0) = \sin(\pi x)\ , \quad 0 < x < 1$$

and Neumann boundary conditions

$$\partial_x v(0,t) = \pi e^{-\pi^2 t}\ , \quad t > 0$$

and

$$\partial_x v(1,t) = -\pi e^{-\pi^2 t}\ , \quad t > 0\ .$$

The exact solution is

$$v(x,t) = e^{-\pi^2 t}\sin(\pi x)\ .$$

The approximate solution at the boundary $x = 0$ is -0.00659. This negative value, even though small and within the accuracy $(O(k) + O(h^2))$ of the method, is the result of the approximation of the derivative in the boundary condition.

As the second test problem, consider the two-dimensional diffusion equation with constant coefficients

$$\partial_t v = \partial_x^2 v + \partial_y^2 v\ , \quad 0 < x,\ y < 1,\ t > 0$$

with initial condition

$$v(x,y,0) = \sin(\pi x)\sin(\pi y)\ , \quad 0 < x,\ y < 1\ .$$

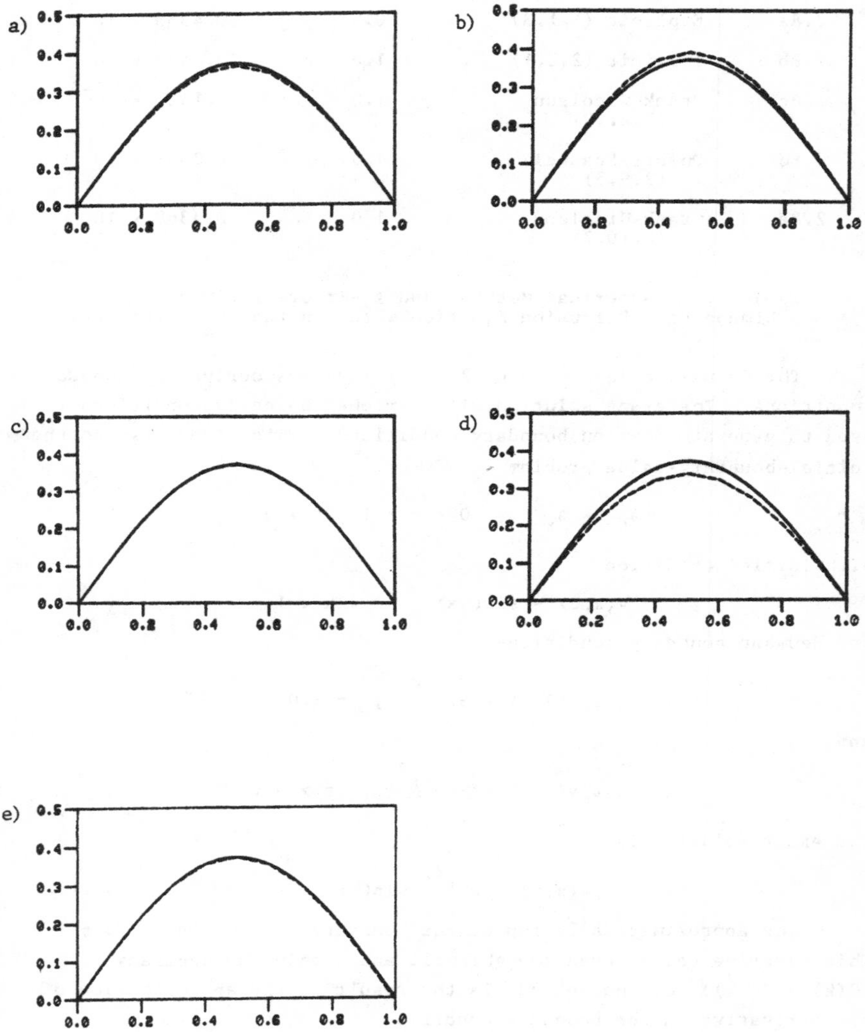

Figure 2.8. Numerical Solution at Time $T = 0.1$ for a) Explicit Method (1.1.3); b) Implicit Method (2.2.1); c) Crank-Nicolson Method (2.2.10); d) Dufort-Frankel method (2.9.5), and e) Crank-Nicolson Method (2.10.7).

and with boundary conditions

$$v(0,y,t) = v(1,y,t) = 0 \ , \quad 0 < y < 1 \ , \quad t > 0$$

$$v(x,0,t) = v(x,1,t) = 0 \ , \quad 0 < x < 1 \ , \quad t > 0 \ .$$

The exact solution is

$$v(x,y,t) = e^{-2\pi^2 t}\sin(\pi x)\sin(\pi y) \ .$$

The grid spacing is taken to be $h = h_x = h_y = 0.1$. In Table 2.2 the numerical methods, values of λ, and the corresonding ℓ_2-errors are presented. Table 3.3 depicts the numerical results for time $T = 1$.

Method	λ	$\|e\|_2$
Explicit (2.3.5)	0.25	0.2278×10^{-2}
Explicit Fractional Step (2.4.7)-(2.4.8)	0.50	0.2361×10^{-2}
Implicit Fractional Step (2.4.10)-(2.4.11)	1.0	0.7780×10^{-2}
Peaceman-Rachford-Douglas (2.5.2)-(2.5.3)	1.0	0.1023×10^{-2}

Table 2.2. Numerical Methods and ℓ_2-Errors for Two-Dimensional Diffusion Equation with Constant Coefficients.

In the third test problem, consider the one-dimensional diffusion equation with variable coefficients

$$\partial_t v = \partial_x(x\partial_x v) - 4v \ , \quad 1 < x < 2 \ , \quad t > 0$$

with initial conditions

$$v(x,0) = \frac{\ln x}{\ln 2} \ , \quad 1 < x < 2 \ ,$$

and boundary conditions

$$v(1,t) = 0 \ , \quad t > 0$$

and

$$v(2,t) = e^{-4t} \ , \quad t > 0 \ .$$

The exact solution is

$$v(x,t) = \frac{e^{-4t}\ln x}{\ln 2} \ .$$

In order to apply the explicit method (2.11.14) the equation must be rewritten as

a)

y \ x	.1	.2	.3	.4	.5	.6	.7	.8	.9
.1	0.128-1	0.244-1	0.336 -01	0.395- 1	0.415-1	0.395-1	0.336-1	0.244-1	0.128-1
.2	0.244-1	0.464-1	0.639 -01	0.751-1	0.790-1	0.751-1	0.639-1	0.464-1	0.244-1
.3	0.336-1	0.639-1	0.879 -01	0.103	0.109	0.103	0.879-1	0.639-1	0.336-1
.4	0.395-1	0.751-1	0.103	0.122	0.128	0.122	0.103	0.751-1	0.395-1
.5	0.415-1	0.790-1	0.109	0.128	0.134	0.128	0.109	0.790-1	0.415-1
.6	0.395-1	0.751-1	0.103	0.122	0.128	0.122	0.103	0.751-1	0.395-1
.7	0.336-1	0.639-1	0.879 -01	0.103	0.109	0.103	0.879-1	0.639-1	0.336-1
.8	0.244-1	0.464-1	0.639 -01	0.751-1	0.790-1	0.751-1	0.639-1	0.464-1	0.244-1
.9	0.128-1	0.244-1	0.336 -01	0.395-1	0.415-1	0.395-1	0.336-1	0.244-1	0.128-1

b)

y \ x	.1	.2	.3	.4	.5	.6	.7	.8	.9
.1	0.128-1	0.244-1	0.336-1	0.395-1	0.415-1	0.395-1	0.336-1	0.244-1	0.128-1
.2	0.244-1	0.464-1	0.639-1	0.751-1	0.790-1	0.751-1	0.639-1	0.464-1	0.244-1
.3	0.336-1	0.639-1	0.879-1	0.103	0.109	0.103	0.879-1	0.639-1	0.336-1
.4	0.395-1	0.751-1	0.103	0.122	0.128	0.122	0.103	0.751-1	0.395-1
.5	0.415-1	0.790-1	0.109	0.128	0.134	0.128	0.109	0.790-1	0.415-1
.6	0.395-1	0.751-1	0.103	0.122	0.128	0.122	0.103	0.751-1	0.395-1
.7	0.336-1	0.639-1	0.879-1	0.103	0.109	0.103	0.879-1	0.639-1	0.336-1
.8	0.244-0	0.464-1	0.639-1	0.751-1	0.790-1	0.751-1	0.639-1	0.464-1	0.244-1
.9	0.128-1	0.244-1	0.336-1	0.395-1	0.415-1	0.395-1	0.336-1	0.244-1	0.128-1

c)

y \ x	.1	.2	.3	.4	.5	.6	.7	.8	.9
.1	0.148-1	0.281-1	0.386-1	0.454-1	0.477-1	0.454-1	0.386-1	0.281-1	0.148-1
.2	0.281-1	0.534-1	0.735-1	0.864-1	0.908-1	0.864-1	0.735-1	0.534-1	0.281-1
.3	0.386-1	0.735-1	0.101	0.119	0.125	0.119	0.101	0.735-1	0.386-1
.4	0.454-1	0.864-1	0.119	0.140	0.147	0.140	0.119	0.864-1	0.454-1
.5	0.477-1	0.908-1	0.125	0.147	0.154	0.147	0.125	0.908-1	0.477-1
.6	0.454-1	0.864-1	0.119	0.140	0.147	0.140	0.119	0.864-1	0.454-1
.7	0.386-1	0.735-1	0.101	0.119	0.125	0.119	0.101	0.735-1	0.386-1
.8	0.281-1	0.534-1	0.735-1	0.864-1	0.908-1	0.864-1	0.735-1	0.534-1	0.281-1
.9	0.148-1	0.281-1	0.386-1	0.454-1	0.477-1	0.454-1	0.386-1	0.281-1	0.148-1

d)

y \ x	.1	.2	.3	.4	.5	.6	.7	.8	.9
.1	0.133-1	0.252-1	0.347-1	0.408-1	0.429-1	0.408-1	0.347-1	0.252-1	0.133-1
.2	0.252-1	0.480-1	0.661-1	0.777-1	0.816-1	0.777-1	0.661-1	0.480-1	0.252-1
.3	0.347-1	0.661-1	0.909-1	0.107	0.112	0.107	0.909-1	0.661-1	0.347-1
.4	0.408-1	0.777-1	0.107	0.126	0.132	0.126	0.107	0.777-1	0.408-1
.5	0.429-1	0.816-1	0.112	0.132	0.139	0.132	0.112	0.816-1	0.429-1
.6	0.408-1	0.777-1	0.107	0.126	0.132	0.126	0.107	0.777-1	0.408-1
.7	0.347-1	0.661-1	0.909-1	0.107	0.112	0.107	0.909-1	0.661-1	0.347-1
.8	0.252-1	0.480-1	0.661-1	0.777-1	0.816-1	0.777-1	0.661-1	0.480-1	0.252-1
.9	0.133-1	0.252-1	0.347-1	0.408-1	0.429-1	0.408-1	0.347-1	0.252-1	0.133-1

Exact Solution

y \ x	.1	.2	.3	.4	.5	.6	.7	.8	.9
.1	0.133-1	0.252-1	0.347-1	0.408-1	0.429-1	0.408-1	0.347-1	0.252-1	0.133-1
.2	0.252-1	0.480-1	0.661-1	0.777-1	0.816-1	0.777-1	0.661-1	0.480-1	0.252-1
.3	0.347-1	0.661-1	0.909-1	0.107	0.112	0.107	0.909-1	0.661-1	0.347-1
.4	0.408-1	0.777-1	0.107	0.126	0.132	0.126	0.107	0.777-1	0.408-1
.5	0.429-1	0.816-1	0.112	0.132	0.139	0.132	0.112	0.816-1	0.429-1
.6	0.408-1	0.777-1	0.107	0.126	0.132	0.126	0.107	0.777-1	0.408-1
.7	0.347-1	0.661-1	0.909-1	0.107	0.112	0.107	0.909-1	0.661-1	0.347-1
.8	0.252-1	0.480-1	0.661-1	0.777-1	0.816-1	0.777-1	0.661-1	0.480-1	0.252-1
.9	0.133-1	0.252-1	0.347-1	0.408-1	0.429-1	0.408-1	0.347-1	0.252-1	0.133-1

Table 2.3. Comparison of Results at Time $t = 1$ for a) Explicit Method (2.3.5), b) Explicit Fractional Step Method (2.4.7)-(2.4.8), c) Implicit Fractional Step Method (2.4.10)-(2.4.11), and d) Peaceman-Rachford-Douglas Method (2.5.2)-(2.5.3).

$$\partial_t v = x\partial_x v^2 + \partial_x v - 4v \ , \quad 1 < x < 2 \ , \quad t > 0 \ .$$

In this case $a(x,t) = x$, $b(x,t) = 0$, $c(x,t) = -4$, and

$$\max_{\substack{0<t<T \\ 1<x<2}} |a(x,t)| = 2 \ .$$

The stability requirement is $\lambda < \frac{1}{4}$.

For the implicit and Crank-Nicolson methods, (2.11.28) and (2.11.34), respectively, $P(x,t) = x$, $r(x,t) = 1$, and $q(x,t) = 4$. Conditions (2.11.29a-c) and satisfied with $p_* = 1$, $r^* = r_* = 1$, and $q^* = 4$. The time-step requirement (2.11.29d) becomes $\frac{1}{k} - 4 > 0$ or $k < \frac{1}{4}$. The grid spacing is taken to be $h = 0.1$. In Table 2.4 the numerical methods, values of λ , and the corresponding ℓ_2-errors are presented for time $T = 0.25$.

Method	λ	$\|e\|_2$
Explicit (2.11.14)	0.2	0.1028×10^{-2}
Implicit (2.11.28)	1.0	0.9717×10^{-3}
Crank-Nicolson (2.11.34)	1.0	0.2584×10^{-4}

Table 2.4. Numerical Methods and ℓ_2-Errors for One-Dimensional Diffusion Equation with Variable Coefficients.

In the final test problem, consider the one-dimensional, quasi-linear diffusion equation

$$\partial_t v = \partial_x^2 v + (1 + v^2)(1 - v) \ , \quad 0 < x < 1 \ , \quad 0 < t < T$$

with initial condition

$$v(x,0) = f(x) = \tan(x) \ , \quad 0 < x < 1 \ ,$$

and boundary conditions

$$v(0,t) = \tan(t) \ , \quad 0 < t < T$$

and

$$v(1,t) = \tan(1 + t) \ , \quad 0 < t < T \ .$$

The exact solution is

$$v(x,t) = f(x + t) = \tan(x + t) \ .$$

This solution represents a traveling wave solution that becomes singular as $x + t \to \pi/2$.

This equation is of the form of (2.13.1) where $r(x,t,v) = 1$, $p(x,t) = \ell$, and $q(x,t,v) = -(1 + v^2)(1 - 2v)$. The grid spacing is taken to be $h = 0.1$. In Table 2.5 the numerical methods, values of λ , and the corresponding ℓ_2-errors are presented for times

$t = 0.1, 0.2, 0.3, 0,4$, and 0.5.

Method	λ	t	$\|e\|_2$
Implicit (2.13.5)	1.0	0.1	0.6857×10^{-2}
		0.2	0.1361×10^{-1}
		0.3	0.2762×10^{-1}
		0.4	0.7097×10^{-1}
		0.5	0.3426
Herative Crank-Nicolson (2.13.17)	1.0	0.1	0.2295×10^{-2}
		0.2	0.5035×10^{-2}
		0.3	0.1172×10^{-1}
		0.4	0.3567×10^{-1}
		0.5	0.2064
Modified Crank-Nicolson (2.13.24)-(2.13.25)	1.0	0.1	0.2292×10^{-2}
		0.2	0.5030×10^{-2}
		0.3	0.1172×10^{-1}
		0.4	0.3566×10^{-1}
		0.5	0.2065

Table 2.5. Numerical Methods and ℓ_2-Errors for One-Dimensional Quasi-Linear Diffusion Equation

In the implementation of the modified Crank-Nicolson method (2.13.24)-(2.13.25) bounds are required on $\partial_v q$, $\partial_v r$, and $\partial_t v$ (see (2.13.28)). In this example, $T = 0.5$ and $v(x,t) < \tan(1.5) = 14.1$. The term a_i in (2.13.28) reduces to $a_i = -\partial_v \bar{q}$, since $\partial_v r = 0$, where $\partial_v q = 6v^2 - 2v + 2$. For $v > 0$, $a_i < 0$ and $a_* = \max(-a_i) = 1166.9$. The bound (2.13.29) becomes $k/h < 0.02$.

For $0 < x < 1$, as $t \to 0.5$, $x + t \to 1.5$. The derivatives of the solution $v(x,t)$ will be very large near the singularity, which results on a large truncation error. For fixed x, as t grows the truncation error is expected to grow. This can be observed in Table 2.5. Figure 2.9 depicts the results of the implicit method (2.13.5) for $T = 0.5$ where the approximate solutions are represented by dashed lines and the exact solutions are represented by solid lines.

A disadvantage of this modified Crank-Nicolson method (2.13.24)-(2.13.25) is that the bound on k/h requires knowing bounds on $\partial_v q$, $\partial_v r$, and $\partial_t v$.

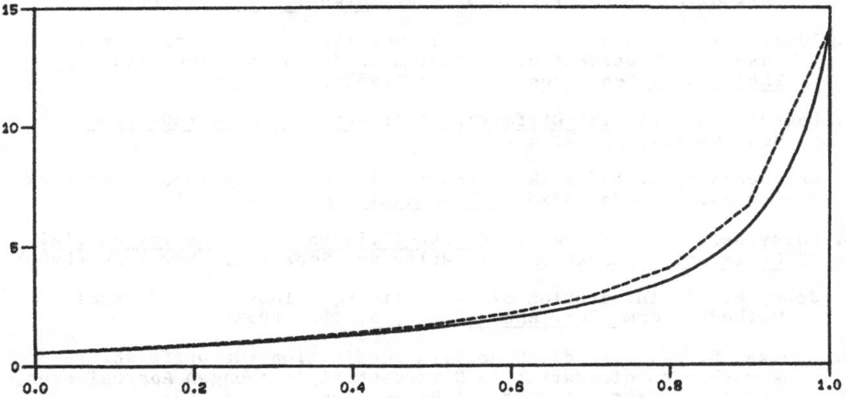

Figure 2.9. Numerical Solution at Time T = 0.5 for the Implicit Method (2.13.5).

References

1. Brian, P. L. T., A Finite-Difference Method of High Order Accuracy for the Solution of Three-Dimensional Transient Heat Conduction Problems, AIChE J., 1, 367 (1961).

2. Cole, J. D., On a Quasi-linear Parabolic Equation Occuring in Aerodynamics, Q. Appl. Math., 9, 225 (1951).

3. Crank, J. and P. Nicolson, A Practical Method for Numerical Evaluation of Solutions of Partial Differential Equations of the Heat-Conduction Type, Proc. Camb. Phil. Soc., 43, 50 (1947).

4. Douglas, J. Jr., On the Numerical Solution of $\partial_x^2 u + \partial_y^2 u = \partial_t u$ by Implicit Methods, SIAM J., 3, 42 (1955).

5. _______________, The Numerical Integration of Quasi-Linear Parabolic Differential Equations, Pacific J. Math., 6, 35 (1956).

6. _______________, The Application of Stability Analysis in the Numerical Solution of Quasi-Linear Parabolic Differential Equations, Trans. Am. Math. Soc., 89, 484 (1958).

7. _______________, Survey of Numerical Methods for Parabolic Differential Equations, Advances in Computers, Vol. 2, F. L. Alt, Ed., Academic Press, New York (1961).

8. _______________, Alternating Direction Methods for Three Spaces Variables, Numer. Math., 4, 41 (1962).

9. ______________ and H. H. Rachford, On the Numerical Solution of Heat Conduction Problems in Two and Three Space Variables, Trans. Am. Math. Soc., 82, 421 (1956).

10. ______________ and B. F. Jones, On Predictor-Corrector Methods for Parabolic Differential Equations, SIAM J., 11, 195 (1963).

11. Dufort, E. C., and S. P. Frankel, Stability Conditions in the Numerical Treatment of Parabolic Differential Equations, Math. Tabl. Natl. Res. Coun., 7, 135 (1953).

12. Epstein, B., Partial Differential Equations, An Introduction, McGraw-Hill, New York (1962).

13. Fairweather, G. and A. R. Mitchell, A New Computational Procedure for ADI Methods, SIAM J. Num. Anal., 4, 163 (1947).

14. Forsythe, G. and W. Wasow, Finite-Difference Methods for Partial Differential Equations, J. Wiley and Sons Inc., New York (1960).

15. John, F., On integration of Parabolic Equations by Difference Methods, Comm. Pure Appl. Math., 5, 155 (1952).

16. Kreiss, H. O., Uber die Stabilitatsdefinition für Differenzeigleichungen die Partielle Differentialgleichungen Approximeren, Nordisk Tidskr Informations-Behandling, 2, 153 (1962).

17. Lees, M., Approximate Solution of Parabolic Equations, SIAM J., 7, 167 (1959).

18. Mitchell, A. R. and G. Fairweather, Improved Forms of Alternating Direction Methods of Douglas, Peaceman, and Rachford for Solving Parabolic and Elliptic Equations, Numer. Math., 6, 285 (1964).

19. Peaceman, D. W. and H. H. Rachford, The Numerical Solution of Parabolic and Elliptic Differential Equations, SIAM J., 3, 28 (1955).

20. Richardson, L. F., The Approximate Arithematical Solution by Finite Differences of Physical Problems Involving Differential Equations, with an Application to the Stresses in a Masonry Dam, Phil. Trans. Roy. Soc., Ser. A., 210, 307 (1910).

21. Tikhonov, A. N., and A. A. Samarskii, Z. Vycisl. Mat. i Mat. Fiz., 1, 5 (1961).

22. Yanenko, N. N., The Method of Fractional Steps, Springer-Verlag (1971).

III. HYPERBOLIC EQUATIONS

III.1. Introduction

The most common hyperbolic equation is the one-dimensional wave equation

$$\partial_t^2 v = c^2 \partial_x^2 v, \quad -\infty < x < +\infty, \quad t > 0, \tag{3.1.1}$$

where c is a constant. By introducing a change of variables

$$\xi = x + ct$$

$$\eta = x - ct,$$

and using the chain rule

$$\partial_x^2 v = \partial_\xi^2 v + 2\partial_{\xi\eta}^2 v + \partial_y^2 v$$

and

$$\partial_t^2 v = c^2(\partial_\xi^2 v - 2\partial_{\xi\eta}^2 v + \partial_\eta^2 v).$$

For $c \neq 0$, equation (3.1.1) reduces to

$$\partial_{\xi\eta}^2 v = 0,$$

which has a general solution of the form

$$\begin{aligned} v(x,t) &= F(\xi) + G(\eta), \\ & F(x + ct) + G(x - ct), \end{aligned} \tag{3.1.2}$$

where F and G are twice differentiable functions.

The change of variable $\xi = x + ct$ is a translation of the coordinate system to the left by an amount ct. This translation is proportional to t so that a point ξ = constant moves to the left with speed c.

The portion of the situation given by

$$v(x,t) = F(x + ct)$$

represents a wave moving to the left with velocity $-c$,

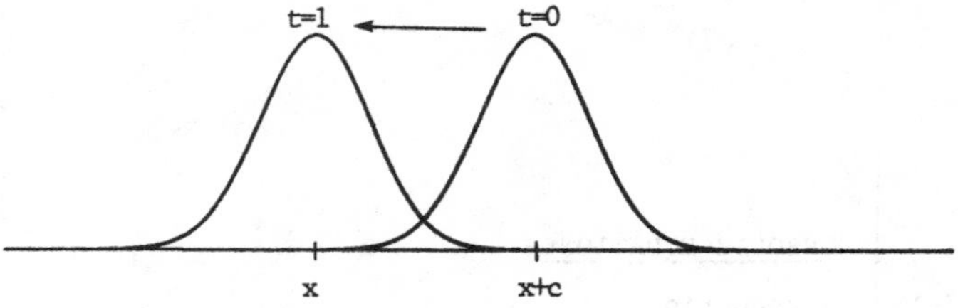

Figure 3.1. Wave Propagating to the Left.

This wave retains its original shape. Similarly, the portion of the solution given by

$$v(x,t) = G(x - ct)$$

represents a wave moving to the right with velocity ct, retaining its original shape as depicted in Figure 3.2. Thus the

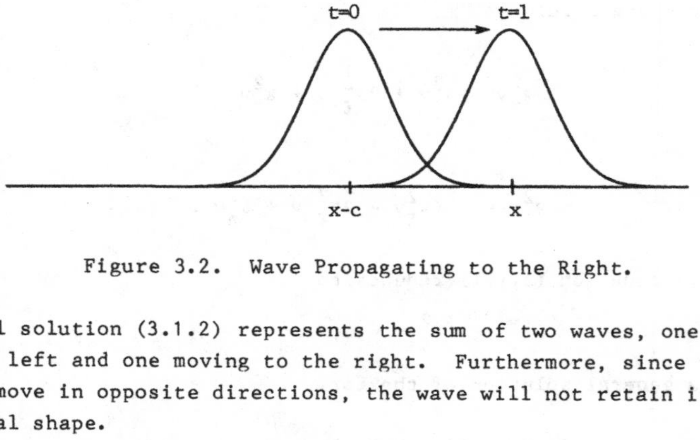

Figure 3.2. Wave Propagating to the Right.

general solution (3.1.2) represents the sum of two waves, one moving to the left and one moving to the right. Furthermore, since the two waves move in opposite directions, the wave will not retain its original shape.

Consider the initial -value problem given by equation (3.1.1) with initial conditions

(3.1.3) $$v(x,0) = f(x), \quad -\infty < x < +\infty$$

and

(3.1.4) $$\partial_t v(x,0) = g(x), \quad -\infty < x < +\infty.$$

In order to obtain a general solution, substitute the initial condition (3.1.3) into the general form of the solution (3.1.2)

(3.1.5) $$F(x) + G(x) = f(x).$$

Differentiating (3.1.2) with respect to t and substituting the initial condition (3.1.4),

$$cF'(x) - cG'(x) = g(x),$$

and differentiating (3.1.5) with respect to x and multiplying by c

$$cF'(x) + cG'(x) = cf'(x).$$

Adding these two expressions gives

$$2cF'(x) = cf'(x) + g(x).$$

$F(\xi)$ is obtained by integrating this relation over the interval $[0,\xi]$,

(3.1.6) $$F(\xi) = \frac{1}{2}(f(\xi) - f(0)) + F(0) + \frac{1}{2c}\int_0^\xi g(s)ds.$$

$G(\eta)$ is obtained by substituting (3.1.6) into (3.1.5),

(3.1.7) $$G(\eta) = \frac{1}{2}f(\eta) + \frac{1}{2}f(0) - f(0) - \frac{1}{2c}\int_0^\eta g(s)ds.$$

Combining (3.1.6) and (3.1.7) in solution (3.1.2)

$$v(x,t) = \frac{1}{2}(f(\xi) + f(\eta)) + \frac{1}{2c}\int_\eta^\xi g(s)ds,$$

which by substitution of the change of variable gives D'Alembert's solution at a point (x_0,t_0)

(3.1.8) $$v(x_0,t_0) = \frac{1}{2}(f(x_0 + ct_0) + f(x_0 - ct_0)) + \frac{1}{2c}\int_{x_0-ct_0}^{x_0+ct_0} g(s)ds\ .$$

From this solution (3.1.8), we see that the solution at a point (x_0,t_0) depends on the initial conditions on the segment of the initial line (x-axis) bounded by the lines $x - ct = x_0 - ct_0$ and $x + ct = x_0 + ct_0$, that is, the interval $[x_0 - ct_0, x_0 + ct_0]$ as depicted in Figure 3.3

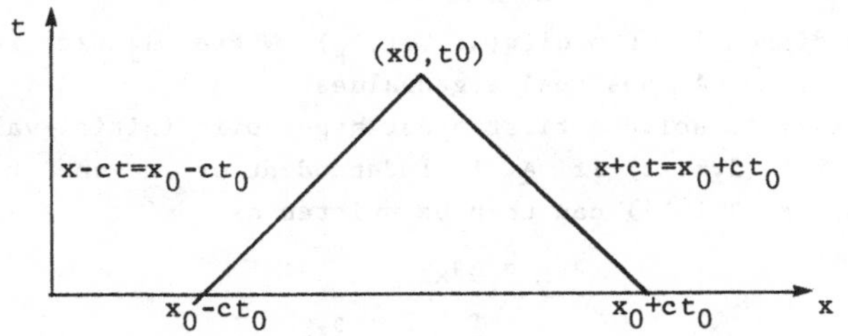

Figure 3.3. Domain of Dependence of (x_0,t_0).

This interval is called the <u>domain of dependence</u> of the point (x_0,t_0). So the solution at a point (x_0,t_0) depends on x such that $x_0 - ct_0 < x < x_0 + ct_0$ means that a perturbation of the initial conditions is propagated at a speed no greater than c.

From the solution (3.1.8), we see that the solution $v(x_0,t_0)$ is altered only if f is perturbed at $x_0 \pm ct$. However, $v(x_0,t_0)$ is altered if the initial condition on $\partial_t v$, g is perturbed anywhere in the interval $x_0 - ct_0 < x < x_0 + ct_0$. This means that a perturbation in the velocity is propagated at all speeds up to c.

The lines $x - ct = \text{constant}$ and $x + ct = \text{constant}$ are called <u>characteristics</u>.

Equation (3.1.1) can be written as a system of first-order equations. Let $v_1 = \partial_x v$ and $v_2 = \partial_t v$, then equation (3.1.1) can be written in the form

$$\partial_t \underline{v} = \underline{A}\partial_x \underline{v} \,, \tag{3.1.9}$$

where $\underline{v} = (v_1,v_2)^T$ and $\underline{A} = \begin{pmatrix} 0 & c \\ c & 0 \end{pmatrix}$. The initial conditions (3.1.3) and (3.1.4) can then be written in the form

$$\underline{v}(x,0) = \underline{f}(x),\ -\infty < x < +\infty, \tag{3.1.10}$$

where $\underline{f} = (f,g)^T$.

Consider a general first-order system of equations of the form

$$\partial_t \underline{v} = \underline{A}\partial_x \underline{v} + \underline{B} \,,\ -\infty < x < +\infty, \quad t > 0 \tag{3.1.11}$$

where $\underline{v}$ $(v_1,\ldots,v_p)^T$ and $\underline{A}$ and $\underline{B}$ are matrices of order $p\times p$ whose elements may depend on x and t. The initial conditions are given by

$$\underline{v}(x,0) = \underline{f}(x), \quad -\infty < x < +\infty. \tag{3.1.12}$$

<u>Definition</u>. The first order system (3.1.11) is said to be hyperbolic if there exists a constant K and a nonsingular matrix $\underline{T}$ with

$$\max(\|\underline{T}\|,\ \|T^{-1}\|) \le K,$$

such that

$$\underline{T}^{-1}\underline{A}\ \underline{T} = \underline{D}$$

where $\underline{D}$ is diagonal, $\underline{D} = \text{diag}(\mu_1,\ldots,\mu_p)$ where μ_j are real for $j = 1,\ldots,p$. Thus $\underline{A}$ has real eigenvalues.

It is easy to solve a first-order hyperbolic initial-value problem (3.1.11)-(3.1.12) if $\underline{A}$ is independent of x and t and $\underline{B} = 0$. Equation (3.1.11) can then be written as

$$\begin{aligned} \partial_t \underline{v} &= \underline{A}\partial_x \underline{v} \\ &= \underline{T}\ \underline{D}\ \underline{T}^{-1}\ \partial_x \underline{v} \end{aligned}$$

or

$$\underline{T}^{-1}\ \partial_t \underline{v} = \underline{D}\ \underline{T}^{-1}\ \partial_x \underline{v}. \tag{3.1.13}$$

Since $\underline{A}$ does not depend on x or t, $\underline{T}$ and, hence $\underline{T}^{-1}$, does not depend on x or t so $\underline{T}^{-1}\partial_t \underline{v} = \partial_t(\underline{T}^{-1}\underline{v})$ and $\underline{T}^{-1}\partial_x \underline{v} = \partial_x(\underline{T}^{-1}\underline{v})$. Define $\underline{w} = \underline{T}^{-1}\underline{v}$, then by substitution into (3.1.13),

(3.1.14) $$\partial_t \underline{w} = \underline{D}\partial_x \underline{w},$$

where $\underline{D}$ is a diagonal matrix whose diagonal elements are the real eigenvalues of $\underline{A}$. The system (3.1.14) represents p uncoupled first-order equations of the form

(3.1.15) $$\partial_t w_j = \mu_j \partial_x w_j \; ,$$

where μ_j is an eigenvalue of $\underline{A}$, for $j = 1,\ldots,p$. The initial condition associated with (3.1.14), by the definition of $\underline{w} = \underline{T}^{-1}\underline{v}$, is

(3.1.16) $$\underline{w}(x,0) = \tilde{\underline{f}}(\underline{x})$$

where $\tilde{\underline{f}}(x) = \underline{T}^{-1}\underline{f}(x)$. Thus (3.1.15)-(3.1.16) reduces to a single (prototype) equation of the form

(3.1.17) $$\partial_t v = c\partial_x v, \; -\infty < x < +\infty, \; t > 0$$

with initial condition

(3.1.18) $$v(x,0) = f(x), \; -\infty < x < +\infty.$$

This initial-value problem has solution

(3.1.19) $$v(x,t) = f(x + ct) \; , \; -\infty < x < +\infty, \; t > 0.$$

As in the case of the wave equation, the solution represents a wave moving to the left with velocity $-c$. The wave retains its original shape. The solution at a point (x_0,t_0) depends on the initial condition evaluated at the point $x_0 + ct_0$.

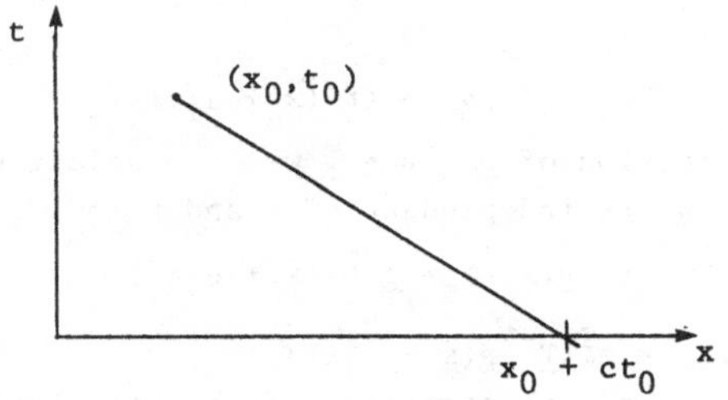

Figure 3.4

In this case, the domain of dependence of the point (x_0,t_0) is the single point $(x_0 + ct_0, 0)$ on the initial line (see Figure 3.4).

The line $x + ct$ = constant is called the characteristic of the equation (3.1.19). The solution is constant along these characteristics. To see this, consider a characteristic line $x + ct = K$, where K is a constant. Write this equation in parametric form

$$t = -\frac{s}{c} + \frac{K}{c}$$

$$x = s,$$

then, by the chain rule

$$\partial_s v = \partial_t v \cdot \frac{dt}{ds} + \partial_x v \cdot \frac{dx}{ds}$$

$$= -\frac{1}{c}\,\partial_t v + \partial_x v = 0.$$

This shows that the derivative of v along the characteristic is 0 and, hence, v is a constant.

Definition. The domain of dependence of a point (x_0, t_0) is the set of points $(x,0)$ which influences the solution at (x_0, t_0).

A property of hyperbolic equations is that the domain of dependence is finite for finite time.

The j-th component of (3.1.16) is

$$w_j(x,0) = \tilde{f}_j(x) = \underline{e}_j^T \tilde{\underline{f}}(x), \tag{3.1.20}$$

where $\underline{e}_j^T = (0,\ldots,0,1,0,\ldots,0)$, that is, a vector with zeros in all elements, except in the j-th element, which is 1. The solution to the j-th component of (3.1.15) and (3.1.20) is

$$w_j(x,t) = \tilde{f}_j(x + \mu_j t)$$

for $j = 1,\ldots,p$, or

$$\underline{w}(x,t) = (\tilde{f}_j(x + \mu_j t)).$$

By using the definition of $\underline{w}$, $\underline{v} = \underline{T}\,\underline{w}$, the solution of (3.1.9)-(3.1.10), where $\underline{A}$ is independent of x and t and $\underline{B} = 0$, is

$$\underline{v}(x,t) = \underline{T}(\tilde{f}_j(x + \mu_j t)) \tag{3.1.21}$$

where $\tilde{f}_j(x + \mu_j t) = \underline{e}_j^T(\underline{T}^{-1}\underline{f}(x + \mu_j t))$.

Considering the j-th component of the solution at the point (x_0, t_0), we see that the domain of dependence is the point $x_0 + \mu_j t_0$ on the initial line. The multiplication of the vector $(\tilde{f}_j(x + \mu_j t))$ by the matrix $\underline{T}$ recouples the solutions of the individual components. Thus the domain of dependence of the system at the point (x_0, t_0) is the set of points $(x_0 + \mu_j t_0, 0)$, $j = 1,\ldots,p$. The lines, $x + \mu_j t =$ constant, gives rise to p families of characteristics, one for each value of $j = 1,\ldots,p$.

Returning to the example of the wave equation (3.1.9)-(3.1.10), define $\underline{T}$ by

$$\underline{T} = \begin{matrix} 1 & -1 \\ 1 & 1 \end{matrix}$$

then

$$\underline{T}^{-1} = \begin{pmatrix} 1/2 & 1/2 \\ -1/2 & 1/2 \end{pmatrix} .$$

This gives

$$\underline{D} = \underline{T}^{-1}\ \underline{A}\ \underline{T} = \begin{pmatrix} c & 0 \\ 0 & -c \end{pmatrix} .$$

Thus $\underline{A}$ has eigenvalues $\pm c$. Defining $\underline{w} = \underline{T}^{-1}\underline{v}$,

$$\begin{pmatrix} w_1 \\ w_2 \end{pmatrix} = \begin{pmatrix} 1/2 & 1/2 \\ -1/2 & 1/2 \end{pmatrix} \begin{pmatrix} v_1 \\ v_2 \end{pmatrix} = \begin{pmatrix} 1/2(v_1 + v_2) \\ 1/2(v_2 - v_1) \end{pmatrix} .$$

The initial condition $\underline{\tilde{f}} = \underline{T}^{-1}\underline{f}$ becomes

$$\begin{pmatrix} w_1(x,0) \\ w_2(x,0) \end{pmatrix} = \begin{pmatrix} f_1 \\ f_2 \end{pmatrix} = \begin{pmatrix} 1/2 & 1/2 \\ -1/2 & 1/2 \end{pmatrix} \begin{pmatrix} f \\ g \end{pmatrix} = \begin{pmatrix} 1/2(f+g) \\ 1/2(g-f) \end{pmatrix} .$$

This gives for $\mu_1 = c$ and $\mu_2 = -c$

$$\begin{pmatrix} w_1(x,t) \\ w_2(x,t) \end{pmatrix} = \begin{pmatrix} (f(x+ct) + g(x+ct)) \\ (g(x-ct) - f(x-ct)) \end{pmatrix} .$$

Finally, $\underline{v}$ is obtained by multiplying $\underline{w}$ by $\underline{T}$ on the left,

$$\underline{v} = \begin{pmatrix} v_1 \\ v_2 \end{pmatrix} = \begin{pmatrix} 1 & -1 \\ 1 & 1 \end{pmatrix} \begin{pmatrix} w_1(x,t) \\ w_2(x,t) \end{pmatrix}$$

$$= \begin{pmatrix} 1/2(f(x+ct) + f(x-ct)) + 1/2(g(x+ct) - g(x-ct)) \\ 1/2(f(x+ct) - f(x-ct)) + 1/2(g(x+ct) + g(x-ct)) \end{pmatrix} .$$

The relation between this and the solution given by D'Alembert's solution (3.1.8) can be seen by observing that $v_1 \equiv \partial_t v$ and $v_2 \equiv \partial_x v$. By differentiating (3.1.8) with respect to t, we obtain the first component in (3.1.20), v_1.

Consider the first order hyperbolic system

$$\partial_t \underline{v} = \underline{A}\ \partial_x\ \underline{v} ,$$

where $\underline{A} = (a_{ij})$ and it is assumed that $|\partial_x a_{ij}| < m$. With the inner product

$$(\underline{v},\underline{w}) = \int_{-\infty}^{+\infty} (\sum_{j=1}^{p} v_j(x,t) w_j(x,t))dx,$$

define the _energy_ $E(t)$ of $\underline{v}(x,t)$ to be

$$E(t) = (\underline{v},\underline{v}).$$

By differentiating with respect to t,

$$\frac{d}{dt} E(t) = \partial_t(\underline{v},\underline{v})$$

$$= (\partial_t\underline{v},\underline{v}) + (\underline{v},\partial_t\underline{v})$$

$$= (\underline{A}\partial_x\underline{v},\underline{v}) + (\underline{v},\underline{A}\partial_x\underline{v})$$

$$= (\underline{v},(\underline{A}\partial_x\underline{v} - \partial_x\underline{A}\ \underline{v})) \ .$$

Since $\partial_x\underline{A}\ \underline{v} = \underline{A}\partial_x\underline{v} + (\partial_x\underline{A})\underline{v}$,

$$\frac{d}{dt} E(t) = (\underline{v},(\partial_x\underline{A})\underline{v})$$

$$\leqslant (\underline{v},m\underline{v})$$

$$= m(\underline{v},\underline{v})$$

$$= mE(t).$$

Define the function $G(t) = E(t)e^{-mt}$, which by differentiating

$$\frac{d}{dt} G(t) = (\frac{d}{dt} E(t))e^{-mt} - mE(t)e^{-mt}$$

$$= e^{-mt}(\frac{d}{dt} E(t) - mE(t)) \leqslant 0.$$

Thus $G(t)$ is a decreasing function of t and $G(t) \leqslant G(0)$ which implies that

$$E(t)e^{-mt} \leqslant E(0)$$
or
$$E(t) \leqslant E(0)e^{mt}.$$

This is the basic energy equality for linear hyperbolic equations.
In the case of constant coefficients, $m = 0$ so we have $E(t) = E(0)$ and the energy is conserved.

III.2. The Courant-Friedrichs-Lewy Condition

The intuitive idea behind the Courant-Friedrichs-Lewy (CFL) condition is that the solution of the finite difference equation should not be independent of any of the data which determines the solution of the partial differential equation, unless it can be shown that the data omitted has a negligible effect.

Definition. The numerical domain of dependence of a finite difference method at a point (ih,nk) on the grid is the set of points $(jh,0)$ on the initial line which influences the solution at the point (ih,nk).

Suppose the domain of dependence at the point (ih,nk) of the finite difference method D_h does not contain the domain of dependence at the point (ih,nk) of the partial differential equation D. Let D_1 denote the set of points in D not contained in D_h. Suppose somewhere in D_1 the initial data is perturbed. The solution of the finite difference equation at the point (ih,nk) will not be affected by this perturbation of the initial data. Thus, it will be impossible for the solution u_i^n of the finite difference method to converge to the solution of the partial differential equation with the perturbed initial data as $h,k \to 0$.

Definition. A finite difference method is said to satisfy the Courant-Friedrichs-Lewy (CFL) condition if the domain of dependence of the solution of the finite difference equation includes the domain of dependence of the solution of the partial differential equation at all points (ih,nk). The CFL condition is necessary for stability. This follows from the above argument.

Typically, a finite difference method that satisfies the CFL condition will have a numerical domain of dependence that contains points not contained in the domain of dependence of the partial differential equation. (This will be exemplified below.) These additional points in the numerical domain of dependence result in additional initial data determining the solution of the finite difference equation. This gives rise to an error. However, the contribution of this additional data tends to 0 as $h,k \to 0$.

Consider the prototype hyperbolic equation

$$\partial_t v = c\partial_x v \ , \quad -\infty < x < +\infty, \quad t > 0 \tag{3.2.1}$$

where c is a constant, with initial condition

$$v(x,0) = f(x), \quad -\infty < x < +\infty. \tag{3.2.2}$$

Approximate the prototype equation (3.2.1) by a finite difference method of the form

$$u_i^{n+1} = \sum_{j=-m_1}^{m_2} A_j S_+^j u_i^n$$

where A_{-m_1} and A_{m_2} are not equal to zero.

The solution at the point $(ih,(n+1)k)$ depends on the solution at $m_1 + m_2 + 1$ points at the previous time level, $((i-m_1)h,nk),\ldots,((i+m_2)h,nk)$. The numerical domain of dependence of the point $(ih,(n+1)k)$ is bounded by the two lines with slopes $-k/m_2h$ and k/m_1h as depicted in the Figure 3.5

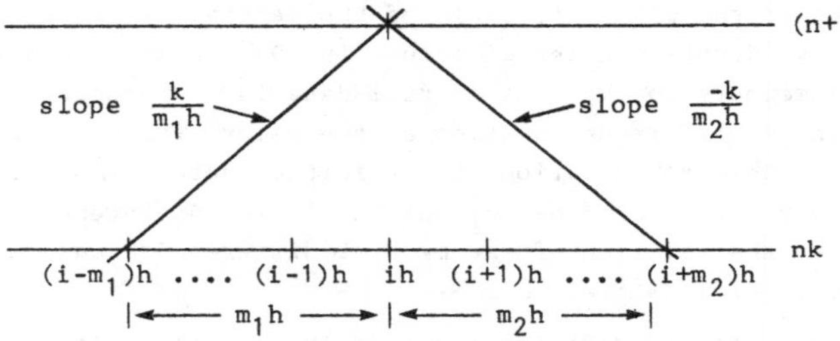

Figure 3.5. Numerical Domain of Dependence

These slopes are negative and positive, respectively. The equations of these lines through the point $(ih,(n+1)k)$ are given by

$$L_1 : t - (n+1)k = \frac{-k}{m_2h}(x - ih)$$

and

$$L_2 : t - (n+1)k = \frac{k}{m_1h}(x - ih)$$

These equations intersect the initial line $(t = 0)$ at the points $((i+m_2(n+1))h,0)$ and $((i-m_1(n+1))h,0)$, respectively. Thus, the numerical domain of dependence of the point $(ih,(n+1)k)$ consists of the $(m_1 + m_2)(n+1) + 1$ points in the interval $[(i-m_1(n+1))h, (i+m_2(n+1))h]$.

The characteristics of the prototype equation (3.2.1) have the form

$$x + ct = K$$

where K is a constant. This represents a family of lines with slope $-1/c$. The domain dependence of the prototype equation (3.2.1) at the point $(ih,(n+1)k)$ is the point $(ih+c(n+1)k,0)$ on the initial line

$(t = 0)$. The CFL condition requires that

$$(ih+c(n+1)k) \in [(i-m_1(n+1))h,(i+m_2(n+1))h].$$

There are two cases to consider corresponding to the sign of c. If $c > 0$, then (3.2.4) is satisfied if the slope of line L_1 is greater than or equal to the slope of the characteristic, that is,

$$\frac{-k}{m_2 h} \geq -\frac{1}{c} .$$

This condition may be written as

$$c \frac{k}{h} \leq m_2 .$$

If $c < 0$, then (3.2.4) is satisfied if the slope of line L_2 is less than or equal to the slope of the characteristic, that is,

$$\frac{k}{m_1 h} \leq -\frac{1}{c} = \frac{1}{|c|} .$$

This condition may be rewritten as

$$|c| \; k/h \leq m_1 .$$

The CFL condition for a finite difference method of the form (3.2.3) approximating the equation (3.2.1) may be summarized as

(3.2.5a) $$c \frac{k}{h} \leq m_2 \quad \text{if} \quad c > 0,$$

(3.5.5b) $$-c \frac{k}{h} \leq m_1 \quad \text{if} \quad c < 0.$$

Suppose we approximate the prototype equation (3.2.1) by

$$\frac{u_i^{n+1} - u_i^n}{k} = cD_+u_i^n ,$$

which may be written in the form of (3.2.3)

(3.2.6) $$\begin{aligned} u_i^{n+1} &= u_i^n + kcD_+u_i^n \\ &= (1-\lambda c)u_i^n + \lambda c u_{i+1}^n \\ &= (1-\lambda c)u_i^n + \lambda c S_+u_i^n \end{aligned}$$

where $\lambda = k/h$. In this case $m_1 = 0$ and $m_2 = 1$. Suppose that c is positive. Then by (3.2.5a), the CFL condition becomes

$$c\lambda \leq 1.$$

The solution at the point $(ih,(n+1)k)$ depends on two points at the previous time level (ih,nk) and $((i+1)h,nk)$, which depend on three points on the next previous time level $(ih,(n-1)k)$, ((i+1)h,(n-1)k), and ((i+2)h,(n-1)k), and so on until the initial line (t = 0) is reached where the solution depends on n+2 points (ih,0),((i+1)h,0),...,((i+n+1)h,0). The numerical domain of dependence of the point (ih,(n+1)k) consists of the n+2 points in the interval [ih,(i+n+1)h] as depicted in Figure 3.6

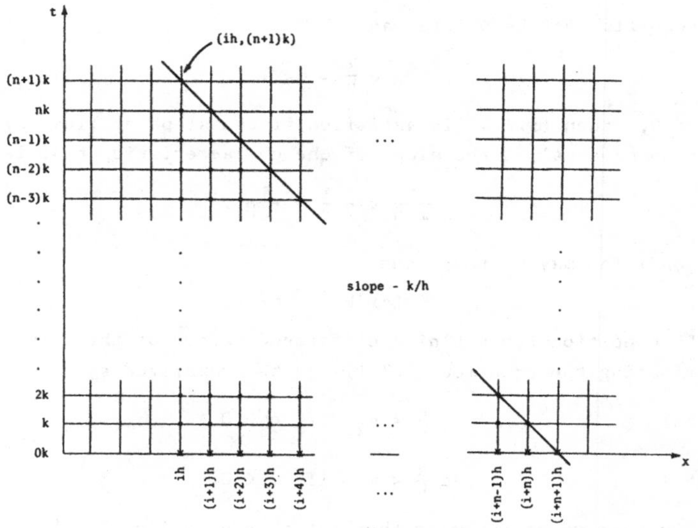

Figure 3.6

The numerical domain of dependence of the point (ih,(n+1)k) is marked by "x".

To consider the stability of method (3.2.6) take the discrete Fourier transform,

$$\hat{u}^{n+1}(\xi) = \rho(\xi)\hat{u}^{n}(\xi)$$

where the symbol

$$\rho(\xi) = 1 - c\lambda + c\lambda e^{-i\xi}.$$

For $c\lambda \leq 1$,

$$\begin{aligned}|\rho(\xi)| &\leq |1 - c\lambda| + |c\lambda e^{-i\xi}| \\ &= 1 - c\lambda + c\lambda|e^{-i\xi}| \\ &= 1.\end{aligned}$$

Thus, for $c\lambda \leq 1$, the von Neumann condition is satisfied and the finite difference method (3.2.6) is stable.

To determine the accuracy of the method, expand v in a Taylor series about the point (ih,nk). Assume that v is twice continuously differentiable so that by equation (3.2.1),

$$\partial_t^2 v = c\partial_t\partial_x v = c\partial_x\partial_t v = c^2\partial_x^2 v,$$

and

$$\frac{v_i^{n+1} - v_i^n}{k} = (\partial_t v)_i^n + \frac{c^2 k}{2}(\partial_x^2 v)_i^n + O(k^2),$$

$$D_+ v_i^n = (\partial_x v)_i^n + \frac{h}{2}(\partial_x v)_i^n + O(h^2).$$

And by using the prototype equation (3.2.1),

$$\text{(3.2.7)}\qquad \frac{v_i^{n+1} - v_i^n}{k} - cD_+ v_i^n = (\frac{c^2 k}{2} - \frac{ch}{2})(\partial_x)_i^n + O(k^2) + O(h^2).$$

Thus the finite difference method (3.2.6) is consistent and accurate of $O(k) + O(h)$.

From (3.2.7) we see that the finite difference method (3.2.6) is also consistent with the equation

$$\text{(3.2.8)}\qquad \partial_t v = c\partial_x v + (\frac{ch}{2} - \frac{c^2 k}{2})\partial_x^2 v$$

and

$$\frac{v_i^{n+1} - v_i^n}{k} - cD_+ v_i^n = O(k^2) + O(h^2).$$

Thus the finite difference method (3.2.6) approximates equation (3.2.8) to $O(k^2) + O(h^2)$. Equation (3.2.8) resembles a diffusion equation.

Since the finite difference method (3.2.6) approximates equation (3.2.8), the solution to (3.2.6) must contain some diffusion. However, this method is used to solve the advection equation (3.2.1). Thus the diffusion introduced is a result of the finite difference method and is "numerically" introduced. This diffusion is called <u>numerical diffusion</u> and the term $ch/2 - c^2k/2$ is called the <u>coefficient of numerical diffusion</u>.

Next, consider the case in which the coefficient c is negative. Then by (3.26b), the CFL condition becomes

$$-ck/h \leq m_1 = 0$$

or

$$\lambda \leq 0,$$

which is an impossible condition. Thus, the CFL condition cannot be satisfied by the finite difference method (3.2.6) with $c < 0$ and, as a result, the method is unstable for $c < 0$. The domain of dependence of the prototype equation (3.2.1) of the point $(ih,(n+1)k)$ with $c < 0$

is the point $((ih + c(n+1)k),0)$. This point lies to the left of point $(ih,0)$ since $c < 0$ and hence, cannot lie in the numerical domain of dependence of (3.2.6) $[ih,(i+n+1)h]$ for any positive choice of k and h. This is depicted in Figure 3.7

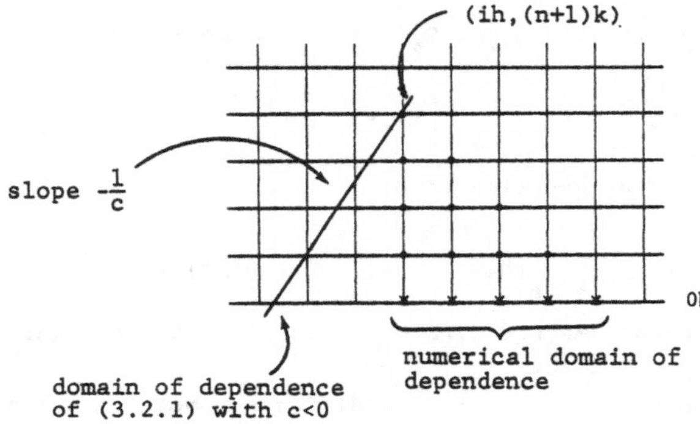

Figure 3.7. Numerical Domain Dependence of $u_i^{n+1} = u_i^n + kcD_+u_i^n$.

A remedy would be to consider a finite difference method whose numerical domain of dependence of $(ih,(n+1)k)$ lies to the left of the point $(ih,0)$. For example, consider the finite difference method

$$\frac{u_i^{n+1} - u_i^n}{k} = cD_-u_i^n \ , \tag{3.2.9}$$

which may be written in the form

$$\begin{aligned} u_i^{n+1} &= u_i^n + kcD_-u_i^n \\ &= (1 + \lambda c)u_i^n - \lambda cS_-u^n \\ &= (1 + \lambda c)u_i^n - \lambda cS_+^{-1}u_i^n \ . \end{aligned}$$

This is in the form of (3.2.3) with $m_1 = 1$ and $m_2 = 0$. With $c < 0$, the CFL condition becomes

$$-c\lambda \leq 1. \tag{3.2.10}$$

The solution at the point $(ih,(n+1)k)$ depends on two points at the previous time level $((i-1)h,nk)$ and (ih,nk). This dependence may be carried to the initial line where we see that the numerical solution at $(ih,(n+1)k)$ depends on the $n+2$ points $((i-(n+1))h,0)$,

$((i-n)h,0),\ldots,(ih,0)$. Thus the umerical domain of dependence of the finite difference method (3.2.9) at $(ih,(n+1)k)$ consists of the n+2 points in the interval $[(i-(n+1))h,ih]$. If the CFL condition (3.2.10) is satisfied, then the domain of dependence of the equation (3.2.1) $((ih + c(n+1)k,0)$ lies in the numerical domain of dependence. For with $c < 0$,

$$ih + c(n+1)k \leq ih,$$

and since, by (3.2.10), $ck \geq -h$,

$$(i - (n+1))h \leq ih + c(n+1)k \leq ih.$$

To consider the stability of the method (3.2.9) take the discrete Fourier transform,

$$\hat{u}^{n+1}(\xi) = \rho(\xi)\hat{u}^n(\xi)$$

where the symbol

$$\rho(\xi) = 1 + c\lambda - c\lambda e^{i\xi}.$$

For $-c\lambda \leq 1$, $0 \leq 1 + c\lambda \leq 1$, and

$$\begin{aligned}|\rho(\xi)| &\leq |1 + c\lambda| + |-c\lambda e^{i\xi}| \\ &= 1 + c\lambda - c\lambda|e^{i\xi}| \\ &= 1 + c\lambda - c\lambda = 1.\end{aligned}$$

Thus, for $-c\lambda \leq 1$, the von Neumann condition is satisfied and the finite difference method (3.2.9) is stable.

The accuracy of the method is determined as in the case of method (3.2.6),

$$\frac{v_i^{n+1} - v_i^n}{k} - cD_-v_i^n = \left(\frac{c^2k}{2} + \frac{ch}{2}\right)(\partial_x^2 v)_i^n + O(h^2) + O(k^2). \tag{3.2.11}$$

Thus, the finite difference method (3.2.9) is consistent and accurate of $O(k) + O(h)$. Again, as in the other method, from (3.2.11), the finite difference method (3.2.9) is also consistent with the diffusion type equation

$$\partial_t v = c\partial_x v - \left(\frac{c^2k}{2} + \frac{ch}{2}\right)\partial_x^2 v \tag{3.2.12}$$

and

$$\frac{v_i^{n+1} - v_i^n}{k} - cD_-v_i^n = O(k^2) + O(h^2).$$

If $c > 0$, then by (3.2.5a) the CFL condition becomes

$$ck/h \leq m_2 = 0.$$

This gives rise to the impossible condition $\lambda \leq 0$. So the CFL condition cannot be satisfied by the finite difference method (3.2.11) with $c > 0$ as a result, the method is unstable for $c > 0$. The domain of dependence of the prototype equation (3.2.1) of the point $(ih,(n+1)k)$ with $c > 0$ is the point $((ih + c(n+1)k),0)$. The point lies to the right of the point $(ih,0)$ since $c > 0$ and, hence, cannot be contained in the numerical domain of dependence of (3.2.11) $[(i-(n+1))h,ih]$ for any positive choice of k and h. This is depicted in the Figure 3.8.

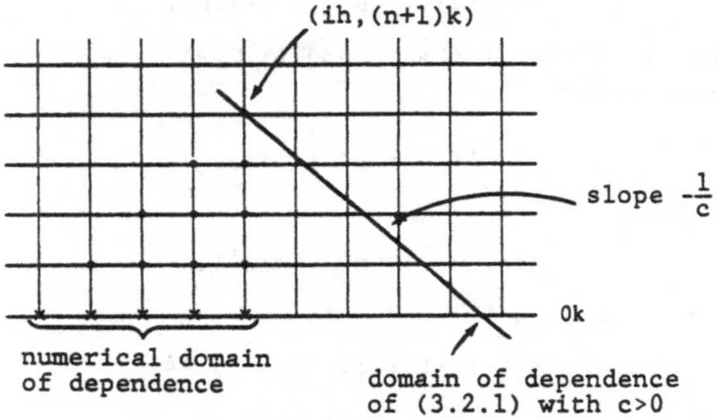

Figure 3.8. Numerical Domain Dependence of
$u_i^{n+1} = u_i^n + kCD_+u_i^n$.

In summary, if $c > 0$, then method (3.2.6) is stable for $c\lambda \leq 1$ and if $c < 0$, then method (3.2.11) is stable for $-c\lambda \leq 1$.

The two methods discussed in this section are first order accurate in space. In an attempt to improve the accuracy in space, consider a centered difference approximation to the term $\partial_x v$. It was shown in Section II.6 that

$$D_0 v_i^n = (\partial_x v)_i^n + O(h^2).$$

This gives rise to the finite difference method

$$\frac{u_i^{n+i} - u_i^n}{k} = cD_0 u_i^n.$$

or

(3.2.13) $$u_i^{n+1} = u_i^n + kcD_0 u_i^n.$$

This finite difference method is consistent with (3.2.1) and accurate of $O(h) + O(h^2)$.

We may rewrite (3.2.13) in the form of (3.2.3)

$$u_i^{n+1} = -c\lambda S_+^{-1} u_i^n + u_i^n + c\lambda S_+ u_i^n \tag{3.2.14}$$

where $m_1 = m_2 = 1$. If $c > 0$ by (3.2.5a) the CFL condition becomes

$$c\lambda \leq 1$$

and if $c < 0$ by (3.2.5b) the CFL condition becomes

$$-c\lambda \leq 1.$$

For arbitrary $c \neq 0$, the CFL condition is

$$\lambda \leq \frac{1}{|c|}.$$

To analyze the stability of the finite difference method (3.2.13) take the discrete Fourier transform

$$\hat{u}^{n+1}(\xi) = \rho(\xi)\hat{u}^n(\xi)$$

where the symbol

$$\begin{aligned}\rho(\xi) &= -c\lambda e^{i\xi} + 1 + c\ e^{-i\xi} \\ &= 1 - ic\lambda \sin \xi.\end{aligned}$$

This gives rise to

$$|\rho(\xi)|^2 = \rho(\xi)\overline{\rho(\xi)} = 1 + c^2\lambda^2\sin^2\xi > 1$$

for any λ not equal to zero. Thus the von Neumann condition is not satisfied and the finite difference method (3.2.13) is unstable for any choice of λ.

This method shows that the CFL condition is not sufficient for stability. For this finite difference method satisfies the CFL condition provided that $\lambda \leq 1/|c|$. However, the method is unconditionally stable.

In order to understand why the method is unstable, consider the individual components. The time derivative depends on the two points $(ih,(n+1)k)$ and (ih,nk) while the spatial derivative depends on the two points $((i-1)h,nk)$ and $((i+1)h,nk)$. The value of the time derivative at $(ih,(n+1)k)$ using the prototype equation (3.2.1) with the space derivative approximated by D_0 is

$$c\ \frac{u_{i+1}^n - u_{i-1}}{2h}$$

and the value of the time derivative at $((i+2)h,nk)$ is

$$c\ \frac{u_{i+3}^n - u_{i+1}}{2h} .$$

These two are coupled by the point u_{i+1}^n. Similarly, the value of the time at $(i-1)h, nk)$ is

$$c \; \frac{u_i^n - u_{i-2}}{2h}$$

and the value of the time derivative at $((i+1)h, nk)$ is

$$c \; \frac{u_{i+2}^n - u_i}{2h} \; .$$

These two are coupled by the point u_i^n. However, those two groups are uncoupled. Thus, the advection occurs independently on two separate grids whose union gives the entire grid, one corresponding to grid points ih where i is even, and the other corresponding to grid points ih where i is odd. As seen in Figure 3.9, the advection on the mesh marked "×" occurs independently of the advection on the mesh marked by "o"

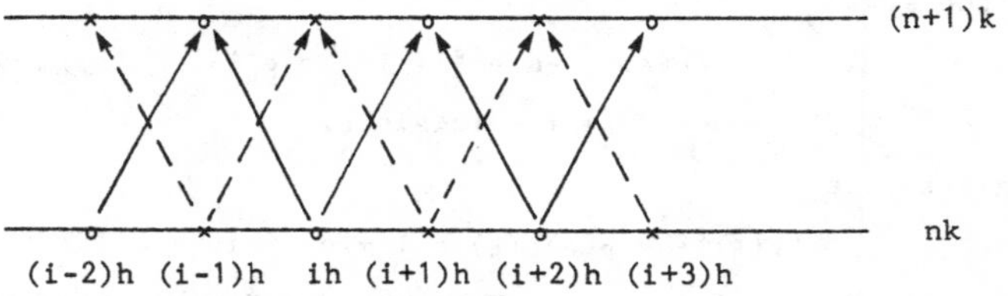

Figure 3.9

A remedy, suggested by Lax [21], is to replace the u_i^n term in approximation to ∂_t in (3.2.13) by an average $\frac{1}{2}(u_{i+1}^n + u_{i-1}^n)$. This gives

(3.2.15) $$u_i^{n+1} = \frac{1}{2}(u_{i+1}^n + u_{i-1}^n) + kcD_0u_i^n \; .$$

Since $m_1 \; m_2 = 1$, the CFL condition remains the same

$$|c|\lambda \leq 1.$$

This method may be written in the form

$$u_i^{n+1} = \frac{1}{2}(S_+ + S_+^{-1})u_i^n + \frac{1}{2} \; c\lambda(S_+ - S_+^{-1})u_i^n$$

so by taking the discrete Fourier transform

$$\hat{u}^{n+1}(\xi) = \rho(\xi)\hat{u}^n(\xi),$$

where

$$\rho(\xi) = \frac{1}{2}(e^{-i\xi} + e^{i\xi}) + \frac{1}{2}\, c\lambda(e^{-i\xi} - e^{i\xi})$$
$$= \cos\xi - ic\lambda\sin\xi .$$

This gives rise to

$$|\rho(\xi)|^2 = \rho(\xi)\overline{\rho(\xi)} = \cos^2\xi + c^2\lambda^2\sin^2\xi$$
$$= 1 - (1 - c^2\lambda^2)\sin^2(\xi).$$

For $|c|\lambda \leq 1$, $0 \leq 1-c^2\lambda^2 \leq 1$ and, hence, $|\rho(\xi)|^2 \leq 1$. Thus, for $|c|\lambda \leq 1$, the von Neumann condition is satisfied and the method (3.2.15), known as the Lax-Friedrichs method, is stable.

We must determine the accuracy of the Lax-Friedrichs method. Expand v in a Taylor series about the point (ih,nk). The average is a second order accurate approximation to v_i^n, that is,

$$\frac{1}{2}(v_{i+1}^n + v_{i-1}^n) = v_i^n + \frac{h^2}{2}(\partial_x^2 v)_i^n + O(h^4).$$

However,

$$\text{(3.2.16)}\quad \frac{v_i^{n+1} - \frac{1}{2}(v_{i+1}^n + v_{i-1}^n)}{k} = (\partial_t v)_i^n + \frac{k}{2}(\partial_t^2 v)_i^n - \frac{h^2}{2k}(\partial_x^2 v)_i^n + O(h^4/k)$$

$$= (\partial_t v)_i^n + \frac{k}{2}(\partial_t^2 v)_i^n - \frac{h}{2\lambda}(\partial_x^2 v)_i^n + O(h^3)$$

$$= (\partial_t v)_i^n + O(k) + O(h).$$

The division by k in the dividend difference is responsible for the reduction from $O(h^2)$ to $O(h)$. By substituting into (3.2.15),

$$\frac{v_i^{n+1} - \frac{1}{2}(v_{i+1}^n + v_{i-1}^n)}{k} - cD_0 v_i^n = (\partial_t v)_i^n - c(\partial_x v)_i^n + O(k) + O(h)$$

Thus, the Lax-Friedrichs method is consistent with the prototype equation (3.2.1) and accurate of $O(k) + O(h)$.

How does the inclusion of the average term stabilize the method (3.2.13)? Observe that the Lax-Friedrichs method (3.2.15) may be written in the form

$$u_i^{n+1} = u_i^n + \frac{h^2}{2}\,\frac{u_{i+1}^n - 2u_i^n + u_{i-1}^n}{h^2} + ckD_0 u_i^n$$

or

$$\text{(3.2.17)}\quad \frac{u_i^{n+1} - u_i^n}{k} = ckD_0 u_i^n + \frac{h^2}{2k}\, D_+D_- u_i^n .$$

By retaining the $\partial_x^2 v$ term in (3.2.16),

$$\frac{v_i^{n+1} - (v_{i+1}^n + v_{i-1}^n)}{k} - cD_0 v_i^n \equiv \frac{v_i^{n+1} - v_i^n}{k} - ckD_0 v_i^n - \frac{h^2}{2k} D_+D_- v_i^n$$

$$= (\partial_t v)_i^n - c(\partial_x v)_i^n - \frac{h^2}{2k}(\partial_x^2 v)_i^n + O(k) + O(h^2).$$

Thus, the Lax-Friedrichs method (3.2.15) or (3.2.17) is also consistent with the diffusion type equation

$$\partial_t v = c\partial_x v + \frac{h^2}{2k}\partial_x^2 v$$

and is accurate to $O(k) + O(h^2)$.

Thus the Lax-Friedrichs method as represented by (3.2.17) involves advection and diffusion. Consider the time derivative at the point $(ih,(n+1)k)$. The diffusion term $\frac{h^2}{2k} D_+D_- u_i^n$ couples the two uncoupled grids because it contains the term u_i^n.

III.3. Algebraic Characterization of Accuracy

Consider the prototype equation (3.2.1)

$$\partial_t v = c\partial_x v, \quad -\infty < x < +\infty, \quad t > 0$$

approximated by the finite difference method (3.2.3).

$$u_i^{n+1} = \sum_{j=-m_1}^{m_2} A_j S_+^j u_i^n$$

where A_j are constants. Recall that by taking the discrete Fourier transform of (3.2.3),

$$\hat{u}^{n+1}(\xi) = \rho(\xi)\hat{u}^n(\xi)$$

where the symbol $\rho(\xi)$ is given by

$$\rho(\xi) = \sum_{j=-m_1}^{m_2} A_j e^{-ij\xi}. \tag{3.3.1}$$

The following theorem was proved by Lax [22].

Theorem 3.1. Consider the advection equation (3.2.1) approximated by a finite difference method of the form of (3.2.3). Suppose the finite difference method is accurate of order (p_1, p_2) and $k = O(h)$, then

$$\rho(\xi) = e^{-ka\xi/h} + O(|\xi|^{p+1}) \tag{3.3.2}$$

where $p = \min(p_1, p_2)$.

Proof: By the definition of accurate of order (p_1, p_2),

$$v_i^{n+1} = \sum_{j=-m_1}^{m_2} A_j v_{i+1}^n + k(O(k^{p_1}) + O(h^{p_2})).$$

Since $k = O(h)$ and $p = \min(p_1, p_2)$. we have

$$v_i^{n+1} = \sum_{j=-m_1}^{m_2} A_j v_{i+1}^n + kO(k^p).$$

By expanding each term in (3.3.3) in a Taylor series in space about th point (ih, nk), we obtain

$$v_i^{n+1} = v_i^n + k\partial_t v_i^n + \ldots + \frac{k^m}{m!} \partial_t^m v_i^n + \ldots + O(k^{p+1}).$$

Using the equation (3.2.1), provided v is sufficiently smooth, $\partial_t^m v = c^m \partial_x^m v$. Substituting into the Taylor series for v_i^{n+1},

(3.3.4) $$v_i^{n+1} = v_i^n + kc\ {}_x v_i^n + \ldots + \frac{(kc)^m}{m!} \partial_x^m v_i^n + \ldots + O(k^{p+1}).$$

Similarly,

(3.3.5) $$v_{i+j}^n = v_i^n + jh\partial_x v_i^n + \frac{(jh)^2}{2!} \partial_x^2 v_i^n + \ldots .$$

By substituting (3.3.5) into (3.3.3),

$$v_i^{n+1} = \sum_{j=-m_1}^{m_2} A_j \left(\sum_{\ell=0}^{p} \frac{(jh)^\ell}{\ell!} \partial_x^\ell v_i^n + O(k^{p+1}) \right)$$

while, interchanging the summation,

(3.6) $$v_i^{n+1} = \sum_{\ell=0}^{p} \left(\sum_{j=-m_1}^{m_2} A_j \frac{(jh)^\ell}{\ell!} \right) \partial_x^\ell v_i^n + O(k^{p+1}).$$

By equating the coefficients of $\partial_x^\ell v_i^n$ in the right hand side of (3.3.6) with those of the Taylor series for v_i^{n+1} given by (3.3.4), we obtain for $\ell = 0$

$$\sum_{j=-m_1}^{m_2} A_j = 1$$

and for $\ell = 1, \ldots, p$

$$\frac{(kc)^\ell}{\ell!}\,\partial_x^\ell v_i^n = \Big(\sum_{j=-m_1}^{m_2} A_j \frac{(jh)^\ell}{\ell!}\Big)\partial_x^\ell v_i^n .$$

Hence, for $0 \le \ell \le p$, by multiplying through by $(\frac{-i\xi}{h})^\ell/\partial_x^\ell v_i^n$

$$\frac{1}{\ell!}\Big(\frac{-i\xi kc}{h}\Big)^\ell = \sum_{j=-m_1}^{m_2} A_j \frac{(-ij\xi)^\ell}{\ell!}. \tag{3.3.7}$$

Expanding e^z in a Taylor series about $z = 0$ and evaluating at $z = -ik\xi c/h$,

$$e^{-ik\xi c/h} = \sum_{\ell=0}^{p} \frac{1}{\ell!}\Big(\frac{-ik\xi c}{h}\Big)^\ell + O\Big(\big(\frac{\xi k}{h}\big)^{p+1}\Big).$$

Since $k = O(h)$ and thus, $k/h = O(1)$,

$$\sum_{\ell=0}^{p} \frac{1}{\ell!}\Big(\frac{-ik\xi c}{h}\Big)^\ell = e^{-ik\xi c/h} + O(\xi^{p+1}) \tag{3.3.8}$$

Similarly, by evaluating the Taylor series for e^z at $z = -ij\xi$,

$$\sum_{\ell=0}^{p} \frac{1}{\ell!}(-ij\xi)^\ell = e^{-ij\xi} + O(\xi^{p+1}). \tag{3.3.9}$$

However, summing (3.3.7) for $\ell = 0,\ldots,p$

$$\begin{aligned}\sum_{\ell=0}^{p} \frac{1}{\ell!}\Big(\frac{-ik\xi c}{h}\Big)^\ell &= \sum_{\ell=0}^{p} \sum_{j=-m_1} A_j \Big(\frac{-ij\xi}{\ell!}\Big)^\ell \\ &= \sum_{j=-m_1}^{m_2} A_j \sum_{\ell=0}^{p} \frac{1}{\ell!}(-ij\xi)^\ell.\end{aligned}$$

But then, by substituting (3.3.9) in the right-hand side,

$$\begin{aligned}\sum_{\ell=0}^{p} \frac{1}{\ell!}\Big(\frac{-ik\xi c}{h}\Big)^\ell &= \sum_{j=-m_1}^{m_2} A_j\big(e^{-ij\xi} + O(\xi^{p+1})\big) \\ &= \sum_{j=-m_1}^{m_2} A_j e^{ij\xi} + O(\xi^{p+1}) \\ &= \rho(\xi) = O(\xi^{p+1}).\end{aligned}$$

By replacing the left-hand side with (3.3.8), the result (3.3.2),

$$e^{-ik\xi c/h} = \rho(\xi) + O(\xi^{p+1}),$$

is obtained.

What is the significance of this result? If a finite difference method is stable, then the von Neumann condition is satisfied and $|\rho(\xi)| \leq 1$. In this case, considering the symbol $\rho(\xi)$ a complex function $\rho(\xi)$ lies inside the unit circle in the complex plane as depicted in Figure 3.10.

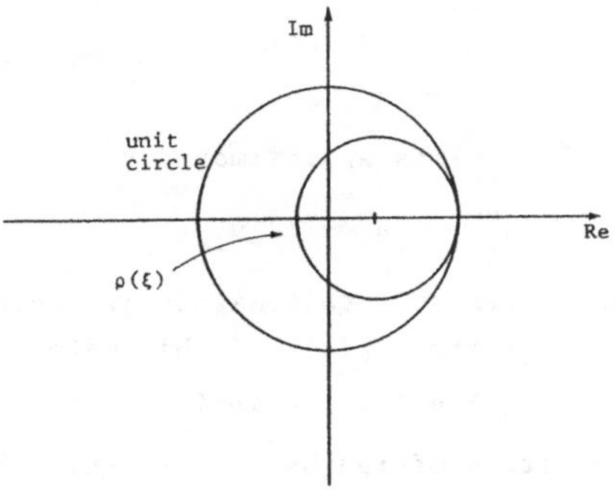

Figure 3.10

Stability indicates how close the entire curve given by $\rho(\xi)$ is to the unit circle. This is a global property. Using (3.3.2), we can determine how close the curve given by $\rho(\xi)$ is to the unit circle in the vicinity of $\xi = 0$, the tangent point $(1,0)$. This is a local property. The more accurate the finite difference method, the smaller $|\xi|^{p+1}$ becomes for ξ near 0 and, hence, the closer the curve given by $\rho(\xi)$ is to the unit circle.

As an example, consider the prototype equation with $c = 1$,

(3.3.10) $$\partial_t v = \partial_x v.$$

Suppose this equation is approximated by

$$u^{n+1} = u_i^n + kD_+u_i^n .$$

It was shown in Section III.2 that this method was stable for $\lambda = k/h \leq 1$ and the symbol $\rho(\xi)$ is given by

$$\rho(\xi) = 1 - \lambda + \lambda e^{-i\xi}.$$

This represents a circle with radius λ and center $1-\lambda$. This is depicted in Figure 3.11.

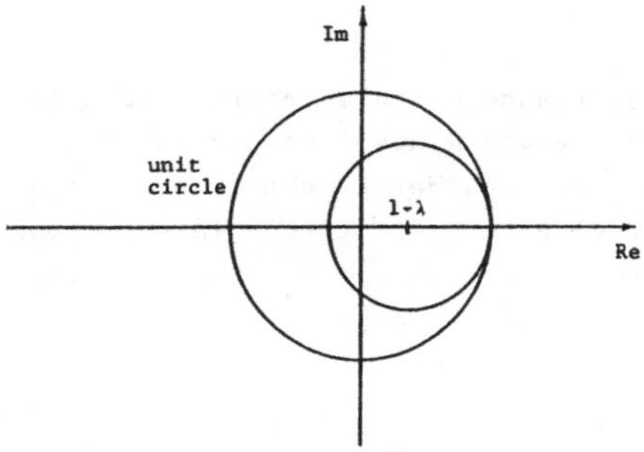

Figure 3.11

Suppose that equation (3.3.10) is approximated by

$$u_i^{n+1} = u_i^n + kD_-u_i^n .$$

It was shown in Section III.2 that this method is unconditionally unstable for $c > 0$. The symbol $\rho(\xi)$ of this method is

$$\rho(\xi) = 1 + \lambda - \lambda e^{i\xi}.$$

This also represents a circle of radius λ. In this case, however, the center of the circle is $1+\lambda$ so that the entire circle lies outside the unit circle as depicted in Figure 3.12.

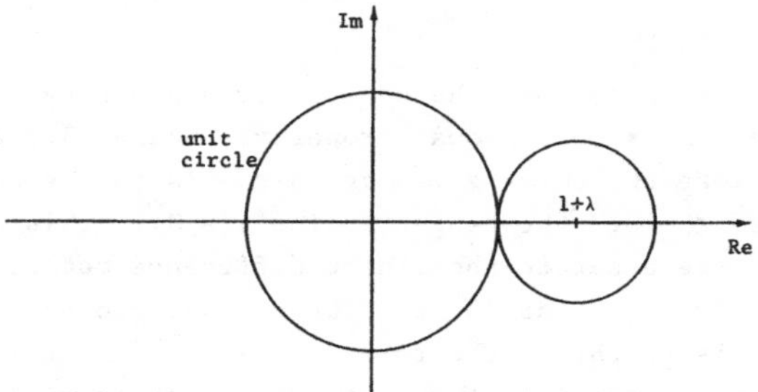

Figure 3.12

The unconditionally unstable approximation to (3.3.10) given by

(3.3.11) $$u_i^{n+1} = u_i^n + kD_0u_i^n$$

has symbol

$$\rho(\xi) = 1 - i\lambda\sin\xi.$$

This represents a vertical line segment as depicted in Figure 3.13.

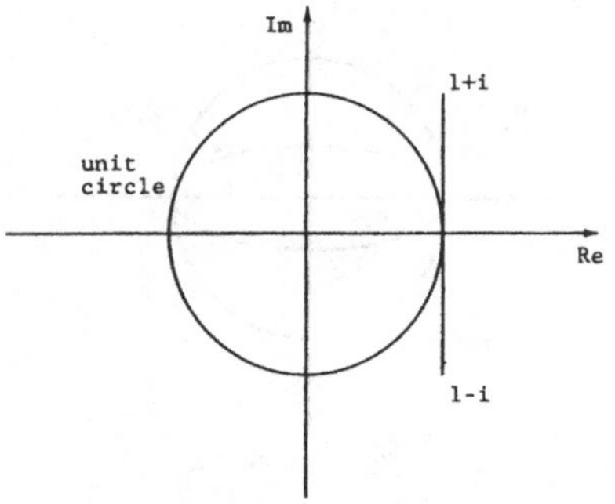

Figure 3.13

In many cases, it may be possible to modify an unstable finite difference method so that it becomes stable. For an unstable finite difference method, the symbol $\rho(\xi)$ lies outside the unit circle for some $0 \leq \xi < 2\pi$. If the CFL condition is satisfied by the finite difference method, then terms of $O(h^q)$ where q is sufficiently large may be added so that the symbol of the modified finite difference method decreases for large values of ξ is an attempt to stablizize the method.

For example, any term which contains terms of the form D_+D_-, $(D_+D_-)^2, \ldots$ can be used to decrease the symbol $\rho(\xi)$ for large values of ξ.

An example of a modified method which is stable is the Lax-Friedrichs method (3.2.21). The finite difference method (3.3.11) is unconditionally stable, however, it satisfies the CFL condition provided that $\lambda \leq 1$. In the Lax-Friedrichs method, the term $\frac{h^2}{2}D_+D_-u_i^n$ was added to the unstable finite difference method (3.3.11) to obtain

$$u_i^{n+1} = u_i^n + kD_0u_i^n + \frac{h^2}{2} D_+D_-u_i^n .$$

This method is stable for $\lambda \leq 1$. The symbol is given by

$$\rho(\xi) = \cos \xi - i\lambda \sin \xi .$$

The term $\frac{h^2}{2} D_+D_-u_i^n$ results in replacing the 1 in the symbol (3.3.11) with $\cos \xi$. The graph of the symbol is an ellipse contained in the unit circle with major axis of length 2 and minor axis of length 2λ as depicted in Figure 3.14.

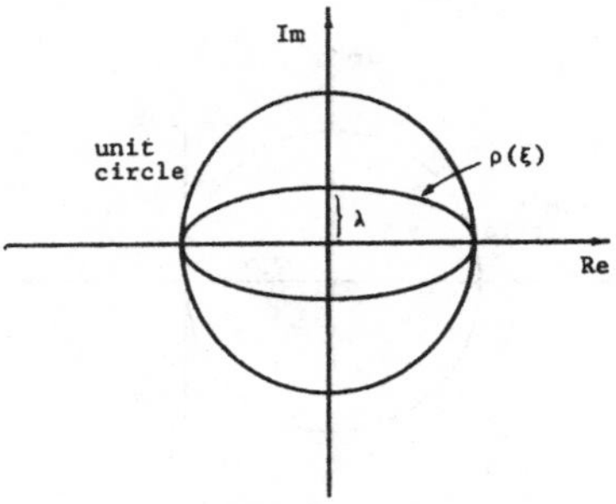

Figure 3.14

III.4. Finite Difference Methods with Positive Coefficients

Consider the differential equation

$$\partial_t v = \partial_x v \quad -\infty < x < +\infty, \quad t > 0$$

with initial condition

$$v(x,0) = f(x), \quad -\infty < x < +\infty,$$

where f is twice differentiable. Let $\frac{k}{h} = t \leq 1$, and consider the grid depicted in Figure 3.15.

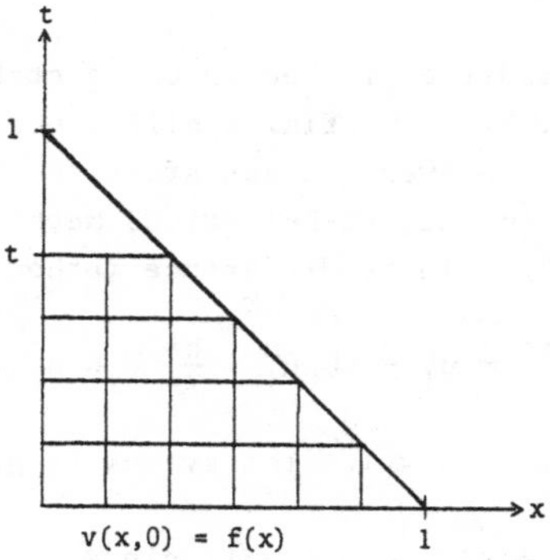

Figure 3.15

Let N be chosen such that $h = 1/N$. Then for a given value of t, $k = t/N$.

Consider the finite difference method

(3.4.1) $$u_i^{n+1} = u_i^n + \frac{k}{h} D_+ u_i^n .$$

It has been shown in Section III.2 that this method satisfies the CFL condition and is stable for $k/h \leq 1$.

Expanding the exact solution $v(x,t)$ in a Taylor series with remainder about the point (ih,nk),

$$(3.4.2)\qquad \frac{v_i^{n+1} - v_i^n}{k} - D_+v_i^n = h\left(\frac{k}{2h}\,\partial_t^2 v(ih,t^*) - \frac{1}{2}\,\partial_x^2 v(x^*,nk)\right)$$

where $ih < x^* < (i+1)h$ and $nk < t^* < (n+1)k$. Since f is twice differentiable, $\partial_t^2 v$ and $\partial_x^2 v$ exist and the Taylor series with remainder exists. Let M be chosen so that $|f''(x)| \leq M$, then $|\partial_t^2 v|$, $|\partial_x^2 v| \leq M$. This follows since the exact solution $v(x,t) = f(x+t)$, so that $\partial_t^2 v = \partial_x^2 v = f''(x+t)$. The truncation error τ is given by

$$(3.4.3)\qquad \tau = h\left(\frac{k}{2h}\,\partial_t^2 v(ih,t^*) - \frac{1}{2}\,\partial_x^2 v(x^*,nk)\right)$$

from which an upper bound on τ is

$$|\tau| \leq h\left(\frac{t}{2}|\partial_t^2 v(ih,t^*)| + \frac{1}{2}|\partial_t^2 v(x^*,nk)|\right)$$

$$\leq h\left(\frac{M}{2}\,t + \frac{M}{2}\right)\ .$$

Since $k/h = t \leq 1$,

$$(3.4.4)\qquad |\tau| \leq hM.$$

The finite difference method (3.4.1) is stable in the maximum norm (it has been shown to be stable in the ℓ_2-norm via the von Neumann condition). By (3.4.1),

$$|u_i^{n+1}| \leq \left(1 - \frac{k}{h}\right)|u_i^n| + \frac{k}{h}|u_{i+1}^n|\ ,$$

since $\frac{k}{h} = t \leq 1$, and

$$|u_i^{n+1}| \leq \left(1 - \frac{k}{h}\right)\max_i|u_i^n| + \frac{k}{h}\max_i|u_{i+1}^n|$$

$$= \|u^n\|_{max}\ .$$

This is true for any i, so

$$\|u^{n+1}\|_{max} \leq \|u^n\|_{max}.$$

Define the error e_i^n by $e_i^n = u_i^n - v_i^n$ where $v_i^n \equiv v(ih,nk)$. By subtracting (3.4.2) from (3.4.1),

$$e_i^{n+1} = \left(1 - \frac{k}{h}\right)e_i^n + \frac{k}{h}\,e_{i+1}^n - k\tau,$$

where τ is given by (3.4.3).

Again, since $k/h = t \le 1$,

$$|e_i^{n+1}| \le (1 - \frac{k}{h})|e_i^n| + \frac{k}{h}|e_{i+1}^n| + k|\tau|$$

$$\le (1 - \frac{k}{h})\|e^n\|_{max} + \frac{k}{h}\|e^n\|_{max} + khM,$$

using inequality (3.4.4). This gives

(3.4.5)
$$\|e^{n+1}\|_{max} \le \|e^n\|_{max} + khM$$
$$\le \|e^{n-1}\|_{max} + 2khM$$
$$\le \ldots \le \|e^0\|_{max} + (n+1)khM$$
$$= (n+1)khM = thM \le hM,$$

since $t \le 1$. Therefore $\|e^{n+1}\|_{max} \to 0$ as $h \to 0$. This shows convergence in the maximum norm. Look at the solution of the finite difference method (3.4.1) at time nk. For $t = k/h$,

$$u_i^n = (1-t)u_i^{n-1} + tu_{i+1}^{n-1}$$
$$= (1-t)^2 u_i^{n-2} + 2t(1-t)u_{i+1}^{n-2} + t^2 u_{i+2}^{n-2}$$
$$= \sum_{j=0}^{n} \binom{n}{j} t^j (1 - t)^{n-j} \; S_+^j u_i^0 \, ,$$

that is, u_i^n is a polynomial of degree n in t with the coefficients depending on $u_i^0 = f(ih)$. We denote the polynomial, called the Bernstein polynomial (see Strang [43]), by $B_n f(t)$. B_n is a linear operator that acts on the initial condition f. Since $B_n f(t) = u_i^n$ and the exact solution is $v(x,t) = f(x+t)$, if $|f''| \le M$, then by (3.4.5),

(3.4.6)
$$|B_n f - f| \le hM.$$

This is an approximation theorem.

Some properties of the polynomial $B_n f(t)$ are:

1) B_n is linear.
2) B_n preserves inequalities, that is, if $f \le g$, then $B_n f \le B_n g$.
3) $B_n a = a$, where a is a constant function.
4) $B_n t = t$.
5) If $f(t) = t^2$ then $|B_n f - f| \le 2/N$.

Property 1 follows from the finite difference method (3.4.1) being linear. Since $k/h = t \le 1$, the finite difference method (3.4.1) has positive coefficients. Thus, if $f \ge 0$, then $B_n f \ge 0$, which proves Property 2. By consistency, the sum of the coefficients of the finite difference method is 1. For the method (3.4.1), $(1 - \frac{k}{h}) + \frac{k}{h} = 1$. Thus $(1 - \frac{k}{h})c + \frac{k}{h}c = c$ from which Property 3 follows. Property 4 follows from the finite difference method (3.4.1) being first order accurate; it is exact for polynomials of degree 1 or less. If $f(t) = t^2$, then $f''(t) = 2$ and hence, $M = \max_t |f''(t) = 2$. Substitution into (3.4.6) with $h = 1/N$ gives Property 5.

Suppose that $f(t)$ is merely continuous on $[0,1]$, then since $[0,1]$ is compact, f is uniformly continuous and uniformly bounded on $[0,1]$ with bound $\bar{M}$. Thus $|f(t)| \leq \bar{M}$ for every $t \in [0,1]$. The definition of uniform continuity states that for any $\varepsilon > 0$, there exists a $\delta > 0$ (independent of ε) such that if $|t - t_0| < \delta$, then $|f(t) - f(t_0)| < \varepsilon$. Suppose $|t - t_0| \geq \delta$, then $\frac{|t-t_0|^2}{\delta^2} \geq 1$. This gives the following bound

$$|f(t) - f(t_0)| \leq |f(t) + |f(t_0)| \leq 2\bar{M}$$
$$\leq 2\bar{M}(t - t_0)^2/\delta^2.$$

Combining this with the bound for $|t - t_0| < \delta$,

$$|f(t) - f(t_0)| \leq \varepsilon + \frac{2\bar{M}(t-t_0)^2}{\delta^2} ,$$

which may be written in the form

$$(3.4.7) \qquad -\varepsilon - \frac{2\bar{M}(t-t_0)^2}{\delta^2} < f(t) - f(t_0) < \varepsilon + \frac{2\bar{M}(t-t_0)^2}{\delta^2}.$$

Since B_n preserves inequalities (Property 2), applying B_n to (3.4.7)

$$(3.4.8) \qquad -\varepsilon - \frac{2\bar{M}}{\delta^2} B_n(t-t_0)^2 \leq B_nf(t) - f(t_0) \leq \varepsilon + \frac{2\bar{M}}{\delta^2} B_n(t-t_0)^2$$

where, by Property 3, $B_nf(t_0) = f(t_0)$. By expanding $(t-t_0)^2$ out, treating t_0 as a constant, using Properties 1), 3), and 4),

$$\frac{2\bar{M}}{\delta^2} B_n(t-t_0)^2 = \frac{2\bar{M}}{\delta^2} [B_nt^2 - 2t_0t + t_0^2]$$
$$= \frac{2\bar{M}}{\delta^2} [B_nt^2 - t^2 + t^2 - 2t_0t - t_0^2]$$
$$= \frac{2\bar{M}}{\delta^2} [(B_nt^2 - t^2) + (t-t_0)^2],$$

which by Property 5),

$$\frac{2\bar{M}}{\delta^2} B_n(t - t_0)^2 < \frac{2\bar{M}}{\delta^2} \cdot \frac{2}{N} + \frac{2\bar{M}}{\delta^2}(t - t_0)^2 .$$

By substitution into (3.4.8),

$$|B_n f(t) - f(t_0)| < \varepsilon + \frac{4\bar{M}}{\delta^2 N} + \frac{2\bar{M}}{\delta^2}(t - t_0)^2 ,$$

and letting $t \to t_0$,

$$|B_n f(t) - f(t)| < \varepsilon + \frac{4\bar{M}}{\delta^2 N} .$$

Thus, as N becomes sufficiently large, $|B_n f(t) - f(t)|$ becomes arbitrarily small, which is the Weierstrass approximation theorem.

We have, in fact, proved that the solution of the finite difference equation converges to the solution of the partial differential equation whenever the initial condition is merely continuous. This is a very important result.

Suppose the finite difference method, which is second order accurate, could be constructed with positive coefficients (that is, $O(h^2)$), then the corresponding B_n would satisfy the additional property $B_n t^2 = t^2$. Inequality (3.4.8) would then become

$$-\varepsilon - \frac{2\bar{M}}{\delta^2}(t - t_0)^2 < B_n f(t) - f(t_0) < \varepsilon + {}_{\delta^2}(t - t_0)^2$$

$$|B_n f(t) - f(t_0)| < \varepsilon + \frac{2\bar{M}}{\delta^2}(t - t_0)^2 ,$$

and letting $t \to t_0$,

$$|B_n f(t) - f(t)| < \varepsilon,$$

for arbitrary ε. Hence, $B_n f(t) = f(t)$. These results are summarized in the following theorem.

Theorem 3.2. If a finite difference method (for linear equations) has positive coefficients, then it has either first-order accuracy or infinite order accuracy.

As an example of such a method, consider the finite difference method

$$u_i^{n+1} = u_i^n + kcD_+ u_i$$

and the Lax-Friedrichs method

$$u_i^{n+1} = \frac{1}{2}(u_{i+1}^n + u_{i-1}^n) + kcD_0 u_i^n .$$

If $c\frac{k}{h} = 1$, the upper limit on the CFL condition, then both methods reduce to

$$u_i^{n+1} = u_{i+1}^n .$$

This method has infinite order accuracy since it follows the characteristic $x + ct = \text{constant}$ from the initial line. The numerical domain of dependence of the point $(ih,(n+1)k)$ in this case is the single point $((i+n+1)h,0)$. The domain of dependence of the prototype equation (3.2.1) of the point $(ih,(n+1)k)$ is the single point $(ih + c(n+1)k,0)$. For $c\frac{k}{h} = 1$ or $ck = h$, these points are the same and the numerical domain of dependence is identical to the domain of dependence of the prototype equation.

This theorem will turn out to have an important consequence. In the case of more than one equation, the accuracy is, in fact, at most one, except in extraordinary cases.

III.5. The Lax-Wendroff Method

Consider a finite difference method that approximates the prototype equation (3.2.1)

$$\partial_t v = c\partial_x v$$

with the following properties

1) u_i^{n+1} depends on u_{i-1}^n, u_i^n, and u_{i+1}^n and

2) the accuracy of the finite difference method is $O(k^2) + O(h^2)$.

In this case the finite difference method is essentially* uniquely determined and is called the Lax-Wendroff method (see Lax-Wendroff [25]). To see this write out the finite difference method so that Property 1) is satisfied. The method takes the form

$$u_i^{n+1} = \sum_{j=-1}^{1} A_j S_+^j u_i^n = A_{-1}u_{i-1}^n + A_0 u_i^n + A_1 u_{i+1}^n. \tag{3.5.1}$$

Upon substitution of the exact solution $v(x,t)$ into (3.2.1),

$$v_i^{n+1} - A_{-1}v_{i-1}^n - A_0 v_i^n - A_1 v_{i+1}^n = k(O(k^2) + O(h^2)) \tag{3.5.2}$$

which satisfies Property 2. Expanding the exact solution v in a Taylor series about the point (ih,nk) and substituting into (3.5.2),

---

*We use "essentially", for no finite difference method is uniquely determined, because a higher order term could be added and still satisfy Properties 1 and 2. For example, consider

$$u^{n+1} = \sum_{j=-1}^{1} A_j S_+^j u_j^n + O(h^6)$$

that is, add on an $O(h^6)$ term and obtain another finite difference method which is still consistent.

$$(v_i^n + k(\partial_t v)_i^n + \frac{k^2}{2}(\partial_t^2 v)^n + O(k^3)) - A_{-1}(v_i^n - h(\partial_x v)_i^n + \frac{h^2}{2}(\partial_x^2 v)_i^n$$
$$+ O(h^3)) - A_0 v_i^n - A_1(v_i^n + h(\partial_x v)_i^n + \frac{h^2}{2}(\partial_x v)_i^n + O(h^3))$$
$$= k(O(k^2) + O(h^2)),$$

which is equivalent to

$$(1 - (A_{-1} + A_0 + A_1))v_i^n + (k(\partial_t v)_i^n - (A_1 - A_{-1})h(\partial_x v)_i^n) + (\frac{k^2}{2}(\partial_t^2 v)_i^n$$
$$- (A_{-1} + A_1)\frac{k^2}{2}(\partial_x^2 v)_i^n) - k(O(k^2) + O(h^2)) = k(O(k^2) + O(h^2)).$$

Using the prototype equation (3.2.1), $\partial_t^2 v = c^2\partial_x^2 v$ which gives by substitution

$$(1 - (A_{-1} + A_0 + A_1))v_i^n + (kc - (A_1 - A_{-1})h)(\partial_x v)_i^n$$
$$+ (\frac{k^2c^2}{2} - (A_{-1} + A_1)\frac{h^2}{2})(\partial_x^2 v)_i^n + k(O(k^2) + O(h^2)) = k(O(k^2) + O(h^2)).$$

In order to satisfy Property 2, the first three terms must be zero which gives rise to 3 equations for the coefficients A_{-1}, A_0, A_1,

$$A_{-1} + A_0 + A_1 = 1\ ,$$
$$A_1 - A_{-1} = c\lambda\ ,$$
$$A_{-1} + A_1 = c^2\lambda^2\ ,$$

where $\lambda = k/h$. This yields $A_{-1} = \frac{1}{2}(c^2\lambda^2 + c\lambda)$, $A_0 = 1 - c^2\lambda^2$, and $A_1 = \frac{1}{2}(c^2\lambda^2 - c\lambda)$. Substitution into (3.5.1)

$$u_i^{n+1} = (\tfrac{1}{2}c^2\lambda^2 - \tfrac{1}{2}c\lambda)u_{i-1}^n + (1 - c^2\lambda^2)u_i^n$$
$$+ (\tfrac{1}{2}c^2\lambda^2 + \tfrac{1}{2}c\lambda)u_{i+1}^n\ ,$$

which may be rewritten in the form

(3.5.3) $$\frac{u_i^{n+1} - u_i^n}{k} = cD_0u_i^n + \frac{kc^2}{2}D_+D_-u_i^n$$

or

$$u_i^{n+1} = u_i^n + \frac{kc}{2h}(u_{i+1}^n - u_{i-1}^n) + \frac{k^2c^2}{2h^2}(u_{i+1}^n - 2u_i^n + u_{i-1}^n)\ .$$

To analyze the stability of the Lax-Wendroff method (3.5.3) take the discrete Fourier transform

$$\hat{u}^{n+1}(\xi) = \rho(\xi)\hat{u}^n(\xi)$$

where the symbol

$$\rho(\xi) = 1 + \tfrac{1}{2}c\lambda(e^{-i\xi} - e^{i\xi}) + \tfrac{1}{2}c^2\lambda^2(e^{-i\xi} - 2 + e^{i\xi})$$
$$= 1 + c^2\lambda^2(\cos\xi - 1) - ic\lambda\sin\xi,$$

so that

$$|\rho(\xi)|^2 = \rho(\xi)\overline{\rho(\xi)} = (1 + c^2\lambda^2(\cos\xi - 1))^2 + c^2\lambda^2\sin^2\xi$$
$$= 1 + c^4\lambda^4(\cos\xi - 1)^2 + 2c^2\lambda^2\cos\xi - c^2\lambda^2 + c^2\lambda^2(\sin^2\xi - 1)$$
$$= 1 + c^4\lambda^4(\cos\xi - 1)^2 + c^2\lambda^2(\cos\xi - 1)^2$$
$$= 1 - c^2\lambda^2(1 - c^2\lambda^2)(\cos\xi - 1)^2$$
$$= 1 - 4c^2\lambda^2(1 - c^2\lambda^2)\sin^4(\xi/2).$$

By (3.5.1) $m_1 = m_2 = 1$ in (3.2.3), so that by (3.2.5a) and (3.2.5b), the CFL condition is

$$|c|\lambda \leq 1.$$

Hence if $|c|\lambda \leq 1$, we have $|\rho(\xi)|^2 \leq 1$ and the von Neumann condition is satisfied. Thus the Lax-Wendroff method is stable if $|c|\lambda \leq 1$.

This method is similar to the Lax-Friedrichs method in that the unstable method

$$u^{n+1} = u_i^n + kcD_0u_i^n$$

was made stable by adding a term of the form $D_+D_-u_i^n$. The only difference is the coefficient of this term. The coefficient used in the Lax-Wendroff method was chosen so that the method is second order accurate.

Observe that the Lax-Wendroff method is also consistent with the diffusion type equation

$$\partial_t v = c\partial_x v + \frac{kc^2}{2}\,\partial_x^2 v\ .$$

The symbol may be written in the form

$$\rho(\xi) = 1 - c^2\lambda^2 + c^2\lambda^2\cos\xi - ic\lambda\sin\xi\ .$$

The term $c^2\lambda^2\cos\xi - ic\lambda\sin\xi$ represents an ellipse with center at the origin in the complex plane. The graph of $\rho(\xi)$ represents an ellipse whose center is translated by $1 - c^2\lambda^2$ as seen in Figure 3.16.

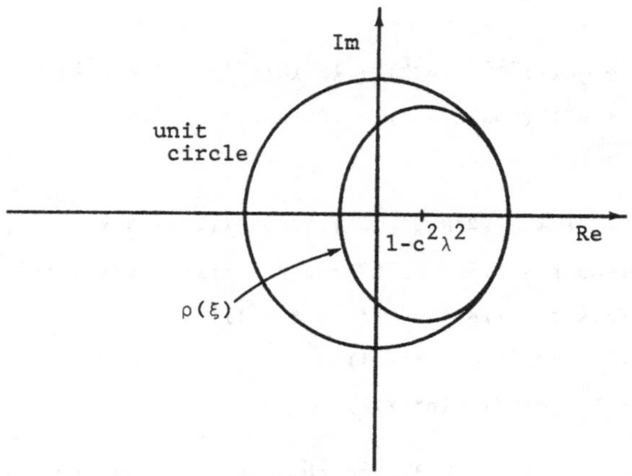

Figure 3.16

Observe that the Lax-Wendroff method reduces to a method with positive coefficients (3.4.9)

$$u_i^{n+1} = u_{i+1}^n$$

where $c\lambda = 1$. It was seen in the previous section that this method has infinite order accuracy.

III.6. Dispersion and Dissipation

A partial differential equation couples points in space and time. The linear properties of a partial differential equation may be described by the action of a wave in space and time. Consider the effect of the partial differential equation on a single wave in space and time called a Fourier mode

$$v(x,t) = \hat{v}e^{i(\omega t+\beta x)} , \tag{3.6.1}$$

where ω is the frequency of the wave, β is the wave number which is related to the wave length Λ by $\beta = 2\pi/\Lambda$. In general, ω is a function of β. Upon substitution of this Fourier mode into the partial differential equation, we obtain an equation which gives $\omega(\beta)$ called the dispersion relation. The frequency $\omega(\beta)$ may be real or imaginary. If ω is real, then oscillations or wave phenomena are being described by the partial differential equation. If ω is imaginary, then the Fourier mode grows or decays depending on the sign of ω.

Consider the diffusion equation

(3.6.2) $$\partial_t v = a\partial_x^2 v \ .$$

By substitution of (3.6.1) into the diffusion equation (3.6.2),

$$i\omega\hat{v}e^{i(\omega t+\beta x)} = -a\beta^2\hat{v}e^{i(\omega t+\beta x)} \ ,$$

which gives rise to the dispersion relation $i\omega = -a\beta^2$, or

(3.6.3) $$\omega = i\beta^2 \ .$$

Since ω is imaginary, the Fourier mode decays with time, for by substitution of (3.6.3) into (3.6.1), we obtain the Fourier mode for the diffusion equation (3.6.2)

$$v(x,t) = \hat{v}e^{-a\beta^2 t+i\beta x}$$

The term $e^{-a\beta^2 t}$ produces decay with time.

Now, consider the advection equation given by the prototype equation (3.2.1)

$$\partial_t v = c\partial_x v \ .$$

By substitution of (3.6.3) into the advection equation (3.2.1),

$$i\omega\hat{v}e^{i(\omega t+\beta k)} = ic\beta\hat{v}e^{i(\omega t+\beta t)} \ ,$$

which gives rise to the dispersion relation

(3.6.4) $$\omega = c\beta.$$

Upon substitution of (3.6.4) into (3.6.1), we obtain the Fourier mode for the advection equation

$$v(x,t) = \hat{v}e^{i\beta(x+ct)} \ .$$

We see from this that each Fourier mode is propagated with unit amplitude (ω real) at a constant speed c independent of the wave number β and, hence, independent of the frequency.

We would like to investigate the corresponding properties of the finite difference methods. We are particularly concerned with the time scales (frequencies) of the problem and their dependence on the wave number (or wave length).

Diffusion or dissipation is where the different Fourier modes decay with time. Dispersion is where Fourier modes of differing wave lengths (or wave numbers) propagate at different speeds.

The von Neumann condition gives vital information concerning the stability of a finite difference method, but it gives no information about the more detailed properties of a finite difference method such as the properties of dissipation and dispersion.

Consider a Fourier mode associated with a finite difference method of the form

$$u_j^n = \hat{u}e^{i(\beta\omega nk+\beta jh)} \tag{3.6.5}$$

where $(x,t) = (ih,nk)$. Upon substitution of this Fourier mode (3.6.5) into the finite difference method, we obtain the discrete frequency ω as a function of β and h and k, so that $\omega = \omega(\beta,h,k)$. This is called the <u>discrete dispersion relation</u>. Using (3.6.5), we obtain

$$u_j^{n+1} = e^{i\beta\omega k}u_j^n , \tag{3.6.6}$$

and

$$u_{j\pm1}^n = e^{\pm i\beta h}u_j^n . \tag{3.6.7}$$

There is a connection between ξ used in the symbol of a finite difference method and the wave number β. By comparing (3.6.7) with the symbol of S_+ and S_-, $\xi = -\beta h$. Thus for a finite difference method of the form

$$u^{n+1} = Qu_1^n \tag{3.6.8}$$

where Q is a polynomial in S_+ and S_-, using (3.6.5)-(3.6.7), gives rise to the discrete dispersion relation $e^{i\beta\omega k} = \rho(-\beta h)$.

We shall now make precise the idea of a finite difference method being dissipative and dispersive. Suppose the finite difference method (3.6.8) approximates the prototype equation (3.2.1) to order q. Furthermore, suppose that the finite difference method (3.6.8) approximates

$$\partial_t v = c\partial_x v + a\partial_x^{q+1} v + b\partial_x^{q+2} v$$

to order $q+2$. If $q+1$ is odd (even), then a is the <u>coefficient of numerical dispersion (dissipation)</u> and b is the <u>coefficient of numerical dissipation (dispersion)</u>. Thus even-order terms in the remainder of the truncation error produce dissipation (or diffusion) and the odd-order terms in the remainder of the truncation error produce dispersion.

<u>Definition</u>. A finite difference method (3.6.8) is called <u>dissipative</u> if for all ξ, $|\rho(\xi)| \leq 1$ and $|\rho(\xi)| < 1$ for some ξ, where $\rho(\xi)$ is the symbol of Q.

<u>Definition</u>. A finite difference method (3.6.8) is called <u>dissipative of order 2s</u> where s is a positive integer if there exists a constant $\delta > 0$ (independent of h, k, and ξ) such that

$$|\rho(\xi)| \leq 1 - \delta|\xi|^{2s} \tag{3.6.9}$$

for $0 \leq \xi \leq \pi$. This restriction on ξ will turn out to be crucial.

Using inequality (3.6.9), we see that solutions of dissipative finite difference methods decay with the number of time steps. Due to the term $|\xi|^{2s} = |-\beta h|^{2s}$ in (3.6.9), the decay of high frequency modes (corresponding to large values of $\xi = -\beta h$) is much greater than the decay of low frequency modes (corresponding to small values of ξ).

<u>Definition</u>. A finite difference method is said to be <u>dispersive</u> if ω depends on β, that is, ω is a function of β.

Consider the finite difference method

$$(3.6.10) \qquad u_j^{n+1} = u_j^n + ckD_+u_j^n = (1 - \lambda c)u_j^n + c\lambda u_{j+1}^n$$

where $\lambda = k/h$. This method is stable for $c\lambda < 1$. From Section III.2, the symbol is

$$\begin{aligned}\rho(\xi) &= 1 - c\lambda + c\lambda e^{-i\xi} \\ &= 1 - c\lambda(1 - \cos\xi) - ic\lambda\sin\xi,\end{aligned}$$

and

$$\begin{aligned}|\rho(\xi)|^2 &= (1 - c\lambda(1 - \cos\xi))^2 + c^2\lambda^2\sin^2\xi \\ &= 1 + 2c^2\lambda^2(1 - \cos\xi) - 2c\lambda(1 - \cos\xi) \\ &= 1 - 2c\lambda(1 - c\lambda)(1 - \cos\xi) \\ &= 1 - 4c\lambda(1 - c\lambda)\sin^2(\xi/2).\end{aligned}$$

Therefore,

$$(3.6.11) \qquad |\rho(\xi)| = [1 - 4c\lambda(1 - c\lambda)\sin^2(\xi/2)]^{1/2}.$$

If $c\lambda < 1$, then $|\rho(\xi)|$ decrease monotonically from $|\rho(\xi)| = 1$ at $\xi = 0$ to $|\rho(\xi)| = [1 - 4c\lambda(1 - c\lambda)]^{1/2}$ at $\xi = \pi$. For small values of ξ, we replace sin by the first term in the Taylor series about $\xi = 0$, that is, $\sin z = z + O(z^2)$. This gives $\sin^2(\xi/2) = \xi^2/4 + O(\xi^4)$. Upon substitution into (3.6.11), for small values of ξ,

$$(3.6.11) \quad |\rho(\xi)| = \left[1 - 4c\lambda(1 - c\lambda)\frac{\xi^2}{4}\right]^{1/2} = \left[1 - c\lambda(1 - c\lambda)\xi^2\right]^{1/2}.$$

By expanding $\sqrt{1-z}$ in a Taylor series about $z = 0$,

$$\sqrt{1-z} = 1 - \tfrac{1}{2}z + O(z^2)\ .$$

Setting $z = c\lambda(1 - c\lambda)\xi^2$,

$$(3.6.12) \qquad |\rho(\xi)| = 1 - \tfrac{1}{2}c\lambda(1 - c\lambda)\xi^2.$$

Thus the order of dissipation cannot be of order less than 2.

By choosing $\delta > \frac{1}{2}c\lambda(1 - c\lambda)$, inequality (3.6.9) is satisfied. So we see that the method is dissipative of order 2.

Upon substitution of (3.6.5) into (3.6.10), using (3.6.6) and (3.6.7),

$$e^{i\beta\omega k}u_j^n = (1 - c\lambda)u_j^n + c\lambda e^{i\beta h}\, u_j^n \;.$$

Dividing through by (3.6.5), we obtain the dispersion relation for (3.6.10)

$$\text{(3.6.13)} \qquad e^{i\omega k} = (1 - c\lambda) + c\lambda e^{i\beta h} = \rho(-\beta h)$$

$$= 1 + c\lambda(\cos \beta h - 1) + ic\lambda \sin(\beta h)$$

$$= 1 - 2c\lambda \sin^2(\tfrac{1}{2}\beta h) + ic\lambda \sin(\beta h).$$

Writing ω as a complex number of the form $\omega = a + ib$,

$$e^{i\beta\omega k} = e^{i\beta(a+ib)k} = e^{-b\beta k}\, e^{ia\beta k}.$$

Substituting into (3.6.13),

$$\text{(3.6.14)} \qquad e^{-b\beta k}\, e^{ia\beta k} = 1 - 2c\lambda \sin^2(\tfrac{1}{2}\,\beta h) + ic\lambda \sin(\beta h)$$

The real part $e^{-b\beta k}$ has already been examined using the definition of dissipation. The real part of ω, a, is of interest as it determines how dispersive a finite difference method is. This may be determined using the relation

$$\text{(3.6.15)} \qquad \tan(a\beta k) = \frac{\operatorname{Im}(\rho(-\beta h))}{\operatorname{Re}(\rho(-\beta h))}$$

Using (3.6.14),

$$\text{(3.6.16)} \qquad \tan(a\beta k) = \frac{c\lambda \sin(\beta h)}{1 - 2c\lambda \sin^2(\frac{1}{2}\beta h)} \;.$$

Considering (3.6.16) first, a depends on β so the method is dispersive. For βh near π (high frequency terms), $\tan(a\beta k)$ is near 0 so that a is near 0. So the high frequency waves are nearly stationary. Consider now small values of βh (low frequency modes). We shall make use of the following Taylor series expansions about the point $z = 0$

$$\sin z = z - \tfrac{1}{6}(z^3) + O(z^5)$$

$$\frac{1}{1 - z} = 1 + z + O(z^2),$$

and

$$\tan^{-1}(z) = z - \tfrac{1}{3}\, z^3 + O(z^5).$$

With this we obtain $c\lambda \sin(\beta h) = c\lambda(\beta h - \frac{1}{6}(\beta h)^3)$, $2c\lambda \sin^2(\frac{1}{2}\beta h) = 2c\lambda(\frac{1}{2}\beta h - \frac{1}{6}(\frac{1}{2}\beta h)^3)^2 = 2c\lambda(\frac{1}{2}\beta h)^2 + O((\beta h)^4)$. By substitution into (3.6.15), using the Taylor series for $\frac{1}{1-z}$,

$$\tan(a\beta k) = c\lambda(\beta h - \frac{1}{6}(\beta h)^3)(1 + 2c\lambda(\frac{1}{2}\beta h)^2)$$

$$= c\lambda\beta h(1 + \frac{1}{2}(\beta h)^2(c\lambda - \frac{1}{3}))$$

or

$$a\beta k = \tan^{-1}(c\lambda\beta h(1 + \frac{1}{2}(\beta h)^2(c\lambda - \frac{1}{3}))).$$

Using the Taylor series for $\tan^{-1}(z)$,

$$a\beta k = c\lambda\beta h(1 + \frac{1}{2}(\beta h)^2(c\lambda - \frac{1}{3} - \frac{2}{3}c^2\lambda^2))$$

$$= c\lambda\beta h(1 + \frac{1}{3}(\beta h)^2(1 - c\lambda)(c\lambda - \frac{1}{2}))$$

or

$$a = c(1 + \frac{1}{3}(\beta h)^2(1 - c\lambda)(c\lambda - \frac{1}{2})). \tag{3.6.17}$$

For $\frac{1}{2} < c\lambda < 1$, $(1 - c\lambda)(c\lambda - \frac{1}{2}) > 0$, so that by (3.6.17), $a > c$. This means for low frequency Fourier modes (small values of β) with $\frac{1}{2} < c\lambda < 1$, the numerical solution leads to the exact solution, that the wave is shifted ahead of the true wave. This is called a <u>phase error</u>. For $0 < c\lambda < \frac{1}{2}$, $(1-c\lambda)(c\lambda-\frac{1}{2}) < 0$, so that by (3.6.17) $a < c$. This means for low frequency Fourier modes with $0 < c\lambda < \frac{1}{2}$, the numerical solution lags behind the exact solution, that is, the wave is shifted behind the true wave. This also is a phase error. Observe that for $c\lambda = \frac{1}{2}$ or $c\lambda = 1$, the second term on the right-hand side of (3.6.17) is 0 so that $a = c$ and the numerical solution is said to be in <u>phase</u> with the exact solution.

Consider the Lax-Friedrichs method

$$u_j^{n+1} = \frac{1}{2}(u_{j+1}^n + u_{j-1}^n) + ckD_0u_j^n .$$

The symbol given in Section III.2 is

$$\rho(\xi) = \cos\xi - ic\lambda\sin\xi ,$$

and

$$|\rho(\xi)|^2 = \cos^2\xi + c^2\lambda^2\sin^2\xi$$

$$= 1 - (1 - c^2\lambda^2)\sin^2\xi .$$

Therefore,

$$|\rho(\xi)| = [1 - (1 - c^2\lambda^2)\sin^2\xi]^{1/2}. \tag{3.6.18}$$

If $|c|\lambda \le 1$, then $|\rho(\xi)|$ decreases monotonically from $|\rho(\xi)| = 1$ at $\xi = 0$ to $|\rho(\xi)| = [1 - (1 - c^2\lambda^2)]^{1/2}$ at $\xi = \pi/2$ and increasing monotonically to $|\rho(\xi)| = 1$ at $\xi = \pi$. For small values of ξ replace $\sin \xi$ by the first term in the Taylor series about $\xi = 0$,

$$|\rho(\xi)| = [1 - (1 - c^2\lambda^2)\xi^2]^{1/2} .$$

Using the Taylor series for $\sqrt{1-z}$, with $z = (1 - c^2\lambda^2)\xi^2$,

$$|\rho(\xi)| = 1 - \tfrac{1}{2}(1 - c^2\lambda^2)\xi^2. \tag{3.6.19}$$

Thus the order of dissipation cannot be of order less than 2. By choosing $\delta > \frac{1}{2}(1 - c^2\lambda^2)$, inequality (3.6.8) is satisfied. So the Lax-Friedrichs method is dissipative of order 2.

Using the general form of the discrete dispersion relation, we obtain for the Lax-Friedrichs method

$$\begin{aligned} e^{i\beta\omega k} &= \rho(-\beta h) \\ &= \cos(\beta h) + ic\lambda \sin(\beta h). \end{aligned}$$

Writing ω as a complex number of the form $\omega = a + ib$, we obtain using (3.6.12)

$$\tan(a\beta k) = \frac{c\lambda \sin(\beta h)}{\cos(\beta h)} = c\lambda \tan(\beta h). \tag{3.6.20}$$

By using a Taylor series about $z = 0$,

$$\tan z = z + \tfrac{1}{3}z^3 + O(z^5)$$

with $z = \beta h$,

$$\tan(a\beta k) = c\lambda\beta h(1 + \tfrac{1}{3}(\beta h)^2)$$

or

$$\begin{aligned} a\beta k &= \tan^{-1}(c\lambda\beta h(1 + \tfrac{1}{3}(\beta h)^2)). \\ &= c\lambda\beta h(1 + \tfrac{1}{3}(\beta h)^2(1 - c^2\lambda^2)) \end{aligned}$$

or

$$a = c(1 + \tfrac{1}{3}(\beta h)^2(1 - c^2\lambda^2)). \tag{3.6.21}$$

For $c\lambda < 1$, by (3.6.21), $a > c$. This means that for low frequency Fourier modes (small values of β), the numerical solution leads the exact solution.

As a final example, consider the Lax-Wendroff method (3.5.3)

$$u_j^{n+1} = u_j^n + kcD_0u_j^n = \frac{k^2c^2}{2} D_+D_-u_j^n .$$

The symbol $\rho(\xi)$ given in Section III.5 is

$$\rho(\xi) = 1 - 2c^2\lambda^2 \sin^2(\tfrac{\xi}{2}) - ic\lambda \sin \xi,$$

and

$$(3.6.22)\qquad |\rho(\xi)| = [1 - 4c^2\lambda^2(1 - c^2\lambda^2)\sin(\tfrac{\xi}{2})]^{1/2}.$$

If $c\lambda \leq 1$, then $|\rho(\xi)|$ decreases monotonically from $|\rho(\xi)| = 1$ at $\xi = 0$ to $|\rho(\xi)| = [1 - 4c^2\lambda^2(1 - c^2\lambda^2)]^{1/2}$ at $\xi = \pi$. For small values of ξ, where $|\rho(\xi)|$ is largest, replace $\sin z$ with the first term of the Taylor series about $z = 0$, so that $\sin^4(\frac{\xi}{2}) = (\frac{\xi}{2})^4$. Substitution into (3.6.22) gives

$$|\rho(\xi)| = [1 - \tfrac{1}{4}c^2\lambda^2(1 - c^2\lambda^2)\xi^4]^{1/2}.$$

Using the Taylor series for $\sqrt{1-z}$, with $z = \frac{1}{4}c^2\lambda^2(1-c^2\lambda^2)\xi^4$,

$$(3.6.23)\qquad |\rho(\xi)| = 1 = \tfrac{1}{8}c^2\lambda^2(1-c^2\lambda^2)\xi^4$$

Thus the order of dissipation cannot be of order less than 4. By choosing $\delta > \frac{1}{8}c^2\lambda^2(1-c^2\lambda^2)$, inequality (3.6.8) is satisfied. So the Lax-Wendroff method is dissipative of order 4.

Using the general form of the discrete dispersion relation, we obtain for the Lax-Wendroff method

$$\begin{aligned} e^{i\beta\omega k} &= \rho(-\beta h) \\ &= 1 - 2c^2\lambda^2 \sin^2(\tfrac{1}{2}\beta h) + ic\lambda \sin(\beta h). \end{aligned}$$

Writing ω as a complex number of the form $\omega = a + ib$, using (3.6.12)

$$(3.6.24)\qquad \tan(a\beta k) = \frac{c\lambda \sin(\beta h)}{1 - 2c^2\lambda^2\sin^2(\frac{1}{2}\beta h)}.$$

Using the Taylor series for $\sin z$, we obtain $c\lambda \sin(\beta h) = c\lambda\beta h(1 - \frac{1}{6}(\beta h)^2)$ and $2c^2\lambda^2\sin^2(\frac{1}{2}\beta h) = 2c^2\lambda^2(\frac{1}{2}\beta h - \frac{1}{6}(\frac{1}{2}\beta h)^3)^2 = 2c\lambda(\frac{1}{2}\beta h)^2$. By substitution into (3.6.24) using (3.6.15) with $z = 2c^2\lambda^2\sin(\frac{1}{2}\beta h) = 2c^2\lambda^2(\frac{1}{2}\beta h)^2$,

$$\begin{aligned} \tan(a\beta k) &= c\lambda\beta h(1 - \tfrac{1}{6}(\beta h)^2)(1 - \tfrac{1}{2}c^2\lambda^2(\beta h)^2) \\ &= c\lambda\beta h(1 - \tfrac{1}{6}(\beta h)^2(1 - 3c^2\lambda^2)), \end{aligned}$$

or

$$\begin{aligned} a\beta k &= \tan^{-1}(c\lambda\beta h(1 - \tfrac{1}{6}(\beta h)^2(1 - 3c^2\lambda^2))) \\ &= c\lambda\beta h(1 - \tfrac{1}{6}(\beta h)^2(1 - 3c^2\lambda^2)) - \tfrac{1}{3}(c\lambda\beta h)^3 \\ &= c\lambda\beta h(1 - \tfrac{1}{6}(\beta h)^2(1 - c^2\lambda^2)) \end{aligned}$$

or

$$(3.6.25) \qquad a = c(1 - \tfrac{1}{6}(\beta h)^2(1 - c^2\lambda^2)).$$

For $c\lambda < 1$, by (3.6.25), $a < c$. This means that for low frequency (small values of β) Fourier modes, the numerical solution lags the exact solution.

III.7. Implicit Methods

In Section III.2, the prototype equation (3.2.1)

$$\partial_t v = c\partial_x v$$

was approximated by

$$u_i^{n+1} = \begin{cases} u_i^n + ckD_+u_i^n, & \text{if } c > 0, \\ u_i^n + ckD_-u_i^n, & \text{if } c < 0, \end{cases}$$

which is stable for $|c|\lambda \leq 1$ where $\lambda = k/h$. It was seen that

$$u_i^{n+1} = u_i^n + ckD_-u_i^n$$

is unconditionally unstable for $c < 0$ because the numerical domain of dependence does not contain the domain of dependence of the prototype equation. Similarly, the finite difference method

$$u_i^{n+1} = u_i^n + ckD_+u_i^n$$

is unconditionally unstable for $c > 0$ for the same reasons.

Consider now the implicit finite difference method approximating the prototype equation with $c > 0$

$$(3.7.1) \qquad u_i^{n+1} = u_i^n + ckD_+u_i^{n+1}.$$

This may be written in the form

$$(3.7.2) \qquad (I - ckD_+)u_i^{n+1} = u_i^n$$

or by expanding out $I - ckD_+$

$$(3.7.2') \qquad (1 + c\lambda)u_i^{n+1} - c\lambda u_{i+1}^{n+1} = u_i^n .$$

Expand the exact solution $v(x,t)$ in a Taylor series about the point $(ih,(n+1)k)$. Assuming that v is twice differentiable, $\partial_t^2 v = c^2\partial_x^2 v$, and

$$(3.7.3) \qquad \frac{v_i^{n+1} - v_i^n}{k} - cD_+v_i^{n+1} = (\partial_t v)_i^{n+1} - c(\partial_x v)_i^{n+1} = \left(\frac{c^2k}{2} + \frac{ch}{2}\right)(\partial_x^2 v)_i^{n+1} + O(k^2) + O(h^2) .$$

From (3.7.3), we see that the finite difference method (3.7.1) is consistent and accurate of $O(k) + O(h)$. Furthermore, (3.7.3) shows that the finite difference method is consistent with the diffusion-type equation

$$\partial_t v = c\partial_x v + \left(\frac{c^2k}{2} + \frac{ch}{2}\right)\partial_x^2 v \tag{3.7.4}$$

accurate to $O(k^2) + O(h^2)$. From Section III.6, this method is dissipative with coefficient of numerical dissipation (or diffusion) given by

$$\frac{c^2k}{2} + \frac{ch}{2} .$$

To analyze the stability of the finite difference method (3.7.2'), take the discrete Fourier transform,

$$(1 + c\lambda - c\lambda e^{-i\xi})\hat{u}^{n+1}(\xi) = \hat{u}^n(\xi)$$

or

$$\hat{u}^{n+1}(\xi) = \rho(\xi)\hat{u}^n(\xi),$$

where the symbol $\rho(\xi)$

$$\rho(\xi) = \frac{1}{i + c\lambda - c\lambda e^{-i\xi}} = \frac{1}{1 + 2c\lambda\sin^2(\frac{\xi}{2}) - ic\lambda\sin\xi} .$$

But then

$$|\rho(\xi)| = \rho(\xi)\overline{\rho(\xi)} = \frac{1}{[1 + 2c\lambda\sin^2(\frac{\xi}{2})]^2 + c^2\lambda^2\sin^2\xi} \leq 1$$

for any $\lambda > 0$. Thus the method is unconditionally stable.

From the discussion of implicit methods in Chapter II, the solution at a point $(ih,(n+1)k)$ depends on all points at the previous time level. Thus the solution at a point $(ih,(n+1)k)$ depends on the entire initial line so the numerical domain of dependence at a point $(ih,(n+1)k)$ consists of the entire initial line so it contains the domain of dependence of the prototype equation. Hence the CFL condition is satisfied for ever $\lambda > 0$.

There is the typical restriction on the time step k due to accuracy considerations. Since the method (3.7.1) is accurate to $O(k) + O(h)$, the optimal choice of k is $O(h)$. However, there is one other consideration. The coefficient of numerical dissipation (3.7.5) grows with the square of k. The larger the coefficient of numerical dissipation, the greater the smearing of the wave.

The coefficient of numerical dissipation for the explicit method (3.2.6) corresponding to (3.7.1) is

$$\frac{ch}{2} - \frac{c^2k}{2},$$

which, due to the negative sign, produces a smaller coefficient for $c\lambda < 1$ than for the implicit method. Another reason for the stability requirement of $c\lambda \leq 1$ on the explicit method (3.2.6) is that if $c\lambda > 1$, then the coefficient of numerical dissipation is negative. The resulting diffusion type equation is not well-posed and so a stable finite difference method is impossible.

Similarly, if $c < 0$ in the prototype equation (3.2.1), the implicit finite difference method

$$u_i^{n+1} = u_i^n + ckD_-u^{n+1} \tag{3.7.5}$$

is unconditionally stable and is accurate of $O(k) + O(h)$.

To solve the implicit method (3.7.1), observe that in (3.7.2') the left-hand side involves only u_i^{n+1} and u_{i+1}^{n+1}, not u_{i-1}^{n+1}. This gives rise to a bidiagonal set of equations. Assume that the real line $-\infty < x < +\infty$ is truncated to $[a,b]$. Let $h = (b - a)/N$ denote the grid spacing. In order to close the system of equations, a boundary condition is required at $x = b$ only. Observe that from the form of (3.7.2') that a boundary condition is not required at $x = a$. This agrees with the prototype equation, for a boundary condition may not be imposed at $x = a$. Assume that at $x = b$, the boundary condition is $v(b,t) = 0$. This is approximated by $u_N^n = 0$.

This leaves a system of equations written in matrix form

$$\underline{A}\ \underline{u}^{n+1} = \underline{u}^n$$

where $\underline{u}^n = (u_1^n, u_2^n, \ldots, u_{N-1}^n)^T$ and A is an $(N-1) \times (N-1)$ bidiagonal matrix which is upper triangular

$$\underline{A} = \begin{pmatrix} 1+c\lambda & -c\lambda & & \\ 0 & 1+c\lambda & -c\lambda & \\ & \ddots & \ddots & \ddots \\ & 0 & 1+c\lambda & -c\lambda \\ & & 0 & -+c\lambda \end{pmatrix}.$$

Since $\underline{A}$ is upper triangular, the forward elimination part of Gaussian elimination is not needed. Only the back substitution part is required.

The finite difference method

$$u_i^{n+1} = u_i^n + ckD_0u_i^n$$

was shown (in Section III.2) to satisfy the CFL condition provided that $|c|\lambda < 1$. However, the method is unconditionally unstable.

Consider the implicit finite difference method based on this unstable method

$$u_i^{n+1} = u_i^n + ckD_0u_i^{n+1} \quad (3.7.6)$$

This may be written in the form

$$(I - ckD_0)u_i^{n+1} = u_i^n . \quad (3.7.7)$$

or by expanding out $I - ckD_0$,

$$\frac{c\lambda}{2} u_{i-1}^{n+1} + u_i^{n+1} - \frac{c\lambda}{2} u_{i+1}^{n+1} = u_i^n . \quad (3.7.7')$$

By expanding the exact solution $v(x,t)$ in a Taylor series in space about the point $(ih,(n+1)k)$,

$$v_{i+1}^{n+1} = v_i^{n+1} + h(\partial_x v)_i^{n+1} + \frac{h^2}{2}(\partial_x^2 v)_i^{n+1} + \frac{h^3}{6}(\partial_x^3 v)_i^{n+1} + \frac{h^4}{24}(\partial_x^4 v)_i^{n+1} + O(h^5),$$

$$v_{i-1}^{n+1} = v_i^{n+1} - h(\partial_x v)_i^{n+1} + \frac{h^2}{2}(\partial_x^2 v)_i^{n+1} - \frac{h^3}{6}(\partial_x^3 v)_i^{n+1} + \frac{h^4}{24}(\partial_x^4 v)_i^{n+1} + O(h^5),$$

and

$$D_0v_i^{n+1} = (\partial_x^3 v)_i^{n+1} + \frac{h^2}{6}(\partial_x^3 v)_i^{n+1} + O(h^4).$$

Observe that by cancellation of pairs of even order derivatives of v, that is, all terms of the form $(\partial_x^{2p} v)_i^{n+1}$ cancel. Thus $D_0v_i^{n+1}$ does not introduce any diffusion. Furthermore,

$$\frac{v_i^{n+1} - v_i^n}{k} = (\partial_t v)_i^{n+1} - \frac{c^2k}{2}(\partial_x^2 v)_i^{n+1} + O(k^2).$$

Substituting into (3.7.6),

$$\frac{v_i^{n+1} - v_i^n}{k} - cD_0v_i^{n+1} = (\partial_t v)_i^{n+1} - c(\partial_x v)_i^{n+1} - \frac{c^2k}{2}(\partial_x^2 v)_i^{n+1} + O(k^2) + O(h^2), \quad (3.7.8)$$

where we assume that v is twice differentiable. The finite difference method (3.7.6) is consistent and accurate of $O(k) + O(h^2)$. Furthermore, (3.7.8) shows that the finite difference method is consistent with the diffusion-type equation

$$\partial_t v = c\partial_x v + \frac{c^2h}{2} \partial_x^2 v$$

accurate to $O(k^2) + O(h^2)$. This method is dissipative due only to the approximation of the time derivative with coefficient of numerical dissipation (of diffusion) given by

$$\frac{c^2k}{2}$$

To analyze the stability of the finite difference method (3.7.7') take the discrete Fourier transform,

$$(1 + ic\lambda \sin \xi)\hat{u}^{n+1}(\xi) = \hat{u}^n(\xi)$$

or

$$\hat{u}^{n+1}(\xi) = \rho(\xi)\hat{u}^n(\xi),$$

where the symbol

$$\rho(\xi) = \frac{1}{1 + ic\lambda \sin \xi},$$

which yields

$$|\rho(\xi)| = \rho(\xi)\rho(\xi) = \frac{1}{1 + c^2\lambda^2 \sin \xi} \leq 1$$

for every $\lambda > 0$. Thus the method is unconditionally stable.

Again the solution at a point $(ih,(n+1)k)$ depends on the solution at every point at the previous time level. Thus the implicit method does not suffer from the "coupling" problems of the explicit version as discussed in Section III.2.

It should, as with the previous implicit method, be noted that the coefficient of numerical dissipation (3.7.9) is proportional to the time step. Thus the larger the time step, the more diffusion that is introduced. However, the amount introduced by the implicit method (3.7.6) is much less than in the implicit method (3.7.1).

The solution of (3.7.7) involves the inversion of a tridiagonal matrix. Thus we need to close the system (3.7.7'). In order to do this, we need to impose two boundary conditions. Truncate the real line to the closed interval $[a,b]$. Since $D_0u_i^{n+1}$ depends on u_{i+1}^{n+1} and u_{i-1}^{n+1}, a boundary condition on the left is necessary to close the system at the left also, unlike the previous implicit method (3.7.1). However, if $c > 0$, then the characteristics, $x + ct =$ constant, are lines with negative slope, so a boundary condition cannot be imposed at $x = a$. The solution at the point (a,t) is obtained by following the characteristic back to the initial line, so $v(a,t) = f(a + ct)$. However, a boundary condition can be imposed at $x = b$.

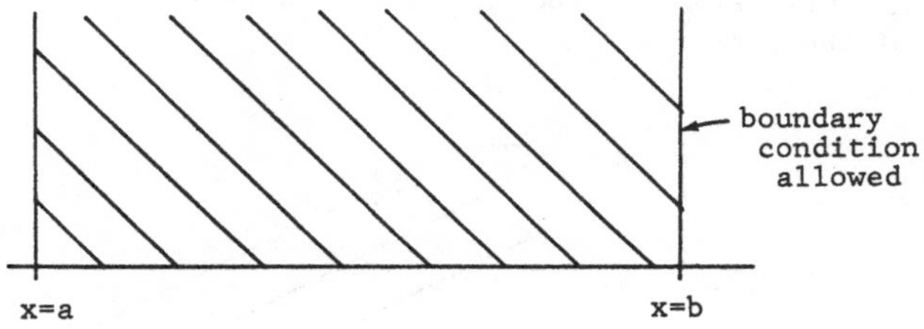

Figure 3.17 $c > 0$

Similarly, if $c < 0$, then the characteristics, $x + ct$ = constants, are lines with positive slope, so a boundary condition cannot be imposed at $x = b$. The solution at the point (b,t) is obtained by following the characteristic back to the initial line, so $v(b,t) = f(b + ct)$. However, a boundary condition can be imposed at $x = a$. This poses a tremendous problem because the method calls for a boundary condition at both points $x = a$ and $x = b$. The discussion of the treatment of boundary conditions is deferred until Chapter V.

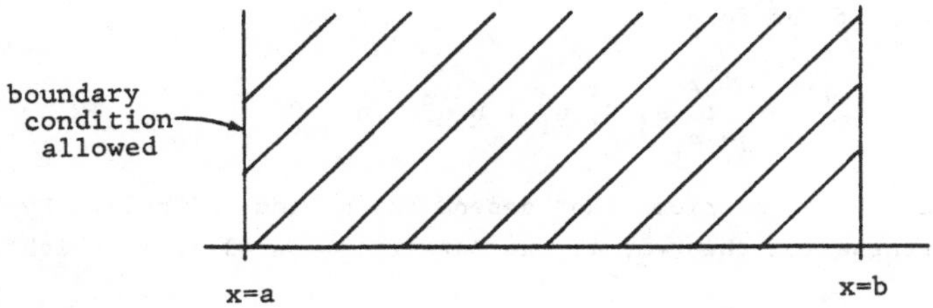

Figure 3.18 $c < 0$

We shall assume that the interval $[a,b]$ is chosen large enough with the initial condition zero everywhere except in a small subinterval of $[a,b]$, so that the solution for $0 < t < T$ is zero at both boundaries. This avoids the boundary conditions. Let the mesh spacing be given by $h = (b - a)/N$.

The system (3.7.7) can be written in the form

$$\underline{A}\ \underline{u}^{n+1} = \underline{u}^{n},$$

where $\underline{u}^n\ (u_1^n,\ldots,u_{N-1}^n)^T$ and $\underline{A}$ is an $(N-1) \times (N-1)$ tridiagonal matrix of the form

$$\underline{A} = \begin{pmatrix} 1 & -\frac{c\lambda}{2} & & & \\ \frac{c\lambda}{2} & 1 & -\frac{c\lambda}{2} & & \\ & \ddots & \ddots & \ddots & \\ & & \frac{c\lambda}{2} & 1 & -\frac{c\lambda}{2} \\ & & & \frac{c\lambda}{2} & 1 \end{pmatrix} .$$

We see that the tridiagonal matrix is diagonally dominant if

$$\left|\frac{c\lambda}{2}\right| + \left|\frac{-c\lambda}{2}\right| = |c|\lambda < 1.$$

III.8. Systems of Hyperbolic Equations in One Space Dimension

Consider the first order system of equations of the form

$$\partial_t\underline{v} = \underline{A}\partial_x\underline{v} + \underline{B}\ \underline{v}\ , \quad -\infty < x < +\infty\ , \quad t > 0 \tag{3.8.1}$$

where $\underline{v} = (v_1,\ldots,v_p)^T$ and $\underline{A}$, $\underline{B}$ are $p \times p$ matrices with constant coefficients. The initial conditions are given by

$$\underline{v}(x,0) = \underline{f}(x)\ , \quad -\infty < x < +\infty\ . \tag{3.8.2}$$

Assume that (3.8.1) is hyperbolic as defined in Section III.1.

Using the notation of Section II.7, consider the approximation of equation (3.8.1) of the form

$$\underline{u}_i^{n+1} = \sum_{j=-m_1}^{m_2} \underline{A}_j\ S_+^j\ \underline{u}_i^n + \underline{B}\ \underline{u}_i^n\ , \quad n > 0 \tag{3.8.3}$$

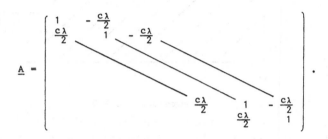

where $\underline{A}_j$ are $p \times p$ matrices that depend on h and k only. By the Strang perturbation theorem, if the matrix $\underline{B}$ in (3.8.3) (which is the same as in equation (3.8.1)) is bounded, then the question of stability reduces to the stability of (3.8.3) without the term $\underline{B}\underline{u}_i^n$. Thus we shall consider the first-order system

(3.8.1') $$\partial_t \underline{v} = \underline{A}\partial_x \underline{v} \ , \ -\infty < x < +\infty \ , \ t > 0$$

approximated by the finite difference method

(3.8.3') $$\underline{u}_i^{n+1} = \sum_{j=-m_1}^{m_2} \underline{A}_j \underline{u}_i^n \ .$$

The concept of stability as well as sufficient conditions for stability have already been developed in Section II.7. These results will be used to show that various methods developed for the scalar equation (3.2.1) in Sections III.2 and III.6 can be extended to first-order hyperbolic systems of the form (3.8.1').

We first need to extend the Courant-Friedrichs-Lewy (CFL) to the case of a hyperbolic system. Let μ_j, $j = 1,..,p$ denote the eigenvalues of $\underline{A}$ in (3.8.1') ordered as

(3.8.4) $$\mu_1 < \mu_2 < \dots < \mu_m < 0 < \mu_{m+1} < \dots < \mu_p$$

There are p-families of characteristics of (3.8.1') given by

$$x + \mu_j t = \text{constant},$$

which represent lines in the (x,t)-plane with slope $-1/\mu_j$. (Refer to Section III.1.)

The definition of numerical domain of dependence for the finite difference method (3.8.3') remains unchanged. The numerical domain of dependence of (3.8.3') of the point $(ih,(n+1)k)$ is bounded by two lines with slopes $-k/m_2h$ and k/m_1h. The numerical domain of dependence of the point $(ih,(n+1)k)$ is the set of $(m_1 + m_2)(n+1) + 1$ points in the interval

$$[(i - m_1(n+1))h,(i + m_2(n+1))h].$$

The domain of dependence of the equation (3.8.1') of the point $(ih,(n+1)k)$ consists of p distinct points corresponding to the points of intersection of the p-characteristics through the point $(ih,(n+1)k)$ with the initial line, that is,

$$(ih + \mu_j(n+1)k,0), \quad j = 1,\dots,p.$$

The CFL condition requires that

(3.8.5) $$(ih + \mu_j(n+1)k) \in [(i - m_1(n+1))h, (i + m_2(n+1)h],$$

for $j = 1,\dots,p$.

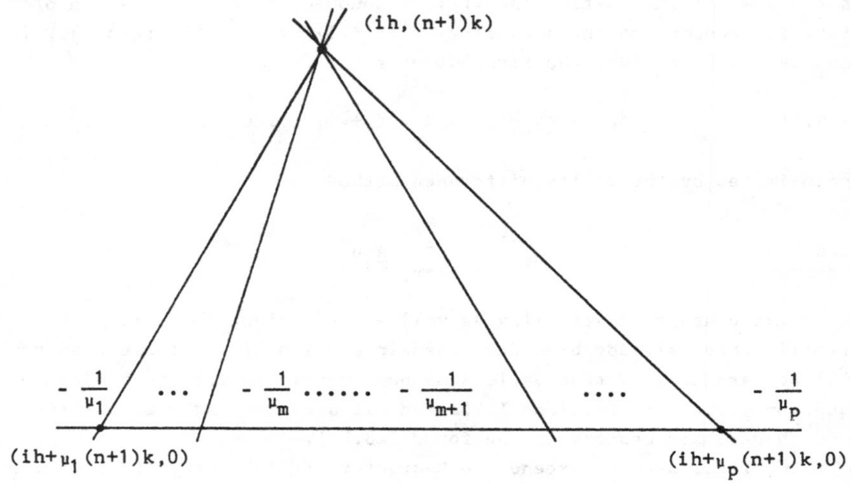

Figure 3.19

From (3.8.4),

$$-\frac{1}{\mu_{m+1}} \leq \ldots \leq -\frac{1}{\mu_p} < 0 < -\frac{1}{\mu_1} \leq \ldots \leq -\frac{1}{\mu_m}$$

The slopes $-\frac{1}{\mu_{m+1}}, \ldots, -\frac{1}{\mu_p}$, corresponding to positive eigenvalues of $\underline{A}$, are less than or equal to the slope $-k/m_2h$, then condition (3.8.5) is satisfied for $j = m+1,\ldots,p$, that is,

$$-\frac{k}{m_2h} \geq -\frac{1}{\mu_j}, \; j = m+1,\ldots,p.$$

Similarly, the slopes $-\frac{1}{\mu_j},\ldots,-\frac{1}{\mu_m}$, corresponding to negative eigenvalues of $\underline{A}$, are greater than or equal to the slope k/m_1h, then condition (3.8.5) is satisfied for $j = 1,\ldots,m$, that is

$$\frac{k}{m_1h} \leq -\frac{1}{\mu_j}, \; j = 1,\ldots,m.$$

The conditions may be combined to give the CFL condition for a finite difference method (3.8.3') approximating equation (3.8.1')

(3.8.6a) $$-\mu_j \frac{k}{h} \leq m_1, \; j = 1,\ldots,m$$

(3.8.6b) $$\mu_j \frac{k}{h} \leq m_2, \; j = m+1,\ldots,p.$$

Consider the Lax-Friedrichs method (3.2.19) (or (3.2.21)) applied to the system (3.8.1'),

(3.8.7)
$$\underline{u}_i^{n+1} = \frac{1}{2}(\underline{u}_{i+1}^n + \underline{u}_{i-1}^n) + \underline{A}kD_0\underline{u}_i^n$$
$$= \sum_{j=-1}^{1} \underline{A}_j S_+^j \underline{u}_i^n$$

where $\underline{A}_{-1} = \frac{1}{2}(\underline{I} - \lambda\underline{A})$, $\underline{A}_0 = 0$, and $\underline{A}_1 = \frac{1}{2}(I + \lambda\underline{A})$. In (3.8.7) $m_1 = m_2 = 1$ so that the CFL condition (3.8.6a) and (3.8.6b) may be combined to yield for the Lax-Friedrichs method

$$|\mu_j|\lambda \le 1 \quad \text{for} \quad j = 1,\ldots,p.$$

In fact, for any finite difference method (3.8.3') where $m_1 = m_2 = 1$, that $\underline{u}_i^{n+1}$ depends on three points $\underline{u}_{i-1}^n$, $\underline{u}_i^n$, and $\underline{u}_{i+1}^n$, the CFL condition (3.8.6a) and (3.8.6b) reduces to

(3.8.8)
$$|\mu_j|\lambda \le 1, \; j = 1,\ldots,p,$$

or

$$\max_{1 \le j \le p} |\mu_j|\lambda \le 1.$$

To analyze the stability of the Lax-Friedrichs method (3.8.7) take the discrete Fourier transform,

$$\hat{\underline{u}}^{n+1}(\xi) = \underline{G}(\xi)\hat{\underline{u}}^n(\xi),$$

where $\underline{G}(\xi)$ is the amplification matrix of order $p \times p$ (see Section II.7) given by

$$\underline{G}(\xi) = \frac{1}{2}(\underline{I}(e^{-i\xi} + e^{i\xi}) + \underline{A}\lambda(e^{-i\xi} - e^{i\xi}))$$
$$= \underline{I}\cos\xi - i\underline{A}\lambda\sin\xi.$$

First, suppose that $\underline{A}$ is symmetric in which case there exists an orthonormal matrix $\underline{T}$ such that $\|\underline{T}\| \le K$, for some constant K, and

$$\underline{A} = \underline{T}\,\underline{D}\,\underline{T}^T,$$

where $\underline{D}$ is a diagonal matrix whose elements are the eigenvalues of $\underline{A}$, that is, $\underline{D} = \mathrm{diag}(\mu_1,\ldots,\mu_p)$. Using the fact that $\underline{T}\,\underline{T}^T = \underline{I}$, since $\underline{T}$ is an orthonormal, the amplification matrix may be written in the form

$$\underline{G}(\xi) = \underline{I}\cos\xi - i\underline{T}\,\underline{D}\,\underline{T}^T\lambda\sin\xi$$
$$= \underline{T}\,\underline{T}^T\cos\xi - i\underline{T}\,\underline{D}\,\underline{T}^T\lambda\sin\xi$$
$$= \underline{T}(I\cos\xi - i\underline{D}\lambda\sin\xi)\underline{T}^T$$
$$= T\,\underline{D}_G(\xi)\underline{T}^T,$$

where $\underline{D}_G(\xi)$ is the diagonal matrix

$$\underline{I}\cos\xi - i\underline{D}\lambda\sin\xi$$

or

$$\underline{D}_G(\xi) = \mathrm{diag}(\cos\xi - i\mu_j\lambda\sin\xi).$$

Since $\underline{A}$ is symmetric, the amplification matrix $\underline{G}(\xi)$ is symmetric and the eigenvalues of $\underline{G}(\xi)$ are the elements of $\underline{D}_G(\xi)$.

Using part i) of Theorem 2.3, it is sufficient to show that $\sigma(G(\xi)) \equiv \sigma(\underline{D}_G(\xi))$ satisfies the von Neumann condition. This reduces the problem of stability to the scalar case.

It suffices to show that

$$|\cos \xi - i\mu_j\lambda \sin \xi| \leq 1 .$$

This has been considered in Section III.2 for the Lax-Friedrichs method, where μ_j is replaced with c. Using this result,

$$(3.8.9) \qquad |\cos \xi - i\mu_j \lambda \sin \xi|^2 = 1 - (1 - \mu_j^2\lambda^2)\sin^2 \xi \leq 1$$

provided $|\mu_j|\lambda \leq 1$, for $j = 1,\ldots,p$. Thus the CFL condition is sufficient for stability of the Lax-Friedrichs method.

Now consider a general $\underline{A}$, not necessarily symmetric. Since $\underline{A}$ is not symmetric, $\underline{G}(\xi)$ is not symmetric and a different approach must be used. Since (3.8.1') is hyperbolic, there exists a nonsingular matrix $\underline{T}$ such that $\max(\|\underline{T}\|, \|\underline{T}^{-1}\|) \leq K$ for some constant K and

$$\underline{A} = \underline{T}\ \underline{D}\ \underline{T}^{-1} ,$$

where $\underline{D}$ is a real diagonal matrix. Rewrite $\underline{G}(\xi)$ in the form

$$\begin{aligned} \underline{G}(\xi) &= \underline{T}(\underline{I} \cos \xi - i\underline{D}\lambda \sin \xi)\underline{T}^{-1} \\ &= \underline{T}\ \underline{D}_G(\xi)\underline{T}^{-1} \end{aligned}$$

where $\underline{D}_G(\xi)$ is the diagonal matrix defined above.

Suppose $\sigma(\underline{D}_G(\xi)) \leq 1 + Ck$ where C is some constant satisfying the restrictions of the von Neumann condition, then

$$\begin{aligned} \|\underline{G}^n(\xi)\| &= \|\underline{T}\ \underline{D}_G^n(\xi)\underline{T}^{-1}\| \\ &\leq \|\underline{T}\|\|\underline{D}_G^n(\xi)\|\|\underline{T}^{-1}\| \\ &\leq K^2\|\underline{D}_G^n(\xi)\| \\ &\leq K^2\|\underline{D}_G(\xi)\|^n. \end{aligned}$$

Since $\underline{D}_G(\xi)$ is symmetric, in fact, diagonal $\sigma(\underline{D}_G(\xi)) = \|\underline{D}_G(\xi)\|$ so that

$$\begin{aligned} \|\underline{G}^n(\xi)\| &\leq K^2\sigma(\underline{D}_G(\xi))^n \\ &\leq K^2(1 + Ck)^n \\ &\leq K^2e^{Cnk} = K^2e^{Ct} . \end{aligned}$$

Thus by Theorem 2.1, the finite difference method is stable. It remains to show that $\sigma(\underline{D}_G(\xi)) \leq 1 + Ck$. Using (3.8.9)

$$\sigma(\underline{D}_G(\xi)) = \max_{1\leq j\leq p} |\cos \xi - i\mu_j\lambda \sin \xi| \leq 1$$

provided $|\mu_j|\lambda \leq 1$ for $j = 1,\ldots,p$ or $\max_{1\leq j\leq p} |\mu_j|\lambda \leq 1$. So, again, the CFL condition is sufficient for stability of the Lax-Friedrichs method. By expanding $\underline{v}$ in a Taylor series about the point $(ih,(nk))$ in the same manner as Section III.2, we see that the Lax-Friedrichs method (3.8.7) is accurate of $O(k) + O(h)$.

Next consider the Lax-Wendroff method (3.5.3) applied to system (3.8.1'),

(3.8.10) $$\underline{u}_i^{n+1} = \underline{u}_i^n + k\underline{A}D_0\underline{u}_i^n + \frac{k^2}{2}\underline{A}^2 D_+D_-\underline{u}_i^n .$$

This may be written in the form

$$\underline{u}^{n+1} = \sum_{j=-1}^{1} \underline{A}_j S_+^j \underline{u}_i^n$$

where $\underline{A}_{-1} = \frac{1}{2}(\underline{A}^2\lambda^2 + \underline{A}\lambda)$, $\underline{A}_0 = \underline{I} - \underline{A}^2\lambda^2$, and $\underline{A}_1 = \frac{1}{2}(\underline{A}^2\lambda^2 - \underline{A}\lambda)$. since $m_1 = m_2 = 1$, the CFL condition is (3.8.8), that is,

$$\max_{1<j<p} |\mu_j|\lambda < 1 .$$

To analyze the stability of the Lax-Wendroff method (3.8.10) take the discrete Fourier transform,

$$\hat{\underline{u}}^{n+1}(\xi) = \underline{G}(\xi)\hat{\underline{u}}^n(\xi),$$

where $\underline{G}(\xi)$ is the amplification matrix of order $p \times p$ given by

$$\begin{aligned}\underline{G}(\xi) &= I + \tfrac{1}{2}\lambda\underline{A}(e^{-i\xi} - e^{i\xi}) + \tfrac{\lambda^2}{2}\underline{A}^2(e^{-i\xi} - 2 + e^{i\xi}) \\ &= I - 2\lambda^2\underline{A}^2 \sin^2(\tfrac{\xi}{2}) - i\underline{A}\lambda \sin \xi .\end{aligned}$$

As with the Lax-Friedrichs method for the general case in which $\underline{A}$ is not necessarily symmetric,

$$\begin{aligned}\underline{G}(\xi) &= \underline{T}((1 - 2\lambda^2\mu_j^2 \sin(\tfrac{\xi}{2}))\underline{I} - i(\mu_j\lambda \sin \xi)\underline{I})\underline{T}^{-1} \\ &= \underline{T}\, \underline{D}_G(\xi)\underline{T}^{-1} ,\end{aligned}$$

where $\underline{D}_G(\xi) = \mathrm{diag}(1 - 2\lambda^2\mu_j^2 \sin^2(\frac{\xi}{2}) - i\mu_j\lambda \sin \xi)$. It suffices to show that $\sigma(\underline{D}_G(\xi)) < 1$, or

$$|1 - 2\lambda^2\mu_j^2 \sin^2(\tfrac{\xi}{2}) - i\mu_j\lambda \sin \xi| < 1$$

for $j = 1,\dots,p$. This has been considered in Section III.5 for the Lax-Wendroff method, when μ_j is replaced with C. Using this result,

(3.8.11) $$|1 - 2\lambda^2\mu_j^2 \sin^2(\tfrac{\xi}{2}) - i\mu_j\lambda \sin \xi| = 1 - 4\mu_j^2\lambda^2(1 - \mu_j^2\lambda^2)\sin^4(\tfrac{\xi}{2}) < 1$$

provided that $|\mu_j|\lambda < 1$ for $j = 1,\dots,p$. Thus if $\max_{1<j<p}|\mu_j|\lambda < 1$, $\sigma(\underline{D}_G(\xi)) < 1$. So, again, the CFL condition is sufficient for the stability of the Lax-Wendroff method.

By expanding $\underline{v}$ in a Taylor series about the point (ih,nk) in the same manner as Section III.5, we see that the Lax-Wendroff method is accurate of $O(k^2) + O(h^2)$.

We may extend the idea of dissipation for the matrix case.

Definiition. A finite difference method (3.8.3') is dissipative of order $2s$, where s is a positive integer, if there exists a constant $\delta > 0$ (independent of h, k, and ξ) such that

$$|\Lambda_j(\xi)| \leq 1 - \delta|\xi|^{2s}$$

with $|\xi| \leq \pi$, where $\Lambda_j(\xi)$ are the eigenvalues of the amplification matrix $\underline{G}(\xi)$ of (3.8.3').

It was seen that the j-th eigenvalue of $\underline{G}(\xi)$ for the Lax-Friedrichs and Lax-Wendroff methods was identical to the symbol of the same methods for the scalar case where c was replaced by μ_j. Thus from the discussion in Section III.6, the Lax-Friedrichs method (3.8.7) is dissipative of order 2 and the Lax-Wendroff method (3.8.10) is dissipative of order 4.

The importance of the idea of dissipation can be seen from the following theorem due to Kreiss [19] and simplified by Parlett [29].

Theorem 3.3. Suppose the finite difference method (3.8.3') is accurate of order $2s-1$ or $2s-2$ and dissipative of order $2s$, then the method is stable.

This theorem applies to both the Lax-Friedrichs method, since it is accurate to order $2s-1$ and dissipative of order $2s$ where $s = 1$ and the Lax-Wendroff method, since it is accurate to order $2s-2$ and dissipative of order $2s$ where $s = 2$.

We close this section with a theorem due to Lax [22] which is an extension of Theorem 3.1 in Section III.3.

Theorem 3.4. Consider the hyperbolic system of equations (3.8.1) approximated by a finite difference method of the form (3.8.3). Suppose the finite difference method is accurate of order (p_1, p_2) and $k = O(h)$, then

$$\underline{G}(\xi) = e^{-k\xi\underline{A}/h} + O(|\xi|^{p+1}),$$

where $p = \min(p_1, p_2)$.

III.9. Systems of Hyperbolic Equations in Several Space Dimensions

Consider the hyperbolic equation

(3.9.1) $$\partial_t v = a\partial_x v + b\partial_y v, \quad -\infty < x,y < +\infty, \quad t > 0,$$

and the initial condition

(3.9.2) $$v(x,y,0) = f(x,y), \quad -\infty < x,y < +\infty$$

As in the case of one-space dimension, the exact solution is given by

(3.9.3) $$v(x,y,t) = f(x+at, y+bt).$$

And the domain of dependence of the point (x,y,t) for equation (3.9.1) is the single point in the (x,y)-plane $(x+at, y+bt)$. The CFL condition requires that the numerical domain of dependence of the point $(x,y,t) \equiv (ihx, jhy, (n+1)k)$ for a finite difference method that approximates (3.9.1)-(3.9.2) must include this point. For a finite difference method of the form

(3.9.4) $$u_{ij}^{n+1} = \sum_{j=-m_1}^{m_2} (A_j S_{+1}^j + B_j S_{+2}^j) u_{ij}^n,$$

where A_{-m_1}, B_{-m_1}, A_{m_2}, and B_{m_2} are not equal to zero, the CFL condition becomes

(3.9.5a) $$\frac{ak}{h_x} \leq m_2 \quad \text{if} \quad a > 0,$$

(3.9.5b) $$\frac{-ak}{h_x} \leq m_1 \quad \text{if} \quad a < 0,$$

(3.9.5c) $$\frac{bk}{h_y} \leq m_2 \quad \text{if} \quad b > 0,$$

(3.9.5d) $$\frac{-bk}{h_y} \leq m_1 \quad \text{if} \quad b < 0.$$

If m_1 and m_2 are each 0 or 1, then these conditions reduce to $\frac{|a|k}{h_x} \leq 1$ and $\frac{|b|k}{h_y} \leq 1$, or

(3.9.6) $$k \leq \min\left(\frac{h_x}{|a|}, \frac{h_y}{|b|}\right).$$

If $h_x = h_y = h$, this condition becomes

(3.9.7) $$k \leq \frac{h}{\max(|a|,|b|)}.$$

Approximate equation (3.9.1) by the two-dimensional Lax-Friedrichs method (see (3.2.19)) which is obtained from the unstable method (see (3.2.17))

$$\frac{u_{ij}^{n+1} - u_{ij}^n}{k} = aD_{01}u_{ij}^n + bD_{02}u_{ij}^n$$

by replacing the u_{ij}^n term in the approximation to ∂_t with the average over the four neighboring points $\frac{1}{4}(u_{i+1,j} + u_{i-1,j}^n + u_{i,j+1}^n + u_{i,j-1}^n)$,

(3.9.8) $$u_i^{n+1} = \frac{1}{4}(u_{i+1,j}^n + u_{i-1,j}^n + u_{i,j+1}^n + u_{i,j-1}^n) + kaD_{01}u_{ij}^n + kbD_{02}u_{ij}^n .$$

To analyze the stability of this method, take the discrete Fourier transform (of (3.9.8),

$$\hat{u}^{n+1}(\xi_1,\xi_2) = (\xi_1,\xi_2)\hat{u}^n(\xi_1,\xi_2)$$

where, for $\lambda_x = k/h_x$ and $\lambda_y = k/h_y$,

$$\rho(\xi 1,\xi 2) = \frac{1}{2}\cos \xi 1 + \frac{1}{2}\cos \xi 2 - ia\lambda_x \sin \xi 1 - ib\lambda_y \sin \xi 2,$$

where, for $\lambda_x = k/h_x$ and $\lambda_y = k/h_y$,

$$\rho(\xi_1,\xi_2) = \tfrac{1}{2}\cos\,\xi_1 + \tfrac{1}{2}\cos\,\xi_2 - ia\lambda_x \sin\,\xi_1 - ib\lambda_y \sin\,\xi_2$$

and

$$\begin{aligned}|\rho(\xi_1,\xi_2)|^2 &= (\tfrac{1}{2}\cos\,\xi_1 + \tfrac{1}{2}\cos\,\xi_2)^2 + (a\lambda_x \sin\,\xi_1 + b\lambda_y \sin\,\xi_2)^2. \\ &= 1 - (\sin^2\xi_1 + \sin^2\xi_2)(\tfrac{1}{2} - (a^2\lambda_x^2 + b^2\lambda_y^2)) \\ &\quad - \tfrac{1}{4}(\cos\,\xi_1 - \cos\,\xi_2)^2 - (a\lambda_x \sin\,\xi_1 - b\lambda_y \sin\,\xi_2)^2.\end{aligned}$$

Observe that the last two terms are always negative so that

$$|\rho(\xi_1,\xi_2)|^2 < 1 - (\sin^2\xi_1 + \sin^2\xi_2)(\tfrac{1}{2} - (a^2\lambda_x^2 + b^2\lambda_y^2)).$$

The von Neumann condition is satisfied if $\frac{1}{2} - (a^2\lambda_x^2 + b^2\lambda_y^2) > 0$, or

$$k < \frac{1}{\sqrt{2}\sqrt{\frac{a^2}{h_x{}^2} + \frac{b^2}{h_y{}^2}}} \tag{3.9.9}$$

In the case that $h_x = h_y = h$

$$k < \frac{h}{\sqrt{2}\,\sqrt{a^2 + b^2}}\,. \tag{3.9.10}$$

The stability condition (3.9.10) is more restrictive than the CFL condition (3.9.7) or the one-dimensional stability condition given in Section III.2. For this reason we turn to other approaches for solving problems in several space dimensions.

As in Section II.4 equation (3.9.1) can be written as the sum of two one-dimensional equations

$$\partial_t v = 2\partial_x v \tag{3.9.11}$$

and

$$\partial_t v = 2\partial_y v\,. \tag{3.9.12}$$

In order to advance the solution from time nk to $(n+1)k$, it is assumed that equation (3.9.11) holds from nk to $(n+\frac{1}{2})k$ and that equation (3.9.12) holds from $(n+\frac{1}{2})k$ to $(n+1)k$. This gives rise to replacing (3.9.1)-(3.9.2) by two initial-value problems, each one-dimensional

$$\begin{cases}\partial_t v' = 2\partial_x v' & nk < t < (n+\frac{1}{2})k \\ v'(x,y,nk) = v''(x,y,nk)\,(= f(x,y) \quad \text{if} \quad n = 0) & \end{cases} \tag{3.9.13}$$

and

$$\begin{cases}\partial_t v'' = 2\partial_y v'' & (n+\frac{1}{2})k < t < (n+1)k \\ v''(x,y,(n+\frac{1}{2})k) = v'(x,y,(n+\frac{1}{2})k). & \end{cases} \tag{3.9.14}$$

Let

$$u^{n+1/2} = u_{ij}^n + kD_1 u_{ij}^n \tag{3.9.15}$$

approximate (3.9.5) and

$$u_{ij}^{n+1} = u_{ij}^{n+1/2} + kD_2 u_{ij}^{n+1/2} \tag{3.9.16}$$

approximate (3.9.14), where D_1 and D_2 are some finite difference operators in x and y, respectively. This two-step method is called a <u>fractional-step method</u> (see Yanenko [48]).

Suppose we use, for example, the Lax-Wendroff method. In this case,

$$D_1 = D_{01} + \frac{k}{2} D_+ D_{-1}$$

and

$$D_2 = D_{02} + \frac{k}{2} D_+ D_{-2}$$

and (3.9.15)-(3.9.16) are each second order accurate in space. Then consider

$$\begin{aligned} u_{ij}^{n+1} &= (I+kD_2)(I+kD_1)u_{ij}^n \\ &= (I+k(D_1+D_2)+k^2 D_1 D_2)u_{ij}^n \ . \end{aligned} \tag{3.9.17}$$

This method is still second order accurate in space. Assume that v is twice differentiable so that $\partial_t^2 v = \partial_x^2 v + 2\partial_y \partial_x v + \partial_y^2 v$, and by expanding the solution v in Taylor series in time about the point (ih_x, jh_y, nk),

$$v_{ij}^{n+1} = v_{ij}^n + k(\partial_x v + \partial_y v)_{ij}^n + \frac{k^2}{2}(\partial_x^2 v + 2\partial_y \partial_x v + \partial_y^2 v)_{ij}^n + O(k^3). \tag{3.9.18}$$

Considering the method (3.9.17), the operator D_{01} in D_1 approximates the $\partial_x v$ term in (3.9.18) to second order in space, the operator $D_+ D_{-1}$ in D_1 approximates the $\partial_x^2 v$ term in (3.9.18) to second order in space. Similarly, D_{02} and $D_+ D_{-2}$ in D_1 approximate the respective terms $\partial_y v$ and $\partial_y^2 v$ in (3.9.18) to second order in space. Finally,

$$\begin{aligned} k^2 D_1 D_2 &= k^2 (D_{01} D_{02} + \frac{k}{2}(D_{01} D_+ D_{-2} + D_{02} D_+ D_{-1}) + \frac{k^2}{4}(D_+ D_{-1} D_+ D_{-2}) \\ &= k^2 D_{01} D_{02} + O(k^3) \end{aligned}$$

and by expanding v in a Taylor series in space about the point (ih_x, jh_y, nk),

$$D_{01} D_{02} v_{ij}^n = (v_{xy})_{ij}^n + O(h_x^2) + O(h_y^2) \ .$$

The operator $D_1 D_2$ approximates the v_{xy} term in (3.9.18) to second order in space. Hence, the product method (3.9.19) is second order in space.

A more direct way to verify that (3.9.17) is second order accurate in space is to use Theorem 3.1 in Section III.3. If (3.9.15) is $O(h_x^2)$, then $e^{-i\lambda_x\xi_1} = \rho_1(\xi_1) + O(\xi_1^3)$ and if (3.9.16) is $O(h_y^2)$, then

$$e^{-i\lambda_y\xi_2} = \rho_2(\xi_2) + O(\xi_2^3)$$

where $\lambda_x = k/h_x$, $\lambda_y = k/h_y$, and $\rho_1(\xi_1)$ denote the symbol of $I+kD_1$ and $\rho_2(\xi_2)$ denotes the symbol of $I+kD_2$. If the combined method (3.9.17) is second order accurate in space, then

$$e^{-i\lambda_x\xi_1-i\lambda_y\xi_2} = \rho(\xi_1,\xi_2) + O(\|\underline{\xi}\|^3),$$

where $\rho(\xi_1,\xi_2)$ denotes the symbol of the combined method (3.9.17) and $\underline{\xi} = (\xi_1,\xi_2)$. However, $\rho(\xi_1,\xi_2) = \rho_1(\xi_1)\rho_2(\xi_2)$ and

$$\begin{aligned} e^{-i\lambda_x\xi_1-i\lambda_y\xi_2} &= e^{-i\lambda_x\xi_1}e^{-i\lambda_y\xi_2} \\ &= (\rho_1(\xi_1) + O(\xi_1^3))(\rho_2(\xi_2) + O(\xi_2^3)) \\ &= \rho_1(\xi_1)\rho_2(\xi_2) + O(\|\underline{\xi}\|^3) \\ &= \rho(\xi_1,\xi_2) + O(\|\underline{\xi}\|^3) . \end{aligned}$$

Thus the fractional step method (3.9.15) and (3.9.16) is second order accurate in space.

Now consider the first order system in two space dimensions

(3.9.19) $\qquad \partial_t\underline{v} = \underline{A}\partial_x v + \underline{B}\partial_y\underline{v}, \quad -\infty < x,y < +\infty, \ t > 0,$

where $\underline{v} = (v_1,\ldots,v_p)$ and $\underline{A}$, $\underline{B}$ are $p \times p$ matrices with constant coefficient. The initial conditions are given by

(3.9.20) $\qquad \underline{v}(x,y,0) = \underline{f}(x,y) \ , \ -\infty < xy < +\infty .$

As in the scalar case, replace (3.9.19) and (3.9.20) with two initial-value problems, each one dimensional

(3.9.21) $$\begin{cases} \partial_t\underline{v}' = 2\underline{A}\partial_x\underline{v}' & nk < t < (n+\frac{1}{2})k \\ \underline{v}'(x,y,nk) = \underline{v}''(x,y,nk)(= \underline{f}(x,y) & \text{if } n = 0), \end{cases}$$

and

(3.9.22) $$\begin{cases} \partial_t\underline{v}'' = 2\underline{B}\partial_y\underline{v}'' & (n+\frac{1}{2})k < t < \frac{1}{2}(n+1)k \\ \underline{v}''(x,y,(n+\frac{1}{2})k) = \underline{v}'(x,y,(n+\frac{1}{2})k) & \end{cases} .$$

Approximate (3.9.21) by

$$\underline{u}_{ij}^{n+1/2} = (\underline{I} + k\underline{A}D_1)\underline{u}_{ij}^n$$

and approximate (3.9.22) by

(3.9.24) $$\underline{u}_{ij}^{n+1} = (\underline{I} + k\underline{B}D_2)\underline{u}^{n+1/2} .$$

Let $\underline{G}_1(\xi_1)$ and $\underline{G}_2(\xi_2)$ denote the amplification matrices of (3.9.23), and (3.9.24), respectively. Furthermore, suppose that (3.9.23) is accurate of $O(h_x^q)$ and (3.9.24) is accurate of $O(h_y^q)$, then by the Theorem 3.4,

$$e^{-i\lambda_x\underline{A}\xi_1} = \underline{G}_1(\xi_1) + O(\xi_1^{q+1})$$

and

$$e^{-i\lambda_y\underline{A}\xi_2} = \underline{G}_2(\xi_2) + O(\xi_2^{q+1}) ,$$

where $\lambda_x = k/h_x$ and $\lambda_y = k/h_y$. For the combined method

(3.9.25) $$\underline{u}_i^{n+1} = (\underline{I} + k\underline{B}D_2)(\underline{I} + k\underline{A}D_1)\underline{u}_i^n ,$$

$$e^{-i\lambda_y\underline{B}\xi_2 - i\lambda_x\underline{A}\xi_1} = \underline{G}(\xi_1,\xi_2) + O(\|\xi\|^{q+1})$$

where $\underline{G}(\xi_1,\xi_2)$ denotes the amplification matrix of the combined method (3.9.25) and $\xi = (\xi_1,\xi_2)$. If the matrices $\underline{A}$ and $\underline{B}$ commute, that is, $\underline{A}\ \underline{B} = \underline{B}\ \underline{A}$, then

$$e^{-i\lambda_y\underline{B}\xi_2 - i\lambda_x\underline{A}\xi_1} = e^{-i\lambda_y\underline{B}\xi_2} e^{-i\lambda_x\underline{A}\xi_1} ,$$

so that

$$\begin{aligned} e^{-i\lambda_y\underline{B}\xi_2 - i\lambda_x\underline{A}\xi_1} &= (\underline{G}_2(\xi_2) + O(\xi_2^{q+1}))(\underline{G}_1(\xi_1) + O(\xi_1^{q+1})) \\ &= \underline{G}_2(\xi_2)\underline{G}_1(\xi_1) + O(\|\underline{\xi}\|^{q+1}) \\ &= \underline{G}(\xi_1,\xi_2) + O(\|\underline{\xi}\|^{q+1}) \end{aligned}$$

from which we see that the same order of accuracy in space $(O(h_x^{q+1}) + O(h_y^{q+1}))$ is achieved.

Now consider the case where the matrices $\underline{A}$ and $\underline{B}$ do not commute, then

$$e^{-i\lambda_y\underline{B}\xi_2 - i\lambda_x\underline{A}\xi_1} \neq e^{-i\lambda_y\underline{B}\xi_2} e^{-i\lambda_x\underline{A}\xi_1} .$$

Expanding $e^{-i\lambda_y\underline{B}\xi_2 - i\lambda_x\underline{A}\xi_1}$ in a Taylor series,

(3.9.26) $$\begin{aligned} e^{-i\lambda_y\underline{B}\xi_2 - i\lambda_x\underline{A}\xi_1} &= \underline{I} - i\lambda_x\underline{A}\xi_1 - i\lambda_y\underline{B}\xi_2 \\ &- \lambda_x^2\underline{A}^2\xi_1^2 - \lambda_y^2\underline{B}^2\xi_1^2 - \lambda_x\lambda_y\xi_1\xi_2(\underline{A}\ \underline{B} + \underline{B}\ \underline{A}) + O(\|\underline{\xi}\|^3), \end{aligned}$$

and expanding $e^{-i\lambda_x\underline{A}\xi_1}$ and $e^{-i\lambda_y\underline{B}\xi_2}$ in Taylor series,

(3.9.27) $$e^{-i\lambda_x\underline{A}\xi_1} = \underline{I} - i\lambda_x\underline{A}\xi_1 - \lambda_x^2\underline{A}^2\xi_1^2 + O(\xi_1^3)$$

and

$$(3.9.28)\qquad e^{-i\lambda_y \underline{B}\xi_2} = \underline{I} - i\lambda_y \underline{B}\xi_2 - \lambda_y^2 \underline{B}^2 \xi_2^2 + O(\xi_2^3)$$

Comparing (3.9.26) with (3.9.27) and (3.9.28),

$$\begin{aligned} e^{-i\lambda_y \underline{B}\xi_2 - i\lambda_x \underline{A}\xi_1} &= e^{-i\lambda_y \underline{B}\xi_2} e^{-i\lambda_x \underline{A}\xi_1} + O(\|\underline{\xi}\|^2) \\ &= (\underline{G}_2(\xi_2) + O(\xi_2^{q+1}))(\underline{G}_1(\xi_1) \\ &\qquad + O(\xi_1^{q+1})) + O(\|\underline{\xi}\|^2) \\ &= \underline{G}_2(\xi_2)\underline{G}_1(\xi_1) + O(\|\underline{\xi}\|^2) \\ &= \underline{G}(\xi_1,\xi_2) + O(\|\xi\|^2) , \end{aligned}$$

which shows that the combined method (3.9.25) is only first order accurate in space independent of the orders of accuracy of the separate steps (3.9.23) and (3.9.24).

By carefully allowing for the neglected terms in each step we can regain the lost accuracies. Comparing the Taylor series (3.9.26)-(3.9.28),

$$e^{-i\lambda_y \underline{B}\xi_2 - i\lambda_x \underline{A}\xi_1} = \tfrac{1}{2}(e^{-i\lambda_x \underline{A}\xi_1} e^{-i\lambda_y \underline{B}\xi_2} + e^{-i\lambda_y \underline{B}\xi_2} e^{-i\lambda_x \underline{A}\xi_1}) + O(\|\underline{\xi}\|^3).$$

This result suggests a way of obtaining second order accuracy in space if $q > 2$. By symmetrically averaging the two combined methods

$$\underline{u}_{ij}^{n+1} = (\underline{I} + k\underline{B}D_2)(\underline{I} + k\underline{A}D_1)\underline{u}_{ij}^n$$

and

$$\underline{u}_{ij}^{n+1} = (\underline{I} + k\underline{A}D_1)(\underline{I} + k\underline{B}D_2)\underline{u}_{ij}^n ,$$

we obtain a combined method

$$(3.9.29)\qquad \underline{u}_{ij}^{n+1} = \tfrac{1}{2}[(\underline{I}+k\underline{B}D_2)(\underline{I}+k\underline{A}D_1) + (\underline{I}+k\underline{A}D_1)(\underline{I}+k\underline{B}D_2)]\underline{u}_{ij}^n$$

which is second order accurate in space.

Another way of preserving second order accuracy in space is to observe that by expanding $e^{-i(\lambda_x/2)\underline{A}\xi_1}$ and $e^{-i(\lambda_y/2)\underline{B}\xi_2}$ in a Taylor series

$$\begin{aligned} & e^{-i(\lambda_x/2)\underline{A}\xi_1} e^{-i(\lambda_y/2)\underline{B}\xi_2} e^{-i(\lambda_y/2)\underline{B}\xi_2} e^{-i(\lambda_x/2)\underline{A}\xi_1} \\ &= [\underline{I} - i\tfrac{\lambda_x}{2}\underline{A}\xi_1 - \tfrac{\lambda_x}{4} A^2\xi_1^2 + O(\xi_1^3)][\underline{I} - i\tfrac{\lambda_y}{2}\underline{B}\xi_2 \\ &\quad - \tfrac{\lambda_y^2}{4}\underline{B}^2\xi^2 + O(\xi_2^3)][\underline{I} - i\tfrac{\lambda_y^2}{2}\underline{B}\xi_2 - \tfrac{\lambda_y}{4}\underline{B}^2\xi_2^2 + O(\xi_2^3)] \\ &\quad [\underline{I} - i\tfrac{\lambda_x}{2}\underline{A}\xi_1 - \tfrac{\lambda_x}{4} A^2\xi_1^2 + O(\xi_1^3)] \\ &= I - i\lambda_x\underline{A}\xi_1 - i\lambda_y\underline{B}\xi_2 - \lambda_x^2\underline{A}^2\xi_1^2 - \lambda_y^2\underline{B}^2\xi_2^2 \\ &\quad - \lambda_x\lambda_y\xi_1\xi_2(\underline{A}\,\underline{B} + \underline{B}\,\underline{A}) + O(\|\underline{\xi}\|^3) . \end{aligned}$$

Hence

$$e^{-i\lambda_y B\xi_2 - i\lambda_x \underline{A}\xi_1} = e^{-i(\lambda_x/2)\underline{A}\xi_1} e^{-i(\lambda_x/2)\underline{B}\xi_2} e^{-i(\lambda_y/2)\underline{B}\xi_2} e^{-i(\lambda_x/2)\underline{A}\xi_1}$$

$$= O(\|\underline{\xi}\|^3).$$

This result shows that second order accuracy in space can be preserved by advancing a solution from nk to (n+1)k in four quarter steps (k/4). The first quarter step (in time) in the x-direction (due to the $e^{-i(\lambda_x/2)\underline{A}\xi_1}$ term) followed by two successive quarter steps in the y-direction (due to the two $e^{-i(\lambda_y/2)\underline{B}\xi_2}$ terms) which is followed by the final quarter step in the x-direction (due to the $e^{-i(\lambda_x/2)\underline{A}\xi_1}$ term). This gives rise to the combined method

$$\underline{u}_{ij}^{n+1} = (\underline{I} + \tfrac{k}{2}\underline{A}D_1)(\underline{I} + \tfrac{k}{2}\underline{B}D_2)(\underline{I} + \tfrac{k}{2}\underline{B}D_2)(\underline{I} + \tfrac{k}{2}\underline{A}D_1)\underline{u}_{ij}^n.$$

However,

$$e^{-i(\lambda_y/2)\underline{B}\xi_2} e^{-i(\lambda_y/2)\underline{B}\xi_2} = e^{-i\lambda_y\underline{B}\xi_2} + O(\xi_1^3),$$

so that

$$e^{-i\lambda_y\underline{B}\xi_2 - i\lambda_x\underline{A}\xi_1} = e^{-i(\lambda_x/2)\underline{A}\xi_1} e^{-i\lambda_y\underline{B}\xi_2} e^{-i(\lambda_x/2)\underline{A}\xi_1}$$

$$+ O(\|\underline{\xi}\|)^3) .$$

This gives rise to the combined method involving two quarter steps and one half step to advance the solution from time nk to (n+1)k

(3.9.30) $$u_{ij}^{n+1} = (\underline{I}+\tfrac{k}{2}\underline{A}D_1)(\underline{I}+k\underline{B}D_2)(\underline{I}+\tfrac{k}{2}\underline{A}D_1)u_{ij}^n$$

or in terms of the three fractional steps

(3.9.31a) $$\underline{u}^{n+1/4} = (\underline{I} + \underline{A}D_1)\underline{u}_{ij}^n ,$$

(3.9.31b) $$\underline{u}_{ij}^{n+3/4} = (\underline{I} + k\underline{B}D_2)\underline{u}_{ij}^{n+1/4} ,$$

and

(3.9.31c) $$\underline{u}_{ij}^{n+1} = (\underline{I} + \tfrac{k}{2}\underline{A}D_1)\underline{u}_{ij}^{n+3/4} .$$

Consider the first order system in three space dimensions

(3.9.32) $\partial_t v = \underline{A}\partial_x v + \underline{B}\partial_y v + \underline{C}\partial_z \underline{v}, \ -\infty < x,y,z < +\infty, \ t > 0,$

where $\underline{v} = (v_1,\ldots,v_p)$ and $\underline{A},\underline{B}$, and $\underline{C}$ are $p \times p$ matrices with constant coefficients. The initial conditions are given by

(3.9.33) $\underline{v}(x,y,z,0) = \underline{f}(x,y,z), \quad -\infty < x,y,z < +\infty .$

Equation (3.9.32) can be written as the sum of three one-dimensional equations

(3.9.34) $$\partial_t\underline{v} = 3\underline{A}\,\partial_x\underline{v} ,$$

(3.9.35) $$\partial_t\underline{v} = 3\underline{B}\,\partial_y v ,$$

(3.9.36) $$\partial_t \underline{v} = 3\underline{C}\partial_z v .$$

In order to advance the solution from time nk to (n+1)k, it is assumed that equation (3.9.34) holds from time nk to $(n+\frac{1}{3})k$, that equation (3.9.35) holds from time $(n+\frac{1}{3})k$ to $(n+\frac{2}{3})k$, and that equation (3.9.36) holds from time $(n+\frac{2}{3})k$ to (n+1)k. This gives rise to replacing (3.9.32)-(3.9.33) with three initial value problems, each one-dimensional

(3.9.37) $$\begin{cases} \partial_t \underline{v}' = 3\partial_x \underline{v}' & nk < t < (n+\frac{1}{3})k \\ \underline{v}'(x,y,t,nk) = \underline{v}'''(x,y,z,nk) & (= \underline{f}(x,y,z) \text{ if } n = 0), \end{cases}$$

(3.9.38) $$\begin{cases} \partial_t \underline{v}'' = 3\partial_y \underline{v}'' & (n+\frac{1}{3})k < t < (n+\frac{2}{3})k \\ \underline{v}''(x,y,z,(n+\frac{1}{3})k) = \underline{v}'(x,y,z,(n+\frac{1}{3})k), \end{cases}$$

and

(3.9.39) $$\begin{cases} \partial_t \underline{v}''' = 3\underline{C}\ {}_z \underline{v}''' & (n+\frac{2}{3})k < t < (n+1)k, \\ \underline{v}'''(x,y,z,(n+\frac{2}{3})k) = \underline{v}''(x,y,z,(n+\frac{2}{3})k). \end{cases}$$

Let

(3.9.40) $$u_{ij\ell}^{n+1/3} = (\underline{I}+k\underline{A}D_1)\underline{u}_{ij\ell}^{n} ,$$

(3.9.41) $$\underline{u}^{n+2/3} = (\underline{I}+k\underline{B}D_2)\underline{u}_{ij\ell}^{n+1/3} ,$$

and

(3.9.42) $$\underline{u}_{ij\ell}^{n+1} = (\underline{I}+k\underline{C}D_3)\underline{u}_{ij\ell}^{n+2/3}$$

approximate (3.9.37)-(3.9.39), respectively, where D_1, D_2, and D_3 are finite difference operators in x, y, and z, respectively. The combined method is

(3.9.43) $$u_{ij\ell}^{n+1} = (\underline{I}+k\underline{C}D_3)(\underline{I}+k\underline{B}D_2)(\underline{I}+k\underline{A}D_1)\underline{u}_{ij\ell}^{n} .$$

If any two of the three matrices $\underline{A}$, $\underline{B}$, or $\underline{C}$ do not commute, then

$$e^{-i\lambda_z\underline{C}\xi_3-i\lambda_y\underline{B}\xi_2-i\lambda_x\underline{A}\xi_1} \neq e^{-i\lambda_z\underline{C}\xi_3}e^{-i\lambda_y\underline{B}\xi_2}e^{-i\lambda_x\underline{A}\xi_1},$$

as in the case of two space dimensions, and

$$e^{-i\lambda_z\underline{C}\xi_3-i\lambda_y\underline{B}\xi_2-i\lambda_x\underline{A}\xi_1} = \underline{G}(\xi_1,\xi_2,\xi_3) + O(\|\underline{\xi}\|^2),$$

where $\underline{G}(\xi_1,\xi_2,\xi_3) = \underline{G}_3(\xi_3)\underline{G}_2(\xi_2)\underline{G}_1(\xi_1)$ is the amplification matrix of the combined method (3.9.43), $\underline{G}_1(\xi_1)$, $\underline{G}_2(\xi_2)$, $\underline{G}_3(\xi_3)$ are the amplification matrices of the fractional steps (3.9.40)-(3.9.42), respectively, and $\underline{\xi} = (\xi_1,\xi_2,\xi_3)$. Thus the combined method (3.9.43) is first order accurate in space even though the fractional steps are each accurate of order of their respective space variables.

Again by carefully allowing for the neglected terms in each step we can regain the lost accuracy. By expanding $e^{-i\lambda_z \underline{C}\xi_3 - i\lambda_y \underline{B}\xi_2 - i\ x\underline{A}\ 1}$, $e^{-i(\lambda_x/2)\underline{A}\xi_1}$, $e^{-i(\lambda_y/2)\underline{B}\xi_2}$, and $e^{-i(\lambda_z/2)\underline{C}\xi_3}$ in Taylor series and comparing terms, we see (as in the case of two space dimensions) that

$$e^{-i\lambda_z \underline{C}\xi_3 - i\lambda_y \underline{B}\xi_2 - i\lambda_x \underline{A}\xi_1} = e^{-i(\lambda_x/2)\underline{A}\xi_1} e^{-i(\lambda_y/2)\underline{B}\xi_2} e^{-i(\lambda_z/2)\underline{C}\xi_3}$$

$$e^{-i(\lambda_z/2)\underline{C}\xi_3} e^{-i(\lambda_y/2)\underline{B}\xi_2} e^{-i(\lambda_x/2)\underline{A}\xi_1} + O(\|\underline{\xi}\|^3) .$$

However,

$$e^{-i(\lambda_z/2)\underline{C}\xi_3} e^{-i(\lambda_z/2)\underline{C}\xi_3} = e^{-i\lambda_z \underline{C}\xi_3} + O(\|\underline{\xi}\|^3) ,$$

so that

$$e^{-i\lambda_z \underline{C}\xi_3 - i\lambda_y \underline{B}\xi_2 - i\lambda_x \underline{A}\xi_1}$$

$$= e^{-i(\lambda_x/2)\underline{A}\xi_1} e^{-i(\lambda_y/2)\underline{B}\xi_2} e^{-i\lambda_z \underline{C}\xi_3} e^{-i(\lambda_y/2)\underline{B}\xi_2} e^{-i(\lambda_x/2)\underline{A}\xi_1} + O(\|\underline{\xi}\|^3).$$

This gives rise to the combined method involving four steps of length k/6 (in time) and one step of length k/3 (in time)

$$u^{n+1}_{ij\ell} = (\underline{I}+\tfrac{k}{2}\underline{A}D_1)(\underline{I}+\tfrac{k}{2}\underline{B}D_2)(I+k\underline{C}D_3)(\underline{I}+\tfrac{k}{2}\underline{B}D_2)(\underline{I}+\tfrac{k}{2}\underline{A}D_1)u^n_{ij\ell}$$

or, in terms of the five fractional steps

(3.9.44a) $$\underline{u}^{n+1/6}_{ij\ell} = (I + \tfrac{k}{2}\underline{A}D_1)\underline{u}^n_{ij\ell} ,$$

(3.9.44b) $$\underline{u}^{n+1/3}_{ij\ell} = (\underline{I} + \tfrac{k}{2}\underline{B}D_2)\underline{u}^{n+1/6}_{ij\ell} ,$$

(3.9.44c) $$\underline{u}^{n+2/3}_{ij\ell} = (\underline{I} + k\underline{C}D_3)\underline{u}^{n+1/3}_{ij\ell} ,$$

(3.9.44d) $$\underline{u}^{n+5/6}_{ij\ell} = (\underline{I} + \tfrac{k}{2}\underline{B}D_2)\underline{u}^{n+2/3}_{ij\ell} ,$$

and

(3.9.44e) $$\underline{u}^{n+1}_{ij\ell} = (\underline{I} + \tfrac{k}{2}\underline{A}D_1)\underline{u}^{n+5/6}_{ij\ell} ,$$

which is second order in space profided that $q < 2$.

For further discussion, see Strang [44-46]. This accuracy analysis carries over to hyperbolic systems with variable coefficients and nonlinear hyperbolic systems. A major problem with fractional step methods arises from boundary conditions. This will be treated in Chapter V.

III.10. Multilevel Methods

In Section III it was shown that the forward method (3.2.6) and the Lax-Friedrichs method (3.2.21) are dissipative of order 2 and the Lax-Wendroff method (3.5.3) is dissipative of order 4. However, the coefficient of numerical dissipation for all of these methods depends on k and/or h. So the amount of dissipation can only be controlled by changing h and, hence, k. This is not a practical method for controlling the amount of dissipation present. It would be more practical and convenient to have a finite difference method that is not dissipative and then to add a dissipation term that does not depend on k or h. In this manner the amount of dissipation can be controlled independent of the mesh spacing or the time step.

Consider the following multilevel method known as the leap-frog method

$$\frac{u_i^{n+1} - u_i^{n-1}}{2k} = cD_0u_i^n \,, \tag{3.10.1}$$

which approximates the prototype equation $\partial_t v = c\partial_x v$. This method which is centered both in space and time depends on two time levels of initial conditions.

Expand the exact solution $v(x,t)$ in a Taylor series in space about the point (ih,nk). This gives

$$v_{i+1}^n = v_i^n + h(\partial_x v)^n + \frac{h^2}{2}(\partial_x^2 v)_i^n + \frac{h^3}{6}(\partial_x^3 v)_i^n + \frac{h^4}{24}(\partial_x^4 v)_i^n + O(h^5)$$

$$v_{i-1}^n = v_i^n - h(\partial_x v)^n + \frac{h^2}{2}(\partial_x^2 v)_i^n - \frac{h^3}{6}(\partial_x^3 v)_i^n + \frac{h^4}{24}(\partial_x^4 v)_i^n + O(h^5),$$

and

$$D_0v_i^n = (\partial_x v)_i^n + \frac{h^2}{6}(\partial_x^3 v)_i^n + O(h^4) \,. \tag{3.10.2}$$

Observe that the even order derivatives, that is, derivatives of v of the form $\partial_x^{2p}v$ cancel in pairs in (3.10.2). Thus $D_0v_i^n$ does not introduce any diffusion. Similarly,

$$\frac{v_i^{n+1} - v_i^{n-1}}{2k} = (\partial_t v)_i^n + \frac{k^2}{6}(\partial_t^3 v)_i^n + O(k^4) \,,$$

where as in (3.10.2) the even order derivatives of v of the form $\partial_t^{2p}v$ cancel in pairs. Thus $(v_i^{n+1} - v_i^{n-1})/2k$ does not introduce any diffusion. Combining these two results

$$\frac{v_i^{n+1} - v_i^{n-1}}{2k} - cD_0v_i^n = (\partial_t v)_i^n - c(\partial_x v)_i^n + O(k^2) + O(h^2)$$

Thus the leap-frog method (3.10.1) is consistent and accurate of $O(k^2) + O(h^2)$.

By using the prototype equation (3.2.1), we obtain $\partial_t^q = c^q\partial_x^q v$, provided that v is sufficiently differentiable. Substitution into (3.1.13) yields

$$\frac{v_i^{n+1} - v_i^{n-1}}{2k} = c(\partial_x v)_i^n + \frac{c^3k^2}{6}(\partial_x^3 v)_i^n + O(k^4)$$

So, as in the case of (3.10.2), by cancellation in pairs of the even order derivatives, $\frac{v_i^{n+1} - v_i^{n+1}}{2k}$ does not introduce any diffusion. This shows that the leap-frog method is not dissipative.

To analyze the stability of the method (3.10.1), we introduce an auxiliary vector

$$\underline{u}^{n+1} = \begin{pmatrix} u^{n+1} \\ u^n \end{pmatrix}.$$

With this, the leap-frog method may be written in the form

$$u_i^{n+1} = 2kcD_0u_i^n + u_i^{n-1}$$
$$u_i^n = u_i^n,$$

which may be written using the auxiliary vector

(3.10.3) $$\underline{u}_i^{n+1} = \begin{pmatrix} 2ckD_0 & 1 \\ 1 & 0 \end{pmatrix} \underline{u}_i^n .$$

By taking the discrete Fourier transform of (3.10.3), we obtain

$$\hat{\underline{u}}^{n+1}(\xi) = \underline{G}(\xi)\hat{\underline{u}}(\xi)$$

where the amplification matrix $\underline{G}(\xi)$ is given by

(3.10.4) $$\underline{G}(\xi) = \begin{pmatrix} -i2c\lambda \sin \xi & 1 \\ 1 & 0 \end{pmatrix},$$

where the symbol D_0 is $-i2 \sin \xi/h$ and $\lambda = k/h$. The eigenvalues of the amplification matrix (3.10.4), denoted by $\varepsilon_\pm$, are

(3.10.5) $$\varepsilon_\pm(\xi) = -ic\lambda \sin \xi \pm\sqrt{1 - c^2\lambda^2\sin^2\xi}.$$

For $|c|\lambda < 1$, ε_+ and ε_- are distinct. Furthermore,

(3.10.6) $$|\varepsilon_\pm|^2 = 1 .$$

Thus both of the eigenvalues lie on the unit circle in the complex plane for $|\xi| \leq \pi$. Furthermore, from the definition of dissipation, it follows (3.10.6) that the leap-frog method is <u>not</u> dissipative. The von Neumann condition is satisfied since $\sigma(\underline{G}(\xi)) = 1$, which is a necessary condition for stability.

In order to prove the stability of the leap-frog method, we shall make use of a version of the Lax-Milgram lemma (see Lax and Milgram [24]).

<u>Lemma (Lax-Milgram)</u>. Consider a finite difference method

$$\underline{u}^{n+1} = \sum_{j=-m_1}^{m_2} \underline{A}_j\, S_+^j\, \underline{u}^n .$$

Suppose there exists a functional $S(\underline{u})$ and positive constants m and M such that

(3.10.7) $$m\|\underline{u}\| \le S(\underline{u}) \le M\|\underline{u}\|,$$

for some norm $\|\ \|$, that is, $S(\underline{u})$ is a quasi-norm. If

(3.10.8) $$S(\underline{u}^n) \le (1 + ck)S(\underline{u}^{n-1}),$$

where c is a constant independent of h, k, ξ, and n, then the finite difference method is stable.

Proof. Using inequality (3.10.8) repeatedly,

$$\begin{aligned} S(\underline{u}^n) &\le (1 + ck)S(\underline{u}^{n-1}) \\ &\le (1 + ck)^2 S(\underline{u}^{n-2}) \\ &\ \ \vdots \\ &\le (1 + ck)^n S(\underline{u}^0) \\ &\le e^{cnk} S(\underline{u}^0) \\ &= e^{ct}\, S(\underline{u}^0)\ . \end{aligned}$$

Using the right half of the inequality (3.10.7),

$$S(\underline{u}^n) \le e^{ct}\, M\|\underline{u}^0\|.$$

Furthermore, using the left half of inequality (3.10.7) combined with the above result

$$m\|\underline{u}^n\| \le S(\underline{u}^n) \le e^{ct}\, M\|\underline{u}^0\|$$

or

$$\|\underline{u}^n\| \le \frac{M}{m}\, e^{ct}\|\underline{u}^0\|\ .$$

Therefore, the finite difference method is stable.

The energy method shall be used to show that the leap-frog method is stable. From the discrete integration by parts (1.2.13), we obtain for $D_0 = \frac{1}{2}(D_+ + D_-)$ and for two grid functions $f = \{f_i\}$ and $g = \{g_i\}$

(3.10.9) $$(g, D_0 f) = -(D_0 g, f)\ .$$

A bound on $\|D_0\|$ is needed. Using the definition of norm of an operator,

$$\begin{aligned} k\|D_0\| &= k \max_{\|\underline{u}_i\| \ne 0} \frac{\|D_0\underline{u}_i\|}{\|\underline{u}_i\|} = k \max_{\|\underline{u}_i\| \ne 0} \frac{\|\underline{u}_{i+1} - \underline{u}_{i-1}\|/2h}{\|\underline{u}_i\|} \\ &\le \frac{k}{2h} \max_{\|\underline{u}_i\| \ne 0} \frac{\|\underline{u}_{i+1}\| + \|\underline{u}_{i-1}\|}{\|\underline{u}_i\|} = \frac{k}{2h} \max_{\|\underline{u}_i\|} \frac{2\|\underline{u}_i\|}{\|\underline{u}_i\|} = \frac{k}{h}\ . \end{aligned}$$

Therefore,

(3.10.10) $$k\|D_0\| \le \lambda\ .$$

Write the leap-frog method (3.10.1) in the form

$$u^{n+1} - u^{n-1} = 2kcD_0u^n\ .$$

Taking the discrete inner product with $u^{n+1} + u^{n-1}$,

$$\begin{aligned}\|u^{n+1}\|_2^2 - \|u^{n-1}\|_2^2 &= (u^{n+1} + u^{n-1},\ 2ckD_0u^n) \\ &= 2kc(u^{n+1} + u^{n-1},\ D_0u^n) \\ &= 2kc[(u^{n+1},\ D_0u^n) + (u^{n-1},\ D_0u^n)] \\ &= 2kc[(u^{n+1},\ D_0u^n) - (u^n,\ D_0u^{n-1})]\ ,\end{aligned}$$

using (3.10.9). Adding $\|u^n\|_2^2$ to both sides we obtain after regrouping terms,

(3.10.11) $$\|u^{n+1}\|_2^2 + \|u^n\|_2^2 - 2kc(u^{n+1}, D_0u^n)$$

$$= \|u^n\|_2^2 + \|u^{n-1}\|_2^2 - 2kc(u^n, D_0u^{n-1}).$$

Define the functional $S(\underline{u}^{n+1})$ by

$$S(\underline{u}^{n+1}) = \|u^{n+1}\|_2^2 + \|u^n\|_2^2 - 2kc(u^{n+1}, D_0u^n),$$

then

(3.10.12) $$S(\underline{u}^n) = \|u^n\|_2^2 + \|u^{n-1}\|_2^2 - 2kc(u^n, D_0u^{n-1})$$

and (3.10.11) may be written in the form

(3.10.13) $$S(\underline{u}^{n+1}) = S(\underline{u}^n)$$

Applying Lemma 3 in Section II.11 to the inner product in (3.10.12) and using the bound on $\|D_0\|$ (3.10.10),

$$\begin{aligned}S(\underline{u}^n) &\le \|u^n\|_2^2 + \|u^{n-1}\|_2^2 + 2k|c|\,\|D_0\|\left(\frac{\|u^n\|_2^2 + \|u^{n-1}\|_2^2}{2}\right) \\ &= (\|u^n\|_2^2 + \|u^{n-1}\|_2^2)(1 + k|c|\,\|D_0\|) \\ &\le (1 + |c|\lambda)(\|u^n\|_2^2 + \|u^{n-1}\|_2^2)\ .\end{aligned}$$

Similarly,

$$\begin{aligned}S(\underline{u}^n) &\ge \|u^n\|_2^2 + \|u^{n-1}\|_2^2 - 2k|c|\,|(u^n, D_0u^{n-1})| \\ &\ge \|u^n\|_2^2 + \|u^{n-1}\|_2^2 - 2k|c|\,\|D_0\|\left(\frac{\|u^n\|_2^2 + \|u^{n-1}\|_2^2}{2}\right) \\ &= (1 - k|c|\,\|D_0\|)(\|u^n\|_2^2 + \|u^{n-1}\|_2^2) \\ &\ge (1 - |c|\lambda)(\|u^n\|_2^2 + \|u^{n-1}\|_2^2)\ .\end{aligned}$$

Combining these two inequalities

(3.10.14) $$(1-|c|\lambda)(\|u^n\|_2^2+\|u^{n-1}\|_2^2) \le S(\underline{u}^n) \le (1+|c|\lambda)(\|u^n\|_2^2+\|u^{n-1}\|_2^2).$$

Let the norm of the auxiliary vector $\underline{u}^n$ be defined by $\|\underline{u}^n\| = \|u^n\|_2^2 + \|u^{n-1}\|_2^2$, then $S(\underline{u}^n)$ satisfies the inequality (3.10.7) of the Lax-Milgram lemma with $m = 1 - |c|\lambda$ and $M = 1 + |c|\lambda$. In order that $m > 0$, $|c|\lambda < 1$. Combining this with (3.10.13), we conclude by the Lax-Milgram lemma that the leap-frog method is stable for $|c| < 1$. Observe the strict inequality. The CFL condition is $|c|\lambda \leq 1$. A natural question arises. Is the leap-frog method stable for $|c|\lambda = 1$? Suppose $|c|\lambda = 1$. The amplification matrix (3.10.4) becomes

$$\underline{G}(\xi) = \begin{pmatrix} 2i & 1 \\ 1 & 0 \end{pmatrix}$$

where $\xi = -\pi/2$. $\underline{G}(\xi)$ may be written in the (Jordan canonical) form

$$\underline{G}(\xi) = \underline{T}\ \underline{V}\ \underline{T}^{-1},$$

where $\underline{T} = \begin{pmatrix} i & 1 \\ 1 & 0 \end{pmatrix}$, $\underline{T}^{-1} = \begin{pmatrix} 0 & 1 \\ 1 & -i \end{pmatrix}$ and $\underline{V} = \begin{pmatrix} i & 1 \\ 0 & i \end{pmatrix} = i\begin{pmatrix} 1 & -i \\ 0 & 1 \end{pmatrix}$,

and

$$\underline{G}^n(\xi) = \underline{T}\ \underline{V}^n\ \underline{T}^{-1} = \begin{pmatrix} i & 1 \\ 1 & 0 \end{pmatrix} i^n \begin{pmatrix} 1 & -ni \\ 0 & 1 \end{pmatrix}\begin{pmatrix} 0 & 1 \\ 1 & -i \end{pmatrix} = i^n \begin{pmatrix} n+1 & -ni \\ -ni & 1-n \end{pmatrix}.$$

So $\underline{G}^n(\xi)$ cannot be bounded independent of the number of timesteps n. Hence, the leap-frog method is <u>not</u> stable for $|c|\lambda = 1$.

To determine the phase error, substitute (3.6.5) into the leap-frog method (3.10.1) to obtain the dispersion relation

$$\frac{e^{i\beta\omega k} - e^{-i\beta\omega k}}{2k} = c\,\frac{e^{i\beta h} - e^{i\beta h}}{2h}$$

or

$$e^{i\beta\omega k} - e^{-i\beta\omega k} = i2c\lambda \sin(\beta h)\ .$$

Multiplying by $e^{i\beta\omega k}$ gives a quadratic equation in $e^{i\omega k}$

$$e^{i2\beta\omega k} - 2ic\lambda \sin(\beta h)e^{i\beta\omega k} - 1 = 0,$$

which has roots

$$e^{i\beta\omega k} = ic\lambda \sin(\beta h) \pm \sqrt{1 - c^2\lambda^2\sin^2(\beta h)} = \varepsilon_{\pm}(-\beta h)\ , \tag{3.10.15}$$

where $\varepsilon_{\pm}$ are the eigenvalues of the amplification matrix of the leap-frog method.

Writing ω as a complex number of the form $\omega = a + ib$,

$$e^{i\beta\omega k} = e^{-b\beta k}e^{ia\beta k}\ .$$

Substituting into (3.10.15),

$$e^{-b\beta k}e^{ia\beta k} = ic\lambda\sin(\beta h) \pm \sqrt{1 - c^2\lambda^2\sin^2(\beta h)}$$

As observed earlier, $e^{-b\beta k} = 1$ so that $b = 0$, and the method is not dissipative. Consider the eigenvalue ε_+,

$$\tan(a\beta k) = \frac{\mathrm{Im}\ \varepsilon_+(-\beta h)}{\mathrm{Re}\ \varepsilon_+(-\beta h)} = \frac{c\lambda\sin(\beta h)}{\sqrt{1 - c^2\lambda^2\sin^2(\beta h)}}$$

For high frequency modes where βh is near π, $\tan(a\beta k)$ is near 0 so that a is near 0. Thus the high frequency waves are nearly stationary. Consider now small values of βh (low frequency modes). We shall make use of the Taylor series for $\sin z$ given by (3.6.14) in Section III.6 and the Taylor series expansion about $z = 0$ of

(3.10.16) $$\frac{1}{\sqrt{1 - z}} = 1 - \frac{1}{2}z + O(z^2) .$$

With this we obtain $c\lambda\sin(\beta h) = c\lambda\beta h(1 - \frac{1}{6}(\beta h)^2)$, $c^2\lambda^2\sin^2(\beta h) = c^2\lambda^2(\beta h)^2$ and $1/\sqrt{1 - c^2\lambda^2\sin^2(\beta h)} = 1 - \frac{1}{2}c^2\lambda^2(\beta h)^2$. This gives by substitution

$$\begin{aligned}\tan(a\beta h) &= c\lambda\beta h(1 - \tfrac{1}{6}(\beta h)^2)(1 - \tfrac{1}{2}c^2\lambda^2(\beta h)^2)\\ &= c\lambda\beta h(1 - \tfrac{1}{6}(1 + 3c^2\lambda^2)(\beta h)^2)\end{aligned}$$

or, using (3.6.16),

$$\begin{aligned}a\beta h &= \tan^{-1}(c\lambda\beta h(1 - \tfrac{1}{6}(1 + 3c^2\lambda^2(\beta h)^2))).\\ &= c\lambda\beta h[1 - \tfrac{1}{6}(1 + 3c^2\lambda^2)(\beta h)^2)] - \tfrac{1}{3}(c\lambda\beta h).\\ &= c\lambda\beta h[1 - \tfrac{1}{6}(1 + 5c^2\lambda^2)(\beta h)^2)]\end{aligned}$$

or

(3.10.17) $$a = c[1 - \frac{1}{6}(1 + 5c^2\lambda^2)(\beta h)^2]$$

From (3.10.17), $a < c$, which means that for low frequency Fourier modes, the numerical solution lags the exact solution. Similar results are obtained with the eigenvalues ε_- in (3.10.15).

In the implementation of the leap-frog method (3.10.1), two time levels of initial conditions are needed. This also means that two time levels of the solution must be stored, which doubles the storage requirement for the method.

Furthermore, as with the method (3.2.17) in Section III.2, the leap-frog method has the "checker board" type problem. Alternate points are connected by the finite difference operator D_0 and, as such, two independent solutions can evolve, one on the even points and one on the odd points.

To damp these high frequency modes, a small diffusion term may be added. This will also serve to couple the odd and even grid points which will represent the formation of two independent solutions.

Observe that by expanding the exact solution $v(x,t)$ in a Taylor series in space about the point $(ih,(n-1)k)$,

$$(3.10.18)\quad (D_+D_-)^2 v_i^{n-1} \equiv \frac{v_{i+2}^{n-1} - 4v_{i+1}^{n-1} + 6v_i^{n-1} - 4v_{i-1}^{n-1} + v_{i-2}^{n-1}}{h^4} = (\partial_x^4 v)_i^{n-1} + O(h^2).$$

Consider the following dissipative leap-frog method using (3.10.18)

$$(3.10.19)\qquad u_i^{n+1} = u_i^{n-1} + 2ckD_0 u_i^n - \sigma\frac{h^4}{16}(D_+D_-)^2 u_i^{n-1} ,$$

which may be written in the form

$$(3.10.20)\qquad \frac{u_i^{n+1} - u_i^{n-1}}{2k} = cD_0 u_i^n - \frac{\sigma h^4}{32k}(D_+D_-)^2 u_i^{n-1} .$$

From (3.10.18) we see that the second term on the right in (3.10.20) is an approximation to $\partial_x^4 v$, in fact, it is an approximation to $\frac{\sigma h^4}{32k}(\partial_x^4 v)_i^{n-1}$. Assume that $k = O(h)$, then $\frac{\sigma h^4}{32k}(\partial_x^4 v)_i^{n-1} = O(h^3)$. Thus the dissipative leap-frog method remains accurate of $O(k^2) + O(h^2)$. It is assumed that σ is a positive constant used to govern the amount of dissipation introduced. The limitations on σ will be determined by the stability condition on (3.10.18).

Using the same auxiliary vector $\underline{u}^n$, the dissipative leap-frog method (3.10.18) may be written in the form

$$\underline{u}_i^{n+1} = \begin{pmatrix} 2ckD_0 & 1 - \sigma\frac{h^4}{16}(D_+D_-)^2 \\ 1 & 0 \end{pmatrix} \underline{u}_i^n .$$

To analyze the stability of the dissipative leap-frog method (3.10.19) take the discrete Fourier transform,

$$\hat{u}^{n+1}(\xi) = \underline{G}(\xi)\hat{u}^n(\xi),$$

where the amplification matrix $\underline{G}(\xi)$ is given by

$$\underline{G}(\xi) = \begin{pmatrix} -i2c\lambda \sin\xi & 1 - \varepsilon \sin^4(\frac{\xi}{2}) \\ 0 & 0 \end{pmatrix},$$

and the symbol of $(D_+D_-)^2$, $\frac{16}{h^4}\sin^4(\frac{\xi}{2})$, is obtained by squaring the symbol of D_+D_-. The eigenvalues of the amplification matrix are

$$(3.10.21)\quad \varepsilon_\pm(\xi) = i2c\lambda \sin\xi \pm \sqrt{1 - c^2\lambda^2\sin^2\xi - \varepsilon \sin^4(\tfrac{\xi}{2})}$$

For $c^2\lambda^2 < 1 - \sigma$, the square root term is real,

(3.10.22) $$|\varepsilon_{\pm}(\xi)|^2 = 1 - \sigma \sin^4(\tfrac{\xi}{2}) \le 1,$$

and so the von Neumann condition is satisfied. Write (3.10.22) as

$$|\varepsilon_{\pm}(\xi)| = \sqrt{1 - \sigma \sin^4(\tfrac{\xi}{2})}\ .$$

If $\sigma < 1$, then $|\varepsilon_{\pm}(\xi)|$ decreases monotonically from $|\xi_{\pm}(\xi)| = 1$ at $\xi = 0$ to $|\xi_{\pm}(\xi)| = \sqrt{1 - \varepsilon}$ at $\xi = \pi$. For small values of ξ, replace $\sin z$ by the first term in the Taylor series about $z = 0$, so that $\sin^4(\frac{\xi}{2}) = \frac{\xi^4}{16}$. Upon substitution into $|\xi_{\pm}(\xi)|$, we obtain for small values of ξ,

$$|\xi_{\pm}(\xi)| = \sqrt{1 - \tfrac{\sigma}{16}\,\xi^4}$$

By expanding $\sqrt{1 - z}$ in a Taylor series about $z = 0$, as in Section III.6,

$$|\varepsilon_{\pm}(\xi)| = 1 - \tfrac{\sigma}{32}\,\xi^4$$

Thus, the order of dissipation cannot be of order less than 4. By choosing $\delta < \frac{\sigma}{32}$, the inequality in the definition of dissipation is satisfied. Hence the dissipative leap-frog method is dissipative of order 4.

It remains to show that the method is stable. Again we shall use the energy method. Write the dissipative leap-frog method (3.10.19) in the form

$$u^{n+1} - u^{n-1} = 2ckD_0u^n - \sigma \tfrac{h^4}{16}(D_+D_-)^2u^{n-1}$$

Taking the discrete inner product with $u^{n+1} + u^{n-1}$,

$$\begin{aligned}\|u^{n+1}\|_2^2 - \|u^{n-1}\|_2^2 &= (u^{n+1}+u^{n-1}, 2ckD_0u^n) + (u^{n+1}+u^{n-1}, -\tfrac{\sigma h^4}{16}(D_+D_-)^2u^{n-1})\\ &= 2ck[(u^{n+1},D_0u^n) + (u^{n-1},D_0u^n)]\\ &\quad - \tfrac{\sigma h^4}{16}[(u^{n+1},(D_+D_-)^2u^{n-1}) + (u^{n-1},(D_+D_-)^2u^{n-1})].\end{aligned}$$

Using the discrete integration by parts (1.2.13) twice, for two grid functions $f = \{f_i\}$ and $g = \{g_i\}$,

(3.10.23) $$(g, D_+D_-f) = (D_+D_-g, f).$$

Using (3.10.23) along with (3.10.9),

(3.10.24) $$\begin{aligned}\|u^{n+1}\|_2^2 - \|u^{n-1}\|_2^2 &= 2ck[(u^{n+1},D_0u^n) - (u^n,D_0u^{n-1})]\\ &\quad - \tfrac{\sigma h^4}{16}[(D_+D_-u^{n+1},D_+D_-u^{n-1}) + \|D_+D_-u^{n-1}\|_2^2]\\ &\le 2ck[(u^{n+1},D_0u^n) - (u^n,D_0u^{n-1})]\\ &\quad + \tfrac{\sigma h^4}{16}|(D_+D_-u^{n+1},D_+D_-u^{n-1})| - \tfrac{\sigma h^4}{16}\|D_+D_-u^{n-1}\|_2^2.\end{aligned}$$

Using Lemma 2 in Section II.11,

$$|(D_+D_-u^{n+1},D_+D_-u^{n-1})| \le \frac{1}{2}(\|D_+D_-u^{n+1}\|_2^2 + \|D_+D_-u^{n-1}\|_2^2),$$

so that

$$\|u^{n+1}\|_2^2 - \|u^{n-1}\|_2^2 \le 2ck[(u^{n+1},D_0u^n) - (u^n,D_0u^{n-1})]$$
$$+ \frac{\sigma h^4}{32}(\|D_+D_-u^{n+1}\|_2^2 + \|D_+D_-u^{n-1}\|_2^2) - \frac{\sigma h^4}{16}\|D_+D_-u^{n-1}\|_2^2$$
$$= 2ck[(u^{n+1},D_0u^n) - (u^n,D_0u^{n-1})]$$
$$+ \frac{\sigma h^4}{32}[\|D_+D_-u^{n+1}\|_2^2 - \|D_+D_-u^{n-1}\|_2^2] .$$

Adding $\|u^n\|_2^2 - \frac{\sigma h^4}{32}\|D_+D_-u^n\|_2^2$ to both sides, after regrouping,

$$\|u^{n+1}\|_2^2 + \|u^n\|_2^2 - 2ck(u^{n+1},D_0u^n) - \frac{\sigma h^4}{32}[\|D_+D_-u^{n+1}\|_2^2 + \|D_+D_-u^n\|_2^2]$$

(3.10.25)

$$\le \|u^n\|_2^2 + \|u^{n-1}\|_2^2 - 2ck(u^n,D_0u^{n-1}) - \frac{\sigma h^4}{32}[\|D_+D_-u^n\|_2^2 + \|D_+D_-u^{n-1}\|_2^2] .$$

Define the functional $S(\underline{u}^n)$, where $\underline{u}^n$ is the auxiliary vector, by

(3.10.26) $$S(\underline{u}^n) = \|u^n\|_2^2 + \|u^{n-1}\|_2^2 - 2ck(u^n,D_0u^{n-1}) - \frac{\sigma h^4}{32}[\|D_+D_-u^n\|_2^2 + \|D_+D_-u^{n-1}\|_2^2] .$$

By substitution into (3.10.25),

$$S(\underline{u}^{n+1}) \le S(\underline{u}^n),$$

so that the inequality (3.10.8) of the Lax-Milgram lemma is satisfied.

Using inequality (1.2.18),

$$\|D_+D_-u^n\|_2^2 \le \frac{16}{h^4}\|u^n\|_2^2$$

and

$$\|D_+D_-u^{n-1}\|_2^2 \le \frac{16}{h^4}\|u^{n-1}\|_2^2,$$

which gives

(3.10.27) $$\frac{\sigma h^4}{32}(\|D_+D_-u^n\|_2^2 + \|D_+D_-u^{n-1}\|_2^2) \le \frac{\sigma}{2}(\|u^n\|_2^2 + \|u^{n-1}\|_2^2) .$$

Using Lemma 3 in Section II.11 and (3.10.10)

(3.10.28) $$2|c|k|(u^n,D_0u^{n-1})| \le 2|c|k\|D_0\|\frac{1}{2}(\|u^n\|_2^2 + \|u^{n-1}\|_2^2) \le |c|\lambda(\|u^n\|_2^2 + \|u^{n-1}\|_2^2).$$

Combining (3.10.27) and (3.10.28) in the definition of $S(\underline{u}^n)$

$$S(\underline{u}^n) \le (1 + |c|\lambda + \frac{\sigma}{2})(\|u^n\|_2^2 + \|u^{n-1}\|_2^2) .$$

Similarly,

$$S(\underline{u}^n) \ge (1 - |c|\lambda - \frac{\sigma}{2})(\|u^n\|_2^2 + \|u^{n-1}\|_2^2)$$

so that by combining

$$(1 - |c|\lambda - \frac{\sigma}{2})(\|u^n\|_2^2 + \|u^{n-1}\|_2^2) \le S(\underline{u}^n) \le (1 + |c|\lambda + \frac{\sigma}{2})(\|u^n\|_2^2 + \|u^{n-1}\|_2^2) .$$

With $m = 1 - |c|\lambda - \frac{\sigma}{2}$ and $M = 1 + |c|\lambda + \frac{\sigma}{2}$ in (3.10.7) of the Lax-Milgram lemma, we conclude that the dissipative leap-frog method is stable provided $m > 0$ and $M > 0$. M is always strictly positive. However, $m > 0$ if $|c|\lambda < 1 - \frac{\sigma}{2}$. Thus the dissipative leap-frog method is stable provided that

$$|c|\lambda < 1 - \frac{\sigma}{2} .$$

Since the dissipation term $(D_+D_-)^2u_i^{n-1}$ depends on five points u_{i-2}^{n-1}, u_i^{n-1}, u_{i+1}^{n-1}, and u_{i+2}^{n-1}, it introduces a complication at the boundary. For suppose the left boundary corresponds to $i = 0$, with this five-point dissipation term, the solution u_2^{n+1} will depend on u_0^{n-1} and the solution u_1^{n+1} will depend on u_{-1}^{n-1}. Something special will have to be done near the boundaries. This will be discussed in a later section.

It should be mentioned that the sign of the dissipation term is negative. This does not result in an ill-posed problem provided that the advection process is the dominant process. If the sign were positive, the method would be unstable, for the eigenvalues of the resulting amplification matrix would be

$$\varepsilon_\pm(\xi) = i2\lambda c \sin \xi \pm\sqrt{1 - c^2\lambda^2 \sin^2\xi + \sigma \sin^2(\frac{\xi}{2})}$$

where

$$|\xi_\pm(\xi)|^2 = 1 + \sigma \sin^4(\frac{\xi}{2}) > 1.$$

From this we see that the von Neumann condition would be violated.

There is another way of introducing dissipation into the leap-frog method (3.10.1). There is a connection between the Lax-Wendroff method and the leap-frog method. To see this, introduce half points $((i + \frac{1}{2})h, (n + \frac{1}{2})k)$ on the grid as depicted in Figure 3.20,

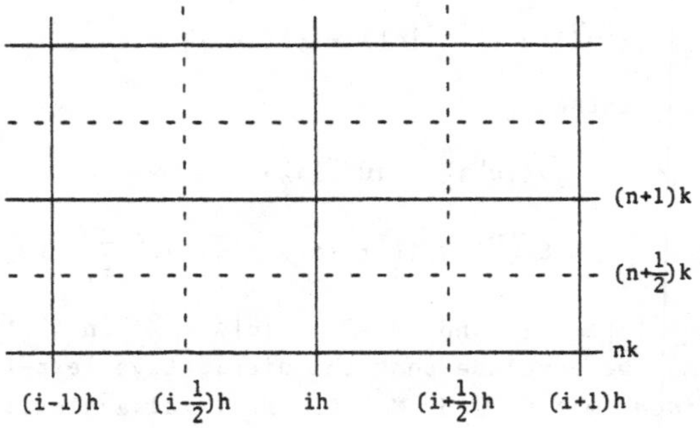

Figure 3.20

The Lax-Wendroff method can be written in a two-step form:

(3.10.30a) $$u_{i+\frac{1}{2}}^{n+\frac{1}{2}} = \frac{1}{2}(u_{i+1}^n + u_i^n) + \frac{1}{2}\ (u_{i+1}^n - u_i^n)$$

(3.10.30b) $$u_i^{n+1} = u_i^n + \ \ (u_{i+\frac{1}{2}}^{n+\frac{1}{2}} - u_{n-\frac{1}{2}}^{n+\frac{1}{2}}) .$$

By substitution of (3.10.30a) into (3.10.30b), one can readily verify that this is the Lax-Wendroff method. The first half-step (3.10.30a) has an average term $\frac{1}{2}(u_{i+1}^n + u_i^n)$, which introduces dissipation. The second half-step is a leap-frog step using the three time levels $(n+1)k$, $(n+\frac{1}{2})k$, and nk in place of $(n+1)h$, nk, and $(n-1)k$. The second half-step (3.10.30b) involves centered differences in time and space about the point $(ih,(n+\frac{1}{2})k)$. This step is not dissipative.

The two methods can also be combined in the form of an m-step Lax-Wendroff method in which there are $(m-1)$ leap-frog steps of the form (3.10.1) followed by one dissipative Lax-Wendroff half-step (3.10.30a). The second half-step (3.10.30b) is used to return to the original form of the leap-frog method. In this way, the Lax-Wendroff half-step is used when the solution needs smoothing.

III.11. Hyperbolic Equations with Variable Coefficients

Consider the first order system of hyperbolic equations

$$\partial_t \underline{v} = \underline{A}(x)\partial_x \underline{v} + \underline{B}(x)\underline{v} \quad , \; -\infty < x < +\infty \; , \; t > 0 \tag{3.11.1}$$

where $\underline{v} = (v_1,\dots,v_p)^T$ and $\underline{A}(x)$, $\underline{B}(x)$ are $p \times p$ matrices whose elements are functions of x.

Definition. The first order system (3.11.1) is said to be hyperbolic if there exists a constant K and a nonsingular matrix $\underline{T}(x)$ with

$$\max(\|\underline{T}(x)\|, \|T^{-1}(x)\|) \le K ,$$

independent of x such that

$$T^{-1}(x)\underline{A}(x)\underline{T}(x) = \underline{D}(x)$$

where $\underline{D}(x)$ is diagonal, $\underline{D}(x) = \operatorname{diag}(\mu_1(x),\dots,\mu_p(x))$ and $\mu_j(x)$ are real for every value of x. The characteristics are no longer straight lines, but the solutions of ordinary differential equations

$$\frac{dx}{dt} = \mu_j(x),$$

for $j = 1,\dots,p$.

One nice result can be obtained; the initial value problem given by equation (3.11.1), along with the initial condition

$$\underline{v}(x,0) = \underline{f}(x) \quad , \; -\infty < x < +\infty$$

is well-posed if and only if all the initial-value problems with frozen coefficients

$$\partial_t \underline{v} = \underline{A}(\bar{x})\partial_x \underline{v} + \underline{B}(\bar{x})\underline{v}$$

for all fixed $\bar{x}$, $-\infty < \bar{x} < +\infty$, are well-posed. This gives some hope for obtaining stability conditions for hyperbolic equations.

Approximate equation (3.11.1) by

$$\underline{u}^{n+1} = \sum_{j=-m_1}^{m_2} (\underline{A}_j(x)S_+^j)\underline{u}^n + \underline{B}(x)u^n .$$

By the Strang perturbation theorem (Theorem 2.4), the undifferentiated term $\underline{B}(x)\underline{v}$ can be eliminated from (3.11.1) and the corresponding term from the finite difference method. Thus we shall consider the first order system of the form

$$\partial_t \underline{v} = \underline{A}(x)\partial_x \underline{v} \; , \; -\infty < x < +\infty \; , \; t > 0 \tag{3.11.2}$$

approximated by

$$\underline{u}^{n+1} = \sum_{j=-m_1}^{m_2} (\underline{A}_j(x)S_+^j)\underline{u}^n, \quad n > 0 , \tag{3.11.3}$$

where $\underline{A}_j(x)$ are $p \times p$ matrices whose elements are functions of x, h, and k. The CFL condition for the finite difference method (3.11.3) is a direct extension of that for the constant coefficient case (compare with (3.8.6) in Section III.8). Let the eigenvalues $\mu_1(x),\dots,\mu_p(x)$ of $\underline{A}(x)$ in (3.11.2) be ordered so that

$$\mu_1(x) \le \mu_2(x) \le \dots \le \mu_m(x) \le 0 \le \mu_{m+1}(x) \le \dots \le \mu_p(x) .$$

The CFL condition for a finite difference method (3.11.2) approximating equation (3.11.2) is for every x

$$-\mu_j(x)\frac{k}{h} \le m_1 \ , \ j = 1,\dots,m \tag{3.11.4a}$$

$$\mu_j(x)\frac{k}{h} \le m_2 \ , \ j = m+1,\dots,p. \tag{3.11.4b}$$

The amplification matrix (or symbol) $\underline{G}(x,\xi)$ of (3.11.3) can be defined by

$$\underline{G}(x,\xi) = \sum_{j=-m_1}^{m_2} \underline{A}_j(x)e^{-ij\xi} \tag{3.11.5}$$

As discussed in Section II.10, it is not true, in general, that

$$\hat{\underline{u}}^{n+1}(\xi) = \underline{G}(x,\xi)\hat{u}^n(\xi)$$

However, stability results do exist. As a first step in developing conditions for stability, we present a theorem due to Friedrichs (see also Widlund [47]).

Theorem 3.5 (Friedrichs). Consider a finite difference method in the form of (3.11.3). Suppose that the matrices $\underline{A}_j(x)$ are hermitian and positive definite independent of k. Furthermore, assume that the matrices $\underline{A}_j(x)$ are Lipschitz continuous in x, that is, there exists a constant K independent of x such that for $j = -m_1,\dots,m_2$

$$\|\underline{A}_j(x) - \underline{A}_j(x_0)\| \le K|x - x_0|.$$

Such a finite difference method is stable. Before this theorem is proved, a few remarks are in order.

It is possible to construct such finite difference methods as described in the theorem. Consider the scalar prototype equation $\partial_t v = c\partial_x v$ where c is a positive constant approximated by

$$u^{n+1} = u_i^n + kcD_+u_i^n$$

$$= (1 - c\frac{k}{h})u_i^n + c\frac{k}{h}\, u_{n+1}^n .$$

If the CFL condition is satisfied, $c\frac{k}{h} \le 1$, then $1 - c\frac{k}{h}$ and $c\frac{k}{h}$ are hermitian and positive definite. It should also be noted that all of these finite difference methods will always be first-order accurate. Why?

The shortest wavelength (highest frequency) that can be resolved on a grid of spacing h is $2h$. For the most part, we are interested in the low frequency (long wavelength) terms, so the lack of resolution of the high frequency terms is not too important. In the case of variable coefficient (or nonlinear) equations, the different Fourier modes will interact in such a way that energy is transferred from the low frequency to the high frequency terms. To reduce the amount of interaction, a smoothness condition will need to be imposed on the coefficients $\underline{A}_j(x)$, such as, requiring the $\underline{A}_j(x)$'s to be Lipschitz continuous in x.

In a turbulent flow, for example, turbulent energy cascades from large eddies to small eddies. This energy at the small eddies is dissipated into internal energy through friction. If a finite difference method is non-dissipative or does not have sufficient dissipation (coefficient of dissipation term too small), energy which has transferred from the longer wavelengths to the short wavelengths will accumulate at the wavelengths of order $2h$. This accumulated energy is transferred back to the longer wavelengths. This result in introducing an error in the Fourier modes of greatest interest. In fact, this solution can grow without bound.

For a finite difference method of the form (3.11.3) that satisfies the conditions of the Friedrichs theorem, $\|\underline{G}(x,\xi)\| \leq 1$, where $\underline{G}(x,\xi)$ denotes the amplification matrix.

Let a_j denote the complex number $e^{-ij\xi}$ so that $|a_j| = 1$ and let $\underline{w}_1$, $\underline{w}_2$ be two vectors. Consider the inner product

$$
\begin{aligned}
|(\underline{w}_1, \sum_{j=-m_1}^{m_2} a_j\underline{A}_j(x)\underline{w}_2)| &= |\sum_{j=-m_1}^{m_2} (\underline{w}_1, a_jA_j(x)\underline{w}_2)| \\
&= |\sum_{j=-m_1}^{m_2} a_j(\underline{w}_1, \underline{A}_j(x)\underline{w}_2)| \\
&\leq \sum_{j=-m_1}^{m_2} |a_j||(\underline{w}_1, \underline{A}_j(x)\underline{w}_2)| \\
&= \sum_{j=-m_1}^{m_2} |(\underline{w}_1, \underline{A}_j(x)w_2)| \, .
\end{aligned}
$$

Since the coefficient matrices $\underline{A}_j(x)$ are hermitian and positive definite, by the Schwartz inequality,

$$
|(\underline{w}_1, \underline{A}_j(x)\underline{w}_2)| \leq \tfrac{1}{2}(\underline{w}_1, \underline{A}_j(x)\underline{w}_1) + \tfrac{1}{2}(\underline{w}_2, \underline{A}_j(x)\underline{w}_2)
$$

and by substitution into the above inequality

$$|(\underline{w}_1, \sum_{j=-m_1}^{m_2} a_j\underline{A}_j(x)\underline{w}_2)| \le \frac{1}{2}\sum_{j=-m_1}^{m_2} (\underline{w}_1, \underline{A}_j(x)\underline{w}_1) + \frac{1}{2}\sum_{j=-m_1}^{m_2} (\underline{w}_2, \underline{A}_j(x)\underline{w}_2)$$

$$= \frac{1}{2}\sum_{j=-m_1}^{m_2} (\underline{w}_1, \underline{A}_j(x)\underline{w}_1) + \frac{1}{2}(\underline{w}_2, \sum_{j=-m_1}^{m_2} \underline{A}_j(x)\underline{w}_2) .$$

However, by consistency $\sum_{j=-m_1}^{m_2} \underline{A}_j(x) = \underline{I}$ (see Section III.3) so that

$$(3.11.6) \qquad |(\underline{w}_1, \sum_{j=-m_1}^{m_2} a_j\underline{A}_j(x)\underline{w}_2)| \le \frac{1}{2}\|\underline{w}_1\| + \frac{1}{2}\|\underline{w}_2\|_2^2$$

Setting $\underline{w}_1 = \sum_{j=-m_1}^{m_2} e^{-ij\xi}\underline{A}_j(x)\underline{w}_2$ and substituting into the left-hand side of (3.11.4),

$$\|\underline{w}_1\|_2^2 \le \frac{1}{2}\|\underline{w}_1\|_2^2 + \frac{1}{2}\|\underline{w}_2^2\|_2^2$$

or

$$\|\underline{w}_1\|_2 \le \|\underline{w}_2\|_2 .$$

By definition of $\underline{w}_1$, $\underline{w}_1 = \underline{G}(x,\xi)\underline{w}_2$, so that

$$\|\underline{G}(x,\xi)\underline{w}_2\|_2 \le \|\underline{w}_2\|_2$$

or

$$\frac{\|\underline{G}(x,\xi)\underline{w}_2\|_2}{\|\underline{w}_2\|_2} \le 1 .$$

Since this is true for any vector $\underline{w}_2$,

$$\|\underline{G}(x,\xi)\| = \max_{\|\underline{w}_2\|_2 \ne 0} \frac{\|\underline{G}(x,\xi)\underline{w}_2\|_2}{\|\underline{w}_2\|_2} \le 1$$

Consider the discrete inner-product of the finite difference method (3.11.3) with $\underline{u}^{n+1}$

$$|(\underline{u}^{n+1}, \underline{u}^{n+1})| = |\sum_{j=-m_1}^{m_2} (\underline{A}_j(x)S_+^j\underline{u}^n, \underline{u}^{n+1})| .$$

Again using the assumption that the coefficient matrices $\underline{A}_j(x)$ are hermitian and positive definite using the Schwartz inequality,

$$|(\underline{A}_j(x)S_+^j\underline{u}^n, \underline{u}^{n+1})| \le \frac{1}{2}(S_+^j\underline{u}^n, \underline{A}_j(x)S_+^j\underline{u}^n) + \frac{1}{2}(\underline{u}^{n+1}, A_j(x)\underline{u}^{n+1}) .$$

Substituting into the preceeding inequality,

$$(3.11.5) \quad \|\underline{u}^{n+1}\|_2^2 \le \frac{1}{2}\sum_{j=-m_1}^{m_2} (S_+^j\underline{u}, \underline{A}_j(x)S_+^j\underline{u}^n) + \frac{1}{2}\sum_{j=-m_1}^{m_2} (\underline{u}^{n+1}, \underline{A}_j(x)\underline{u}^{n+1}).$$

There is a difficulty in bounding $(S_+^j\underline{u}^n, \underline{A}_j(x)S_+^j\underline{u}^n)$ because the $\underline{A}_j(x)$'s are functions of x. If the $\underline{A}_j$'s were independent of x, then by discrete integration by parts,

$$\begin{aligned}
S_+^j\underline{u}^n, A_j S_+^j\underline{u}^n) &= (\underline{u}^n, S_-^j\ \underline{A}_j\ S_+^j\underline{u}^n) \\
&= (\underline{u}^n, \underline{A}_j\ \underline{S_-^j S_+^j}\ \underline{u}^n) \\
&= (\underline{u}^n, \underline{A}_j\underline{u}^n) \\
&= \underline{A}_j \|\underline{u}^n\|_2^2
\end{aligned}$$

so that

$$\sum_{j=-m_1}^{m_2} (S_+^j\ \underline{u}^n, \underline{A}_j\ S_+^j\ \underline{u}^n) = \sum_{j=-m_1}^{m_2} \underline{A}_j \|\underline{u}^n\|_2^2 = \|\underline{u}^n\|_2^2\ .$$

However, by discrete integration by parts

$$\begin{aligned}
(S_+^j\underline{u}^n, \underline{A}_j(x)S_+^j\underline{u}^n) &= (\underline{u}^n, S_-^j(\underline{A}_j(x)S_+^j\underline{u}^n)) \\
&= (\underline{u}^n, (S_-^j\ \underline{A}_j(x))S_-^j S_+^j\underline{u}^n) \\
&= (\underline{u}^n, (S_-^j\ \underline{A}_j(x))\underline{u}^n)\ .
\end{aligned}$$

Since the $\underline{A}_j(x)$'s are Lipschitz continuous in x,

$$S_-^j A_j(x) \equiv \underline{A}_j(x - jh) = \underline{A}_j(x) + O(h)$$

and hence

$$\begin{aligned}
(u^n, (S_-^j\ \underline{A}_j(x))\underline{u}^n) &= (\underline{u}^n, \underline{A}_j(x)\underline{u}^n) + (\underline{u}^n, O(h)\underline{u}^n) \\
&= (\underline{u}^n, \underline{A}_j(x)\underline{u}^n) + O(h)\|\underline{u}^n\|_2^2\ .
\end{aligned}$$

Summing over j,

$$\begin{aligned}
(3.11.7)\quad \sum_{j=-m_1}^{m_2} (\underline{u}^n, (S_-^j\ \underline{A}_j(x))\underline{u}^n) &= \sum_{j=-m_1}^{m_2} (\underline{u}^n, \underline{A}_j(x)\underline{u}^n) + O(h)\|\underline{u}^n\|_2^2 \\
&= (\underline{u}^n, \sum_{j=-m_1}^{m_2} \underline{A}_j(x)\underline{u}^n) + O(h)\|\underline{u}^n\|_2^2 \\
&= (1 + O(h))\|\underline{u}^n\|_2^2\ ,
\end{aligned}$$

because of the consistency condition, $\sum_{j=-m_1}^{m_2} \underline{A}_j(x) = \underline{I}$.

Similarly,

$$\sum_{j=-m_1}^{m_2} (\underline{u}^{n+1}, \underline{A}_j(x)\underline{u}^{n+1}) = \|\underline{u}^{n+1}\|_2^2 \sum_{j=-m_1}^{m_2} \underline{A}_j(x) = \|u^{n+1}\|_2^2.$$

By substitution into the inequality (3.11.7)

$$\|u^{n+1}\|_2^2 \le \tfrac{1}{2}(1 + O(h))\|\underline{u}^n\|_2^2 + \tfrac{1}{2}\|\underline{u}^{n+1}\|_2^2$$

or

$$\|\underline{u}^{n+1}\|_2^2 \le (1 + O(h))\|\underline{u}^n\|_2^2$$

from which stability follows, provided that $k = O(h)$.

We extend the idea of dissipation to the case of a finite difference method with variable coefficients.

Definition. A finite difference method (3.11.3) is dissipative of order 2s, where s is a positive integer, if there is a constant $\delta > 0$ so that

$$|\mu_j(x,\xi)| \le 1 - \delta|\xi|^{2s}$$

for each x and all ξ with $|\xi| \le \pi$, where the $\mu_j(x,\xi)$ are the eigenvalues of the amplification matrix $\underline{G}(x,\xi)$ associated with (3.11.3).

The importance of the idea of dissipation can be seen from the variable coefficient version of the main stability theorem in Section III.8, again due to Kreiss [19]. A proof of this theorem is given in Appendix B.

Theorem 3.6 (Kreiss). Suppose that the hyperbolic system (3.11.2) is approximated by the finite difference method (3.11.3), each having hermitian coefficient matrices that are Lipschitz continuous and uniformly bounded. If the finite difference method is accurate of order $2s-1$ and dissipative of order $2s$, then the method is stable.

It has yet to be shown that, in general, the condition of accuracy to order $2s-1$ can be overcome, as in the case of constant coefficients. Parlett [29], however, has proved under slightly stronger conditions a theorem that will now be given. This theorem treats the special case in which the hyperbolic system (3.11.2) is regular hyperbolic, that is, the matrix $\underline{A}(x)$ has distinct eigenvalues.

Theorem 3.7. Suppose that the regular hyperbolic system (3.11.2) is approximated by the finite difference method (3.11.3), each having hermitian coefficient matrices that are Lipschitz continuous and uniformly bounded. If the finite difference method (3.11.3) is accurate to order $2s-2$ and dissipative to order $2s$, then the method is stable.

Theorems 3.6 and 3.7 remain valid for systems of hyperbolic equations with variable coefficients in several space dimensions (see Appendix B).

Kreiss's theorem, as well as the theorem due to Parlett, requires a great deal of information about the structure of the finite difference method. The main theorem of this section, the Lax-

Nirenberg theorem, requires little information about the structure of the finite difference method.

As an introduction to the Lax-Nirenberg theorem, a weaker theorem, due to Lax [22], for a single scalar equation shall be presented. Proofs of both theorems are given in Appendix C. Consider the hyperbolic equation (3.11.2) where $p = 1$, which may be written in the form

(3.11.8) $$\partial_t v = a(x)\partial_x v$$

approximated by

(3.11.9) $$u_i^{n+1} = \sum_{j=-m_1}^{m_2} a_j(x) S_+^j u_i^n .$$

Let the symbol of the difference operator (3.11.8) be given by

$$\rho(x,\xi) = \sum_{j=-m_1}^{m_2} a_j(x) e^{-ij\xi}$$

Theorem 3.8 (Lax). The finite difference method (3.11.9) which approximates the scalar hyperbolic equation (3.11.8) is stable provided the symbol $\rho(x,\xi)$ given by (3.11.10) is Lipschitz continuous in x and

(i) $|\rho(x,\xi)| < 1$ if $\xi \neq 0$,

(ii) $\rho(x,\xi) = 1 - Q(x)\xi^{2p} + O(\xi^{2p+1})$, where ξ is near 0 and $Q(x) > 0$ for every x.

Observe by consistency (see Section III.3),

$$\rho(x,0) = \sum_{j=-m_1}^{m_2} a_j(x) = 1 .$$

Since $|\rho(x,\xi)| < 1$ for $\xi \neq 0$, condition (i) implies that there is some dissipation. Condition (ii) indicates how much dissipation is present in the finite difference method (3.11.8).

We now state the main result, the Lax-Nirenberg theorem [26].

Theorem 3.9 (Lax-Nirenberg). Consider the hyperbolic system (3.11.2) approximated by the finite difference method (3.11.3). Suppose that the coefficient matrices $\underline{A}_j(x)$ are real and symmetric, independent of t, and have bounded second derivatives. If the amplification matrix $\underline{G}(x,\xi)$ of the finite difference operator (3.11.3) satisfies

$$\|\underline{G}(x,\xi)\| \leq 1$$

for every x and ξ, then the finite difference method (3.11.3) for real vector grid functions $\underline{u}^n$ is stable.

In the Lax-Nirenberg theorem, the coefficient matrices $\underline{A}_j(x)$ are symmetric, so that the amplification matrix $\underline{G}(x,\xi)$ associated with the finite difference method (3.11.3) is symmetric. As a result, the matrix norm of $\underline{G}(x,\xi)$ can be readily evaluated,

$$\|\underline{G}(x,\xi)\| = \sigma(\underline{G}(x,\xi)) .$$

The Lax-Nirenberg theorem requires less information about the structure of the finite difference method than Kreiss's theorem. However, the Lax-Nirenberg theorem requires a stronger smoothness condition to be satisfied by the coefficient matrices. The coefficient matrices are required to have bounded second derivatives. Kreiss's theorem requires that the coefficient matrices be Lipschitz continuous. This condition is satisfied if the coefficient matrices have bounded first derivatives.

The Lax-Nirenberg theorem remains valid for systems of hyperbolic equations with variable coefficients in several space variables (see Appendix C).

Consider the Lax-Friedrichs method (3.8.7) (or (3.2.19)) applied to the system (3.11.2) with variable coefficients

$$\text{(3.11.11)} \qquad \underline{u}_i^{n+1} = \tfrac{1}{2}(\underline{u}_{i+1}^n + \underline{u}_{i-1}^n) + \underline{A}(ih)kD_0\underline{u}_i^n = \sum_{j=-1}^{1} \underline{A}_j(ih)S_+^j\underline{u}_i^n ,$$

where $\underline{A}_{-1}(x) = \frac{1}{2}(\underline{I} - \lambda\underline{A}(x))$, $\underline{A}_0(x) = 0$, and $\underline{A}_1(x) = \frac{1}{2}(\underline{I} + \lambda\underline{A}(x))$. In (3.11.11), $m_1 = m_2 = 1$ so that the CFL condition (3.11.4a) and (3.11.4b) may be combined to yield

$$|\mu_j(x)|\lambda \le 1 \quad \text{for} \quad j = 1,\ldots,p$$

In fact, for any finite difference method (3.11.3) where $m_1 = m_2 = 1$, that is, $\underline{u}_i^{n+1}$ depends on $\underline{u}_{i-1}^n$, $\underline{u}_i^n$ and $\underline{u}_{i+1}^n$, the CFL condition (3.11.4a) and (3.11.4b) reduces to (3.11.12), which may be rewritten as

$$\max_{\substack{1\le j\le p \\ x}} |\mu_j(x)|\lambda \le 1$$

If $\underline{A}(x)$ in (3.11.2) is Lipschitz continuous in x, then by the definition of the three coefficient matrices, the $\underline{A}_j(x)$, $j = -1,0,1$ in (3.11.11) are Lipschitz continuous in x. The Taylor series expansion, as described in Section III.2, can be used for the variable coefficient case where c is replaced with $\underline{A}(ih)$, the Taylor series expansion of $A(x)$ about the point ih evaluated at ih. With this the variable coefficient Lax-Friedrichs method (3.11.11) is accurate to $O(k) + O(h)$.

The amplification matrix $\underline{G}(x,\xi)$ associated with (3.11.11) is

$$\underline{G}(x,\xi) = \sum_{j=-1}^{1} \underline{A}_j(x)e^{-ij\xi} = \underline{I}\cos\xi - i\underline{A}(x)\lambda\sin\xi .$$

Following the same procedure of Section III.8, where $\underline{A}(x)$ is hermitian, the eigenvalues of $\underline{G}(x,\xi)$ are $\varepsilon_j(x,\xi)$, $j = 1,\ldots,p$ given by

$$\varepsilon_j(x,\xi) = \cos\xi - i\mu_j(x)\lambda\sin\xi$$

and following the same procedure of Section III.6, we may write for small ξ

$$|\varepsilon_j(x,\xi)| = 1 - \frac{1}{2}(1 - \mu_j^2(x)\lambda^2)\xi^2 .$$

Thus, the Lax-Friedrichs method is dissipative of order 2. As such, by Kreiss' theorem, the Lax-Friedrichs method (3.11.11) is stable.

Next consider the Lax-Wendroff method (3.8.10) (or (3.5.3)) applied to the system (3.11.2) with variable coefficients. An alternate, yet equivalent, way to derive the Lax-Wendroff method is to consider the Taylor series expansion in time of $\underline{v}(x,t)$ about the point (ih,nk),

$$\underline{v}_i^{n+1} \equiv \underline{v}(ih,(n+1)k) = \underline{v}_i^n + k(\partial_t\underline{v})_i^n + \frac{k^2}{2}(\partial_t^2\underline{v})_i^n + O(k^3).$$

Using equation (3.11.2), $\partial_t\underline{v} = \underline{A}(x)\partial_x\underline{v}$ and

$$\partial_t^2\underline{v} = \partial_t(\underline{A}(x)\partial_x\underline{v}) = \underline{A}(x)\partial_x(\partial_t\underline{v}) = \underline{A}(x)\partial_x(\underline{A}(x)\partial_x\underline{v}).$$

Substituting into the Taylor series

$$(3.11.14) \qquad \underline{v}_i^{n+1} = v_i^n + kA(ih)(\partial_x\underline{v})_i^n + \frac{k^2}{2}\underline{A}(ih)(\partial_x(\underline{A}(x)\partial_x\underline{v}))_i^n + O(k^3).$$

Approximate the two space derivatives in (3.11.14) by second order differences (see Section II.11)

$$D_0\underline{v}_i^n = (\partial_x\underline{v})_i^n + O(h^2)$$

and

$$D_+(\underline{A}_{i-\frac{1}{2}}D_-\underline{v}_i^n) = (\partial_x(\underline{A}(x)\partial_x\underline{v}))_i^n + O(h^2),$$

where $\underline{A}_{i-\frac{1}{2}} = \underline{A}(i+\frac{1}{2})h)$. Substituting into (3.11.14) gives the Lax-Wendroff method

$$(3.11.15) \qquad \underline{u}_i^{n+1} = \underline{u}_i^n + k\underline{A}_iD_0\underline{u}_i^n + \frac{k^2}{2}\underline{A}_iD_+(\underline{A}_{i-\frac{1}{2}}D_-\underline{u}_i^n),$$

which is accurate of $O(k^2) + O(h^2)$. This may be written in the form

$$\underline{u}_i^{n+1} = \sum_{j=-1}^{1}\underline{A}_j(ih)S_+^j\,\underline{u}_i^n$$

where $\underline{A}_{-1}(x) = \frac{\lambda^2}{2}\underline{A}(x)\underline{A}(x-\frac{h}{2}) - \frac{\lambda}{2}\underline{A}(x)$, $\underline{A}_0(x) = 1 - \frac{\lambda^2}{2}\underline{A}(x)(\underline{A}(x+\frac{h}{2}) + \underline{A}(x-\frac{h}{2}))$, and $\underline{A}_1(x) = \frac{\lambda^2}{2}\underline{A}(x)\underline{A}(x+\frac{h}{2}) + \frac{\lambda}{2}\underline{A}(x)$. Since $m_1 = m_2 = 1$, the CFL condition is given by (3.11.12).

If $\underline{A}(x)$ in (3.11.2) is Lipschitz continuous in x and uniformly bounded, then $\underline{A}(x)\underline{A}(x \pm \frac{h}{2})$ is Lipschitz continuous in x and uniformly bounded, and hence the three coefficient matrices $\underline{A}_j(x)$, $j = -1,0,1$ are Lipschitz continuous in x and uniformly bounded.

To complete the stability analysis, the terms $\underline{A}(x \pm \frac{h}{2})$ must be replaced. Expanding $\underline{A}(x)$ in a Taylor series about the point x

$$\underline{A}(x \pm \tfrac{h}{2}) = \underline{A}(x) \pm \tfrac{h}{2}\underline{A}'(x^{\pm}),$$

where $x < x^+ < x + \frac{h}{2}$ and $x - \frac{h}{2} < x^- < x$. Substituting into (3.11.15),

$$(3.11.16) \quad u_i^{n+1} = [\underline{u}_i^n + k\underline{A}_i D_0 \underline{u}_i^n + \tfrac{k^2}{2}\underline{A}_i^2 D_+ D_- \underline{u}_i^n]$$

$$+ [\tfrac{\lambda^2 h}{4}\underline{A}_i\underline{A}'(x^+)(\underline{u}_{i+1}^n - \underline{u}_i^n) + \tfrac{\lambda^2 h}{4}\underline{A}_i\underline{A}'(x)(\underline{u}_i^n - \underline{u}_{i-1}^n).$$

The first term in the bracket on the right hand side of (3.11.16) may be written in the form

$$(3.11.17) \qquad \underline{u}_i^{n+1} = \sum_{j=-1}^{1} \tilde{\underline{A}}_j(ih) S_+^j \underline{u}_i^n,$$

where $\tilde{\underline{A}}_{-1}(x) = \frac{\lambda^2}{2}\underline{A}^2(x) - \frac{\lambda}{2}\underline{A}(x)$, $\tilde{A}_0(x) = 1 - \lambda^2\underline{A}^2(x)$, and $\tilde{\underline{A}}_1(x) = \frac{\lambda^2}{2}\underline{A}^2(x) + \frac{\lambda}{2}\underline{A}(x)$. The amplification matrix associated with (3.11.17) is (cf. Section III.8)

$$\underline{G}(x,\xi) = \underline{I} - 2\lambda^2\underline{A}^2(x)\sin(\tfrac{\xi}{2}) - i\underline{A}(x)\lambda\sin(\xi)$$

If $\underline{A}(x)$ is hermitian, then the coefficient matrices $\tilde{\underline{A}}_j(x)$ are hermitian, and hence $\underline{G}(x,\xi)$ is hermitian. Following the same procedure of Section III.8, the eigenvalues of $\underline{G}(x,\xi)$, $\varepsilon_j(x,\xi)$, $j = 1,\ldots,p$ are

$$\varepsilon_j(x,\xi) = 1 - 2\lambda^2\mu_j^2(x)\sin^2(\tfrac{\xi}{2}) - i\mu_j(x)\lambda\sin\xi .$$

Again, following the same procedure of Section III.6, for small ξ

$$|\varepsilon_j(x,\xi)| = 1 - \tfrac{1}{8}\mu_j^2(x)\lambda^2(1 - \mu_j^2(x)\lambda^2)\xi^4 .$$

Thus, the finite difference method is dissipative of order 4. As such Parlett's theorem applies, provided that $\underline{A}(x)$ has distinct eigenvalues, and the method (3.11.17) is stable.

If $\underline{A}(x)$ does not have distinct eigenvalues but is real and symmetric with bounded second derivatives, the Lax-Nirenberg theorem may be used. Since $\underline{G}(x,\xi)$ is symmetric

$$\|\underline{G}(x,\xi)\| = \sigma(\underline{G}(x,\xi))$$

$$= \max_{1<j<p} |\varepsilon_j(x,\xi)| < 1$$

provided that the CFL condition is satisfied. Hence, the finite difference method (3.11.17) is stable.

The second term in brackets on the right hand side of (3.11.16) may be written in the form $k\underline{B}$, where

$$\underline{B} = \tfrac{\lambda}{4}\underline{A}_i\underline{A}'(x^+)(\underline{u}_{i+1}^n - u_i^n) + \tfrac{\lambda}{4}\underline{A}_i\underline{A}'(x^-)(\underline{u}_i^n - u_{i-1}^n),$$

which is uniformly bounded. By the Strang perturbation theorem (Theorem 2.4) the Lax-Wendroff method (3.11.15), written as,

$$\underline{u}_i^{n+1} = \sum_{J=-1}^{1} \underline{A}_j(ih)S_+^j\underline{u}_i^n + k\underline{B}$$

is stable.

Consider the scalar hyperbolic equation

$$\partial_t v = a(x)\partial_x v \ . \qquad (3.11.18)$$

A method, due to Courant, Isaacson, and Rees [8] (see also Lelevier [27], called the upwind difference method, will be described. In Section III.2, it was shown that

$$u_i^{n+1} = u_i^n + kcD_+u_i^n \ .$$

approximates the scalar equation with constant coefficients

$$\partial_t v = c\partial_x v \qquad (3.11.19)$$

where c is a positive constant. This method is stable provided the CFL condition is satisfied, $c\lambda \leq 1$.

Also, the finite difference method

$$u_i^{n+1} = u_i^n + kcD_-u_i^n$$

approximates (3.11.19) where c is a negative constant. This method is stable provided the CFL condition is satisfied, $-c\lambda \leq 1$.

These two methods can be combined in the case of the variable coefficient equation (3.11.18) to yield

$$u^{n+1} = u_i^n + \begin{cases} ka(ih)D_+u_i^n \ , & \text{if } a(ih) > 0 \ , \\ ka(ih)D_-u_i^n \ , & \text{if } a(ih) < 0 \ . \end{cases} \qquad (3.11.20)$$

Since each method is accurate to $O(k) + O(h)$, the combined method is accurate to $O(k) + O(h)$. The symbol of the upwind difference method, using the symbols of the two individual methods in Section III.2, is

$$\rho(x,\xi) = \begin{cases} 1 - 2a(x)\lambda\sin^2(\tfrac{\xi}{2}) - ia(x)\lambda\sin\xi \ , & a(x) > 0 \\ 1 + 2a(x)\lambda\sin^2(\tfrac{\xi}{2}) - ia(x)\lambda\sin\xi \ , & a(x) < 0 \end{cases}$$

or

$$\rho(x,\xi) = 1 - 2|a(x)|\lambda\sin(\tfrac{\xi}{2}) - i|a(x)|\lambda\sin\xi \ .$$

Following the procedure of Section III.6, for $|\xi| \leq \pi$,

$$|\rho(x,\xi)| = 1 - 4|a(x)|\lambda(1 - |a(x)|\lambda)\sin^2(\tfrac{\xi}{2})$$

and for small ξ,

$$|\rho(x,\xi)| = 1 - \frac{1}{2}|a(x)|\lambda(1 - |a(x)|\lambda)\xi^2.$$

Thus the upwind method is dissipative of order 2. As such, by Kreiss's theorem, the upwind difference method (3.11.20) is stable.

We close this section with the most general version of the theorem of Lax [5]. This is an extension of Theorem 3.1 in Section III.3 and Theorem 3.4 in Section III.9.

Theorem 3.10. Consider the hyperbolic system of equations (3.11.2) approximated by a finite difference method of the form (3.11.1). Suppose the finite difference method is accurate of order (p_1, p_2) and $k = O(h)$. Then

$$\underline{G}(x,\xi) = e^{-k\xi\underline{A}(x)/h} + O(|\xi|^{p+1}),$$

where $p = \min(p_1, p_2)$.

III.12. Random Choice Method

The random choice method (RCM) was introduced by Glimm [12] for the construction of solutions of systems of nonlinear hyperbolic conservation laws (see Chapter IV). This construction was the basis for the existence theorem in the large, with restrictions on the type of systems allowed and on the size and variation of the initial data. The random choice method was developed for hydrodynamics by Chorin [2-3] and further developed by Albright et al [1], Colella [4-5], Concus et al [6], Glaz and Liu [11-12], Glimm et al [13-17] and Sod [31-42].

Our discussion of the random choice method requires a few basic ideas from probability theory. The set of all possible outcomes of an experiment is called a sample space. A random variable is a mapping from a sample space to the real numbers. As an example, consider the tossing of a coin. The sample space consists of two elements, the two outcomes of the experiment, heads and tails. One random variable assigns the outcome "heads" the number one and the outcome "tails" the number zero. If $f(x)$ is a function and ξ is a random variable, then $f(\xi)$ is a random variable. If ξ_1 and ξ_2 are two random variables, then $\xi_1+\xi_2$ is a random variable.

Let $\bar{x}$ denote an element of the range of ξ, that is, $\bar{x}$ is one of the values that ξ assumes. All points in the sample space in which ξ assumes the value $\bar{x}$ forms the event that $\xi = \bar{x}$ with its probability denoted by $P\{\xi = \bar{x}\}$. For all x in the range, the function

$$f(x) = P\{\xi = x\} \tag{3.12.1}$$

is called the probability distribution function (of the random variable ξ).

Let ξ_1 and ξ_2 denote two random variables defined on the same sample space. Let x and y denote elements in the ranges of ξ_1 and ξ_2, respectively. All points in the sample space in which ξ_1 assumes the value x and ξ_2 assumes the value y forms the event that $\xi_1 = x$ and $\xi_2 = y$ with probability $P\{\xi_1 = x, \xi_2 = y\}$. For all x in the range of ξ_1 and y in the range of ξ_2, the function

$$h(x,y) = P\{\xi_1 = x, \xi_2 = y\}$$

is called the joint probability distribution function (of the random variables ξ_1 and ξ_2). Let f(x) and g(y) denote the probability distribution functions of ξ_1 and ξ_2, respectively. If for all x in the range of ξ_1 and for all y in the range of ξ_2, $h(x,y) = f(x)g(y)$, then the random variables ξ_1 and ξ_2 are called independent. This is saying that two random variables ξ_1 and ξ_2 are independent if the outcome of the experiment that determines ξ_1 has no effect on the experiment that determines ξ_2, and vice versa.

Consider a random variable ξ with probability distribution function f(x). If the range of ξ is a countable set $\{x_j\}$, the mean or expected value of ξ, denoted by $E[\xi]$, is

(3.12.2) $$E[\xi] = \sum_j x_j f(x_j).$$

More generally,

(3.12.3) $$E[\xi] = \int_{-\infty}^{+\infty} x\, f(x)dx.$$

If ξ_1 and ξ_2 are two random variables, then $E[\xi_1+\xi_2] = E[\xi_1] + E[\xi_2]$, which can be seen directly from the definition of expected value and the linearity of the integral. If, in addition, ξ_1 and ξ_2 are independent random variables, then $E[\xi_1\xi_2] = E[\xi_1]E[\xi_2]$.

The second moment of a random variable ξ is $E[\xi^2]$ which for a countable range is

$$E[\xi^2] = \Sigma\, x_j^2 f(x_j)$$

and more generally,

$$E[\xi^2] = \int_{-\infty}^{+\infty} x^2 f(x)dx.$$

The deviation of a random variable from its mean, $\xi - E[\xi]$, is a random variable. The second moment of $\xi - E[\xi]$ is the variance of ξ, denoted by $\mathrm{Var}(\xi)$,

(3.12.4) $$\begin{aligned}\mathrm{Var}(\xi) &= E[(\xi - E[\xi])^2] \\ &= E[\xi^2 - 2\xi E[\xi] + E[\xi]^2] \\ &= E[\xi^2] - E[\xi]^2.\end{aligned}$$

If ξ_1 and ξ_2 are independent random variables, then $\mathrm{Var}(\xi_1 + \xi_2) = \mathrm{Var}(\xi_1) + \mathrm{Var}(\xi_2)$. Since the mean of a set of nonnegative numbers is always nonnegative, the variance is positive. With this define the

<u>standard deviation</u> of a random variable ξ as

(3.12.5) $$\sigma(\xi) = (\text{Var}(\xi))^{1/2}$$

For further details see Feller [9-10].

Two versions of the random choice method will be presented, the first version is a two-step method using a staggard grid and the second version is a two-step method using a single grid. Consider the prototype equation (3.2.1)

(3.2.1) $$\partial_t v = c\partial_x v \quad -\infty < x < +\infty, \quad t > 0$$

with initial condition (3.2.2)

(3.2.2) $$v(x,0) = f(x), \quad -\infty < x < +\infty.$$

For the two step version of the random choice method, let u_i^n and $u_{i+\frac{1}{2}}^{n+\frac{1}{2}}$ denote approximations to $v(ih,nk)$ and $v((i+\frac{1}{2})h, (n+\frac{1}{2})k)$, respectively. Rather than considering discrete points in space, assume that the solution is constant on intervals of length h, that is, piecewise constant. The goal is to find the solution $u_{i+\frac{1}{2}}^{n+\frac{1}{2}}$ given u_i^n and u_{i+1}^n; and to find the solution u_i^{n+1} given $u_{i-\frac{1}{2}}^{n+\frac{1}{2}}$ and $u_{i+\frac{1}{2}}^{n+\frac{1}{2}}$.

To find the solution $u_{i+\frac{1}{2}}^{n+\frac{1}{2}}$, consider equation (3.2.1) along with the piecewise constant initial condition

(3.12.6) $$v(x,nk) = u_i^n, \ (i - \tfrac{1}{2})h < x < (i + \tfrac{1}{2})h.$$

Such a problem is called a <u>Riemann problem</u>. On the interval $ih < x < (i+i)h$ (3.12.6) represents a step function (a discontinuity if $u_i^n \neq u_{i+1}^n$).

$$v(x,nk) = \begin{cases} u_i^n, & x < (i + \frac{1}{2})h \\ u_{i+1}^n, & x > (i + \frac{1}{2})h. \end{cases}$$

The value u_i^n is called the left state and the right value u_{i+1}^n is the right state of the Riemann problem. This discontinuity, considered as a wave, is propagated with speed C. If the CFL condition

$$|c|k/h \leq 1$$

is satisfied, the discontinuity will not propagate beyond the interval $[ih,(i+1)h]$, for each integer i. The basis of the random choice method is that it is easier to solve a sequence of Riemann problems (3.2.1) and (3.12.7) on the interval $[ih,(i+1)h]$ for each i than it is to solve (3.2.1) with the more general initial data (Cauchy data).

If the CFL condition is satisfied, the waves (discontinuities) generated by the individual Riemann problems, one for each i, will not interact. Hence, the solution can be combined by superposition into a single exact solution $v^e(x,t)$ for $nk < t < (n + \frac{1}{2})k$ (see Figure 3.21).

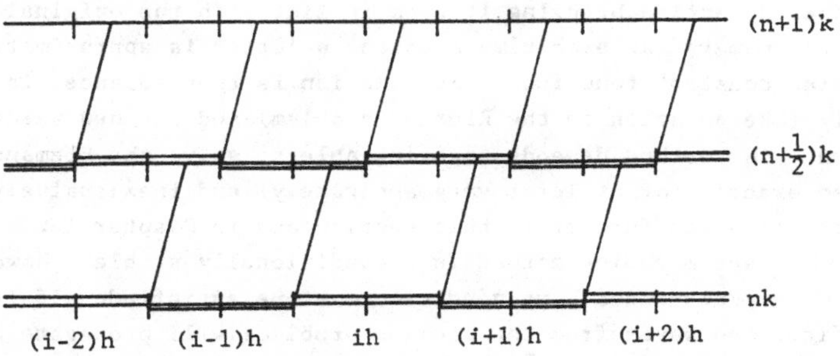

Figure 3.21. Sequence of Riemann Problems for Two-Step Version of Random Choice Method

Let ξ_n denote a uniformly distributed random variable in the interval $(-\frac{1}{2}, \frac{1}{2})$, that is, ξ_n has probability density function which takes the value 1 in $(-\frac{1}{2}, \frac{1}{2})$ and 0 elsewhere. Define

$$u_{i+\frac{1}{2}}^{n+\frac{1}{2}} = v^e((i + \frac{1}{2} + \xi_n)h, k/2)$$

(see Figure 3.22).

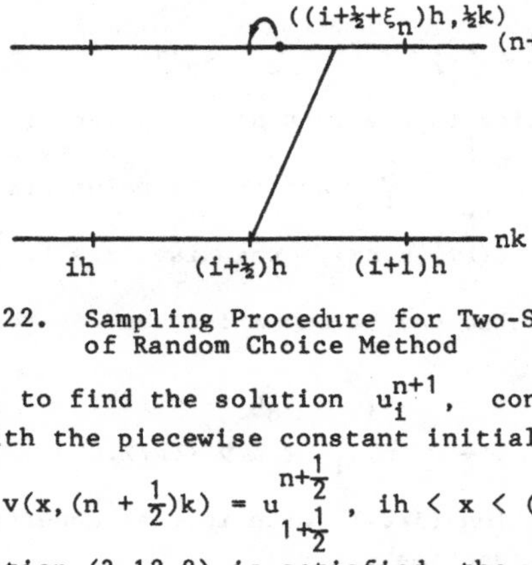

Figure 3.22. Sampling Procedure for Two-Step Version of Random Choice Method

Similarly, to find the solution u_i^{n+1}, consider equation (3.2.1) along with the piecewise constant initial condition

$$v(x,(n + \frac{1}{2})k) = u_{1+\frac{1}{2}}^{n+\frac{1}{2}}, \quad ih < x < (i+1)h.$$

If the CFL condition (3.12.8) is satisfied, the waves (discontinuities) generated by the individual Riemann problems, one for each i, will not interact. Hence, the solution can be combined by superposition into a single exact solution $v^e(x,t)$ for $(n+\frac{1}{2})k < t < (n+1)k$. Let $\xi_{n+\frac{1}{2}}$ denote a uniformly distributed random variable in the interval $(-\frac{1}{2}, \frac{1}{2})$. Define

$$u_i^{n+1} = v^e((i+\xi_{n+\frac{1}{2}})h, \frac{k}{2}).$$

The grid at the end of one half step is offset by h/2. However, at the end of the next half step, the grid will be offset by h/2 in the other direction bringing it back in line with the original grid.

In summary, at each time step the solution is approximated by a piecewise constant function. The solution is then advanced in time exactly (the solution to the Riemann problem) and the new values are sampled. The method depends on being able to solve the Riemann problem exactly (or at least very accurately) and inexpensively. This will be discussed further in this section and in Chapter IV.

The random choice method is unconditionally stable. However, the Courant-Friedrichs-Lewy condition must be satisfied. If it is not satisfied, the waves from the Riemann problems will propagate beyond the sampling interval $(-\frac{h}{2}, \frac{h}{2})$, and result in incorrect sampling probabilities. This results in changing the Riemann problem being solved.

Consider the Riemann problem (3.2.1) and (3.12.6) on the interval $[ih,(i+1)h]$. The solution $v(x,t)$ is constant along the characteristics

$$x + ct = \text{const.}$$

By following the characteristic passing through the point $((i + \frac{1}{2} + \xi_n)h, (n + \frac{1}{2})k)$ to the point of intersection with the line $t = nk$, denoted by $\bar{x},nk)$, we obtain

$$x = \frac{ck}{2} + (i + \frac{1}{2} + \xi_n)h.$$

If the point $(\bar{x},nk)$ lies to the left of the point $((i+\frac{1}{2})h,nk)$, that is, $\bar{x} < (i+\frac{1}{2})$, then $u_{i+\frac{1}{2}}^{n+\frac{1}{2}} = u_i^n$, and if the point $(\bar{x},nk)$ lies to the right of the point $((i+\frac{1}{2})h,nk)$, that is, $\bar{x} > (i+\frac{1}{2})h$ then $u_{1+\frac{1}{2}}^{n+\frac{1}{2}} = u_{i+1}^n$. These results may be summarized as

(3.12.10)
$$u_{i+\frac{1}{2}}^{n+\frac{1}{2}} = \begin{cases} u_i^n & \xi_n h < -ck/2 \\ u_{i+1}^n & \xi_n h > -ck/2. \end{cases}$$

Consider the equation (3.2.1) with initial condition (3.2.2) given by a single step function

(3.12.11)
$$f(x) = \begin{cases} 0, & x < 0 \\ 1, & x > 0, \end{cases}$$

the solution is

$$v(x,t) = f(x+ct) = \begin{cases} 0, & x < -ct \\ 1, & x > -ct. \end{cases}$$

In solving this initial-value problem, which happens to be a Riemann problem, using the random choice method there are only three possible combinations of values of the left and right states. First, both u_i^n and u_{i+1}^n are 0 and rather than a step function, the initial condition is a single constant state with value 0. In this case, the sampled value for $u_{1+\frac{1}{2}}^{n+\frac{1}{2}}$ can only be 0. Second, both u_i^n and u_{i+1}^n are 1 and rather than a step function the initial condition is a single constant state with value 1. In this case, the sampled value for $u_{1+\frac{1}{2}}^{n+\frac{1}{2}}$ can only be 1. Finally, $u_i^n = 0$ and $u_{i+1}^n = 1$ which is a discontinuity. If the sampled value satisfies $\xi_n h < -ck/2$, then $u_{i+\frac{1}{2}}^{n+\frac{1}{2}} = 0$. In this case, the initial discontinuity (3.12.11) move a distance of h/2 to the right. If the sampled value satisfies $\xi_n h > -ck/2$, then $u_{i+\frac{1}{2}}^{n+\frac{1}{2}} = 1$. In this case, the initial discontinuity (3.12.11) moves a distance of h/2 to the left. Thus the discontinuity remains perfectly sharp. The random choice method has no numerical diffusion. The position of the discontinuity obtained by the random choice method fluctuates about the true position of the discontinuity (see Figure 3.23).

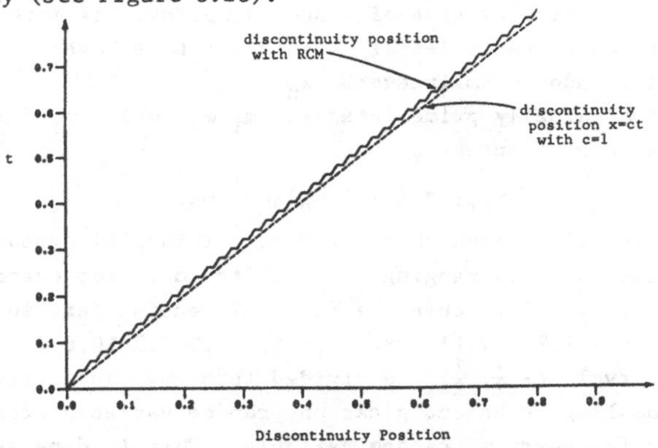

Figure 3.23. Discontinuity Position with
c > 0 for Random Choice Method

It should be noted that some randomness (or some means of choosing an equipartition of the interval $(-\frac{1}{2}, \frac{1}{2})$) is needed. Suppose the more natural choice of obtaining $u_{i+\frac{1}{2}}^{n+\frac{1}{2}}$ is made, that is, the solution to the Riemann problem is sampled at the midpoint of the interval $\xi_n = 0$ and

$$u_{i+\frac{1}{2}}^{n+\frac{1}{2}} = v^e((i + \tfrac{1}{2})h, \tfrac{k}{2}).$$

If the initial-value problem (3.2.1) and (3.12.11) is considered with $c < 0$, then the value $(i + \frac{1}{2})h$ will always be to the left of the discontinuity and $u_{i+\frac{1}{2}}^{n+\frac{1}{2}} = u_i^n$. The initial discontinuity (3.12.11) will move to the right with speed h/k independent of the true speed c. A natural question arises: How much randomness is needed?

The choice of the sequence of random variables ξ_n determines the behavior of the solution. Consider the Riemann problem (3.2.1) and (3.12.6) on the interval $[ih,(i+1)h]$. Suppose ξ_n is close to $-\frac{1}{2}$ then the value of the left state u_i^n will propagate to the right a distance of $\frac{h}{2}$. If ξ_n is chosen close to $\frac{1}{2}$, then the value of the right state u_{i+1}^n will propagate to the left a distance of $\frac{h}{2}$. Suppose a new random variable is chosen for each i (in space) and n (in time), there is a finite probability that a given state will propagate in both directions and create a spurious constant state. To this end only one random variable is chosen per time step (hence the subscript n in ξ_n). This sequence can be approximated by means of a pseudorandom number generator (see Hammersley and Handscomb [18]).

An improvement introduced by Chorin [3] is a method by which the sequence of random variables ξ_n reaches an equidistribution over $(-\frac{1}{2}, \frac{1}{2})$ at a faster rate. This method, known as stratified random sampling, is a particular type of random sampling. The method combines a sequence of random variables ξ_n equidistributed over $(-\frac{1}{2}, \frac{1}{2})$ and a sequence of pseudorandom integers k_n.

Given two mutually prime integers m_1, m_2 with $m_1 < m_2$, consider the sequence of integers

$$k_{n+1} = (m_1 + k_n)(\text{mod } m_2) \tag{3.12.12}$$

when k_0 is specified such that $k_0 < m_2$. This will produce a series of pseudorandom integers ranging from 0 to m_2. For example, if $k_0 = 2$, $m_1 = 7$, $m_2 = 11$, then the first eleven integers in the sequence k_i, $i = 1,2,\ldots,11$ are 9,5,1,8,4,0,7,3,10,6,2.

The interval $(-\frac{1}{2}, \frac{1}{2})$ is divided into m_2 subintervals. The stratified sampling technique picks one random variable from each subinterval once for every m_2 random variable. This is done as follows. Let ξ_n' denote a sequence of random variables equidistributed over the interval $(-\frac{1}{2}, \frac{1}{2})$, then consider the stratified sequence

$$\xi_n = \frac{k_n + \xi_n' + \frac{1}{2}}{m_2} - \frac{1}{2}. \tag{3.12.13}$$

A recent result of Liu [28] has shown that it is not necessary to use random numbers, all that is required is that the sequence approach equidistribution. The first nonrandom sampling technique known as the Richtmeyer-Ostrowski sequence was considered in Lax [23].

One nonrandom type sampling technique has been studied by Colella [45]. This quasirandom sampling known as van der Corput sampling. Consider the binary representation of the integer n,

$$n = \sum_{k=0}^{m} i_k 2^k, \tag{3.12.14}$$

where $i_k = 0,1$. Then the van der Corput sequence is given by

$$\xi_n = \sum_{k=0}^{m} i_k 2^{-(k+1)}, \tag{3.12.15}$$

where the values of m and i_k for $k = 0,\ldots,m$ is determined by the binary representation of n.

For two-dimensional problems two independent sequences are used. To this end the van der Corput sampling must be modified. Let k_1, k_2 be two positive, mutually prime integers with $k_1 > k_2$. Consider the base k_1 expansion of n

$$n = \sum_{\ell=0}^{m} i_\ell k_1^\ell \tag{3.12.15}$$

where $i_\ell = 0,1,\ldots,k_1 - 1$. Define the (k_1,k_2) van der Corput sequence ξ_n by

$$\xi_n = \sum_{\ell=0}^{m} q_\ell k_1^{-(\ell+1}, \tag{3.12.16}$$

where $q_\ell = k_2 i_\ell (\text{mod } k_1)$. The binary van der Corput sampling discussed above is a special case when $k_1 = 2$ and $k_2 = 1$.

Partition the interval $(-\frac{1}{2}, \frac{1}{2})$ into k_1^{m+1} subintervals $q_j = [jk_1^{-m-1}, (j+1)k_1^{-m-1}]$ for $j = 0,1,\ldots,k_1^{m+1}$. The (k_1,k_2) van der Corput sequence $\{\xi_n \mid n = 0,1,\ldots,k_1^{m+1}\}$ is such that for every subinterval g_j, there is one and only one element n for which $\xi_n \in q_j$.

An interesting phenomenon occurs as a result of the random choice method being a two-step method and the sampling in the interval $(-\frac{1}{2}, \frac{1}{2})$. It is possible for a stationary discontinuity (or a solid boundary) to move at the end of a full time step. For example, consider the equation $\partial_t v = 0$ with discontinuous initial data. The solution consists of the discontinuous initial data for all time. However, should the random variable forthe first half step be chosen in the interval $(-\frac{1}{2},0)$ and, again, for the second half step in the interval $(-\frac{1}{2},0)$, the discontinuity (and solution) will be translated over by one grid point.

Consider the Riemann problem given by (3.2.1) and (3.12.6) on the interval $[ih,(i+1)h]$. The solution is constant along the characteristic $x + ct = \text{const}$. The initial discontinuity propagates along the characteristic passing through the point $((i+\frac{1}{2})h,nk)$. Choose a uniformly distributed random variable ξ_n in the interval $(-\frac{1}{2},\frac{1}{2})$. Define the point $Q = ((i+\frac{1}{2}+\xi_n)h,(n+1)k)$. If the point Q lies to the left of the characteristic, then $u_{i+\frac{1}{2}}^{n+\frac{1}{2}} = u_i^n$, and if the point Q lies to the right of the characteristic, then $u_{i+\frac{1}{2}}^{n+\frac{1}{2}} = u_{i+1}^n$. Let P_i and P_{i+1} denote the points $(ih,(n+1)k)$ and $((i+1)h,(n+1)k)$, respectively. The characteristic intersects the line $t = (n+1)k$ at $P_{i+\frac{1}{2}} = ((i+\frac{1}{2})h - ck/2,(n+1)k)$. The point Q lies to the left of the characteristic with probability $(h-ck)/2h$ (the length of the segment $\overline{P_iP_{i+\frac{1}{2}}}$ normalized by the length of the interval h, that is, $\overline{P_iP_{i+\frac{1}{2}}}/h$) and Q lies to the right of the characteristic with probability $(h+ck)/2h$ (the length of the segment $\overline{P_{i+\frac{1}{2}}P_{i+1}}$ normalized by the length of the interval h, that is, $\overline{P_{i+\frac{1}{2}}P_{i+1}}/h$) (see Figure 3.24).

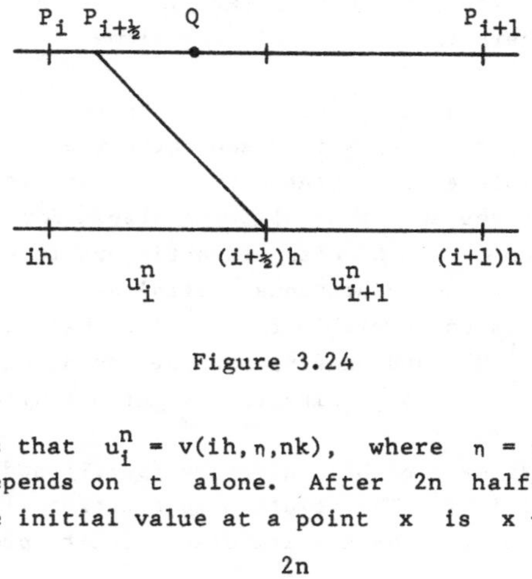

Figure 3.24

It follows that $u_i^n = v(ih,\eta,nk)$, where $\eta = \eta(t)$ is a random variable that depends on t alone. After 2n half-steps the displacement of the initial value at a point x is $x + \eta$,

$$\eta = \sum_{j=1}^{2n} \eta_j,$$

where η_j are independent, identically distributed random variables with probability distribution

$$P\{\eta = -\frac{h}{2}\} = \frac{h+ck}{2h}$$

and

$$P\{\eta = \frac{h}{2}\} = \frac{h-ck}{2h}$$

The expected value, second moment, and variance of η_j are

$$E[\eta_j] = (\frac{h+ck}{2h})(-\frac{h}{2}) + (\frac{h-ck}{2h})(\frac{h}{2}) = -\frac{ck}{2},$$

$$E[\eta_j^2] = (\frac{h+ck}{2h})(-\frac{h}{2})^2 + (\frac{h-ck}{2h})(\frac{h}{2})^2 = \frac{h^2}{4},$$

and

$$\mathrm{Var}[\eta_j] = E[\eta_j^2] - E[\eta_j]^2 = \frac{h^2}{4} - (\frac{ck}{2})^2 = \frac{h^2-c^2k^2}{4}.$$

From which,

$$E[\eta] = \sum_{j=1}^{2n} E[\eta_j] = -c\eta k = -ct$$

$$\mathrm{Var}[\eta] = \sum_{j=1}^{2n} \mathrm{Var}[\eta_j] = (h^2 - c^2k^2)$$
$$= \frac{c^2kt}{2}((\frac{h}{ck})^2 - 1).$$

If the ratio h/k is held constant, then $\mathrm{Var}[\eta] \to 0$ as $h \to 0$ for fixed t. Thus the computed solution converges to the exact solution as h tends to zero.

The second version of the random choice method (see Colella [5]) is a two-step method using a single grid. To find u_i^{n+1} given u_{i-1}^n, u_i^n, and u_{i+1}^n consider the sequence of Riemann problems given by equation (3.2.1) along with the piecewise constant initial condition (3.12.6). The solution u_i^{n+1} is based on the solution of two Riemann problems with initial conditions

(3.12.17) $$v_{i-}(x,nk) = \begin{cases} u_{i-1}^n, & x < (i-\frac{1}{2})h \\ u_i^n, & x > (i-\frac{1}{2})h, \end{cases}$$

and

(3.12.18) $$v_{i+}(x,nk) = \begin{cases} u_i^n, & x < (i+\frac{1}{2})h \\ u_{i+1}^n, & x > (i+\frac{1}{2})h. \end{cases}$$

The CFL condition becomes

$$\frac{|c|k}{2h} < 1, \tag{3.12.19}$$

where the factor of 1/2 comes from replacing k with $k/2$ in (3.12.18). If the CFL condition is satisfied the discontinuities (waves) in each of the intervals $[ih,(i+1)h]$ will remain in their respective intervals and hence, the solution can be combined by superposition into a single exact solution $v^e(x,t)$ for $nk \in t \in (n+1)k$.

Let ξ_n denote an equidistributed random variable in the interval $(-\frac{1}{2}, \frac{1}{2})$. Sample the Riemann problems associated with (3.12.17) and (3.12.18),

$$\hat{u}_{i-\frac{1}{2}}^{n+\frac{1}{2}} = v^e((i-\tfrac{1}{2}+\xi_n)h, k/2) \tag{3.12.20}$$

and

$$\hat{u}_{i+\frac{1}{2}}^{n+\frac{1}{2}} = v^e((i+\tfrac{1}{2}+\xi_n)h, k/2). \tag{3.12.21}$$

The second step of the random choice method is

$$u_i^{n+1} = \begin{cases} \hat{u}_{i-\frac{1}{2}}^{n+\frac{1}{2}}, & \xi_n > 0, \\ \hat{u}_{i+\frac{1}{2}}^{n+\frac{1}{2}}, & \xi_n < 0 \end{cases} \tag{3.12.22}$$

(see Figure 3.25).

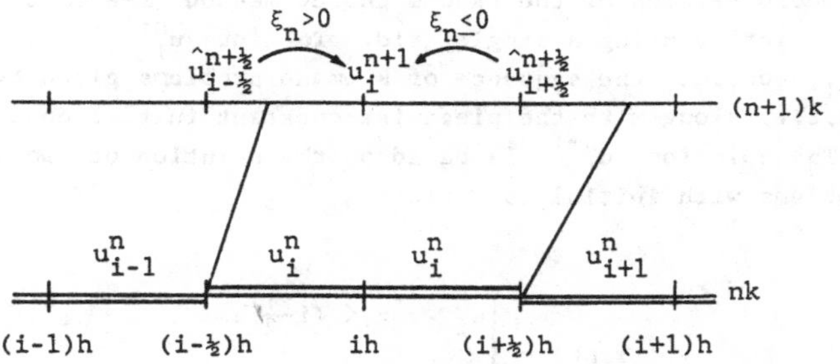

Figure 3.25. Sampling Procedure for Random Choice Method on a Single Grid

Both versions of the random choice method can be extended to the case of a scalar hyperbolic equation with variable coefficients. The random choice method described above for the constant coefficient case remains unchanged for the variable coefficient. The only difference is in the solution of the Riemann problem. The basic random choice method assumes the solution to the Riemann problems is available.

What remains there to be described is the solution to the Riemann problem. To this end, consider the equation

$$\partial_t v = a(x)\partial_x v$$

along with the piecewise constant initial condition (3.12.6) on the interval $[ih,(i+1)h]$. The solution $v(x,t)$ is constant along the characteristics defined by

$$\frac{dx}{dt} = -a(x) \tag{3.12.23}$$

or

$$\int \frac{dx}{a(x)} - t = \text{const.} \tag{3.12.23'}$$

Suppose the integral in (3.12.23') can be explicitly evaluated

$$A(x) = \int \frac{dx}{a(x)},$$

where $A(x)$ is invertible. In this case the characteristic curves are

$$A(x) - t = \text{const.} \tag{3.12.24}$$

Follow the characteristic passing through the sampled point $((i+\frac{1}{2}+\xi_n)h,(n+1)k)$ to the point of intersection with the line $t = nk$, denoted by $(\bar{x},nk)$. Using (3.12.24),

$$\bar{x} = A^{-1}(A((i+\tfrac{1}{2}+\xi_n)h) - k). \tag{3.12.25}$$

If the point $(\bar{x},nk)$ lies to the left of the point $((i+\frac{1}{2})h,nk)$, that is, $\bar{x} < (i+\frac{1}{2})h$, then $\hat{u}^{n+\frac{1}{2}}_{i+\frac{1}{2}} = u^n_i$. Similarly, if the point $(\bar{x},nk)$ lies to the right of the point $((i+\frac{1}{2})h,nk)$, that is, $\bar{x} > (i+\frac{1}{2})h$, then $\hat{u}^{n+\frac{1}{2}}_{i+\frac{1}{2}} = u^n_{i+1}$.

In the event that $A(x)$ is (3.12.24) is not invertible, a nonlinear equation in $\bar{x}$ is obtained

$$A(\bar{x}) - A((i+\tfrac{1}{2}+\xi_n)h) + k = 0, \tag{3.12.26}$$

which can be solved by standard techniques, such as Newton's method.

Finally, suppose that the indefinite integral in (3.12.23') cannot be explicitly evaluated. In this case the characteristic curve through the sampled point can be replaced by the tangent line to the characteristic at the sampled point. From (3.12.23) the slope of the tangent line is $-\frac{1}{a(x)}$ and the equation of the tangent line is

$$x = a_i(\xi_n)(t - (n+1)k) + (i+\tfrac{1}{2}+\xi_n)h,$$

where $a_i(\xi_n) \equiv a((i+\frac{1}{2}+\xi_n)h)$. Let $(\bar{x},nk)$ denote the point of intersection of the tangent line with the line $t = nk$, then

$$\bar{\bar{x}} = (i+\tfrac{1}{2}+\xi_n)h - a_i(\xi_n)k \tag{3.12.27}$$

(see Figure 3.26). This reduces the equation (locally) to the constant

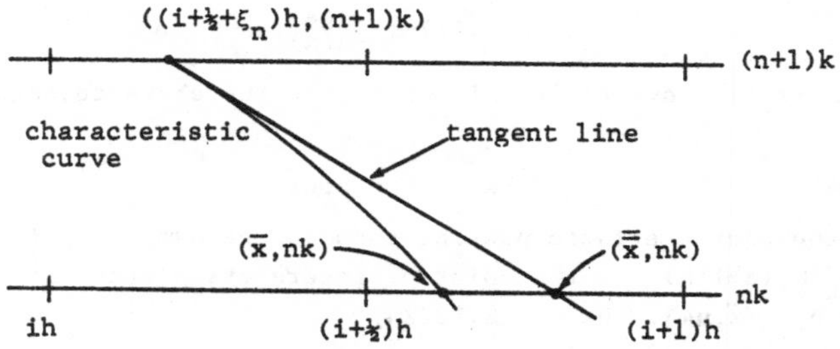

Figure 3.26.

coefficient case, where $c = a_i(\xi_n)$. In this case

$$\hat{u}_{i+\frac{1}{2}}^{n+\frac{1}{2}} = \begin{cases} u_i^n, & \xi_n h < a_i(\xi_n)k \\ u_{i+1}^n, & \xi_n h > a_i(\xi_n)k \end{cases} \tag{3.12.28}$$

Both versions of the random choice method extend directly to hyperbolic systems with constant or variable coefficients. Again, the solution of the Riemann problem for the system of equations is assumed to be available.

III.13. Asymptotic Error Expansions

Suppose a hyperbolic partial differential equation is approximated by a finite difference method that is accurate of $O(k^{q_1}) + O(h^{q_2})$. Let $k = O(h)$ then the finite difference method is accurate of $O(h^q)$ where $q = \min(q_1, q_2)$. Suppose that the error can be written in the asymptotic form

$$\tau = \sum_{j=q}^{\infty} h^j v^j(x,t) ,$$

where the $v^j(x,t)$ are smooth functions which depend on the exact solution $v(x,t)$ but are independent of h and $v^q(x,t)$ is not identically zero. In this case the approximate solution u_i^n may be written in the form

$$u_i^n = v_i^n + \sum_{j=q}^{\infty} h^j v^j(ih,nk) . \tag{3.13.1}$$

Consider the prototype equation (3.2.1)

$$\partial_t v = c\partial_x v, \quad -\infty < x < +\infty , \quad t > 0 \tag{3.2.1}$$

where, without loss of generality, we assume that $c > 0$, along with the initial condition

$$v(x,0) = f(x), \quad -\infty < x < +\infty . \tag{3.2.2}$$

Approximate this equation by

$$u_i^{n+1} = u_i^n + ckD_+u_i^n , \tag{3.13.2}$$

which was shown in Section III.2 to be first order accurate. Thus $q = 1$ in (3.13.1) and u_i^n, given by (3.13.2) may be written as

$$\begin{aligned} u_i^n &= v_i^n + \sum_{j=1}^{\infty} h^j v^j(ih,nk) \\ &= v_i^n + h(v^1)_i^n + \sum_{j=2}^{\infty} h^i v^j(ih,nk) \\ &= v_i^n + h(v^1)_i^n + O(h^2) . \end{aligned}$$

How is this function $v^1(x,t)$ defined? Formally expanding $v + hv^1$ in a Taylor series about the point (ih,nk) gives

$$\begin{aligned} (v + hv^1)_i^{n+1} &= (v_i^n + k(\partial_t v)_i^n + \tfrac{k^2}{2}(\partial_t^2 v)_i^n + O(k^3)) \\ &\quad + h((v^1)_i^n + k(\partial_t v^1)_i^n + \tfrac{k^2}{2}(\partial_t^2 v^1)_i^n + O(k^3)) \end{aligned}$$

and

$$\begin{aligned} (v + hv^1)_{i+1}^n &= (v_i^n + h(\partial_x v)_i^n + \tfrac{h^2}{2}(\partial_x^2 v)_i^n + O(h^3)) \\ &\quad + h((v^1)_i^n + h(\partial_t v^1)_i^n + \tfrac{h^2}{2}(\partial_x^2 v^1)_i^n + O(h^3)) . \end{aligned}$$

Substitution of these Taylor series expansions in the finite difference method (3.13.2)

$$\frac{(v + hv^1)_i^{n+1} - (v + hv^1)_i^n}{k} - cD_+(v + hv^1)_i^n$$

$$= (\partial_t v - c\partial_x v)_i^n + (\tfrac{k}{2}\partial_t^2 v - \tfrac{ch}{2}\partial_x^2 v + h\partial_t v^1 - ch\partial_x v^1)_i^n$$

$$+ O(k^2) + O(kh) + O(h^2) .$$

If v is twice continuously differentiable, $\partial_t^2 v = c^2\partial_x^2 v$, and using (3.2.1)

$$\text{(3.13.3)} \quad \frac{(v + hv^1)_i^{n+1} - (v + hv^1)_i^n}{k} - cD_+(v + hv_1)_i^n$$

$$= (\tfrac{c^2k}{2} - \tfrac{ch}{2})(\partial_x^2 v)_i^n + (h\partial_t v^1 - ch\partial_x v^1)_i^n + O(h^2) ,$$

where $k = O(h)$. In order to obtain accuracy of $O(h^2)$, the sum of the two terms on the right-hand side must vanish. This gives an equation for v^1

$$\text{(3.13.4)} \quad \partial_t v^1 - c\partial_x v^1 + \tfrac{1}{2}(c-c^2\lambda)\partial^2 v$$

where $\lambda = \frac{k}{h}$. The initial condition for this equation is obtained from the initial condition (3.2.2) of v and u

$$u^0 = f(x) = v(x,0) + hv^1(x,0) = f(x) = hv^1(x,0) .$$

This gives rise to the initial condition

$$\text{(3.13.5)} \quad v^1(x,0) = 0 .$$

The initial-value problem (3.13.4)-(3.13.5) has a unique solution. This verifies the existence of such an asymptotic expansion provided the exact solution v is sufficiently smooth, for $\partial_x^2 v$ is needed to determine v^1 in (3.12.4).

With v^1 being the solution of the initial-value problem (3.13.4)-(3.13.5), we see the aid of (3.13.2) and (3.13.3) that

$$\frac{(u-v-hv^1)_i^{n+1} - (u-v-hv^1)_i^n}{k} - cD_+(u-v-hv^1)_i^n$$

$$= O(k^2) + O(h^2) .$$

One application of this technique is Richardson's extrapolation. Write (3.13.1) in the form

$$\text{(3.13.6)} \quad v_i^n = u_i^n + \sum_{j=q}^{\infty} h^j v^j(ih,nk) .$$

Suppose the approximate solution is completed for two different values of h, that is, on two grids with spacing h and rh where $r \neq 0,1$.

Then (3.13.6) may be written as

$$v = u^h + \sum_{j=q}^{\infty} h^j v^j(x,t) \tag{3.13.7}$$

and

$$v = u^{rh} + \sum_{j=q}^{\infty} (rh)^j v^j(x,t) \ ,$$

where u^h and u^{rh} denote the approximate solutions on the grids with spacings h and rh, respectively. Multiplying (3.13.7) by $(rh)^q$ and (3.13.8) by h^q, subtracting and solving for v yields

$$v = \frac{r^q u^h - u^{rh}}{r^q - 1} + \sum_{j=q+1}^{\infty} \frac{r^q - r^j}{r^q - 1} h^j v^j(x,t)$$

$$= \frac{r^q u^h - u^{rh}}{r^q - 1} + O(h^{q+1}) \ .$$

Thus Richardson's extrapolation allows us to combine two approximate solutions that are accurate of $O(h^q)$ to obtain an approximate solution that is accurate of $O(h^{q+1})$. However, without a careful examination of the error term in (3.13.9), we cannot say that (3.13.9) is more accurate than either (3.13.7) or (3.13.8). If v is sufficiently smooth, then (3.13.9) is more accurate for h sufficiently small. This new approximate solution given by (3.13.9) is defined on the coarser of the two grids only.

One feature of Richardson's extrapolation is that higher order approximations are obtainable without explicitly finding v^1.

This process can be repeated successively to cancel higher and higher order terms in the truncation error, provided the solution v is sufficiently smooth. Each application of Richardson's extrapolation calls for an approximate solution of an ever finer grid. This can become expensive, particularly for problems in two- and three-space dimensions.

Another application of asymptotic error expansions that does not require recomputing the solution on a finer grid is the method known as a deferred correction.

In general, $v^j(x,t)$ in (3.13.1) is not known since it is the solution of an equation similar to the one that we are trying to solve. We can, however, find an approximate solution to the equation that determines v^j.

For example, consider the equation (3.13.4) for v^1. This equation can be approximated by employing the same finite difference method (with suitable changes) used to solve the original equation

$$(u^1)_i^{n+1} = (u^1)_i^n + ckD_+(u^1)_i^n + \frac{k}{2}(c - c^2\lambda)D_+D_-u_i^n \ .$$

Then by using $(u^1)_i^n$ in place of v^1 in (3.13.1),

$$u_i^n + h(u^1)_i^n = v_i^n + O(h^2) ,$$

where $k = O(h)$.

A natural question arises. Do we really gain anything from using higher order finite difference methods? This question will be considered in the next section.

III.14. <u>How Accurate Should a Finite Difference Method Be?</u>

Using the techniques developed in Section III.12, we can construct high order accurate finite difference methods, provided the exact solution is smooth enough. Those methods may be either dissipative or nondissipative.

For example, consider the leap-frog method

(3.10.1) $$\frac{u_i^{n+1} - u_i^{n-1}}{2k} = cD_0u_i^n$$

which is accurate of $O(k^2) + O(h^2)$. It was shown that

$$D_0v_i^n = (\partial_x v)_i^n + \frac{h^2}{6}(\partial_x^3 v)_i^n + \frac{h^4}{120}(\partial_x^5 v)_i^n + O(h^6)$$

or

$$(\partial_x v)_i^n = D_0v_i^n - \frac{h^2}{6}(\partial_x^3 v)_i^n - \frac{h^4}{120}(\partial_x v)_i^n + O(h^6) .$$

Let $D_0(h)$ denote the dependence of D_0 on h, then with a grid spacing of ah, where a is a position integer,

$$D_0(ah)v_i^n = (\partial_x v)_i^n + \frac{(ah)^2}{6}(\partial_x^3 v)_i^n + \frac{(ah)^4}{120}(\partial_x^5 v)_i^n + O(h^6)$$

or

$$(\partial_x v)_i^n = D_0(ah)v_i^n - \frac{(ah)^2}{6}(\partial_x^3 v)_i^n - \frac{(ah)^4}{120}(\partial_x^5 v)_i^n + O(h^6) .$$

In particular, for $a = 2$ and 3, we have

$$(\partial_x v)_i^n = D_0(2h)v_i^n - \frac{2}{3}h^2(\partial_x^3 v)_i^n - \frac{2}{15}h^4(\partial_x^5 v)_i^n + O(h^6)$$

and

$$(\partial_x v)_i^n = D_0(3h)v_i^n - \frac{3}{2}h^2(\partial_x^3 v)_i^n + \frac{81}{120}h^4(\partial_x^5 v)_i^n + O(h^6) .$$

Applying Richardson's extrapolation to the two second order accurate approximations to ∂_x, $D_0(h)$ and $D_0(2h)$,

(3.14.1) $$(v_x)_i^n = \frac{1}{3}(4D_0(h) - D_0(2h))v_i^n + O(h^4) ,$$

which results in a fourth order accurate approximation to ∂_x. Replacing the $D_0(h)$ term in the leap-frog method with $\frac{1}{3}(4D_0(h) - D_0(2h))$,

(3.14.2) $$\frac{u_i^{n+1} - u_i^{n-1}}{2k} - \frac{c}{3}(4D_0(h) - D_0(2h))u_i^n ,$$

which is accurate of $O(k^2) + O(h^4)$.

Applying Richardson's extrapolation to the three second order accurate approximations to ∂_x, $D_0(h)$, $D_0(2h)$, and $D_0(3h)$, we have

(3.14.3) $$(\partial_x v)_i^n = \frac{1}{10}(15D_0(h) - 6D_0(2h) + D_0(3h))v_i^n + O(h^6),$$

which results in a sixth order accurate approximation to ∂_x. Replacing the $D_0(h_0)$ term in the leap-frog method with $\frac{1}{10}(15D_0(h) - 6D_0(2h) + D_0(3h))$,

(3.14.4) $$\frac{u_i^{n+1} - u_i^{n-1}}{2k} = \frac{c}{10}(15D_0(h) - 6D_0(2h) + D_0(3h))u_i^n,$$

which is accurate of $O(k^2) + O(h^6)$.

This approach can be generalized. Applying Richardson's extrapolation to the m second order accurate approximations to ∂_x, $D_0(h), D_0(2h),\dots,D_0(mh)$ gives

(3.14.5) $$(\partial_x v)_i^n = D^{[2m]}(h)v_i^n + O(h^{2m}),$$

where

$$D^{[2m]}(h) = \sum_{j=1}^{m} \frac{2(-1)^{j+1}(m!)^2}{(m+j)!(m-j)!} D_0(jh) .$$

Observe that for $m = 1,2,$ and 3 (3.14.6) reduces to the second, fourth, and sixth order accurate approximations to ∂_x just described.

Replacing $D_0(h)$ in the leap-frog method (3.10.1) with $D^{[2m]}(h)$ gives

(3.14.7) $$\frac{u_i^{n+1} - u_i^{n-1}}{2k} = cD^{[2m]}(h)u_i^n ,$$

which is accurate of $O(k^2) + O(h^{2m})$.

It can be seen by the same technique used for the original leap-frog method that the fourth-order multilevel method (3.14.2), the sixth-order multilevel method (3.14.4), and the 2m-th-order multilevel method (3.14.7) are stable for $\lambda = \frac{k}{h} < 1$.

A question arises--how far should we go in increasing the order of accuracy of a finite difference method or in refining the grid?

Consider the prototype equation (3.2.1.)

$$\partial_t v = c\partial_x v , \quad -\infty < x < +\infty , \quad t > 0$$

along with the initial condition

$$v(x,0) = e^{i\beta x}, \quad -\infty < x < +\infty ,$$

which has the exact solution

(3.14.8) $$v(x,t) = e^{i\beta(x + ct)} .$$

This initial-value problem can be treated as an ordinary differential equation by discretizing in space only.

Approximate the space derivative $\partial_x v$ by $D_0(h)$, which is accurate of $O(h^2)$, giving the ordinary differential equation

$$\partial_t u = cD_0(h)u \ . \tag{3.14.9}$$

Consider the ordinary differential equation (3.13.9) along with the initial condition

$$u(x,0) = e^{i\beta x} \ . \tag{3.14.10}$$

The solution to the ordinary differential equation has the form

$$u(x,t) = e^{i\beta(x+c_2(\beta)t)} \ .$$

Differentiating with respect to t,

$$\partial_t u(x,t) = i\beta c_2(\beta)u(x,t)$$

and substituting with D_0,

$$D_0(h)u = u(x,t)\frac{(e^{i\beta h} - e^{-i\beta h})}{2h}$$

$$= u(t,x)\frac{2\ \sin(\beta h)}{h} \ .$$

Combining these two results equation (3.14.9),

$$i\beta c_2(\beta)u(x,t) = cu(x,t)\frac{i\ \sin(\beta h)}{h} \ .$$

or

$$c_2(\beta) = \frac{c\ \sin(\beta h)}{\beta h} \ . \tag{3.14.11}$$

This is consistent for $c_2(\beta) \to c$ as $h \to 0$. For $h > 0$, since $|\sin(\beta h)/\beta h| < 1$, $|c_2(\beta)| < c$. This produces dispersion. Define the <u>phase error</u>, $e_2(\beta)$ to be

$$e_2(\beta) = \beta t(c - c_2(\beta)) \ . \tag{3.14.12}$$

Observe that $e_2(\beta)$ is an increasing function of time.

Expanding $\sin(\beta h)$ in a Taylor series about zero,

$$c_2(\beta) = c(1 - \frac{(\beta h)^2}{6} + O((\beta h)^4)) \ . \tag{3.14.13}$$

Thus

$$c - c_2(\beta) = O(\beta^2 h^2) \ . \tag{3.14.14}$$

Using (3.14.12) along with (3.14.13),

$$e_2(\beta) = O(h^2) \ .$$

Suppose we must include all frequencies $\leq \beta_0$, and we want to integrate (3.14.9) up to time T keeping the phase error $e_2(\beta)$ less than some tolerance ε. We must determine how many points are needed for the computation. Let $N_2 = 2\pi/\beta h$ denote the number of grid points per wave length for the second order accurate leap-frog method (3.10.1). Write time T as $\frac{2\pi p}{\beta c}$, where p denotes the number of periods to be computed.

Since $e_2(\beta)$ is an increasing function of time, the requirement that $e_2(\beta) < \varepsilon$ is satisfied if $\beta T(c - c_2(\beta)) = \varepsilon$. However,

$$\beta T(c - c_2(\beta)) = 2\pi p(1 - \frac{\sin(\beta h)}{\beta h})$$
$$= 2\pi p(1 - \frac{\sin(\frac{2\pi}{N_2})}{\frac{2\pi}{N_2}})$$

Using the first two terms of the Taylor series for $e_2(\beta)$ in (3.14.13), replacing βh with $1/N_2$,

$$\beta T(c - c_2(\beta)) \approx \frac{p}{6N_2^2}(2\pi)^3 .$$

Thus for a given value of the largest allowable phase error ε,

$$N_2 \approx 2\,(\frac{2\pi p}{6\varepsilon})^{1/2}. \tag{3.14.15}$$

Suppose the space derivative in the prototype equation (3.2.1) is approximated by $\frac{1}{3}(4D_0(h) - D_0(2h))$, which is accurate of $O(h^4)$, giving the ordinary differential equation

$$\partial_t u = \frac{c}{3}(4D_0(h) - D_0(2h))u . \tag{3.14.16}$$

Consider the ordinary differential equation (3.14.16), along with the initial condition (3.14.10). The exact solution has the form

$$u(x,t) = e^{i\beta(x+c_4(\beta)t)} ,$$

where following the same procedure as with $c_2(\beta)$,

$$c_4(\beta) = c\,(\frac{8\sin(\beta h) - \sin(2\beta h)}{6\beta h}) . \tag{3.14.17}$$

This is consistent for $c_4(\beta) \to c$ as $h \to 0$. For $h \to 0$, as in the case of $c_2(\beta)$, $|c_4(\beta)| < c$ so there is dispersion. The phase error $e_4(\beta)$ is defined to be

$$e_4(\beta) = \beta t(c - c_4(\beta)) , \tag{3.14.18}$$

which is an increasing function of t.

Again, expanding $\sin(\beta h)$ and $\sin(2\beta h)$ in a Taylor series about zero,

$$c_4(\beta) = c(1 - \frac{1}{30}(\beta h)^4 + O((\beta h)^6)) \tag{3.14.19}$$

Thus

$$c - c_4(\beta) = O(\beta^4 h^4)$$

and

$$e_4(\beta) = O(h^4) .$$

Requiring $e_4(\beta)$ to be $< \varepsilon$ gives $\beta T(c - c_4(\beta)) = \varepsilon$. However,

$$\beta T(c - c(\beta)) = p\left(1 - \frac{8\sin(\frac{2\pi}{N_4}) - \sin(\frac{4\pi}{N_4})}{\frac{12\pi}{N_4}}\right),$$

where $N_4 = \frac{2\pi}{\beta h}$. Using the first two terms of the Taylor series for $c_4(\beta)$ in (3.14.19), replacing βh with $1/N_4$,

$$\beta T(c - c(\beta)) \approx \frac{(2\pi)^5 p}{30 N_4^4}.$$

Thus for a given value of the largest allowable phase error ε,

(3.14.20) $$N_4 \approx 2\pi\left(\frac{2\pi p}{30\varepsilon}\right)^{1/4}.$$

Suppose now the space derivative in the prototype equation (3.2.1) is approximated by $\frac{1}{10}(15D_0(h) - 6D_0(2h) + D_0(3h))$, which is accurate of $O(h^6)$, giving the ordinary differential equation

(3.14.21 $$\partial_t u = \frac{c}{10}(15D_0(h) - 6D_0(2h) + D_0(3h))u.$$

Consider the ordinary differential equation (3.14.21) along with the initial condition (3.13.10). The exact solution has the form

$$u(x,t) = e^{i\beta(x+c_6(\beta)t)},$$

where $c_6(\beta)$ is given by

$$c_6(\beta) = c\left(\frac{45\sin(\beta h) - 9\sin(2\beta h) + \sin(3\beta h)}{30\beta h}\right).$$

This is consistent for $c_6(\beta) \to c$ as $h \to 0$. For $h \to 0$, $|c_6(\beta)| < c$, so there is dispersion. The phase error $e_6(\beta)$ is defined to be

(3.14.22) $$e_6(\beta) = \beta t(c - c_6(\beta)),$$

which is an increasing function of t.

Expanding $\sin(\beta h)$, $\sin(2\beta h)$, and $\sin(3\beta h)$ in a Taylor series about zero,

(3.14.23) $$c_6(b) = c\left(1 - \frac{1}{140}(bh)^6 + O((bh)^8)\right).$$

Thus

$$c - c_6(\beta) = O(\beta^6 h^6)$$

and

$$e_6(\beta) = O(h^6).$$

Requiring $e_6(\beta)$ to be $\leq \varepsilon$ gives $\beta T(c - c_6(\beta)) = \varepsilon$. However,

$$\beta T(c - c_6(\beta)) = 2\pi p\left(1 - \frac{45\sin(\frac{2\pi}{N_6}) - 9\sin(\frac{4\pi}{N_6}) + \sin(\frac{6\pi}{N_6})}{\frac{60\pi}{N_6}}\right)$$

where $N_6 = \frac{2\pi}{\beta h}$. Using the first two terms of the Taylor series for

$c_6(\beta)$ in (3.14.23), replacing h with $1/N_6$,

$$\beta T(c - c_6(\beta)) \approx \frac{(2\pi)^7 p}{140 N_6^6} .$$

Thus for a given value of the largest allowable phase error ε,

$$N_6 \approx 2\pi\left(\frac{2\pi p}{140\varepsilon}\right)^{1/6} . \tag{3.14.24}$$

Finally, the space derivative in the prototype equation (3.2.1) is approximated by $D^{[2m]}(h)$, which is accurate of $O(h^{2m})$, giving the ordinary differential equation

$$\partial_t u = cD^{[2m]}(h)u . \tag{3.14.25}$$

Consider the ordinary differential equation (3.14.25) along with the initial condition (3.14.10). As in the three other cases, the solution of the ordinary differential equation has the form

$$u(x,t) = e^{i\beta(x+c_{2m}(\beta)t)} , \tag{3.14.26}$$

where $c_{2m}(\beta)$ will be determined below.

Kreiss and Oliger [20] introduced the formal expansion of ∂_x in terms of the difference operators D_+D_-,

$$\partial_x = D_0(h) \sum_{j=0}^{\infty} (-1)^j \alpha_{2j} (h^2 D_+ D_-)^j . \tag{3.14.27}$$

Observe that

$$h^2 D_+ D_- e^{i\beta x} = (e^{i\beta h} - 2 + e^{-i\beta h})e^{i\beta x}$$

$$= -4 \sin^2\left(\frac{\beta h}{2}\right)e^{i\beta x}$$

Applying (3.14.27) to the initial condition (3.14.10), using the result just obtained as well as the previous results of this section,

$$i\beta e^{i\beta x} = e^{i\beta x} \frac{i \sin(\beta h)}{h} \sum_{j=0}^{\infty} (-1)^j \alpha_{2j} (-4)^j \sin^{2j}\left(\frac{\beta h}{2}\right)$$

or

$$\beta h = \sin(\beta h) \sum_{j=0}^{\infty} \alpha_{2j}\, 2^{2j} \sin^{2j}\left(\frac{\beta h}{2}\right). \tag{3.14.28}$$

Since $\sin(\beta h) = 2\sin(\frac{\beta h}{2})\cos(\frac{\beta h}{2})$ and $\cos^{-1}(\frac{\beta h}{2}) = (1 - \sin^2(\frac{\beta h}{2}))^{-1/2}$, by substitution into (3.14.28)

$$\frac{\beta h}{2}\left(1 - \sin^2\left(\frac{\beta h}{2}\right)\right)^{-1/2} = \sin\left(\frac{\beta h}{2}\right) \sum_{j=0}^{\infty} \alpha_{2j} 2^{2j} \sin^{2j}\left(\frac{\beta h}{2}\right) . \tag{3.14.29}$$

Introduce the change of variable $\tau = \sin(\frac{\beta h}{2})$ or $\frac{\beta h}{2} = \text{arc} \sin \tau$ and observe that

$$\frac{1}{2}\frac{d}{d\tau}(\text{arc sin } \tau)^2 = \text{arc sin } \tau\cdot(1-\tau^2)^{-1/2} = \frac{\beta h}{2}(1-\sin(\frac{\beta h}{2}))^{-1/2}.$$

Substituting into (3.14.29),

$$\frac{1}{2}\frac{d}{d\tau}(\text{arc sin } \tau) = \tau\sum_{j=0}^{\infty} \alpha_{2j}2^{2j}\tau^{2j},$$

which, upon integration term-by-term, gives

$$\begin{aligned}(\text{arc sin } \tau)^2 &= \sum_{j=0}^{\infty} \alpha_{2j}\frac{2^{2j+1}}{2j+2}\tau^{2j+2} \\ &= 2\tau^2\sum_{j=0}^{\infty}\frac{\alpha_{2j}2^{2j}}{2j+2}\tau^{2j}.\end{aligned}$$

Thus the coefficients α_{2j} of the formal expansion (3.14.27) are obtained from the Taylor series for $(\text{arc sin } \tau)^2$.

The usefulness of this expansion can be seen from the observation that

$$D^{[2m]}(h) = D_0(h)\sum_{j=0}^{m-1}(-1)^j\alpha_{2j}(h^2D_+D_-)^j. \tag{3.14.30}$$

Substituting (3.14.26) into the ordinary differential equation (3.14.25), where $D^{[2m]}(h)$ is given by (3.14.30),

$$c_{2m}(\beta) = c\,\frac{\sin(\beta h)}{\beta h}\sum_{j=0}^{m-1}\alpha_{2j}2^{2j}\sin^{2j}(\frac{\beta h}{2}).$$

This is consistent for $c_{2m}(\beta) \to c$ as $h \to 0$. Define the phase error $e_{2m}(\beta)$ to be

$$e2m(\beta) = \beta t(c - c2m(\beta)).$$

Multiplying (3.14.28) by c and dividing by βh gives

$$c = c_{2m}(\beta) + \frac{c\sin(\beta h)}{\beta h}\sum_{j=m}^{\infty}\alpha_{2j}2^{2j}\sin^{2j}(\frac{\beta h}{2}),$$

so that

$$e_{2m}(\beta) = \beta ct\,\frac{\sin(\beta h)}{\beta h}\sum_{j=m}^{\infty}\alpha_{2j}2^{2j}\sin^{2j}(\frac{\beta h}{2}). \tag{3.14.31}$$

Requiring that $e_{2m}(\beta) < \varepsilon$ gives

$$2\pi p\,\frac{\sin(\frac{2\pi}{N_{2m}})}{\frac{2\pi}{N_{2m}}}\sum_{j=m}^{\infty}\alpha_{2j}2^{2j}\sin^{2j}(\frac{\pi}{N_{2m}}) = \varepsilon, \tag{3.14.32}$$

where $N_{2m} = \frac{2\pi}{\beta h}$.

We would like to observe the behavior of N_{2m} as m becomes large. It was shown by Kreiss and Oliger [20] that

$$\text{(3.14.33)} \quad \lim_{m\to\infty} \left(\sum_{j=m}^{\infty} \alpha_{2j} 2^{2j} \sin^{2j}\left(\frac{\beta h}{2}\right)\right)^{1/m} = \sin\left(\frac{\beta h}{2}\right) = \lim_{m\to\infty} \sin\left(\frac{\pi}{N_{2m}}\right)$$

By raising (2.14.32) to the 1/m power,

$$\lim_{m\to\infty}(2\pi p)^{1/m} \left(\frac{\sin(\frac{2\pi}{N_{2m}})}{\frac{2\pi}{N_{2m}}}\right)^{1/m} \left(\sum_{j=m}^{\infty} {}_{2j} 2^{2j} \sin^{2j}\left(\frac{\pi}{N_{2m}}\right)\right)^{1/m} = \lim_{m\to\infty} \varepsilon^{1/m}$$

or by using (3.14.33),

$$\lim_{m\to\infty} \sin\left(\frac{\pi}{N_{2m}}\right) = 1 .$$

Thus as $m \to \infty$, $\frac{\pi}{N_{2m}} \to \frac{\pi}{2}$ or $N_{2m} \to 2$. This means that there must be at least 2 points per wavelength.

Comparing the fourth order method (3.14.16) with the second order method (3.14.9), it is seen that the amount of computational labor involved in the fourth order method is about two times that of the second order method. Similarly, comparing the sixth order method (3.14.21) with the second order method (3.14.9), it is seen that the amount of computational labor involved in the sixth order method is about three times that of the second order method. And, in general, comparing the 2m-th order method (3.14.25) with the second order method (3.14.9), it is seen that the amount of computational labor involved in the 2m-th order method is about m times that of the second order method.

In Table 3.1 the approximate number of points required for the second order, fourth order, and sixth order methods is given for maximum errors of $\varepsilon = 0.1$ and $\varepsilon = 0.01$.

ε	0.1	0.01
N_2	$20P^{1/2}$	$64p^{1/2}$
N_4	$7P^{1/4}$	$13p^{1/4}$
N_6	$5P^{1/6}$	$8p^{1/6}$

Table 3.1

From the results in the table, the fourth and sixth order methods require substantially fewer points for a given error ε than the second order method. For large times T (or large values of p), the higher order methods are superior because of the $P^{1/2j}$. It is also observed that the sixth order method is superior to the fourth order method for small errors ε.

III.15. Numerical Examples

Here we present results from two test problems. First, consider a hyperbolic equation with constant coefficients

$$\partial_t v = c\partial_x v \qquad -\infty < x < +\infty, \quad t > 0$$

with initial condition

$$v(x,0) = f(x) = \begin{cases} 1, & 0.05 < x < 0.15, \\ 0, & \text{otherwise} . \end{cases}$$

The exact solution is

$$v(x,t) = f(x + ct).$$

Choose the constant c to be -1. To avoid dealing with boundary conditions restrict t to $0 \leq t \leq 0.75$ and x to $-1 \leq x \leq 2$. The grid spacing is taken to be $h = 0.01$. Figures 3.27-3.34 depict results for 50, 100, and 150 time steps for $|c|\frac{k}{h} = 0.5$ where the approximate solutions are represented by dashed lines and the exact solutions are represented by solid lines.

In Table 3.2 the numerical methods and the corresponding coefficients of numerical diffusion are presented.

Figures 3.27 and 3.28 depict the results of the (1,1) explicit method (3.2.10) and the Lax-Friedrichs method. The large amount of dissipation in both methods is apparent by the smearing of the waves. It is also apparent that the smearing increases with time. The phase errors of both methods appears very small.

Figure	Method	Diffusion Term	Coefficient
3.27	(1,1) explicit (3.2.10)	$-(\frac{c^2k}{2} + \frac{ch}{2})\partial_x^2 v$	$.25 \times 10^{-2}$
3.28	Lax-Friedrichs (3.2.15)	$\frac{h^2}{2k}\partial_x^2 v$	$.10 \times 10^{-3}$
3.29	Lax-Wendroff (3.5.3)	$\frac{c^2k}{2}\partial_x^2 v$	$.25 \times 10^{-2}$
3.30	(1,1) implicit (3.7.5)	$(\frac{c^2k}{2} - \frac{ch}{2})\partial_x^2 v$	$.75 \times 10^{-2}$
3.31	(1,2) implicit (3.7.6)	$\frac{c^2k}{2}\partial_x^2 v$	$.25 \times 10^{-2}$
3.32	leap frog (3.10.1)	———	———
3.33	dissipative (3.10.19)	$\frac{\sigma h^4}{32k}\partial_x^4 v$	$.25 \times 10^{-7}$ where $\sigma = .4$
3.34	random choice method (3.12.22)	———	———

Table 3.2. Methods and Numerical Diffusion Coefficients

The oscillatory type behavior of the results obtained by the Lax-Friedrichs method is due to the D_0 operator and decomposing the wave into two waves of differing phases. The average term used to produce the Lax-Friedrichs method from the unstable centered method (3.2.13) does not completely cure this problem.

Figure 3.29 depicts the results of the Lax-Wendroff method. The large wave number modes are not damped by the small numerical dissipation and as a result a trail of oscillations is left behind the wave.

Figures 3.30 and 3.31 depict the results from the (1,1) and (1.2) implicit methods, respectively. The (1,1) implicit method has three times the numerical damping of the (1,2) implicit method. This is obvious upon comparison of the two sets of results. The large wave number modes are not completely damped as there is a small trail of oscillations behind the wave at time step $n = 50$.

Figures 3.32 and 3.33 depict the results fro the leap frog and the dissipative leap frog methods, respectively. The second level of initial conditions u_i^1 was obtained by using the exact solution. In this way the errors in the results represent the errors in the leap frog methods and not a combination of the errors in the leap from methods and the errors in the method used to obtain the solution at the first time level. The leap frog method has no numerical dissipation to damp the high frequency Fourier modes that are poorly represented by a finite difference method. The introduction of the high order dissipation term damps much of the high frequency Fourier modes. The results, in this case, are very similar to those of the Lax-Wendroff method.

Figure 3.34 depicts the results of the random choice method. The structure of the wave is perfect. However, at $n = 50$ and 150, the wave lags the true wave and at $n = 100$ the wave leads the true wave. This is not a phase error in the sense of Section III.6, but, rather a fluctuation due to the randomness. In the mean the wave speed is correct

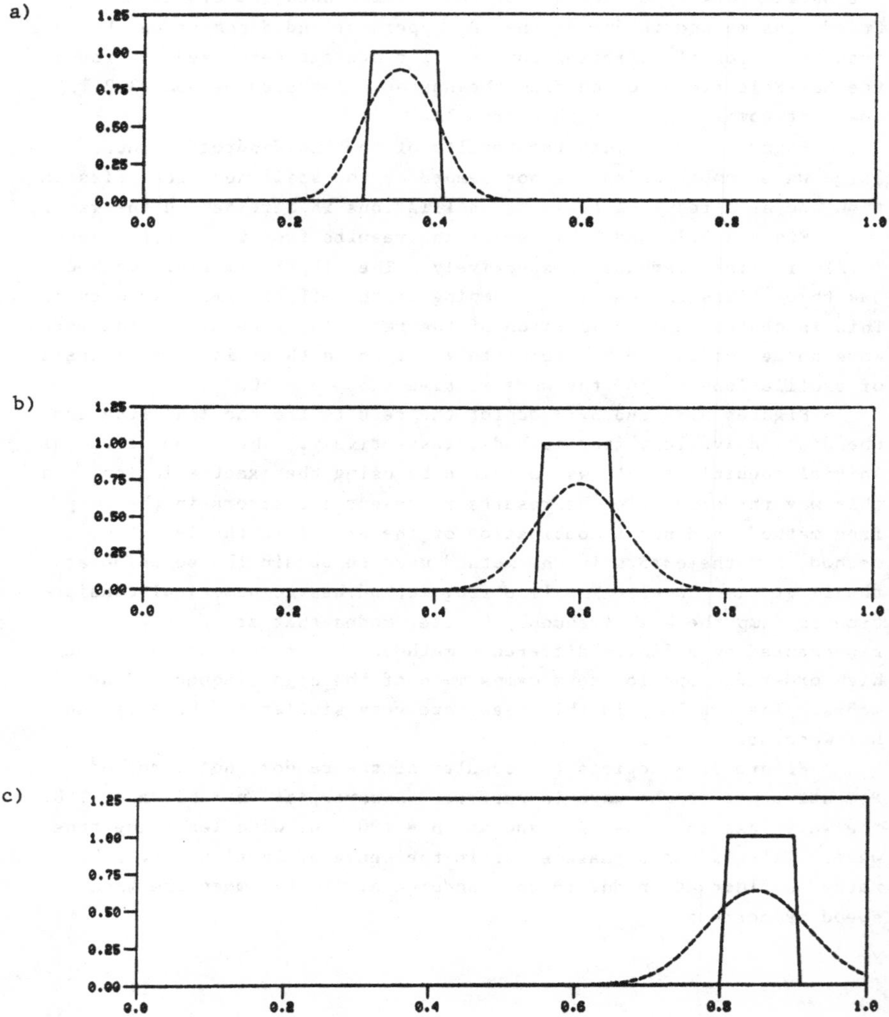

Figure 3.27. (1,1) Explicit Method (3.2.10) for $|c|\frac{k}{h} = 0.5$ at time steps a) n = 50, b) n = 100, and c) n = 150.

a)

b)

c)

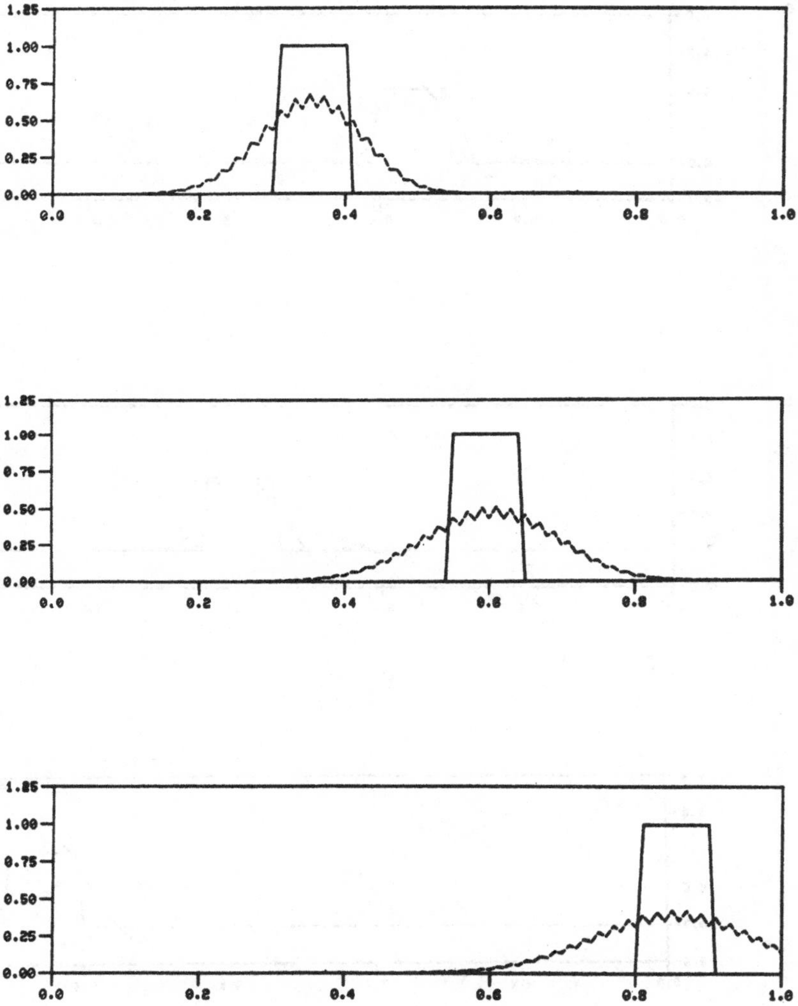

Figure 3.28. Lax-Friedrichs Method (3.2.15) for $|c|\frac{k}{c} = 0.5$ at time steps a) $n = 50$, b) $n = 100$, and c) $n = 150$.

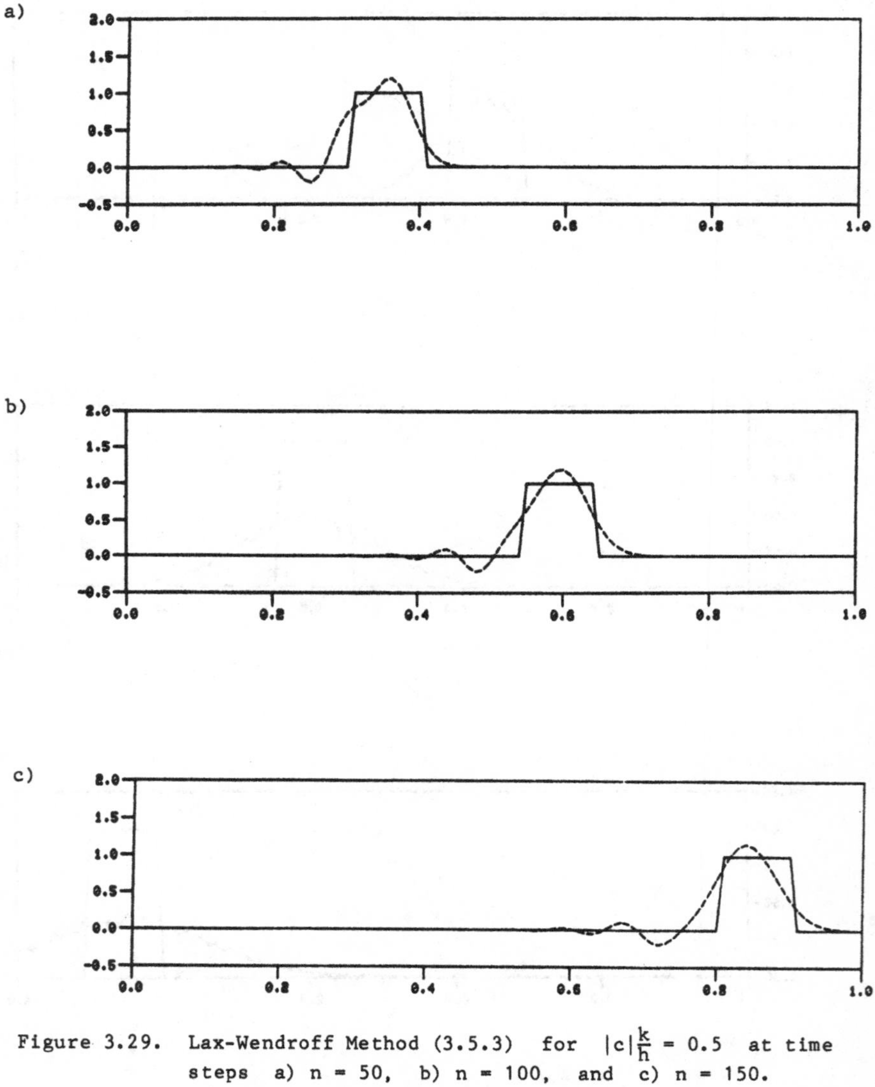

Figure 3.29. Lax-Wendroff Method (3.5.3) for $|c|\frac{k}{h} = 0.5$ at time steps a) n = 50, b) n = 100, and c) n = 150.

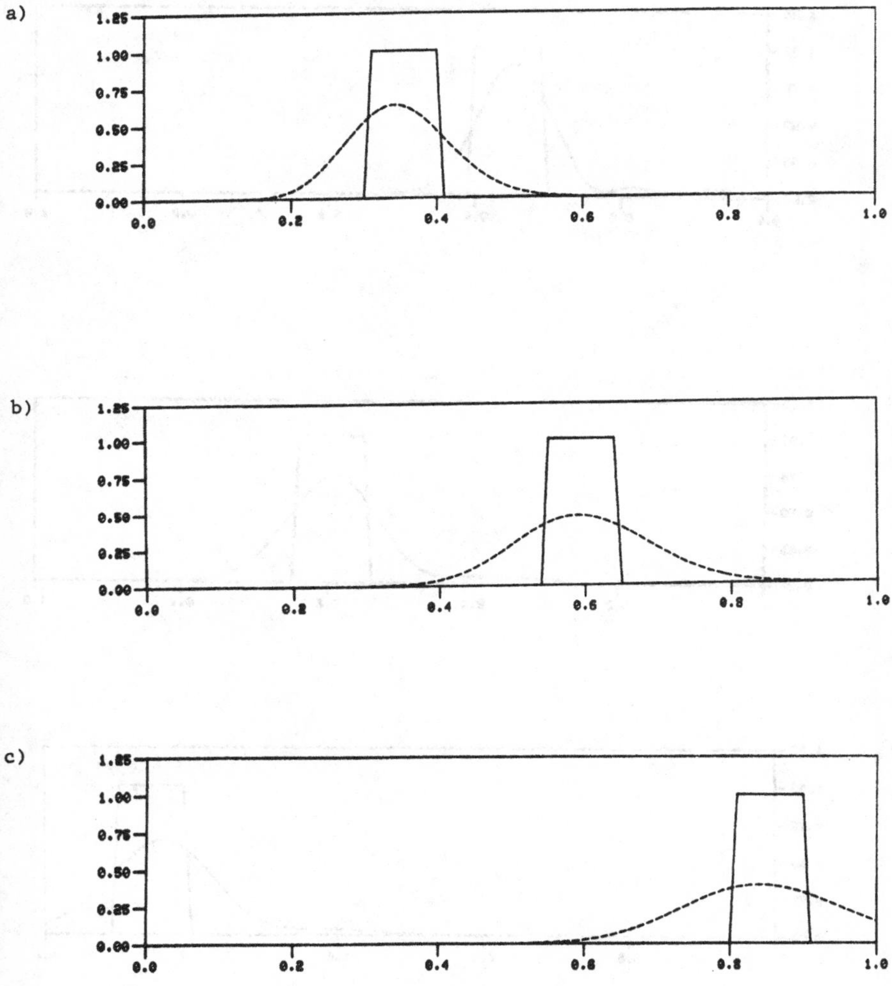

Figure 3.30. (1,1) Implicit Method (3.7.5) for $|c|\frac{k}{h} = 0.5$ at time steps a) $n = 50$, b) $n = 100$, and c) $n = 150$.

a)

b)

c)

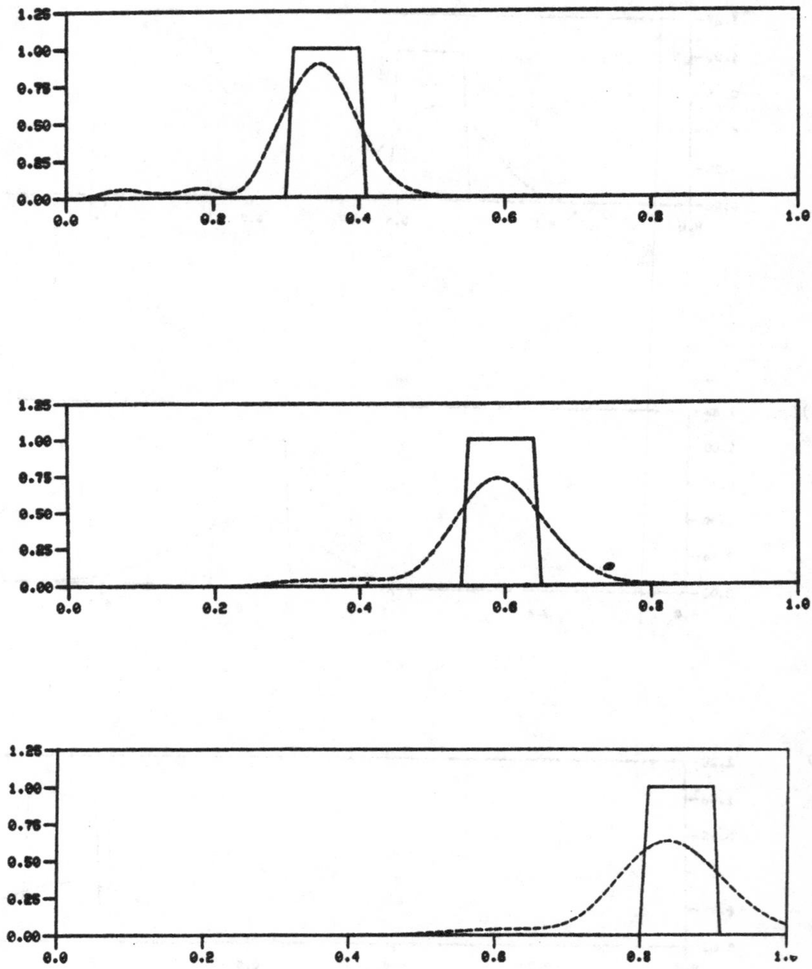

Figure 3.31. (1,2) Implicit Method (3.7.6) for $|c|\frac{k}{h} = 0.5$ at time steps a) n = 50, b) n = 100, and c) n = 150.

a)

b)

c)

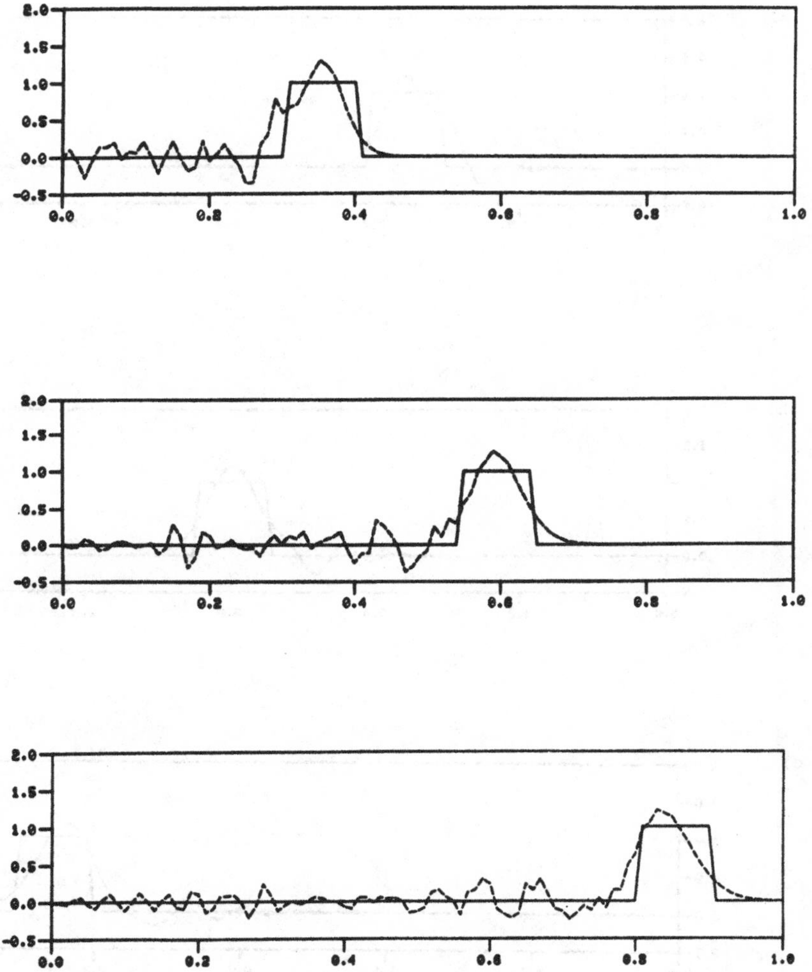

Figure 3.32. Leap Frog Method (3.10.14) for $|c|\frac{k}{h} = 0.5$ at time steps a) n = 50, b) n = 100, and c) n = 150.

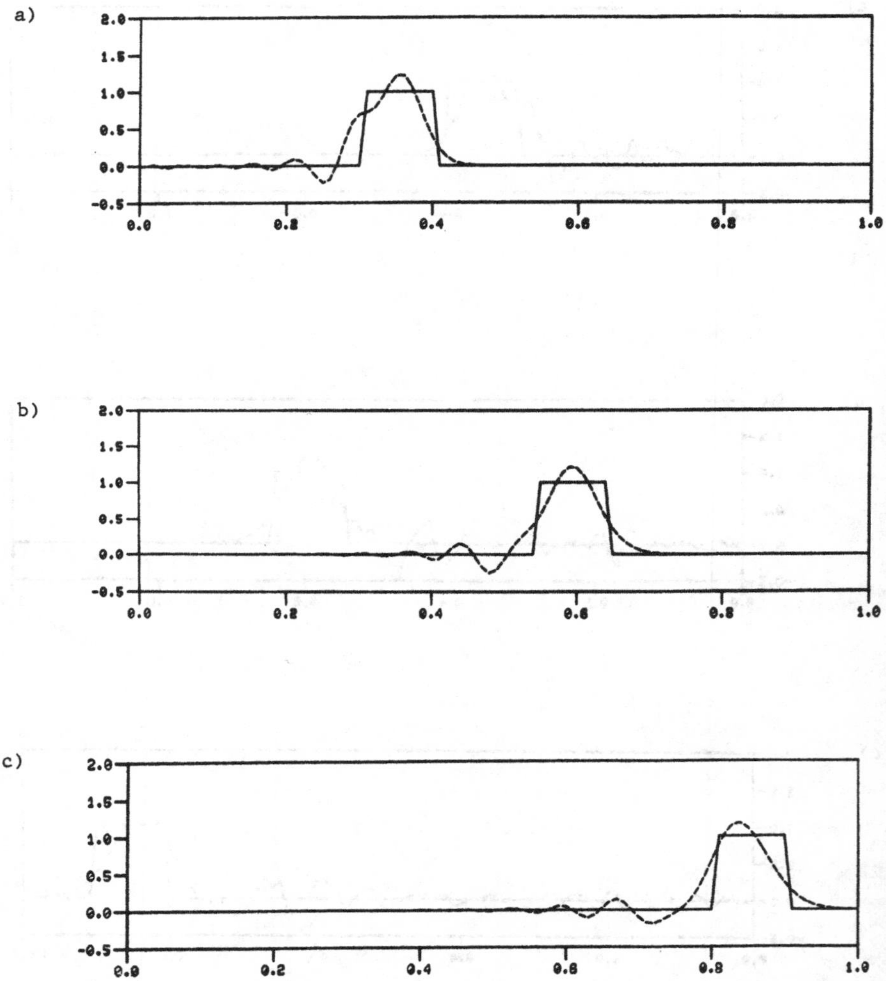

Figure 3.33. Dissipative Leap Frog Method (3.7.19) for $|c|\frac{k}{h} = 0.5$ at time steps a) n = 50, b) n = 100, and c) n = 150.

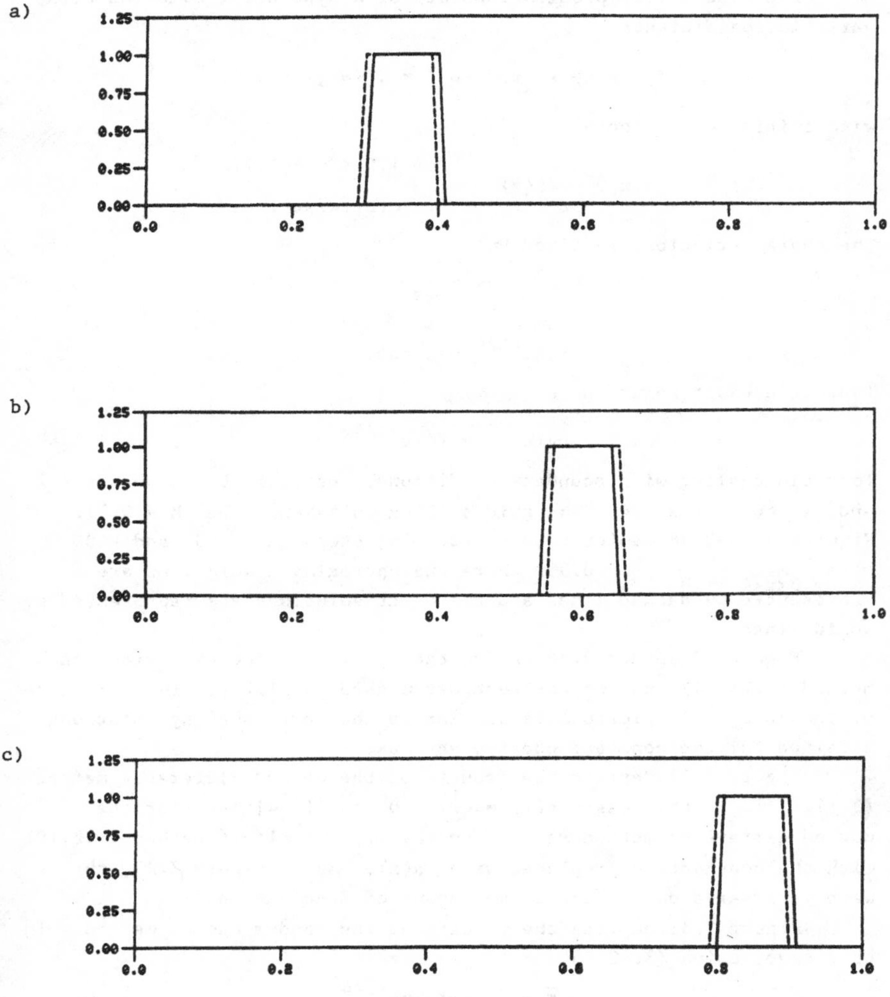

Figure 3.34. Random Choice Method for $|c|\frac{k}{h} = 0.5$ at time steps a) n = 50, b) n = 100, and c) n = 150.

The second test problem consists of a hyperbolic equation with variable coefficients

$$\partial_t v = -\frac{1}{4} x \partial_x v, \quad -\infty < x < +\infty , \quad t > 0$$

with initial condition

$$v(x,0) = f(x) = \begin{cases} 1, & 1.0 < x \leq 1.2, \\ 0, & \text{otherwise} . \end{cases}$$

The characteristics are given by

$$\frac{dx}{dt} = \frac{1}{4}x$$

or

$$xe^{-t/4} = \text{const.}$$

Thus, the exact solution is

$$v(x,t) = f(xe^{-t/4}).$$

To avoid dealing with boundary conditions, restrict t to $0 \leq t \leq 2$ and x to $0 \leq x \leq 4$. The grid spacing is taken to be $h = 0.01$. Figures 3.35-3.38 depict results for time steps $n = 200$ and 400 with $\max_{0 \leq x \leq 4} |-\frac{x}{4}|\frac{k}{h} \equiv \frac{k}{h} = 0.5$, where the approximate solutions are represented by dashed lines and the exact solutions are represented by solid lines.

Figures 3.35 and 3.36 depict the results of the Lax-Friedrichs method (3.11.11) and the Lax-Wendroff method (3.11.14). The structure of the numerical solutions is similar to the corresponding solutions obtained for the constant coefficient case.

Figure 3.37 depicts the results of the upwind difference method (3.11.18). In this case $a(x) = -\frac{1}{4}x \leq 0$ on $[0,4]$ so that the upwind difference method reduces to the (1,1) explicit method (3.2.10) with the constant c replaced with $a(x)$. As in Figure 3.27, the wave is smeared due to the large amount of dissipation.

Figure 3.38 depicts the results of the random choice method. In this case, using (3.12.25

$$\bar{x} = (i+\frac{1}{2}+\xi_n)he^{-k/4}.$$

Again, the structure of the wave is perfect. The position of the wave is not, however, exact due to the randomness.

An additional test was made of the random choice method. The characteristic curve was replaced by its tangent line through the sampling point, as outlined in Section III.12. The results were identical to those in Figure 3.38.

Using these two tests as a basis, it would appear that the random choice method is superior to the finite difference methods. An advantage of the random choice method is that in solving the Riemann problem, information about the wave is used that is not used by the finite difference methods. Finite difference methods are constructed by approximating derivativea and by trying to take combinations of differences to cancel leading terms in the truncation error. This is done formally, often without regard to the physical meaning of the equation being approximated.

A problem associated with the random choice method that will be made clear in Chapter IV is that the basis of the method is the Riemann problem. The method assumes an exact (or at least a very accurate) solution to the Riemann problem. This may not always be available.

a)

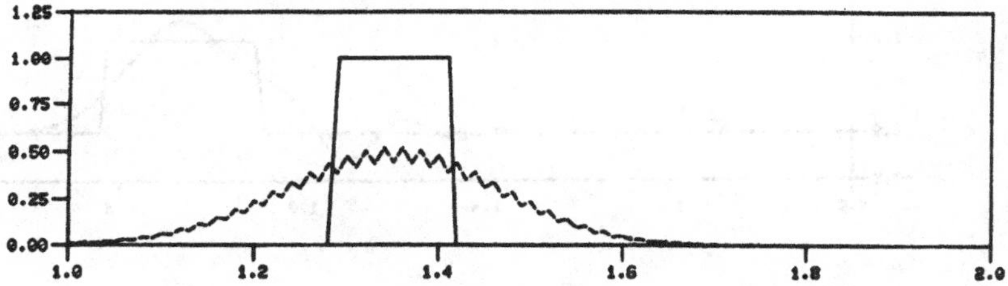

b)

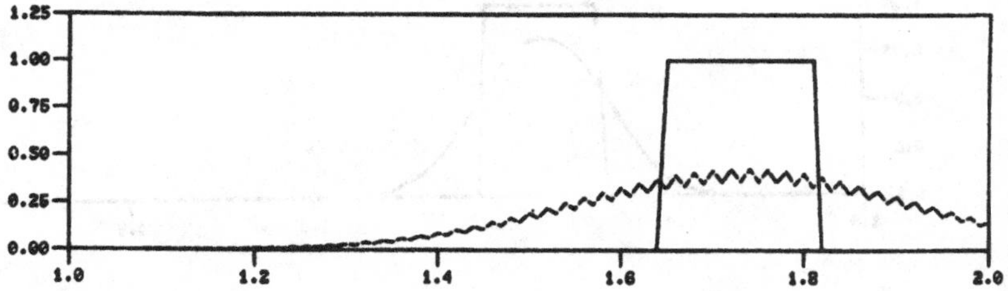

Figure 3.35. Variable Coefficient Lax-Friedrichs Method (3.11.11) for $\max|a(x)|\frac{k}{h} = 0.5$ at time steps a) n = 200 and b) n = 400.

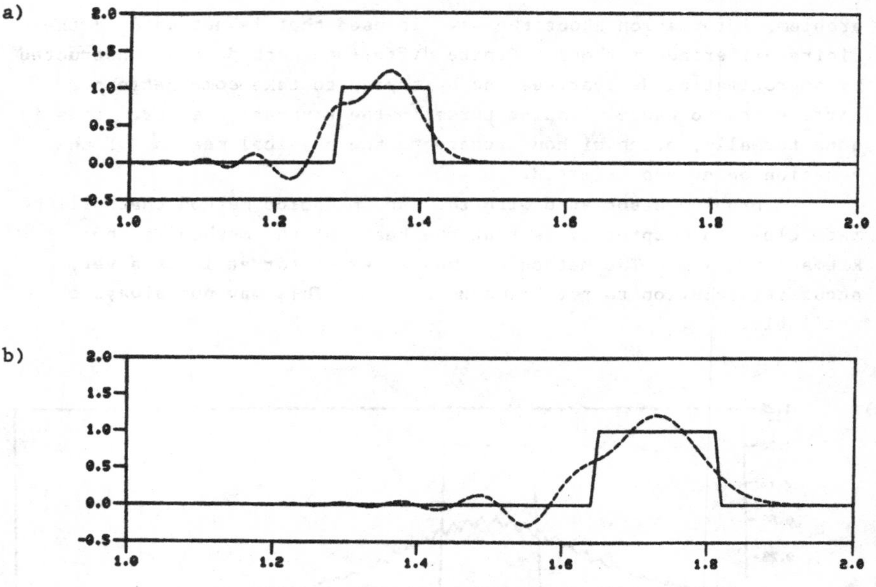

Figure 3.36. Variable Coefficient Lax-Wendroff Method (3.11.14) for $\max|a(x)|\frac{k}{h} = 0.5$ at time steps a) n = 200 and b) n = 400.

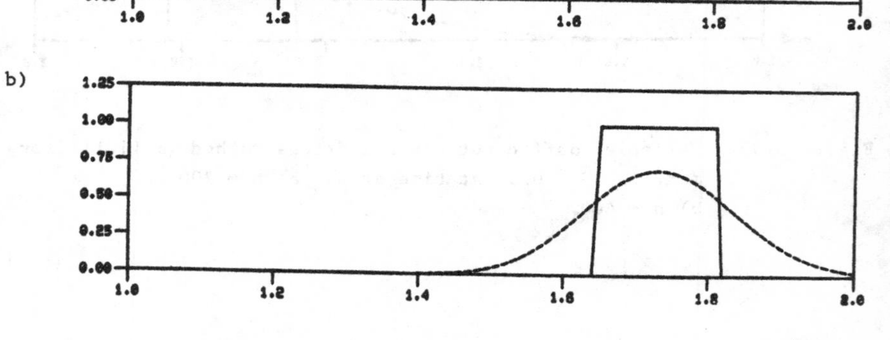

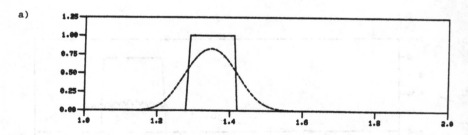

Figure 3.37. Upwind Difference Method (3.11.18) for $\max|a(x)|\frac{k}{h} = 0.5$ at time steps a) n = 100 and b) n = 400.

a)

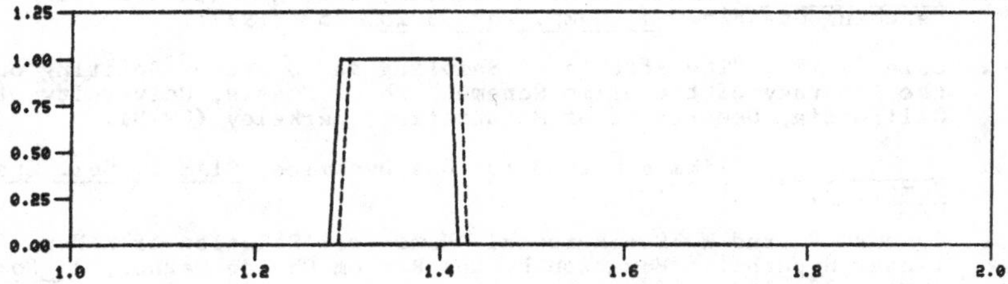

b)

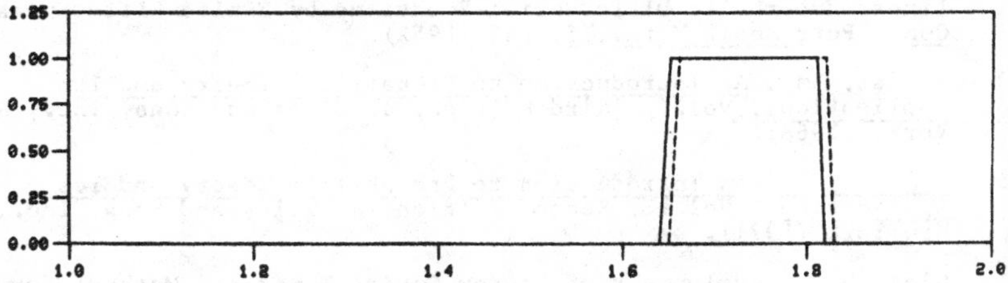

Figure 3.38. Variable Coefficient Random Choice Method $\max|a(x)|\frac{k}{h} = 0.5$ at time steps a) n = 100 and b) n = 400.

III.16. References

1. Albright, N., P. Concus, and W. Proskurowski, Numerical Solution of the Multi-dimensional Buckley-Leverett Equations by a Sampling Method. SPE Paper No.7681 (1979).

2. Chorin, A. J., Random Choice Solution of Hyperbolic Systems, J. Comp. Phys., 22, 517 (1976).

3. ____________, Random Choice Methods, with Applications to Reaching Gas Flow, J. Comp. Phys., 25, 253 (1977).

4. Colella, P., "The Effects of Sampling and Operator Spliting on the Accuracy of the Glimm Scheme," Ph.D. Thesis, University of California, Department of Mathematics, Berkeley (1978).

5. __________, Glimm's Method for Gas Dynamics, SIAM J. Sci. Stat. Comp., 3, 76 (1982).

6. Concus, P. and W. Proskurowski, Numerical Solution of a Non-linear Hyperbolic Equation by the Random Choice Method, J. Comp. Phys., 30, 153 (1979).

7. Courant, R., K. O. Friedrichs, and H. Lewy, Über die Partiellen Differenzengleichungen der Mathematischen Physic, Math. Ann., 100, 32 (1928). English translation: AEC Research and Development Report NYO-7689, New York University, New York (1956).

8. __________, E. Isaacson, and M. Rees, On the Solution of Non-linear Hyperbolic Differential Equations by Finite Differences, Comm. Pure Appl. Math., 5, 243 (1952).

9. Feller, Wm., An Introduction to Probability Theory and Its Applications, Vol.I, Third Edition, J. Wiley and Sons, Inc., New York, (1968).

10. __________, An Introduction to Probability Theory and Its Applications, Vol.II, Second Edition, J. Wiley and Sons, Inc., New York (1971).

11. Glaz, H., "Development of Random Choice Numerical Methods for Blast wave Models," NSWC Report NSWC/WDL TR 78-211, Silver Spring (1979).

12. ________ and T.-P. Liu, The Asymptotic Analysis of Wave Interactions and Numerical Calculations of Transonic Nozzle Flow, Adv. Appl. Math., 5, 111 (1984).

13. Glimm, J., Solutions in the Large for Nonlinear Hyperbolic Systems of Conservation Laws, Comm. Pure Appl. Math., 18, 697 (1965).

14. ________, J. Marchesin, and D. D. McBryan, Statistical Fluid Dynamics: Unstable Fingers, Comm. Math. Phys., 74, 1 (1980).

15. ________________________________, A Numerical Method for Two Phase Flow with an Unstable Interface, J. Comp. Phys., 38, 179 (1981).

16. ________________________________, Unstable Fingers in Two Phase Flow, Comm. Pure Appl. Math., 24, 53 (1981).

17. Glimm, J., G. Marshall, and B. Plohr, A Generalized Riemann Problem for Quasi-One-Dimensional Gas Flows, Adv. Appl. Math., 5, 1 (1984).

18. Hammersley, J. M. and D. C. Handscomb, Monte Carlo Methods, Metheuen and Co., London (1967).

19. Kreiss, H. O., On Different Approximations of the Dissipative Type for Hyperbolic Differential Equations, Comm. Pure Appl. Math., 17, 335 (1964).

20. Kreiss, H. O. and J. Oliger, Comparison of Accurate Methods for the Integration of Hyperbolic Equations, Tellus, 24, 199 (1972).

21. Lax, P. D., Weak Solutions of Nonlinea Hyperbolic Equations and Their Numerical Computations, Comm. Pure Appl. Math., 7, 159 (1954).

22. _________, On the Stability of Difference Approximations to the Solution of Hyperbolic Equations with Variable Coefficients, Comm. Pure Appl. Math., 14, 497 (1967).

23. _________, Nonlinear Partial Differential Equations and Computing, SIAM Review, 11, 7 (1969).

24. Lax, P. D. and A. N. Milgram, Parabolic Equations, Contributions to the Theory of Partial Differential Equations, Annals of Math. Studies, 33, 167 (1954).

25. Lax, P. D. and B. Wendroff, Systems of Conservation Laws, Comm. Pure Appl. Math., 13, 217 (1960).

26. Lax, P. D. and L. Nirenberg, On Stability of Difference Schemes: A Sharp Form of Garding's Inequality, Comm. Pure Appl. Math., 19, 473 (1966).

27. Lelevier, R., Lectures on Hydrodynamics and Shock waves, Lawrence Livermore Laboratory Report UCRL-4333 Rev. 1, University of California, Livremore (1965).

28. Liu, T.-P., The Deterministic Version of the Glimm Scheme, Comm. Math. Phys., 57, 135 (1977).

29. Parlett, B. N., Accuracy and Dissipation in Difference Schemes, Comm. Pure Appl. Math., 19, 111 (1966).

30. Richtmyer, R. D., and K. W. Morton, Difference Methods for Initial-Value Problems, 2nd ed., Interscience, New York (1967).

31. Sod, G. A., A Numerical Study of a Converging Cylindrical Shock, J. Fluid Mech., 83, 785 (1977).

32. _________, "A Study of Cylindrical Implosion," Lecture on Combustion Theory, Samuel Z. Burstein, Peter D. Lax, and Gary A. Sod, editors, Department of Energy Research and Development Report COO-3077-153, New York University, (1978).

33. _________, A Survey of Several Finite Difference Methods for Systems of Nonlinear Hyperbolic Conservation Laws, J. Comp. Phys., 27, 1 (1978).

34. _________, "A Hybrid Random Choice Method with Application to Combustion Engines," Proceedings Sixth International Conference on Numerical Methods in Fluid Dynamics, Lecture Notes in Physics, Vol.90, H. Cabannes, M. Holt, and V. Rusanov, Editors. Springer-Verlag (1979).

35. Sod, G. A., "Automotive Engine Modeling with a Hybrid Random Choice Method," SAE paper 790242, (1979).

36. __________, "A Hybrid Random Choice Method with Applications to Automotive Engine Modeling," Proceedings of the Third IMACS International Symposium on Computer Methods for Partial Differential Equations, Bethlehem, Pennsylvania, (1979).

37. __________, Computational Fluid Dynamics with Stochastic Techniques, Proceedings of the von Karman Institute of Fluid Dynamics Lecture Series on Computational Fluid Dynamics, Brussels (1980).

38. __________, "Automotive Engine Modeling with a Hybrid Random Choice Method, II," "SAE paper 800288, (1980).

39. __________, Numerical Modeling of Turbulent Combustion on Reciprocating Inter Combustion Engines, SAE paper 820041, (1982).

40. __________, A Flame Dictionary Approach to Unsteady Combustion Phenomena, Mat. Aplic. Comp., 3, 157 (1984).

41. __________, A Random Choice Method with Application to Reaction-Diffusion Systems in Combustion, Comp. and Maths. Appls., 11, (1985).

42. __________, A Numerical Study of Oxygen-Diffusion in a Spherical Cell with the Michaelis-Menten Oxygen Uptake Kinetics, to appear in Advances in Hyperbolic Partial Differential Equations, Vol. III, (1985).

43. Strang, G., Polynomial Approximations of Bernstein Type, Trans. Amer. Math. Soc., 105, 525 (1962).

44. __________, Accurate Partial Difference Methods I: Linear Cauchy Problems, Arch. Rational Mech. Anal., 12, 392 (1963).

45. __________, Accurate Partial Difference Methods II: Nonlnear Problems, Numerische Mathematik, 6, 37 (1964).

46. __________, On the Construction and Comparison of Difference Schemes, J. Soc. Indust. Appl. Math. Num. Anal., 5, 506 (1968).

47. Widlund, O., On Lax's Theorem, on Friedrichs Type Finite Difference Schemes, Comm. Pure Appl. Math., 24, 117 (1971).

48. Yanenko, N. N., The Method of Fractional Steps, Springer-Verlag, New York (1971).

IV. HYPERBOLIC CONSERVATION LAWS

IV.1. Introduction to Hyperbolic Conservation Laws

Consider the nonlinear equation written in the form

(4.1.1) $$\partial_t v + \partial_x F(v) = 0 .$$

Such a differential equation is said to be in conservation form. A differential equation that can be written in conservation form is a conservation law.

For example, consider the inviscid Burgers' equation

$$\partial_t v + v\partial_x v = 0 .$$

If the solution v is smooth, this equation may be written in conservation form (4.1.1) where $F(v) = \frac{1}{2} v^2$, that is,

(4.1.2) $$\partial_t v + \partial_x(\frac{1}{2} v^2) = 0 .$$

A conservation law states that the time rate of change of the total amount of a substance contained in some region, for example the closed interval $[a,b]$, is equal to the inward flux of that substance across the boundaries of that region. Suppose $v(x,t)$ denotes the density of the substance at the point (x,t). Then the total amount of the substance in $[a,b]$ is given by the integral

$$\int_a^b v(x,t)dx .$$

Let the flux be denoted by $F(v)$, then the flux across the boundaries of $[a,b]$ is $-F(v(x,t))|_a^b$. Thus the conservation law may be written in the form

$$\frac{d}{dt} \int_a^b v(x,t)dx = -F(v(x,t))|_a^b$$

or by interchanging the integration and differential signs

$$\int_a^b \partial_t v(x,t)dx = -F(v(x,t))|_a^b$$

Observe that $F(v(x,t))|_b^a = \int_a^b \partial_x F(v)dx$ so that the integral form of the conservation law becomes

(4.1.3) $$\int_a^b (\partial_t v + \partial_x F(v))dx = 0 .$$

At every point where all partial derivatives of v and $F(v)$ exist, we obtain the differential form of the conservation law (4.1.1).

Carrying out the differentiation with respect to x in (4.1.1) yields the quasilinear equation

$$\partial_t v + a(v)\partial_x v = 0 \tag{4.1.4}$$

where $a(v) = \partial F/\partial v$. Equation (4.1.4) is <u>genuinely nonlinear</u> provided that a is a non-constant function of $\underline{v}$, that is, $\partial a/\partial v \neq 0$.

The characteristic (or characteristic curve) is given by the equation

$$\frac{dx}{dt} = a(v) \ . \tag{4.1.5}$$

The left-hand side of (4.1.4) can be viewed as the derivative of v in the characteristic direction. So that v is constant along the characteristic. Since v is constant along the characteristic, it follows that $a(v)$ is constant along the characteristic. Thus, by (4.1.5), the characteristics are straight lines. However, the slope depends on the solution v and, therefore, may intersect.

Consider equation (4.1.1) (or (4.1.3)) along with the initial condition

$$v(x,0) = f(x) \ . \tag{4.1.6}$$

Let x_1 and x_2 be two points on the initial line. The slopes of the characteristics at these points are $a_1 = a(v(x_1,0))$ and $a_2 = a(v(x_2,0))$. If the two characteristics with slopes a_1 and a_2 intersect at some point (x,t), then the solution v at this point (x,t) must be equal to both $v(x_1,0)$ and $v(x_2,0)$ (see Figure 4.1).

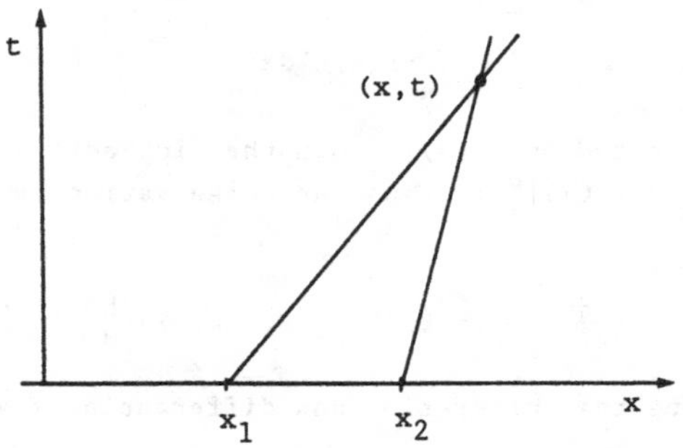

Figure 4.1

Something has gone awry; the solution becomes multi-valued at the point (x,t). This means the differential form of the conservation law (4.1.1) ceases to be valid. Physical (experimental) experience discloses that a discontinuous solution appears beyond this point (x,t) in time. To understand the discontinuous solution, we return to the integral form of the conservation law (4.1.3).

Define the operator $\underline{\nabla} \equiv (\partial_t, \partial_x)$ and the vector $\underline{w}_v \equiv (v, F(v))$. Then equation (4.1.1) may be written in the form

$$\underline{\nabla} \cdot \underline{w}_v = 0 .$$

Let $\phi(x,t)$ be a smooth scalar function that vanishes for $|x|$ sufficiently large, then $\phi \underline{\nabla} \cdot \underline{w}_v = 0$ and by integrating over $-\infty < x < +\infty$ and $t > 0$,

$$\int_0^{\infty} \int_{-\infty}^{+\infty} \phi \underline{\nabla} \cdot \underline{w}_v \, dx dt = 0 .$$

Using the divergence theorem,

$$\int_0^{\infty} \int_{-\infty}^{+\infty} (\underline{\nabla}\phi) \cdot \underline{w}_v \, dx dt + \int_0^{\infty} \int_{-\infty}^{+\infty} \phi(x,0) f(x) dx = 0 . \tag{4.1.7}$$

If $\underline{w}_v$ is smooth, the steps can be reversed. Equation (4.1.7) is called the <u>weak form of the conservation law</u>. Any function $\underline{w}_v$ that satisfies equation (4.1.7) for every smooth function ϕ that vanishes for $|x|$ sufficiently large is called a <u>weak solution</u> of the conservation law. If $\underline{w}_v$ is smooth, then it is called a <u>strong solution</u> of the conservation law.

It is important to note that equation (4.1.7) remains valid even if $\underline{w}_v$ is not smooth because $\underline{w}_v$ can still be integrated.

Consider the weak solution $\underline{w}_v$ of (4.1.7), which is discontinuous along some line L whose equation is $x = \psi(t)$ in the (x,t)-plane (see Figure 4.2).

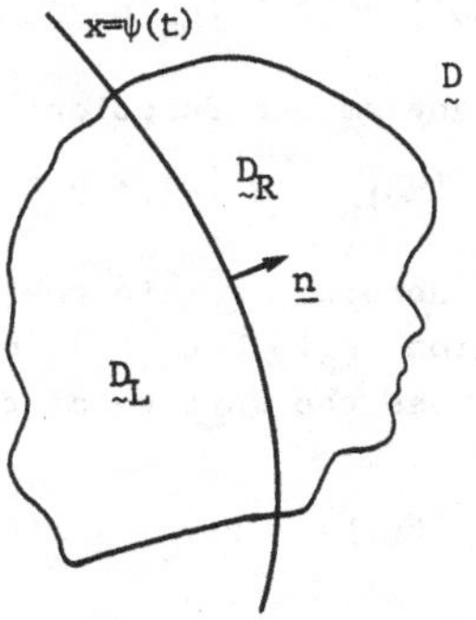

Figure 4.2

The region $\underset{\sim}{D}$ in the upper half plane is cut into two regions $\underset{\sim}{D}_L$ and $\underset{\sim}{D}_R$ by the line L. Since $\underline{w}_v$ is a weak solution of (4.1.7),

(4.1.8) $\iint_{\underset{\sim}{D}} (\underline{\nabla}\phi)\cdot\underline{w}_v dx\, dt = \iint_{\underset{\sim}{D}_L} (\underline{\nabla}\phi)\cdot\underline{w}_v\, dx\, dt + \iint_{\underset{\sim}{D}_R} (\underline{\nabla}\phi)\cdot\underline{w}_v dx\, dt = 0.$

ϕ can be any smooth function, so let ϕ be a smooth function that vanishes on all of the boundaries of $\underset{\sim}{D}_L$ and $\underset{\sim}{D}_R$ except for the line L. By applying the divergence theorem to the first integral on the right hand side of (4.1.8)

$$\iint_{\underset{\sim}{D}_L} (\underline{\nabla}\phi)\cdot\underline{w}_v dx\, dt = \iint_{\underset{\sim}{D}_L} \underline{\nabla}\cdot(\phi\underline{w}_v) dx\, dt - \iint_{\underset{\sim}{D}_L} \phi(\nabla\cdot\underline{w}_v) dx\, dt$$

$$= \int_{L_L} \phi\underline{w}_v\cdot n\, ds - \iint_{\underset{\sim}{D}_L} \phi(\underline{\nabla}\cdot\underline{w}_v) dx\, dt,$$

where $\underline{n}$ is a unit normal vector to L with orientation as depicted in Figure 4.2. $\underline{w}_v$ is smooth in $\underset{\sim}{D}_L$, so $\underline{\nabla}\cdot\underline{w}_v = 0$ and

$$\iint_{\underset{\sim}{D}_L} \phi(\underline{\nabla}\cdot\underline{w}_v) dx\, dt = 0\ .$$

Therefore,

(4.1.9) $$\iint_{\underset{\sim}{D}_L} (\underline{\nabla}\phi)\cdot\underline{w}_v\, dx\, dt = \int_{L_L} \phi\underline{w}_v\cdot \underline{n}\, ds\ .$$

Similarly, by applying the divergence theorem to the second integral on the right hand side of (4.1.8)

(4.1.10) $$\iint_{\underset{\sim}{D}_R} (\underline{\nabla}\phi)\cdot\underline{w}_v\, dx\, dt = -\int_{L_R} \phi\underline{w}_v \cdot \underline{n}\, ds\ ,$$

where the negative sign in (4.1.10) is a result of the orientation of the unit normal $\underline{n}$ (see Figure 4.2). By substitution of (4.1.8) and (4.1.10) in (4.1.7), we obtain the condition across the discontinuity along L,

$$\int_{L_L} \phi\underline{w}_v \cdot \underline{n}\, ds - \int_{L_R} \phi\underline{w}_v \cdot \underline{n}\, ds = 0\ .$$

Since this is true for any smooth function ϕ,

(4.1.11) $$(\underline{w}_v|_L - \underline{w}\,|_R) \cdot \underline{n} = 0$$

across L. Here $\underline{w}_v|_L$ denotes $\underline{w}_v$ in the region $\underset{\sim}{D}_L$ and $\underline{w}_v|_R$ denotes $\underline{w}_v$ in the region $\underset{\sim}{D}_R$. Let [ ] denote the jump in the quantity in the brackets across the surface of the discontinuity, for example,

$$[f(v)] = f(v_L) - f(v_R)\ .$$

With this, (4.1.11) becomes

(4.1.12) $$[\underline{w}_v\cdot n] = 0\ ,$$

called the <u>Rankine-Hugoniot (R-H) condition</u> for the jump condition. Let S denote the speed of propagation of the discontinuity, then $S = d\psi/dt$. From the equation of the curve L, $x - \psi(t) = 0$, we see that the normal vector $\underline{n}$ has components

$$\underline{n} = (-\frac{d\psi}{dt}, 1) = (-s,1) .$$

But then, the R-H condition (4.1.12) becomes

(4.1.13a) $$S[v] = [f(v)]$$

across the discontinuity. This may be expanded to yield across the discontinuity,

(4.1.13b) $$S(v_R - v_L) = F(v_R) - F(v_L) ,$$

where v_L and v_R are the states on the left and right of the discontinuity (see Figure 4.3).

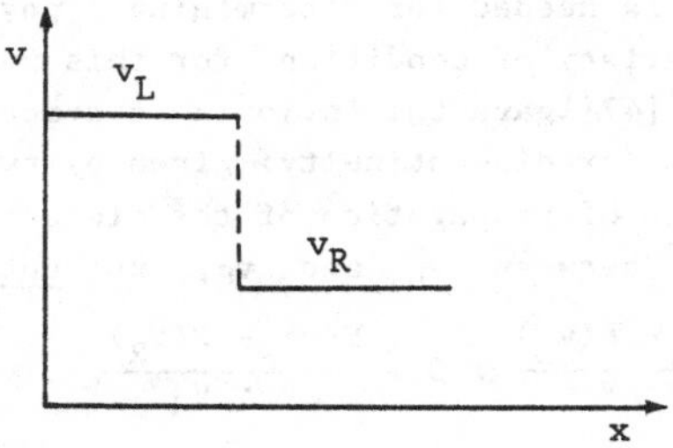

Figure 4.3. Profile of v at a given time.

From (4.1.13b),

(4.1.14) $$S = \frac{F(v_R) - F(v_L)}{v_R - v_L} .$$

The class of weak solutions of weak form of the conservation law (4.1.6) is too large in the sense that there is no uniqueness in general. To see this, consider the stationary Burgers' equation

$$\partial_x(v^2) = 0$$

which, in the class of smooth solutions, has only one $v \equiv$ constant. However, the class of weak solutions gives rise to $v^2 = c > 0$ from which $v(x) = \{\begin{smallmatrix} c, & 0 < x < p \\ -c, & p < x < 2p \end{smallmatrix}$, $v(x+2p) = v(x)$, and p is a positive constant, as depicted in Figure 4.4.

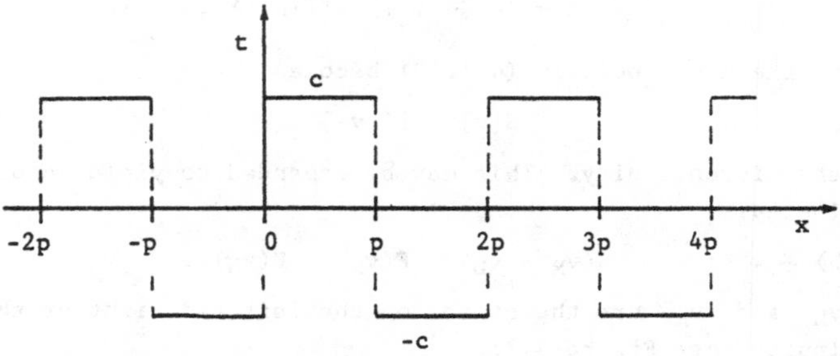

Figure 4.4.

An additional principle is needed for determining a physically relevant solution. There is a variety of conditions for this purpose. In a classic paper, Oleinick [47] gave the following characterization of an admissible weak solution (or discontinuity) given by two states v_L and v_R, along with the speed of propagation of the discontinuity S (4.1.14). For every v between v_L and v_R, the <u>entropy condition</u>

(4.1.15) $$\frac{F(v) - F(v_L)}{v - v_L} \geq S \geq \frac{F(v) - F(v_R)}{v - v_R}$$

must be satisfied. Weak solutions satisfying the entropy condition (4.1.15) are uniquely determined by the initial data.

A discontinuity is called a <u>shock</u> if the inequality signs in (4.1.15) are strict inequalities for all v between v_L and v_R. A discontinuity is called a <u>contact discontinuity</u> if equality in (4.1.15) holds identically.

In the case of a contact discontinuity, (4.1.15) yields for v between v_L and v_R,

$$F(v) = S(v - v_L) + F(v_R),$$

which is linear. Thus the conservation law (4.1.1) is linear in the range of v between v_L and v_R.

By the definition of a(v),

(4.1.16a) $$a(v_L) = \frac{\partial F}{\partial v}(v_L) = \lim_{v \to v_L} \frac{F(v) - F(v_L)}{v - v_L}$$

and

(4.1.16b) $$a(v_R) = \frac{\partial F}{\partial v}(v_R) = \lim_{v \to v_R} \frac{F(v) - F(v_R)}{v - v_R} .$$

Thus, by the entropy condition (4.1.15),

$$a(v_L) \geq S \geq a(v_R) .$$

Consider the conservation law (4.1.1) with initial conditions given by the step function

$$(4.1.17) \qquad v(x,0) = f(x) = \begin{cases} v_L , & x < 0 , \\ v_R , & x > 0 . \end{cases}$$

Such a problem is called a <u>Riemann problem</u>. A property of the solution to a Riemann problem is that it is <u>self-similar</u>, that is,

$$v(x,t) = h(x/t) ,$$

where h is a piecewise continuous function of one variable.

Another property of the solution of a Riemann problem is an additivity property. Let v_m denote the function $v(\xi,t\xi)$. Then the function $v_1(x,t)$,

$$v_1(x,t) = \begin{cases} v(x,t) & , x/t < \xi \\ v_m(x,t) & , x/t > \xi \end{cases}$$

is the solution of the Riemann problem given by the conservation law (4.1.1), along with the initial condition

$$v_1(x,0) = \begin{cases} v_L , & x < 0 \\ v_m , & x > 0 . \end{cases}$$

The function $v_2(x,t)$,

$$v_2(x,t) = \begin{cases} v_m(x,t) , & x/t < \xi \\ v(x,t) , & x/t > \xi \end{cases}$$

is the solution of the Riemann problem given by the conservation law (4.1.1), along with the initial condition

$$v_2(x,0) = \begin{cases} v_m , & x < 0 \\ v_L , & x > 0 . \end{cases}$$

Thus the two solutions v_1 and v_2 combine to form the solution v of the original Riemann problem (see Figure 4.5).

Let S be given by (4.1.14). The solution v has the form of a wave (shock wave) propagation with speed S, that is,

$$(4.1.18) \qquad v(x,t) = f(x-St) = \begin{cases} v_L , & x/t < S , \\ v_R , & x/t > S , \end{cases}$$

(see Figure 4.6).

Suppose now v_L and v_R are two states that violate the entropy condition (4.1.15), that is,

$$a(v_L) < a(v_R) .$$

Clearly, (4.1.18) is a weak solution that violates the entropy condition (4.1.15).

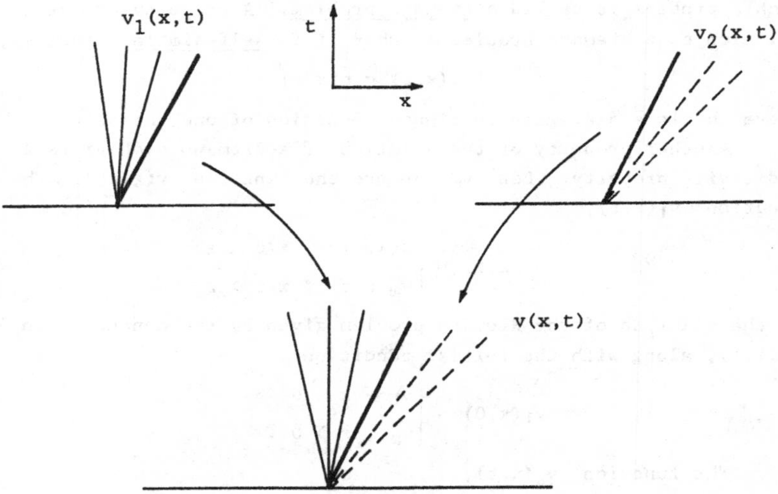

Figure 4.5.
Additivity Property of a Riemann Problem

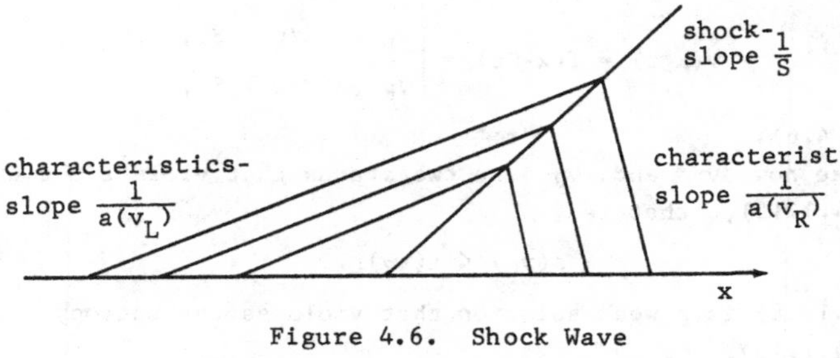

Figure 4.6. Shock Wave

In this case, the solution is an <u>expansion</u> or <u>rarefaction</u> wave. The solution has the general form

$$(4.1.19)\qquad v(x,t) = w(x/t) = \begin{cases} v_L\ , & x/t < a(v_L) \\ h(x/t)\ , & a(v_L) < x/t < a(v_R) \\ v_R\ , & x/t > a(v_R)\ , \end{cases}$$

where h is a function such that $a(h(z)) = z$. If a is an increasing function of v $(da/dv > 0)$ for every v between v_L and v_R, then the solution (4.1.19) is continuous for all time $t > 0$.

The portion of the profile for $a(v_L) < x/t < a(v_R)$ is called the <u>fan</u> (see Figure 4.7).

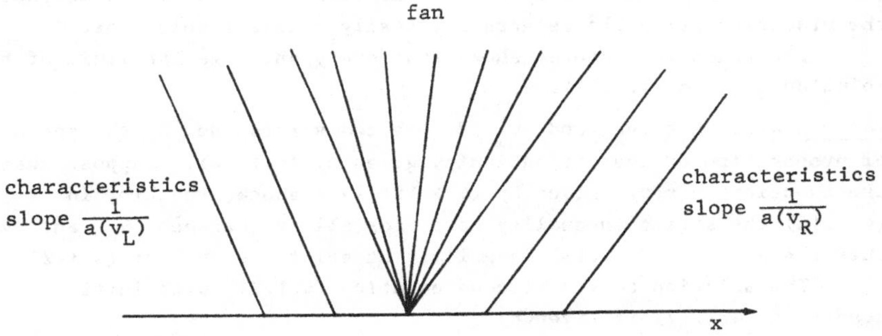

Figure 4.7. Rarefaction Wave

As an example of the solution of a Riemann problem, consider Burgers' equation (4.1.2) with initial condition (4.1.17). In this case $F(v) = \frac{1}{2}v^2$ and $a(v) = v$. For $a(v_L) > a(v_R)$, that is $v_L > v_R$ the wave is a shock with speed

$$S = \frac{F(v_L)-F(v_R)}{v_L-v_R} = \frac{v_L+v_R}{2}\ .$$

Substitution of S in (4.1.18) gives the solution for a shock wave.

Suppose $a(v_L) < a(v_R)$, that is, $v_L < v_R$. In this case the solution is a rarefaction wave with solution (4.1.19). It remains to determine the function $h(\frac{x}{t})$. Let $P = (x,t)$ denote a point in the fan, that is, $v_L < x/t < v_R$. Equate the slope of the characteristic $\frac{dx}{dt} = a(v) = v$ to the line segment between P and the origin. This gives $\frac{1}{a(v)} \equiv \frac{1}{v} = \frac{t}{x}$ or $v = x/t$. The solution (4.1.19) becomes

$$v(x,t) = \begin{cases} v_L\ , & x/t < v_L \\ x/t, & v_L < x/t < v \\ v_R\ , & x/t > v_R\ . \end{cases}$$

Another approach used to determine a physically relevant solution is the viscosity principle. In this approach, those solutions which

which are given by

(4.1.20) $$v(x,t) = \lim_{\substack{\nu \to 0 \\ \nu > 0}} v_\nu(x,t) ,$$

where $v_\nu(x,t)$ satisfies the parabolic equation

(4.1.21) $$\partial_t v_\nu + \partial_x F(v_\nu) = \nu \partial_x(\beta(v_\nu)\partial_x v_\nu) ,$$

where $\beta(w) > 0$, are admissible.

By the maximum principle (see Section II.1), this solution $v_\nu(x,t)$ is unique. The solution $v_\nu(x,t)$ of (4.1.21) exists for all time $t > 0$. Furthermore, v_ν converges to the limit v as $\nu \to 0$. Keyfitz [35] has shown that the discontinuities of v given by (4.1.20) satisfy the entropy condition (4.1.16). This verifies that the viscosity principle selects physically relevant solutions.

The following theorem shows that every shock is the limit of the solution v_ν to (4.1.21).

<u>Theorem 4.1</u>. Let v_L and v_R denote two states and S the speed of propagation of the discontinuity given by (4.1.14). Suppose that the function $v(x,t)$ given by (4.1.18) is a shock, that is, in (4.1.15) the strict inequality holds for all v between v_L and v_R. Then the solution (4.1.18) is a limit of solutions v_ν of (4.1.21).

The solution to the viscous equation (4.1.21) with initial condition (4.1.17) is given by

$$v_\nu(x,t) = w(\frac{x-st}{\nu}),$$

where $w(z)$ satisfies the ordinary differential equation

(4.1.22) $$-S\partial_z w + \partial_z F(w) = \partial_z(\beta(w)\partial_z w)$$

This solution represents a traveling wave, called a <u>shock profile</u>. Oleinik [47] has shown that every weak solution whose discontinuities are shocks is the limit of viscous solutions v_ν of (4.1.21). The viscous solutions v_ν are continuous. The effect of the viscosity is to smear (or diffuse) the continuity.

Let $g(v)$ denote the flux function in a coordinate system moving with the shock,

(4.1.23) $$g(v) = f(v) = Sv.$$

The Rankine-Hugoniot condition (4.1.13) becomes

$$g(v_L) = g(v_R) = C,$$

where C is a constant. Thus $g(v)$ is continuous across the shock. In terms of $g(v)$ the entropy condition (4.1.15) may be written in the form

$$\frac{g(v)-C}{v-v_L} > 0 > \frac{g(v)-C}{v-v_R}$$

for $v_L < v < v_R$. The strict inequality follows as the discontinuity is a shock.

Substituting g into (4.1.22),

$$\partial_z g(w) = \partial_z(\beta(w)\partial_z w)$$

or

$$g(w) - C = \beta(w)\partial_z w. \tag{4.1.25}$$

Since $g(w) - C \neq 0$, (4.1.25) may be solved for z as a function of w,

$$z(w) = \int_{w_0}^{w} \frac{\beta(w)}{g(w)-C}\, dw$$

where $v_L < w_0, w < v_R$. Let $W(v_-,v_+)$ denote the length of the y-interval in which $v_\nu(x,t) = w(\frac{y}{\nu})$ assumes values between v_- and v_+, where $v_L < v_-, v_+ < v_R$. Let $w(\frac{y_-}{\nu}) = v_-$ and $w(\frac{y_\pm}{\nu}) = v_+$ then

$$W(v_-,v_+) = y_+ - y_- = \nu\int_{v_-}^{v_+} \frac{\beta(w)}{g(w)-C}\, dw. \tag{4.1.26}$$

Thus the width of the transition region across the shock is $O(\nu)$.

Now consider a system of conservation laws

$$\partial_t \underline{v} + \partial_x \underline{F}(\underline{v}) = 0, \tag{4.1.27}$$

where $\underline{v} = (v_1,\ldots,v_p)^T$ and $\underline{F}$ is a vector-valued function of $\underline{v}$. Let $\underline{A}(\underline{v})$ denote the Jacobian of $\underline{F}(\underline{v})$, that is,

$$\underline{A}(\underline{v}) = \frac{\partial \underline{F}}{\partial \underline{v}} = (a_{ij}(\underline{v}))$$

where $a_{ij} = \partial F_j/\partial v_i$. The system (4.1.20) is <u>hyperbolic</u> if $\underline{A}(\underline{v})$ has real and distinct eigenvalues $\mu_1(v), \mu_2(v),\ldots,\mu_p(v)$ for all values of $\underline{v}$. Thus hyperbolicity not only depends on the equation but on the solution.

Without loss of generality, assume that the eigenvalues $\mu_j(\underline{v})$ are arranged in increasing order. Let $\underline{r}_j(\underline{v})$ denote the right eigenvector of $\underline{A}(\underline{v})$ corresponding to $\mu_j(\underline{v})$, that is,

$$\underline{A}(\underline{v})\underline{r}_j(\underline{v}) = \mu_j(\underline{v})\underline{r}_j(\underline{v}).$$

As in the scalar case, in order that equation (4.1.27) be genuinely nonlinear, such eigenvalue $\mu_j(v)$ must be a nonconstant function of $\underline{v}$, that is, $\underline{\nabla}_v \mu_j(v) \neq 0$, where $\underline{\nabla}_v = (\partial_{v_1},\ldots,\partial_{v_p})$. Furthermore, it is required that $\nabla_v \mu_j(v)$ not be orthogonal to $\underline{r}_j(v)$. Thus equation (4.1.20) is <u>genuinely nonlinear</u> if for $1 \leq j \leq p$ and all $\underline{v}$

$$\underline{r}_j(\underline{v}) \cdot \underline{\nabla}_v \mu_j(\underline{v}) \neq 0. \tag{4.1.28}$$

In this case, in order to solve the initial-value problem consisting of equation (4.1.20) along with the initial condition

$$\underline{v}(x,0) = \underline{f}(x)$$

for all time, we must admit weak solutions. Let $\phi(x,t)$ be a smooth scalar function that vanishes for $|x|$ sufficiently large, $\underline{v}(x,t)$ is a weak solution if it satisfies

$$\int_0^{\infty}\int_{-\infty}^{+\infty}(\partial_t\phi\underline{v} + \partial_x\phi F(\underline{v}))dx\,dt - \int_{-\infty}^{+\infty}\phi(x,0)\underline{f}(x)dx = 0.$$

This is equivalent to requiring that for all rectangles $[a,b] \times [t_1,t_2]$ equation obtained by integrating (4.1.17) over the rectangle holds

(4.1.29) $$\int_a^b\int_{t_1}^{t_2} (\partial_t\underline{v} + \partial_x F(v))dt\,dx$$

$$= \int_a^b\underline{v}(x,t_2)dx - \int_a^b\underline{v}(x,t_1)dx + \int_{t_1}^{t_2}\underline{F}(\underline{v}(b,t))dt - \int_{t_1}^{t_2}\underline{F}(v(a,t))dt = 0.$$

For a piecewise smooth weak solution, each of the p conservation laws must satisfy the Rankine-Hugoniot condition,

(4.1.30) $$S[v_j] = [F_j] \;,\; j = 1,\ldots,p,$$

across each discontinuity where S denotes the speed of propagation of the discontinuity.

As in the case of a scalar conservation law, a weak solution is not uniquely determined by the initial condition. For a scalar conservation law, the entropy condition (4.1.16) states that the characteristics on either side of the discontinuity given by $x = \psi(t)$ intersects the discontinuity. The discontinuity is a shock if in (4.1.16) we have strict inequality,

$$a(v_L(\psi(t),t)) > S(t) > a(v_R(\psi(t),t)) \;.$$

For systems of conservation laws (4.1.27), there are p families of characteristics given by the solution of the ordinary differential equation

$$\frac{dx}{dt} = \mu_j(\underline{v}) \;,\; j = 1,\ldots,p \;.$$

Let $\underline{v}_L$ and $\underline{v}_R$ denote the states to the left and right, respectively, of a discontinuity given by $x = \psi(t)$. Lax [39] proposed the following shock condition. For each j, $1 \leq j \leq p$, we require that

(4.1.31a) $$\mu_j(\underline{v}_L(\psi(t),t)) > S(t) > \mu_j(\underline{v}_R(\psi(t),t))$$

and

(4.1.31b) $$\mu_{j-1}(\underline{v}_L(\psi(t),t)) < S(t) < \mu_{j+1}(\underline{v}_R(\psi(t),t)) \;.$$

Inequalities (4.1.31a) and (4.1.31b) may be combined to obtain

$$\mu_1(\underline{v}_L) \leq \mu_2(\underline{v}_L) \leq \cdots \leq \mu_{j-1}(\underline{v}_L) < S(t) < \mu_j(\underline{v}_L)$$

and

$$\mu_j(\underline{v}_R) < S(t) < \mu_{j+1}(\underline{v}_R) \leq \mu_{j+2}(\underline{v}_R) \leq \cdots \leq \mu_p(\underline{v}_R) \;,$$

using the fact that the eigenvalues are arranged in increasing order. The number of characteristics drawn in the positive t direction with respect to the state $\underline{v}_L$ that remain to the left of the discontinuity is $j-1$. Similarly, the number of characteristics drawn in the positive t direction with respect to the state $\underline{v}_R$ that remain to the right of the discontinuity is $p-j$. Thus the total number of characteristics that do not intersect the discontinuity is $p-1$. See Figure 4.8, where the numbers refer to the different families of characteristics.

The number j is called the index of the shock. Furthermore, a discontinuity that satisfies both conditions (4.1.30) and (4.1.31) is called a j-shock.

Consider the case where the j-th family of characteristics are linearly degenerate, that is, for all solutions $\underline{v}$

(4.1.32) $$\underline{r}_j(\underline{v}) \cdot \nabla_v \mu_j(\underline{v}) = 0.$$

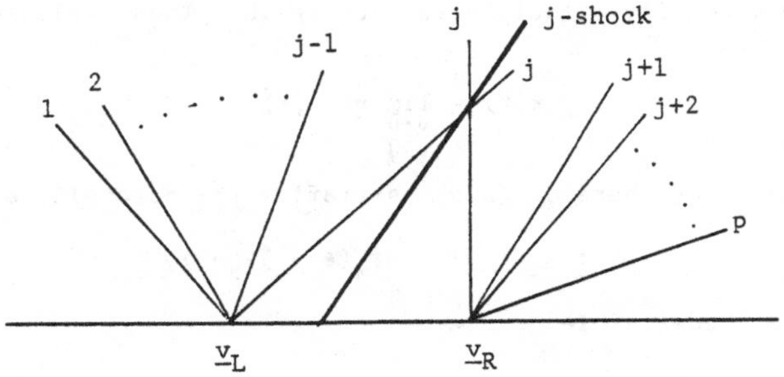

Figure 4.8. j-Shock

(4.1.32) results in the existence of a traveling wave solution, that is, a solution of the form

$$\underline{v}(x,t) = \underline{w}(x-ct) \equiv \underline{w}(t) \ .$$

By substitution into the conservation law (4.1.27),

$$-c\underline{w}'(z) + \underline{A}(\underline{w})\underline{w}'(z) = 0 \ .$$

Thus $\underline{w}'(z) = K\underline{r}_j(\underline{w}(z))$, where K is a constant, and $c = \mu_j(\underline{w}(z))$. From this, using (4.1.32)

$$\begin{aligned}\frac{dc}{dt} &= \underline{\nabla}_v c \cdot \underline{w}'(z)\\ &= K\underline{\nabla}_v \mu_j \cdot \underline{r}_j\\ &= 0 \ .\end{aligned}$$

This gives the condition that c is a constant and a traveling wave solution exists.

Suppose a function $R_j(\underline{v})$ satisfies

$$\underline{r}_j(\underline{v}) \cdot \nabla_{\underline{v}} R_j(\underline{v}) = 0$$

for all $\underline{v}$. The function $R_j(\underline{v})$ is called a <u>j-Riemann invariant</u>. Thus if the j-th family of characteristics is linearly degenerate, by (4.1.32), the eigenvalue $\mu_j(\underline{v})$ is a j-Riemann invariant.

A theorem due to Lax will now be stated. This will be very useful in a later chapter.

<u>Theorem 4.2</u>. Consider a discontinuity separated by two states $\underline{v}_L$ and $\underline{v}_R$ with the same j-Riemann invariant, then the Rankine-Hugoniot conditions (4.1.30) are satisfied with

$$S = \mu_j(\underline{v}_L) = \mu_j(\underline{v}_R) .$$

Such a discontinuity is called a <u>contact discontinuity</u>.

As in the scalar case weak solutions of (4.1.27) are not uniquely determined by the initial condition. To determine a physically relevant solution the viscosity principle is used again. Those solutions given by

$$\underline{v}(x,t) = \lim_{\substack{\nu \to 0 \\ \nu > 0}} \underline{v}_\nu(x,t), \tag{4.1.33}$$

boundedly and a.e., where $\underline{v}_\nu(x,t)$ satisfies the parabolic equation

$$\partial_t \underline{v} + \partial_x \underline{F}(\underline{v}_\nu) = \nu \partial_x (\beta(\underline{v}_\nu) \partial_x \underline{v}_\nu), \tag{4.1.34}$$

$\beta(\underline{v}_\nu) > 0$, are admissible.

<u>Definition</u>. An <u>entropy function</u> $U(\underline{v})$ is a scalar function defined by the two conditions

1) U is a convex function of $\underline{v}$, that is, $\partial^2_{\underline{v}} U > 0$

and

2) U satisfies the equation

$$\partial_{\underline{v}} U \partial_{\underline{v}} \underline{F} = F_E, \tag{4.1.35}$$

where F_E is some function called the <u>entropy flux</u>.

Consider hyperbolic systems of conservation laws (4.1.27) that have an entropy function. Multiplying (4.1.27) by $\partial_{\underline{v}} U$ and using (4.1.35), a weak solution of (4.1.27) satisfies

$$\partial_t U + \partial_x F_E = 0.$$

Lax [40] showed that if $\underline{v}(x,t)$ is given by (4.1.33) then, in the weak sense, $\underline{v}(x,t)$ satisfies

$$\partial_t U(\underline{v}) + \partial_x F_E(\underline{v}) < 0. \tag{4.1.36a}$$

This is equivalent to requiring that for all rectangles $[a,b] \times [t_1,t_2]$ the inequality obtained by integrating (4.1.36b) over the rectangle holds

(4.1.36b) $\int_a^b U(\underline{v}(x,t_2))dx - \int_a^b U(\underline{v}(x,t\ ))dx + \int_{t_1}^{t_2} F_E(\underline{v}(b,t))dt$

$$- \int_{t_1}^{t_2} F_E(\underline{v})a,t))dt \le 0.$$

If the weak solution $\underline{v}(x,t)$ is piecewise smooth with discontinuity, then across the discontinuity

$$F_E(\underline{v}_R) - F_E(\underline{v}_L) \le S[U(\underline{v}_R) - U(\underline{v}_L)],$$

where S is the speed of the discontinuity and $\underline{v}_L$ and $\underline{v}_R$ are states on either side of the discontinuity. Conditions (4.1.36) are called <u>entropy conditions</u>.

How do we construct an entropy function $U(\underline{v})$ and the corresponding entropy flux? Equation (4.1.35) represents a system of p equations. If $p \le 2$, then there are many possible solutions. For example, in the scalar case $(p = 1)$, equation (4.1.35) may be written as

$$\partial_v U(v) = \frac{F_e(v)}{a(v)},$$

or

$$U(v) = \int \frac{F_e(v)}{a(v)} dv,$$

where $a(v) = \partial_v F$, the Jacobian of $F(v)$. $F_e(v)$ can be chosen so that $U(v)$ is convex, one possibility is $F_e(v) = 2va(v)$ so that $U(v) = v^2$.

For $p > 2$, the existence of an entropy function becomes a special property of the conservation law (4.1.27). Suppose (4.1.27) can be written as a <u>symmetric hyperbolic system</u>, that is,

$$\underline{P}\partial_t \underline{w} + S\partial_x \underline{w} = 0,$$

where $\underline{P}$ is a symmetric, positive definite matrix and $\underline{S}$ is a symmetric matrix. This can be accomplished by replacing $\underline{v}$ with a new dependent variable $\underline{w}$, $\underline{v} = \underline{v}(\underline{w})$. Substitution in (4.1.27 gives a new equation

$$\partial_{\underline{w}}\underline{v}\partial_t \underline{w} + \partial_{\underline{w}}\underline{F}\partial_x \underline{w} = 0.$$

This system is symmetric hyperbolic if $\partial_{\underline{w}}\underline{v}$ is symmetric and positive definite and $\partial_{\underline{w}}\underline{F}$ is symmetric. If $r(\underline{w})$ is a smooth function of $\underline{w}$, then $\partial_{\underline{w}}^2 r$ is a symmetric matrix. Thus if $\partial_{\underline{w}}\underline{v}$ and $\partial_{\underline{w}}\underline{F}$ are symmetric, then there exist functions $r(\underline{w})$ and $s(\underline{w})$ such that

$$\partial_{\underline{w}} r = \underline{v}^T.$$

and

$$\partial_{\underline{w}} s = \underline{F}^T.$$

The final condition, that $\partial_{\underline{w}}\underline{v}$ is positive definite, is equivalent to the function $r(\underline{w})$ being convex.

The following theorem due to Godunov (see Harten and Lax [31]) shows how to construct an entropy function and entropy flux if the system of conservation laws can be written as a symmetric hyperbolic system.

Theorem 4.3. Suppose that the hyperbolic system of conservation laws (4.1.27) can be written as a symmetric hyperbolic system by introducing a new dependent variable $\underline{w}$. Suppose $r(\underline{w})$ and $s(\underline{w})$ are functions such that $r(\underline{w})$ is convex and

$$\partial_{\underline{w}} r = \underline{v}^T$$

and

$$\partial_{\underline{w}} s = \underline{F}^T,$$

then (4.1.27) has an entropy function $U(\underline{v})$ and entropy flux $F_e(\underline{v})$ defined by

$$U(\underline{v}) = \underline{v}^T\underline{w} - r(\underline{w})$$

and

$$F_e(\underline{v}) = F^T\underline{w} - s(\underline{w})$$

IV.2. Conservative Finite-Difference Methods

In this section we begin the discussion of the numerical solution of the system of hyperbolic conservation laws (4.1.1)

$$\partial_t\underline{v} + \partial_x\underline{F}(\underline{v}) = 0 .$$

A finite difference approximation of this system of conservation laws must be consistent. A weak solution of this system of conservation laws satisfies the weak (or integral) form of the conservation law. It seems natural to expect the approximate solution to satisfy the discrete analogue. This concept was introduced by Lax and Wendroff [ ].

Consider the finite difference method

$$\underline{u}_i^{n+1} = Q\underline{u}_i^n , \tag{4.2.1}$$

where Q is some finite difference operator, not necessarily linear. The finite difference operator Q has the divergence property if the solutions $\underline{u}_i^{n+1}$ and $\underline{u}_i^n$ of (4.2.1) satisfy

$$\frac{1}{k}\Big(\sum_{i=I_1}^{I_2} \underline{u}_i^{n+1}h - \sum_{i=I_1}^{I_2} \underline{u}_i^n h\Big) = -(\underline{F}(\underline{u}_{I_2}^n) - \underline{F}(\underline{u}_{I_1}^n)),$$

for any integers $I_2 > I_1$. Compare this with (4.1.2), where the first and second sums on the left hand side of (4.2.2) approximate

$$\int_{x_{I_1}}^{x_{I_2}} v(x,(n+1)k)dx \quad \text{and} \quad \int_{x_{I_1}}^{x_{I_2}} v(x,nk)dx,$$

respectively, and the divided difference on the left hand side of (4.2.2) aproximates

$$\partial_t \int_{x_{I_1}}^{x_{I_2}} v(x,t)dx \ .$$

As an example, consider the scalar conservation law

$$\partial_t v + \partial_x(a(x)v) = 0 \ ,$$

where $a(x) < 0$. In this case $F(v) = a(x)v$. Approximate this conservation law by the finite difference method

$$u_i^{n+1} = u_i^n - kD_+(a_i u_i^n)$$

or

$$\frac{u_i^{n+1} - u_i^n}{h} = -D_+(a_i u_i^n),$$

where $a_i \equiv a(ih)$. Multiply both sides by h and summing over i, $I_1 < i < I_2$,

$$\frac{1}{k}\left(\sum_{i=I_1}^{I_2} u_i^{n+1}h - \sum_{i=I_1}^{I_2} u_i^n h\right) = -\sum_{i=I_1}^{I_2} D_+ a_i u_i^n h = -\sum_{i=I_1}^{I_2} (a_{i+1}u_{i+1}^n - a_i u_i^n) \ .$$

All but two terms in the sum of the right hand side cancel in pairs resulting in $-(a_{I_2}u_{I_2}^n - a_{I_1}u_{I_1}^n)$, which corresponds to the right hand side of (4.2.2). Thus, the finite difference operator $Q = I + kD_+a$ has the divergence property.

Consider a point x and let $\underline{u}_i^n = \underline{u}(x+ih,nk)$. A finite difference method is said to be in <u>conservation form</u> if it can be written in the form

$$\text{(4.2.3)} \qquad \frac{\underline{u}(x,(n+1)k) - \underline{u}(x,nk)}{k} = -\frac{\underline{G}(x+\frac{h}{2}) - \underline{G}(x-\frac{h}{2})}{h} \ ,$$

where $\underline{G}(x+\frac{h}{2}) = \underline{G}(\underline{u}_{-q+1}^n, \underline{u}_{-q+2}^n, \ldots, \underline{u}_q^n)$ and $\underline{G}(x-\frac{h}{2}) = \underline{G}(\underline{u}_{-q}^n, \underline{u}_{-q+1}^n, \ldots, \underline{u}_{q-1}^n)$, that is, $\underline{G}$ is a vector-valued function of $2q$ arguments called a <u>numerical flux</u>. Furthermore, in order that (4.2.3) be consistent with (4.1.21), the numerical flux must be consistent with the physical flux in the sense that $\underline{G}(v,\ldots,\underline{v}) = \underline{F}(v)$.

From these two definitions, it is clear that a finite difference method that is in conservation form also satisfies the divergence property.

We now present the main theorem of this section due to Lax and Wendroff [42] that shows why it is essential that a finite difference approximation to a conservation law be in conservation form.

Theorem 4.4. Suppose that the solution $\underline{u}(x,nk)$ of a finite difference method in conservation form converges boundedly almost everywhere to some function $\underline{v}(x,t)$ as h and k approach zero. Then $\underline{v}(x,t)$ is a weak solution of (4.1.21).

Proof: Multiply (4.2.3) by a smooth scalar function $\phi(x,t)$ that vanishes for $|x|$ sufficiently large. Integrate over x and sum over all values of t that are integer multiples of k

$$(4.2.4)\qquad \sum_n \int_{-\infty}^{+\infty} \phi(x,nk)\left(\frac{\underline{u}(x,(n+1)k)-\underline{u}(x,nk)}{k}\right)dxk + \sum_n \int_{-\infty}^{+\infty} \phi(x,nk)\left(\frac{\underline{G}(x+\frac{h}{2}) - \underline{G}(x-\frac{h}{2})}{k}\right)dxk = 0.$$

By summation by parts

$$\sum_n \int_{-\infty}^{+\infty} \phi(x,nk)\left(\frac{\underline{u}(x,(n+1)k)-\underline{u}(x,nk)}{k}\right)dxk$$
$$= -\int_{\infty}^{\infty} \phi(x,0)\underline{u}(x,0)dx - \sum \int_{-\infty}^{+\infty} \left(\frac{\phi(x,nk)-\phi(x,(n-1)k)}{k}\right)\underline{u}(x,nk)dxk$$

By introducing a change of variable

$$\sum_n \int_{-\infty}^{+\infty} \frac{1}{h}\,\phi(x,nk)\underline{G}(x \pm \frac{h}{2})dxk = \sum_n \int_{-\infty}^{+\infty} \frac{1}{h}\,\phi(x \mp \frac{h}{2},nk)\underline{G}(x)dxk,$$

where $\underline{G}(x) = \underline{G}(\underline{u}^n_{-q},\dots,\underline{u}^n_{-1},\underline{u}^n_1,\dots,\underline{u}^n_q)$ and the values $\underline{u}^n_{-q},\dots,\underline{u}^n_{-1}$, $\underline{u}^n_1,\dots,\underline{u}^n_q$ are the values of $\underline{u}$ at the $2q$ points symmetrically distributed about the point (x,nk). Substitution of these two results in (4.2.4) yields

$$\sum_n \int_{-\infty}^{+\infty} \left(\frac{\phi(x,nk)-\phi(x,(n-1)k)}{k}\right)\underline{u}(x,nk)dxk + \sum_n \int_{-\infty}^{+\infty} \left(\frac{\phi(x+\frac{h}{2},nk)-(x-\frac{h}{2},nk)}{h}\right)\underline{G}(x)dxk + \int_{-\infty}^{+\infty} \phi(x,0)\underline{u}(x,0)dx = 0\ .$$

If $\underline{u}(x,nk)$ tends boundedly almost everywhere to a function $\underline{v}(x,t)$, then so do $\underline{u}^n_{-q},\dots,\underline{u}^n_{-1},\underline{u}^n_1,\dots,\underline{u}^n_q$ and $\underline{G}(x)$ tends to $\underline{G}(\underline{v},\dots,\underline{v})$, which by consistency equals $\underline{F}(\underline{v})$. Thus as $h,k \to 0$ in (4.2.5), the first term tends to $\int_0^\infty \int_{-\infty}^{+\infty} \partial_t\phi(x,t)\underline{v}(x,t)dxdt$ and the second term tends to $\int_0^\infty \int_{-\infty}^{+\infty} \partial_x\phi(x,t)\underline{F}(\underline{v})dxdt$, from which we obtain the desired limit

$$\int_0^\infty \int_{-\infty}^{+\infty} (\partial_t\phi\underline{v} + \partial_x\phi\underline{F}(\underline{v}))dxdt + \int_{-\infty}^{+\infty} \phi(x,0)\underline{v}(x,0)dx = 0 ,$$

that is, $\underline{v}(x,t)$ is a weak solution of (4.1.21).

We need to extend the Courant-Friedrich-Lewy condition to the case of a system of hyperbolic conservation laws (4.1.21). Let $\underline{A}(\underline{v})$ denote the Jacobian of $\underline{F}(\underline{v})$, that is

$$\underline{A}(\underline{v}) = \frac{\partial \underline{F}}{\partial \underline{v}} = (a_{ij}(v)) ,$$

where $a_{ij} = \partial F_j/\partial v_i$. Since the system is hyperbolic, $\underline{A}(\underline{v})$ has real and distinct eigenvalues $\mu_1(\underline{v}),\mu_2(\underline{v}),\ldots,\mu_p(\underline{v})$ for all values of $\underline{v}$. As in the linear case, the eigenvalues of $\underline{A}(\underline{v})$ represent signal speeds. The CFL condition again states that a necessary condition for convergence of a finite difference method is that the numerical signal speed must be at least as fast as the maximum signal speed. The maximum signal speed is given by

$$\max_{\substack{1<j<p \\ \underline{v}}} |\mu_j(\underline{v})| .$$

Consider a finite difference method in conservation form

$$u_i^{n+1} = Q\underline{u}_i^n = \underline{u}_i^n - \lambda(\underline{G}_{i+\frac{1}{2}} - G_{i-\frac{1}{2}}), \tag{4.2.6}$$

where $\lambda = k/h$, $\underline{G}_{i+\frac{1}{2}} = \underline{G}(\underline{u}^n_{i-q+1},\underline{u}^n_{i-q+2},\ldots,\underline{u}^n_{i+q})$, $\underline{G}_{i-\frac{1}{2}} = \underline{G}(u^n_{i-q},\underline{u}^n_{i-q+1},\ldots,\underline{u}^n_{i+q-1})$, and q is an integer > 1. In general, Q is a nonlinear finite difference operator.

The numerical domain of dependence of the point $(ih,(n+1)k)$ is the set of $2q(n+1)+1$ points on the x-axis in the interval $[(i-q(n+1))h,(i+q(n+1))h]$. Thus the numerical signal speed is qh/k and the CFL condition becomes

$$\max_{\substack{1<j<p \\ \underline{v}}} |\mu_j(v)| < \frac{qh}{k}$$

or

$$k < \frac{qh}{\max_{\substack{1<j<p \\ \underline{v}}} |\mu_j(\underline{v})|} . \tag{4.2.7}$$

As with parabolic equations, the notion of stability for the linear case has not been extended to the nonlinear case. Von Neumann, however, observed that when instabilities develop in a finite difference method, they appear as oscillations of short wave-length and of initially small amplitude superimposed on a smooth solution. The first variation of this nonlinear finite difference operator Q, which is a linear finite difference operator with variable coefficients, can often be used to determine the conditions on λ to which these instabilities will develop. It is further observed that, since these instabilities first appear in a small neighborhood, if the coefficients of the first variation of Q are sufficiently smooth in this neighbor-hood, then the variable coefficients can be replaced with their values at some point in the neighborhood, resulting in a reduction to constant coefficient linear finite difference operator. This is called a local linearization of Q and allows the use of the discrete Fourier transform and the von Neumann condition.

To develop this idea we present a theorem due to Strang [72]. Consider the system of hyperbolic conservation laws (4.1.21) with smooth solutions, which allows this system to be written as a quasi-linear system

(4.2.8) $$\partial_t \underline{v} + \underline{A}(\underline{v})\partial_x \underline{v} = 0,$$

where $\underline{A}(\underline{v})$ is the Jacobian of $\underline{F}(\underline{v})$ which, in general, is a nonlinear function of $\underline{v}$. Approximate this system (4.1.21) by the finite difference method

(4.2.9) $$\underline{u}^{n+1}(x) = H\underline{u}^n(x) = \underline{H}(\underline{u}^n(x-qh),\ldots,\underline{u}^n(x+qh)),$$

where $\underline{H}(\underline{u}^n(x-qh),\ldots,\underline{u}^n(x+qh)) = \underline{u}^n(x) - \lambda(\underline{G}(x+h/2) - \underline{G}(x-h/2))$. The function $\underline{H}$ is a nonlinear function of $2q+1$ arguments and H is a nonlinear finite difference operator acting on $\underline{u}^n(x)$.

Comparing this with (4.2.3), we see that consistency requires that for any smooth function $\underline{w}(x,t)$

$$H(\underline{w}^n(x),\ldots,\underline{w}^n(x)) = \underline{w}^n(x) + k\partial_x F(\underline{w}^n(x)) + O(k) ,$$

or by using (4.2.8)

(4.2.10) $$\underline{H}(\underline{w}^n(x),\ldots,\underline{w}^n(x)) = \underline{w}^n(x) + kA(\underline{w}^n(x))\partial_x \underline{w}^n(x) + O(k) ,$$

where $\underline{w}^n(x) = \underline{w}(x,nk)$.

The first variation of the nonlinear finite difference operator H in (4.2.9) is a linear finite difference operator with variable coefficients M, defined by

(4.2.11) $$M\underline{u}^n(x) = \sum_{j=-q}^{q} \underline{C}_j(\underline{u}^n(x))\underline{u}^n(x+jh)$$

$$= (\sum_{j=-q}^{q} \underline{C}_j(\underline{u}^n(x))S_+^j)\underline{u}^n(x),$$

where $\underline{C}_j(\underline{u}^n(x))$ denotes the Jacobian of $\underline{H}(\underline{u}^n(x-qh),\ldots,u^n(x+qh))$ with respect to the j-th argument $\underline{u}^n(x+jh)$ evaluated with each argument equal to $\underline{u}^n(x)$.

Since M is a linear finite difference operator with variable coefficients, we can consider its stability. In the following theorem due to Strang [72], the stability of the first variation of H is used to prove convergence of the finite difference method (4.2.9), provided that the solution is sufficiently smooth.

Theorem 4.5. Suppose that the finite difference method (4.2.9) is consistent and accurate $O(k^n)$ (where $k = O(h)$). If the first variation of the finite difference operator H is stable in the ℓ_2-norm, and if the solution $\underline{v}(x,t)$, the Jacobian $\underline{A}(\underline{v})$, and the function $\underline{H}$ in (4.2.9) are sufficiently smooth, then the approximate solution $\underline{u}_i^n$ converges to $\underline{v}(ih,nk)$ as $k \to 0$, that is $\underline{u}_i^n = v(ih,nk) + O(k^r)$.

Consider the Lax-Friedrichs method applied to the system of hyperbolic conservation laws (4.1.1)

(4.2.12)
$$\begin{aligned}\underline{u}_i^{n+1} &= \tfrac{1}{2}(\underline{u}_{i+1}^n + \underline{u}_{i-1}^n) - kD_0\underline{F}(\underline{u}_i^n)\\ &= \tfrac{1}{2}(\underline{u}_{i+1}^n + \underline{u}_{i-1}^n) - \tfrac{\lambda}{2}(\underline{F}_{i+1}^n - \underline{F}_{i-1}^n),\end{aligned}$$

where $\lambda = k/h$ and $\underline{F}_i^n \equiv \underline{F}(\underline{u}_i^n)$. With $q = 1$ in (4.2.6) define

(4.2.13a)
$$\underline{G}_{i+\frac{1}{2}} \equiv \underline{G}(\underline{u}_i^n,\underline{u}_{i+1}^n) = -\tfrac{1}{2\lambda}(\underline{u}_{i+1}^n - \underline{u}_i^n) + \tfrac{1}{2}(\underline{F}_{i+1}^n + \underline{F}_i^n)$$

and

(4.2.13b)
$$\underline{G}_{i-\frac{1}{2}} \equiv \underline{G}(\underline{u}_{i-1}^n,\underline{u}_i^n) = -\tfrac{1}{2\lambda}(\underline{u}_i^n - \underline{u}_{i-1}^n) + \tfrac{1}{2}(\underline{F}_i^n + \underline{F}_{i-1}^n)$$

The Lax-Friedrichs method (4.2.12) can be written in conservation form (4.2.6) with this choice of G. The form of $\underline{G}$ a function of two variables is

(4.2.14)
$$\underline{G}(\underline{a},\underline{b}) = -\tfrac{1}{2\lambda}(\underline{a}-\underline{b}) + \tfrac{1}{2}(\underline{F}(\underline{a}) + \underline{F}(\underline{b})),$$

so that $\underline{G}(\underline{v},\underline{v}) = \underline{F}(\underline{v})$ and, hence, the Lax-Friedrichs method is consistent. $\underline{H}$ in (4.2.9) is given by

(4.2.15)
$$\begin{aligned}\underline{H}(\underline{u}_{i-1}^n,\underline{u}_i^n,\underline{u}_{i+1}^n) &= \underline{u}_i^n - \lambda(\underline{G}_{i+\frac{1}{2}} - \underline{G}_{i-\frac{1}{2}})\\ &= \tfrac{1}{2}(\underline{u}_{i+1}^n + \underline{u}_{i-1}^n) - \tfrac{\lambda}{2}(\underline{F}_{i+1}^n - \underline{F}_{i-1}^n).\end{aligned}$$

The form of $\underline{H}$ as a function of three variables is

(4.2.16)
$$\underline{H}(\underline{a},\underline{b},\underline{c}) = \tfrac{1}{2}(\underline{c} + \underline{a}) - \tfrac{\lambda}{2}(\underline{F}(\underline{c}) - \underline{F}(\underline{a})).$$

Consider the first variation of h and the linear finite difference operator M (4.2.11),

(4.2.17) $$M\underline{u}^n(x) = (\sum_{j=-1}^{1} \underline{C}_j(\underline{u}^n(x))S_+^j)\underline{u}^n(x),$$

where

$$\underline{C}_{-1}(\underline{w}) = \partial_{\underline{a}}\underline{H}(\underline{w},\underline{w},\underline{w}) = (\tfrac{1}{2} + \tfrac{\lambda}{2}\,\partial_{\underline{a}}\underline{F})(\underline{w}) = \tfrac{1}{2} + \tfrac{\lambda}{2}\underline{A}(\underline{w}),$$

$$\underline{C}_0(w) = \partial_{\underline{b}}\underline{H}(\underline{w},\underline{w},\underline{w}) = 0,$$

$$\underline{C}_1(\underline{w}) = \partial_{\underline{c}}\underline{H}(\underline{w},\underline{w},\underline{w}) = (\tfrac{1}{2} - \tfrac{\lambda}{2}\,\partial_{\underline{a}}\underline{F})(\underline{w}) = \tfrac{1}{2} - \tfrac{\lambda}{2}\underline{A}(\underline{w}),$$

and $\underline{A}(\underline{w})$ denote the Jacobian of $\underline{F}$ evaluated at $\underline{w}$. Setting $\underline{w} = \underline{u}^n(x)$ and substituting into (4.2.17),

$$\underline{u}^{n+1}(x) = [(\tfrac{1}{2} + \tfrac{\lambda}{2}\underline{A}(\underline{u}^n(x)))S_+^{-1} + (\tfrac{1}{2} - \tfrac{\lambda}{2}\underline{A}(\underline{u}^n(x)))S_+]\underline{u}^n(x)$$

$$= \tfrac{1}{2}(\underline{u}^n(x+h) + \underline{u}^n(x-h)) - \tfrac{\lambda}{2}\underline{A}(\underline{u}^n(x))(\underline{u}^n(x+h) - \underline{u}^n(x-h)).$$

With $x = ih$ this reduces to

(4.2.18) $$\underline{u}_i^{n+1} = \tfrac{1}{2}(\underline{u}_{i+1}^n + \underline{u}_{i-1}^n) - \tfrac{\lambda}{2}\underline{A}(\underline{u}^n(ih))(\underline{u}_{i+1}^n - \underline{u}_{i-1}^n),$$

which is the Lax-Friedrichs method for hyperbolic systems with variable coefficients (3.11.11). Assuming that $\underline{A}(\underline{u}^n(x))$ is sufficiently smooth, by Kreiss' theorem (Theorem 3.6), (4.2.18) is stable provided that the CFL condition (4.2.7) is satisfied. This completes the linearized stability analysis.

An alternate, yet equivalent, way to derive the Lax-Wendroff method is to consider the Taylor series expansion in terms of $v(x,t)$ about the (ih,nk),

(4.2.19) $$\underline{v}_i^{n+1} \equiv \underline{v}(ih,(n+1)k) = \underline{v}_i^n + k(\partial_t\underline{v})_i^n + \tfrac{k^2}{2}(\partial^2\underline{v})_i^n + O(k^3).$$

Using the conservation law (4.1.1),

$$\partial_t\underline{v} = -\partial_x F = -\underline{A}(\underline{v})\partial_x\underline{v}$$

and

$$\partial_t^2\underline{v} = -\partial_t\partial_x\underline{F} = -\partial_x(\partial_t\underline{F}) = -\partial_x(\underline{A}(\underline{v})\partial_t\underline{v})$$

$$= \partial_x(\underline{A}^2(\underline{v})\partial_x\underline{v}),$$

where $\underline{A}(\underline{v})$ is the Jacobian of $\underline{F}(\underline{v})$. Substituting into the Taylor series

(4.2.20) $$\underline{v}_i^{n+1} = \underline{v}_i^n - k\underline{A}(\underline{v}_i^n)(\partial_x\underline{v})_i^n + \frac{k^2}{2}[\partial_x(\underline{A}^2(\underline{v})\partial_x\underline{v})]_i^n + O(k^3)$$

Approximate the two space derivatives in (4.2.20) by second order difference (see Section III.11)

$$D_0\underline{v}_i^n = (\partial_x\underline{v})_i^n + O(h^2)$$

and

(4.2.21) $$D_+(\underline{A}^2_{i-\frac{1}{2}}D_-\underline{v}_i^n) = [\partial_x(\underline{A}^2(\underline{v})\partial_x\underline{v})]_i^n + O(h^2),$$

where $\underline{A}^2_{i-\frac{1}{2}}$ $\underline{A}^2(v^n_{i-\frac{1}{2}}) = \underline{A}^2(\frac{\underline{v}_i^n + \underline{v}_{i+1}^n}{2}) + O(h^2)$. Substituting into (4.2.20) gives the Lax-Wendroff method

(4.2.22) $$\underline{u}_i^{n+1} = \underline{u}_i^n - k\underline{A}_iD_0\underline{u}_i^n + \frac{k^2}{2}D_+(\underline{A}^2_{i-\frac{1}{2}}D_-\underline{u}_i^n),$$

where $\underline{A}_i = \underline{A}(\underline{u}_i^n)$ and $\underline{A}^2_{i-\frac{1}{2}} = \underline{A}^2(\frac{\underline{u}_i^n + \underline{u}_{i+1}^n}{2})$, which is accurate of $O(k^2) + O(h^2)$.

Clearly this finite difference method is not in conservation form. It can be written in the form

$$\underline{u}_i^{n+1} = H\underline{u}_i^n = H(\underline{u}_{i-1}^n,\underline{u}_i^n,\underline{u}_{i+1}^n),$$

where

$$H(\underline{a},\underline{b},\underline{c}) = \underline{b} - \frac{\lambda}{2}A(\underline{b})(\underline{c}-\underline{a}) + \frac{\lambda^2}{2}(\underline{A}^2(\frac{b+c}{2})c-b)$$
$$- \underline{A}^2(\frac{a+b}{2})(b-a)).$$

Consider the first variation of H and the linear finite difference operator M (4.2.11)

(4.2.23) $$M\underline{u}^n(x) = (\sum_{j=-1}^{1}\underline{C}_j(\underline{u}^n(x))S_+^j)\underline{u}^n(x)$$

where

$$\underline{C}_{-1}(\underline{w}) = \partial_{\underline{a}}\underline{H}(\underline{w},\underline{w},\underline{w}) = \frac{\lambda}{2}\underline{A}(\underline{w}) + \frac{\lambda^2}{2}\underline{A}^2(\underline{w}),$$

$$\underline{C}_0(\underline{w}) = \partial_{\underline{b}}\underline{H}(\underline{w},\underline{w},\underline{w}) = 1 - \lambda^2\underline{A}^2(\underline{w}),$$

$$\underline{C}_1(\underline{w}) = \partial_{\underline{c}}\underline{H}(\underline{w},\underline{w},\underline{w}) = -\frac{\lambda}{2}\underline{A}(\underline{w}) + \frac{\lambda^2}{2}\underline{A}^2(\underline{w}).$$

Setting $\underline{w} = \underline{u}^n(x)$ and substituting into (4.2.20),

$$\underline{u}^{n+1}(x) = [(\tfrac{\lambda}{2}\underline{A}(\underline{u}^n(x)) + \tfrac{\lambda^2}{2}\underline{A}^2(\underline{u}^n(x)))S_+^{-1} + (1-\lambda^2\underline{A}^2(\underline{u}^n(x)))I$$

$$+ (-\tfrac{\lambda}{2}\underline{A}(\underline{u}^n(x)) + \tfrac{\lambda^2}{2}\underline{A}^2(\underline{u}^n(x)))S_+]\underline{u}^n(x)$$

$$= u^n(x) + \tfrac{\lambda}{2}\underline{A}(\underline{u}^n(x))D_0\underline{u}^n(x) + \tfrac{\lambda^2}{2}\underline{A}^2(\underline{u}^n(x))D_+D_-\underline{u}^n(x).$$

This is the linear finite difference method with variable coefficients (3.11.17). If $\underline{A}(\underline{u}^n(x))$ is hermitian, Lipschitz continuous, and uniformly bounded, then the linearized finite difference method (4.2.23) is stable, provided the CFL condition (4.2.7) is satisfied.

It should be mentioned that there is another choice for the form of $\underline{A}^2_{i-\frac{1}{2}}$ in (4.2.22). We could also take

$$\underline{A}_{i-\frac{1}{2}} = \tfrac{1}{2}(\underline{A}_i + \underline{A}_{i-1}),$$

since

$$\underline{A}_{i-\frac{1}{2}} \equiv \underline{A}(\underline{v}^n_{i-\frac{1}{2}}) = \frac{\underline{A}(\underline{v}^n_i) + \underline{A}(\underline{v}^n_{i-1})}{2} + O(h^2)$$

Ignoring the fact that this finite difference method is not in conservation form, a disadvantage of the method is the expense of evaluating the Jacobian of $\underline{F}$, $\underline{A}(\underline{v})$. The former version of $\underline{A}_{i-\frac{1}{2}}$ in (4.2.22) requires that $\underline{A}$ be evaluated at values $\underline{u}^n_i$ while the later version of $\underline{A}_{i-\frac{1}{2}}$ in (4.2.22) requires that $\underline{A}$ be evaluated at values $\underline{u}^n_i$ and $\frac{1}{2}(\underline{u}^n_i + \underline{u}^n_{i+1})$. The later form of $\underline{A}_{i-\frac{1}{2}}$ reduces the number of evaluations of $\underline{A}$ by half, yet the method is still inefficient.

Replace $\partial_t v$ with $-\partial_x F$ and $\partial_t^2 v$ is replaced with $\partial_x(A(v)\partial_x F)$ in the Taylor series (4.2.19),

(4.2.24) $$\underline{v}^{n+1}_i = \underline{v}^n_i - k(\partial_x\underline{F})^n_i + \frac{k^2}{2}[\partial_x(\underline{A}\partial_x\underline{F})]^n_i + O(k^3).$$

Approximate the two space derivatives in (4.2.24) by the second order differences

$$D_0\underline{F}^n_i = (\partial_x\underline{F})^n_i + O(h^2)$$

and

$$D_+(\underline{A}_{i-\frac{1}{2}}D_-\underline{F}^n_i) = [\partial_x(\underline{A}\partial_x\underline{F})]^n_i + O(h^2),$$

where $\underline{A}_{i-\frac{1}{2}} = \underline{A}(\underline{v}^n_{i-\frac{1}{2}}) = \underline{A}(\frac{1}{2}(\underline{v}_i^n + \underline{v}_{i-1}^n)) + O(h^2)$. Substituting into (4.2.24) gives the Lax-Wendroff method in conservation form

$$\underline{u}_i^{n+1} = \underline{u}_i^n - kD_0\underline{F}_i^n + \frac{k^2}{2}D_+(\underline{A}_{i-\frac{1}{2}}D_-\underline{F}_i^n),$$

which is accurate of $O(k^2) + O(h^2)$. In this case the function $\underline{G}$ in (4.2.6) is defined by

$$\underline{G}_{i+\frac{1}{2}} = \underline{F}_{i+1}^n + \frac{\lambda}{2}\underline{A}_{i+\frac{1}{2}}(\underline{F}_{i+1}^n - \underline{F}_i^n).$$

This version of the Lax-Wendroff method is still inefficient because it requires the evaluation of $\underline{A}(v)$ once for each grid point. Can the Lax-Wendroff method be constructed so that it does not require the evaluation of $\underline{A}(\underline{v})$?

A modification due to Richtmyer [52] was to consider the two-step method, called the two-step Lax-Wendroff method,

(4.2.26a) $$\underline{u}_{i+\frac{1}{2}}^{n+\frac{1}{2}} = \frac{1}{2}(\underline{u}_{i+1}^n + \underline{u}_i^n) - \frac{\lambda}{2}(\underline{F}_{i+1}^n - \underline{F}_i^n)$$

(4.2.26b) $$\underline{u}_i^{n+1} = \underline{u}_i^n - \lambda(\underline{F}_{i+\frac{1}{2}}^{n+\frac{1}{2}} - \underline{F}_{i-\frac{1}{2}}^{n+\frac{1}{2}}),$$

where $\underline{F}_{i\pm\frac{1}{2}}^{n+\frac{1}{2}} = \underline{F}(\underline{u}_{i\pm\frac{1}{2}}^{n+\frac{1}{2}})$. This method is in conservation form. The first step (4.2.26a) computes intermediate values at the midpoint of the grid points; this being accurate of $O(k) + O(h)$. The second step (4.2.26b) corrects these values and realigns the grid; this being accurate of $O(k^2) + O(h^2)$.

In the constant coefficient case, $F(v) = \underline{A}\,\underline{v}$, where $\underline{A}$ is a constant matrix, the two-step Lax-Wendroff method reduces to the one-step Lax-Wendroff method (3.8.10) for systems of hyperbolic equations with constant coefficients.

The stability of the two-step Lax-Wendroff method is analyzed by the local linearization of (4.2.26). This is achieved by expanding $\underline{F}(\underline{v})$ in a Taylor series in $\underline{v}$ about the point $\underline{v}_0$, retaining only the first two terms,

$$\underline{F}(\underline{v}) = \underline{F}(\underline{v}_0) + \underline{A}(\underline{v}_0)(\underline{v}-\underline{v}_0) + \cdots .$$

Differentiating with respect to x,

(4.2.27) $$\partial_x\underline{F}(\underline{v}) = \underline{A}(\underline{v}_0)\partial_x\underline{v}.$$

Substitution of (4.2.27) in (4.2.26) results in the linear two-step method with constant coefficients which reduces to the one-step Lax-Wendroff method (3.8.10), which is stable provided the CFL condition is satisfied. The stability analysis where there is the further reduction from variable to constant coefficients is called local linearization.

As seen in the linear case (in Table 3.2), the Lax-Wendroff method has a small amount of numerical dissipation. This lack of numerical dissipation is responsible for the poor representation of shock waves. From the views of Fourier analysis, a steep gradient in the solution, such as a shock is built of high frequency (short wave length) components. Dissipation terms damp the high frequency terms more strongly than the low frequency (long wave length) terms. This results in smoothing (or smearing) shocks and contacts. High frequency components move more slowly than the low frequency waves. Thus if there is a small amount of numerical dissipation these high frequency components will lag the low frequency components. This is represented by oscillations trailing the shock wave. These oscillations will often remain in the region behind the shock and eventually merge with the large scale structure.

To explain the oscillations in the Lax-Wendroff method, consider the system of hyperbolic equations with constant coefficients

(4.2.28) $$\partial_t \underline{v} - \underline{A}\partial_x \underline{v} = 0.$$

The Lax-Wendroff method (3.8.10) is consistent with (4.2.28) to $O(k^2) + O(h^2)$. $\underline{Q}(\underline{v})$ can be constructed so that the Lax-Wendroff method is consistent with the equation

(4.2.29) $$\partial_t \underline{v} - \underline{A}\partial_x \underline{v} = \underline{Q}(\underline{v})$$

to $O(k^3) + O(h^3)$. Expanding the terms in the Lax-Wendroff method (3.8.10) in a Taylor series about the point (ih,nk), as in Section III.5, retaining terms up to $\partial_t^3 v$ and $\partial_x^3 v$,

(4.2.30) $$\underline{Q}(\underline{v}) = -\tfrac{1}{6}\underline{A}(h^2 - \underline{A}^2 k^2)\partial_x^3 \underline{v},$$

which represents a dispersion term because of the odd derivative. This has components traveling at different speeds resulting in oscillations $\underline{Q}(\underline{v})$ given by (4.2.30) is the first term in the truncation error of the Lax-Wendroff method.

The Lax-Wendroff method has natural dissipation, as it is dissipative of order 4 (see Section III.6). To find this natural dissipation, construct $\underline{Q}'(\underline{v})$ so that the Lax-Wendroff method is consistent with the equation

(4.2.31) $$\partial_t \underline{v} - \underline{A}\partial_x \underline{v} = \underline{Q}(\underline{v}) + \underline{Q}'(\underline{v})$$

to $O(k^4) + O(h^4)$, where $\underline{Q}(\underline{v})$ is given by (4.2.30). Expanding the terms in the Lax-Wendroff method in a Taylor series about the point (ih,nk), as in Section III.5, retaining terms up to $\partial_t^4\underline{v}$ and $\partial_x^4\underline{v}$,

(4.2.32) $$\underline{Q}'(\underline{v}) = -\tfrac{1}{8}\underline{A}^2k(h^2 - \underline{A}k^2)\partial_x^4\underline{v}$$

which is a dissipative term because of the even derivative.

Von Neumann and Richtmyer [80] developed an artificial viscosity term which was introduced in the Lagrangian form of the equations of gas dynamics (see Chapter III, Volume II). The goal of the artificial viscosity was to reduce the oscillations while allowing the shock transition to occupy only a few grid points and having negligible effect in smooth regions. All other forms of artificial viscosity are variations of the one introduced by von Neumann and Richtmyer. The one presented here due to Lax and Wendroff [42] is no exception.

The artificial viscosity should not resemble a realistic viscosity because it will dampen frequencies that one wants to see. It should only damp the high frequency terms. The resulting finite difference method should also be in conservation form.

Suppose $\underline{Q}$ is a positive vector-valued function of $2q$ arguments such that $\underline{Q}(\underline{v},\underline{v},\ldots,\underline{v}) = 0$. Define $\underline{Q}(x+\frac{h}{2}) = \underline{Q}(\underline{u}^n_{-q+1},\underline{u}^n_{-q+2},\ldots,\underline{u}^n_q)$ and $\underline{Q}(x-\frac{h}{2}) = \underline{Q}(\underline{u}_{-q},\underline{u}_{-q+1},\ldots,\underline{u}^n_{q-1})$. If a term of the form

$$\frac{\underline{Q}(x+\frac{h}{2}) - \underline{Q}(x-\frac{h}{2})}{h}$$

is added to a finite difference method in conservation form, then the resulting method is in conservation form. To see this, suppose the original finite difference method is in the form (4.2.3), then the modified method is

$$\frac{\underline{u}(x,(n+1)k) - \underline{u}(x,nk)}{k} = \frac{\underline{G}(x+\frac{h}{2}) - \underline{G}(x-\frac{h}{2})}{h}$$
$$+ \frac{\underline{Q}(x+\frac{h}{2}) - \underline{Q}(x-\frac{h}{2})}{2},$$

where $\underline{G}(x+\frac{h}{2}) = \underline{G}(\underline{u}^n_{-q+1},\underline{u}^n_{-q+2},\ldots,\underline{u}^n_{-q})$. Define a new vector-valued function $\underline{G}'$ of $2q$ arguments, where

$$\underline{G}'(x \pm \tfrac{h}{2}) = \underline{G}(x \pm \tfrac{h}{2}) + \underline{Q}(x \pm \tfrac{h}{2}).$$

The consistency requirement on $\underline{G}'$ is satisfied since, $\underline{G}'(\underline{v},\ldots,\underline{v}) = \underline{G}(\underline{v},\ldots,\underline{v}) + \underline{Q}(\underline{v},\ldots,\underline{v}) = \underline{F}(\underline{v}) + 0 = \underline{F}(\underline{v})$. The finite difference method (4.2.31) may then be written in the conservation form

$$\frac{\underline{u}(x,(n+1)k) - \underline{u}(x,nk)}{k} = \frac{\underline{G}'(x+\frac{h}{2}) - \underline{G}'(x-\frac{h}{2})}{h} .$$

We shall consider the function Q of two arguments $(q=1)$ so that $\underline{Q}((i+\frac{1}{2})h) = \underline{Q}(\underline{u}_i^n, \underline{u}_{i+1}^n)$ and $\underline{Q}((i-\frac{1}{2})h) = \underline{Q}(\underline{u}_{i-1}^n, \underline{u}_i^n)$. Since $\underline{Q}(\underline{v},\underline{v}) = 0$, $\underline{Q}(\underline{u}_i^n, \underline{u}_{i+1}^n)$ will be close to 0 when $\underline{u}_i^n$ and $\underline{u}_{i+1}^n$ are close. Under these conditions, where there are no osciallations $\underline{Q}$ will be negligible.

In the paper by Lax and Wendroff [42] the function $\underline{Q}$ was constructed so that the artificial viscosity term (4.2.33) with $q = 1$ can be written in the form

(4.2.35)
$$\frac{h}{2}D_+(\underline{Q}_{i-\frac{1}{2}}D_-\underline{u}_i^n) = \frac{1}{2h}[\underline{Q}_{i+\frac{1}{2}}(\underline{u}_{i-1}^n - \underline{u}_i^n) - \underline{Q}_{i-\frac{1}{2}}(\underline{u}_i^n - \underline{u}_{i-1}^n)],$$

where $\underline{Q}_{i+\frac{1}{2}} \equiv \underline{Q}_{i+\frac{1}{2}}(\underline{u}_i^n, \underline{u}_{i+1}^n)$ and $\underline{Q}_{i-\frac{1}{2}} \equiv \underline{Q}_{i-\frac{1}{2}}(\underline{u}_{i-1}^n, \underline{u}_i^n)$ are matrices that are 0 when $\underline{u}_{i+1}^n - \underline{u}_i^n$ and $\underline{u}_i^n - \underline{u}_{i-1}^n$ are 0, respectively. In this case

$$\underline{Q}(\underline{u}_i^n, \underline{u}_{i+1}^n) = \frac{1}{2}\underline{Q}_{i-\frac{1}{2}}(\underline{u}_{i+1}^n - \underline{u}_i^n)$$

and

$$\underline{Q}(\underline{u}_{i-1}^n, \underline{u}_i^n) = \frac{1}{2}\underline{Q}_{i-\frac{1}{2}}(\underline{u}_i^n - \underline{u}_{i-1}^n).$$

For the case where (4.1.1) is a scalar conservation law, Lax and Wendroff choose

(4.2.36)
$$Q_{i+\frac{1}{2}}(u_i^n, u_{i+1}^n) = \frac{\varepsilon}{2}|a(u_{i+1}^n) - a(u_i^n)|$$

where $a(v) = \partial_v F$ and $\varepsilon = O(1)$.

Adding the artificial viscosity term (4.2.35)-(4.2.36) to the Lax-Wendroff method (4.2.22),

(4.2.37)
$$u_i^{n+1} = u_i^n - ka_i D_0 u_i^n + \frac{k^2}{2} D_+(a^2_{i-\frac{1}{2}} D_- u_i^n) + \frac{kh}{2} D_+(Q_{i-\frac{1}{2}} D_- u_i^n),$$

where $a_i = a(u_i^n)$ and $a_{i-\frac{1}{2}} = a(\frac{1}{2}(u_i^n + u_{i-1}^n))$. To analyze the stability of this method, consider the local linearization of (4.2.37) where $a(v)$ is replaced with a typical value a_0, that is, the value of $a(v)$ at some point, and $Q_{i\pm\frac{1}{2}}$ in (4.2.36) is replaced with a typical difference in $a(v)$, denoted by Q_0. The resulting linearized finite difference method is

(4.2.38) $$u_i^{n+1} = u_i^n - ka_0 D_0 u_i^n + \left(\frac{k^2 a_0^2 + khQ_0}{2}\right) D_+ D_- u_i^n.$$

Taking the discrete Fourier transform of (4.2.38), the symbol

$$\rho(\xi) = 1 + (a_0^2\lambda^2 + \lambda Q_0)(\cos\xi - 1) + ia_0\lambda \sin\xi$$

is obtained so that

$$|\rho(\xi)|^2 = 1 - 4[Q_0\lambda - ((a_0^2\lambda^2 + Q_0\lambda)^2 - a_0^2\lambda^2)\sin^2\tfrac{\xi}{2}]\sin^2\tfrac{\xi}{2}\ .$$

If $a_0^2\lambda^2 + Q_0\lambda \leq 1$ and the CFL condition $|a_0|\lambda \leq 1$ is satisfied, then $Q_0\lambda - ((a_0^2\lambda^2 + Q_0\lambda)^2 - a_0^2\lambda^2) \geq 0$ and $|\rho(\xi)|^2 \leq 1$. This verifies the stability of linearized finite difference method (4.2.38) by the von Neumann condition. Let $\alpha = \max_v |a(v)|$ and V = variation of $a(v)$ then $Q_0 \leq \frac{\varepsilon}{2} V$. If $a(v)$ does not change sign, then $V \leq \alpha$ and $Q_0 \leq \frac{\varepsilon\alpha}{2}$. If

(4.2.39) $$\alpha^2\lambda^2 + \tfrac{\varepsilon}{2}\alpha\lambda \leq 1$$

then

$$a_0^2\lambda^2 + Q_0\lambda \leq 1.$$

Inequality (4.2.39) may be rewritten as

$$\alpha\lambda \leq (1+(\tfrac{\varepsilon}{4})^2)^{1/2} - \tfrac{\varepsilon}{4} < 1.$$

This is a stronger condition than the CFL condition.

We now return to the case of a system of conservation laws and the artificial viscosity term (4.2.35). The form of $Q_{i\pm\frac{1}{2}}$ (4.2.36) in the scalar case will be used to construct the matrices $\underline{Q}_{i\pm\frac{1}{2}}$ in (4.2.35). Let $\mu_{i,j}^n$ and $\mu_{i+1,j}^n$ denote the j-th eigenvalues of $\underline{A}(\underline{u}_i^n)$ and $\underline{A}(\underline{u}_{i+1}^n)$, respectively, for $j = 1,\ldots,p$. The matrix $\underline{Q}_{i+\frac{1}{2}}(\underline{u}^n, \underline{u}_{i+1}^n)$ in (4.2.35) is defined by having eigenvalues

$\frac{\varepsilon_j}{2}|\mu_{i+1,j} - \mu_{i,j}|$, where $\varepsilon_j = O(1)$, $1 \leq j \leq p$. The eigenvalues of $\underline{Q}_{i-\frac{1}{2}}(\underline{u}_{i-1}^n, \underline{u}_i^n)$ are similarly defined. This does not uniqely determine the matrices $\underline{Q}_{i\pm\frac{1}{2}}$. In Lax and Wendroff [42] the matrices $\underline{Q}_{i\pm\frac{1}{2}}$ were required to commute with $\underline{A}$. Thus the matrices $\underline{Q}_{i\pm\frac{1}{2}}$ are functions of $\underline{A}$. The matrix $\underline{Q}_{i+\frac{1}{2}}$ may be written as a polynomial of degree p-1 in $\underline{A}$,

$$\underline{Q}_{i+\frac{1}{2}}(\underline{u}_i^n,\underline{u}_{i+1}^n) = g_0\underline{I} + g_1\underline{A} + \ldots + g_{p-1}\underline{A}^{p-1},$$

where the coefficients g_j, $0 \le j \le p-1$ are determined by prescribed eigenvalues of $\underline{Q}_{i+\frac{1}{2}}$.

Specific examples of the construction of $\underline{Q}_{i\pm\frac{1}{2}}$ will be given in Chapter III, Volume II.

IV.3. Monotone Difference Methods

In Theorem 4.4 it is shown that if the solution of a finite difference method in conservation form converges boundedly almost everywhere to a function as h and k approach zero, then the limit is a weak solution. However, this theorem does not indicate whether this limit satisfies the entropy condition (4.1.5). In order to insure that the entropy condition is satisfied, the finite difference method will have to meet an additional requirement. To this end, we introduce the notion of a monotone finite difference approximation to a scalar conservation law (4.1.1) (see Jennings [33]).

Definition. A finite difference method of the form

$$u^{n+1} = Qu_i^n \equiv H(u_{i-q}^n, u_{i,q+1}^n, \ldots, u_{i+q}^n), \tag{4.3.1}$$

is monotone if H is a monotone increasing function of each of its $2q+1$ arguments, that is,

$$H_j \equiv \frac{\partial H}{\partial u_{i-j}^n} \ge 0 \quad \text{for} \quad -q \le j \le q\,. \tag{4.3.2}$$

If the function H in (4.3.1) satisfies the consistency condition

$$H(v,v,\ldots,v) = F(v)$$

and can be written in the form

$$H(u_{i-q}^n,\ldots,u_{i+q}^n) = u_i^n - \lambda(G(u_{i-q+1}^n,u_{i-q+2}^n,\ldots,u_{i+q}^n) - G(u_{i-q}^n,u_{i-q+1}^n,\ldots,u_{i+q-1}^n)), \tag{4.3.3}$$

then the finite difference method (4.3.1) is in conservation form.

In Harten, Hyman, and Lax [30], the main theorem of this section is proved.

Theorem 4.6. Let (4.3.1) and (4.3.3) be a monotone finite difference method in conservation form. If the solution of this finite difference method u_i^n converges boundedly almost everywhere to some function $v(x,t)$ as k and h approach zero with $\lambda = k/h$ fixed, then $v(x,t)$ is a weak solution and the entropy condition (4.1.15) is satisfied at all discontinuities of v.

Writing out the truncation error for the monotone finite difference method (4.3.1)

$$\frac{1}{k}(v_i^{n+1} - H(v_{i-q}^n, v_{i-q+1}^n, \ldots, v_{i+q}^n)) = \partial_t v + \partial_x F(v) \tag{4.3.4}$$

$$= -k\partial_x(\beta(v,\lambda)\partial_x v) + O(k^2),$$

where

$$\beta(v,\lambda) = \frac{1}{2\lambda^2}\sum_{j=-q}^{q} j^2 H_j(v,\ldots,) - \frac{1}{2}a^2(v) \tag{4.3.5}$$

and H_j denotes the partial derivative of H with respect to the $(p+j-1)$-th argument (cf. (4.3.2)).

As an example, consider the Lax-Friedrichs method (4.2.12). Define H by

$$H(u_{i-1}^n, u_i^n, u_{i+1}^n) = \frac{1}{2}(u_{i+1}^n + u_{i-1}^n) - (F_{i+1}^n - F_{i-1}^n) \tag{4.3.6}$$

Differentiating H with respect to its 3 arguments

$$\frac{\partial H}{\partial u_{i-1}^n}(v,v,v) = \frac{1}{2} + \frac{\lambda}{2}a(v) > 0,$$

$$\frac{\partial H}{\partial u_i^n}(v,v,v) = 0,$$

$$\frac{\partial H}{\partial u_{i+1}^n}(v,v,v) = \frac{1}{2} - \frac{\lambda}{2}a(v) > 0,$$

provided the CFL condition is satisfied. Thus the Lax-Friedrichs method is monotone. By substituting these derivatives of H in (4.3.5), we obtain the $\beta(v,\lambda)$ for the Lax-Friedrichs method

$$\beta(v,\lambda) = \frac{1}{2}(\frac{1}{\lambda^2} - a^2(v)),$$

which is non-negative provided the CFL condition is satisfied.

<u>Theorem 4.7</u>. A monotone finite difference method in conservation form is first order accurate.

<u>Proof</u>: By direct calculation we obtain the identities

$$\sum_{j=-q}^{q} H_j(v,v,\ldots,v) = 1$$

and

$$\sum_{j=-q}^{q} jH_j(v,v,\ldots,v) = -\lambda a(v).$$

Since the finite difference is monotone, $H_j > 0$, $\sqrt{H_j}$ is real for $-q < j < q$. Using the Schwartz inequality and the two identities

$$\lambda^2 a^2(v) = (\sum_j jH_j)^2 = (\sum_j j\sqrt{H_j}\,\sqrt{H_j})^2 \tag{4.3.7}$$

$$< (\sum_j j^2 H_j)(\sum_j H_j)$$

$$= \sum_j j^2 H_j .$$

Substituting this inequality in (4.3.5) gives $\beta(v,\lambda) > 0$. The only way in which $\beta(v,\lambda) \equiv 0$ is for (4.3.7) to be an equality which is possible if and only if $H_j(v,\ldots,v) = 0$ for all j, $-q < j < q$, except one. This case gives rise to a finite difference operator that is a translation, which is a trivial case. If $\beta(v,\lambda) \not\equiv 0$, then by (4.3.4), the monotone finite difference method is first order accurate.

Finite difference methods, particularly those with accuracy greater than one, may produce oscillations when applied across a discontinuity. These oscillations can not only be responsible for the poor representation of the discontinuity but can also induce nonlinear instabilities (instabilities not predicted by the linearized stability analysis) and may result in convergence to a weak solution that does not satisfy the entropy condition. (An example of the later effect of oscillations will be given below.) From this we conclude that it is desirable to try to construct nonoscillatory methods, that is, finite difference methods that do not produce oscillations when applied across a discontinuity. It is for this reason that we are interested in monotone finite difference methods.

Definition. A finite difference operator Q is monotonicity preserving if for any monotone grid function u, $w = Qu$ is also monotone.

Theorem 4.8. A monotone finite difference method (4.3.1) is monotonicity preserving.

Proof. Let u denote any monotone grid function and define

$$w_i = H(u_{i-q}, u_{i-q+1}, \ldots, u_{i+q}).$$

Using the mean value theorem, there exists a number $0 < \alpha < 1$ such that

$$\begin{aligned} w_{i+1} - w_i &= H(u_{i-q+1}, u_{i-q+2}, \ldots, u_{i+q+1}) \\ &\quad - H(u_{i-q}, u_{i-q+1}, \ldots, u_{i+q}) \\ &= \sum_{j=-q}^{q} H_j(\alpha u_{i-q+1} + (1-\alpha)u_{i-q}, \ldots, \alpha u_{i+q+1} + (1-\alpha)u_{i+q}) \\ &\quad \cdot (u_{i+j+1} - u_{i+j}). \end{aligned}$$

Since $H_j > 0$ for $-q < j < q$, by the definition of monotone, if u is a monotone increasing (decreasing) grid function of i, $u_{i+1} - u_i > 0$ (< 0) then $w_{i+1} - w_i > 0$ (< 0). Thus w is a monotone grid function of the same kind as u. This theorem shows that monotone finite difference methods are nonoscillatory methods.

Gudonov [24] showed that in the linear constant coefficient case monotonicity is equivalent to monotonicity preserving. However, in the nonlinear case, the converse of Theorem 4.8 is not true. The set of all monotone preserving finite difference methods is larger than the set of all monotone finite difference methods. It was shown above that a monotone finite difference method is first order accurate. However, it is possible to construct monotonicity preserving finite difference methods that are second order accurate (see van Leer [75]). This will be demonstrated later in this section.

The notion of a monotone finite difference method can be extended to the vector case. Inequailty (4.3.2) is taken to mean all eigenvalues of the matrix $\underline{H}_j$ are nonnegative for every j, $-q \leq j \leq q$. However, the notion of monotonicity preserving does not extend to the vector case.

In Harten [27] a necessary condition for a finite difference method to be monotonicity preserving is established. In inequality (4.3.8) set $u_{i-q} = u_{i-q+1} = \cdots = u_{i+j} = u_L$ and $u_{i+j+1} = u_{i+j+2} = \cdots = u_{i+q+1} = u_R$, for each j, $-q \leq j \leq q$ then

$$w_{i+1}-w_i = H_j(u_L,\ldots,u_L,u_L+\alpha(u_R-u_L),u_R,\ldots,u_R)\,(u_R-u_L).$$

From this we obtain the set of necessary conditions

$$(4.3.9) \qquad H_j(u_L,\ldots,u_L,u_L+\alpha(u_R-u_L),u_R,\ldots,u_R) \geq 0$$

for all $-q \leq j \leq q$ and $0 < \alpha < 1$.

In Harten, Hyman, and Lax [30] an example is given of a nonmonotone finite difference method (the Lax-Wendroff method (4.2.25)) whose solution converges to a weak solution that does not satisfy the entropy condition. Consider the scalar hyperbolic conservation law (4.1.1) where $F(v)$ is given by

$$F(v) = v - 3\sqrt{3}v^2(v-1)^2$$

and initial condition

$$v(x,0) = f(x) = \begin{cases} 1, & x \leq 0.5 \\ 0, & x > 0.5. \end{cases}$$

The function

$$v(x,t) = f(x-t)$$

solves the initial-value problem and satisfies the RH-condition (4.1.13a) and the entropy condition. Therefore, $v(x,t)$ is the unique solution.

Approximate this initial-value problem by the Lax-Wendroff method (4.2.25) which is in conservation form. Since the Lax-Wendroff method is second order accurate it is not monotone. At the beginning of each time step a new time step is chosen according to $\max|a(u_i^n)|\lambda = 0.9$.

In Figure 4.9a-c the numerical solution (represented by a solid line) and the exact solution (represented by a dashed line) are depicted at time steps $n = 5,15,$ and $150,$ respectively.

The numerical results indicated that the approximate solution converges as $n \to \infty$ to

$$u(x,t) = \begin{cases} u_L = 1 & x < 0.5 - 3.3t, \\ u_0 = 1.41, & 0.5 - 3.3t < x < 0.5 \\ u_1 = -0.17, & 0.5 < x < 0.5 + 2.2t, \\ u_R = 0, & 0.5 + 2.2t < x. \end{cases}$$

The approximate solution has one. The R-H condition (4.1.13a) is satisfied at the three discontinuities of $u(x,t)$ (within the accuracy of the method). Thus, by Theorem 4.3, $u(x,t)$ is a weak solution of the initial value problem. Also the entropy condition is satisfied at the discontinuities between u_L and u_0 and between u_1 and u_R. However, it is not satisfied at the third discontinuity between u_0 and u_1.

The nonmonotonicity of the Lax-Wendroff method (4.2.25) is responsible for the development of the beginning of the oscillations represented by the overshoot forming the discontinuity between u_L and u_0 in Figure 4.9a and by the undershoot forming the discontinuity

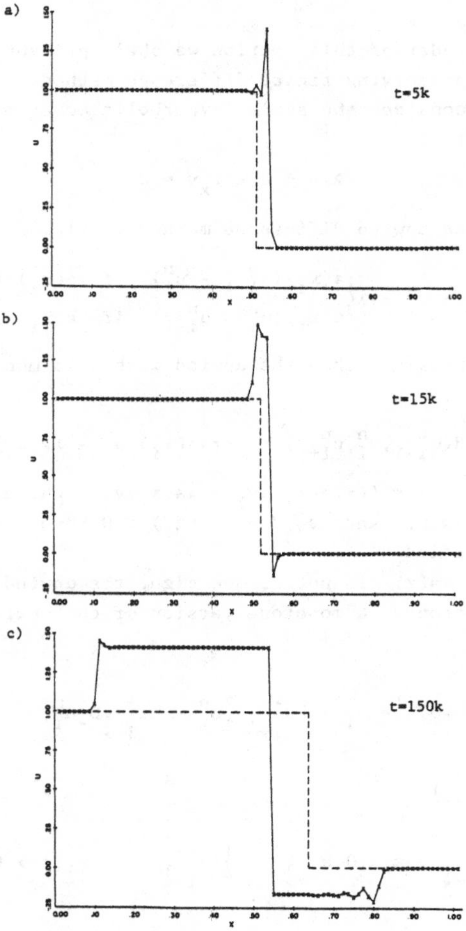

Figure 4.9. Example of Non-Monotone Lax Wendroff Method in Conservation Form (4.2.25) Whose Solution Converges to a Weak Solution that Biolates the Entropy Condition at Time Steps a) n = 5, b) n = 15, and c) n = 150. (Courtesy of A. Harten, J. M. Hyman, and P. D. Lax.)

between u_1 and u_R in Figure 4.9b. However, the flux function $F(v)$ is not concave or convex so that $a(v) = \partial F/\partial v$ satisfies $a(u_0) < a(u_L)$ and $a(u_1) > a(u_R)$. This results in the backward propagating discontinuity between u_L and u_0 and the forward propagating discontinuity between u_1 and u_R.

In the remainder of this section we shall present some monotone and monotonicity preserving finite difference methods. As an introductory example, consider the scalar hyperbolic equation with variable coefficients (3.11.19)

$$\partial_t v + a(x)\partial_x v = 0$$

approximated by the upwind difference method (3.11.20)

$$u_i^{n+1} = u_i^n - \lambda\begin{cases} a(x_i)(u_{i+1}^n - u_i^n) & \text{if } a(x_i) < 0 \\ a(x_i)(u_i^n - u_{i-1}^n) & \text{if } a(x_i) > 0 \end{cases}$$

If $a(x)$ is of one sign, then the upwind method is monotone. Suppose $a(x) < 0$, then

$$u_i^{n+1} = H(u_{i-1}^n, u_i^n, u_{i+1}^n) = (1+\lambda a(x_i))u_i^n - \lambda a(x_i)u_{i+1}^n,$$

that is, $H(v_1,v_2,v_3) = (1+\lambda a(x_i))v_2 - \lambda a(x_i)v_3$, where $\partial_{v_1}H = 0$, $\partial_{v_2}H = 1 + \lambda a(x_i) \geq 0$, and $\partial_{v_3}H = -a(x_i) \geq 0$ provided the CFL condition is satisfied.

However, if $a(x)$ is not of one sign, the upwind difference method is not monotone. A monotone version of the upwind difference method is

(4.3.10)
$$u_i^{n+1} = u_i^n - ka^-_{i+\frac{1}{2}}D_+u_i^n - ka^+_{i-\frac{1}{2}}D_-u_i^n,$$

where $a_{i\pm\frac{1}{2}} = a(x_{i\pm\frac{1}{2}})$ and

(4.3.11a)
$$a^+_{i+\frac{1}{2}} \equiv \max(0, a_{i+\frac{1}{2}}) = \frac{1}{2}(a_{i+\frac{1}{2}} + |a_{i+\frac{1}{2}}|) \geq 0$$

and

(4.3.11b)
$$a^-_{i+\frac{1}{2}} \equiv \min(0, a_{i+\frac{1}{2}}) = \frac{1}{2}(a_{i+\frac{1}{2}} - |a_{i+\frac{1}{2}}|) \leq 0.$$

If $\max_i |a_{i+\frac{1}{2}}|\lambda \leq 1$, then the upwind method (4.3.10) is monotonicity preserving. Without loss of generality assume u_i^n is monotone increasing grid function. We shall establish that u_i^{n+1} is monotone increasing grid function. Since u_i^n is monotone increasing $u_{i-1}^n \leq u_i^n \leq u_{i+1}^n \leq u_{i+2}^n$. From (4.3.10)

$$u_i^{n+1} = \lambda a^+_{i-\frac{1}{2}} u^n_{i-1} + (1+\lambda(a^-_{i+\frac{1}{2}} - a^+_{i-\frac{1}{2}}))u^n_i - \lambda a^-_{i+\frac{1}{2}} u_{i+1}.$$

Replacing u^n_{i-1} with u^n_i, since $a^+_{i-\frac{1}{2}} \geq 0$,

$$u_i^{n+1} \leq (1 + \lambda a^-_{i+\frac{1}{2}})u^n_i - \lambda a^-_{i+\frac{1}{2}} u^n_{i+1}. \tag{4.3.12}$$

Similarly

$$u_{i+1}^{n+1} = \lambda a^+_{i\pm\frac{1}{2}} u^n_i + (1 + \lambda(a^-_{i+\frac{3}{2}} - a^+_{i+\frac{1}{2}}))u^n_{i+1} - \lambda a^-_{i+\frac{3}{2}} u_{i+1}$$

which by replacing u^n_{i+2} with u^n_{i+1}, since $a^-_{i+\frac{3}{2}} \leq 0$,

$$u_{i+1}^{n+1} \geq \lambda a^+_{i+\frac{1}{2}} u^n_i + (1 - \lambda a^+_{i+\frac{1}{2}})u^n_{i+1}. \tag{4.3.13}$$

Subtracting (4.3.12) from (4.3.13)

$$u^{n+1}_{i+1} - u^{n+1}_i \geq (1 - (a^+_{i+\frac{1}{2}} - a^-_{i+\frac{1}{2}}))(u^n_{i+1} - u^n_i).$$

Using (4.3.11), $a^+_{i+\frac{1}{2}} - a^-_{i+\frac{1}{2}} = |a_{i+\frac{1}{2}}|$ so that, $(1 - \lambda(a^+_{i+\frac{1}{2}} - a^-_{i+\frac{1}{2}})) \geq 0$, provided the CFL condition is satisfied. Thus $u^{n+1}_{i+1} - u^{n+1}_i \geq 0$ for every i, that is, u^{n+1}_i is a monotone increasing grid function of i.

We shall construct a monotone upwind difference method for the scalar conservation law (4.1.1) (see Harten [27]). Define $\delta(v,w)$ to be

$$\delta(v,w) = \begin{cases} 1, & \frac{f(w)-f(v)}{w-v} < 0 \\ 0, & \frac{f(w)-f(v)}{w-v} \geq 0 \end{cases}$$

and $\delta^n_{i+\frac{1}{2}} = \delta(u^n_i, u^n_{i+1})$. Consider the nonlinear analogue of (4.3.10)

$$\begin{aligned} u^{n+1}_i &= u^n_i - k\delta^n_{i+\frac{1}{2}} D_+ F^n_i - k(1 - \delta^n_{i-\frac{1}{2}})D_- F^n_i \\ &= u^n_i - \lambda\delta^n_{i+\frac{1}{2}}(F^n_{i+1} - F^n_i) - \lambda(1 - \delta^n_{i-\frac{1}{2}})(F^n_i - F^n_{i-1}). \end{aligned}$$

This method may be written in the conservation form

$$u^{n+1}_i = u^n_i - \lambda(G(u^n_i, u^n_{i+1}) - G(u^n_{i-1}, u^n_i)),$$

where

$$G(u^n_i, u^n_{i+1}) = F^n_i + \delta^n_{i+\frac{1}{2}}(F^n_{i+1} - F^n_i).$$

Writing $G(v_1, v_2) = F(v_i) = F(v_1) + \delta(v_1, v_2)(F(v_2) - F(v_1))$, the consistency condition is satisfied since $G(v,v) = F(v)$.

To establish that the method is monotonicity preserving we shall show that (4.3.14) may be reduced to (4.3.6). Define

$$a_{i+\frac{1}{2}} = \begin{cases} \dfrac{F_{i+1}^n - F_i^n}{u_{i+1}^n - u_i^n}, & u_{i+1}^n \neq u_i^n, \\ 0, & u_{i+1}^n = u_i^n \end{cases}$$

$$a^-_{i+\frac{1}{2}} = \delta_{i+\frac{1}{2}} a_{i+\frac{1}{2}},$$

and

$$a^+_{i-\frac{1}{2}} = (1 - \delta_{i-\frac{1}{2}})a_{i-\frac{1}{2}}.$$

Substitution of these values in (4.3.14) gives (4.3.10) which is monotonicity preserving provided the CFL condition

$$\max_i |a_{i+\frac{1}{2}}|\lambda = \max\left|\frac{F_{i+1}^n - F_i^n}{u_{i+1}^n - u_i^n}\right|\lambda < 1.$$

The upwind difference method can be extended to systems of hyperbolic conservation laws. As a first step, consider the system of hyperbolic equations with constant coefficients,

$$\partial_t \underline{v} + \underline{A}\partial_x \underline{v} = 0,$$

where $\underline{A}$ is a constant $p\times p$ matrix. By the definition of hyperbolic, $\underline{A}$ can be diagonalized by a similarity transformation, that is, there exists a nonsingular matrix $\underline{T}$ such that, $\underline{A} = \underline{T}\,\underline{D}\,\underline{T}^{-1}$, where $\underline{D} = \mathrm{diag}(\mu_1,\ldots,\mu_p)$, where μ_j denote the eigenvalues of $\underline{A}$. As in Section III.1, substituting $\underline{w} = \underline{T}^{-1}\underline{v}$ in (4.3.15) results in p uncoupled equations of the form

(4.3.16) $$\partial_t w_j + \mu_j \partial_x w_j = 0,$$

each being a special case of (3.11.19). Each component equation can be approximated by the upwind method (4.3.10),

$$\tilde{u}_{ij}^{n+1} = \tilde{u}_{ij}^n - \tfrac{k}{2}(\mu_j - |\mu_j|)D_+\tilde{u}_{ij}^n - \tfrac{k}{2}(\mu_j + |\mu_j|)D_-\tilde{u}_{ij}^n,$$

for $j = 1,\ldots,p$. This may be written in vector form

(4.3.17) $$\underline{u}_i^{n+1} = \underline{u}_i^n - \tfrac{k}{2}(\underline{D} - |\underline{D}|)D_+\tilde{\underline{u}}_i^n - \tfrac{k}{2}(\underline{D} + |\underline{D}|)D_-\tilde{\underline{u}}_i^n,$$

where $\tilde{\underline{u}}_i^n$ is an approximation to $\underline{w}(ih,nk)$ and

(4.3.18) $$|\underline{D}| = \mathrm{diag}(|\mu_1|,\ldots,|\mu_p|).$$

Multiplying (4.3.17) on the left by $\underline{T}$ transforms the upwind method for $\underline{w}$ into an upwind method for $\underline{v}$,

(4.3.19) $$\underline{u}_i^{n+1} = \underline{u}_i^n - k\underline{A}^- D_+\underline{u}_i^n - k\underline{A}^+ D_-\underline{u}_i^n,$$

where $\underline{u}_i^n = \underline{T}\,\tilde{\underline{u}}_i^n$,

(4.3.20a) $$\underline{A}^- = \tfrac{1}{2}(\underline{A} - |\underline{A}|),$$

(4.3.20b) $$\underline{A}^+ = \tfrac{1}{2}(\underline{A} + |\underline{A}|),$$

and

(4.3.21) $$|A| = \underline{T}|\underline{D}|\underline{T}^{-1}.$$

Clearly the upwind method (4.3.19) is stable provided the CFL condition $\max_{1<j<p} |\mu_j|\lambda < 1$ is satisfied.

To consider systems of hyperbolic conservation laws we shall restrict our attention to "three point" methods in conservation form, that is, methods of the form (4.2.3) where the numerical flux is a function of two arguments $(q=1)$, $\underline{G}(\underline{v}_1,\underline{v}_2)$. We now make precise the definition of a general "three point" upwind method (see Harten, Lax, van Leer [32]).

<u>Definition</u>. A finite difference method in conservation form (4.2.3) is an <u>upwind method</u> if:

1) For $\underline{v}_1$ and $\underline{v}_2$ any two states near some reference state $\underline{v}^*$, the numerical flux $\underline{G}$ may be written in the form

(4.3.22) $$\underline{G}(\underline{v}_1,\underline{v}_2) = F(\underline{v}^*) + \underline{A}^+(\underline{v}^*)(\underline{v}_1-\underline{v}^*) + \underline{A}^-(\underline{v}^*)(\underline{v}_2-\underline{v}^*) + O(|\underline{v}_1-\underline{v}^*| + |\underline{v}_2-\underline{v}^*|),$$

where $A(\underline{v})$ denotes the Jacobian of $\underline{F}(\underline{v})$ and $\underline{A}^+$ and $\underline{A}^-$ are defined by (4.3.20) and (4.3.21).

2) When all signal speeds associated with the numerical flux $G(\underline{v}_1,\underline{v}_2)$ are positive, $\underline{G}(\underline{v}_1,\underline{v}_2) = \underline{F}(\underline{v}_1)$. When all signal speeds associated with the numerical flux are negative, $\underline{G}(\underline{v}_1,\underline{v}_2) = \underline{F}(\underline{v}_2)$.

A natural choice for the reference state $\underline{v}^*$ is $\frac{\underline{v}_1+\underline{v}_2}{2}$ so that (4.3.22) reduces to

(4.3.23) $$\underline{G}(\underline{v}_1,\underline{v}_2) = \tfrac{1}{2}(\underline{F}(\underline{v}_1) + \underline{F}(\underline{v}_2)) - \tfrac{1}{2}|\underline{A}(\tfrac{\underline{v}_1+\underline{v}_2}{2})|(\underline{v}_2 - \underline{v}_1) + O(|\underline{v}_2 - \underline{v}_1|),$$

where using (4.3.20), $A^+ - A^- = |\underline{A}|$, and

$$\underline{F}(\tfrac{\underline{v}_1+\underline{v}_2}{2}) = \tfrac{1}{2}(\underline{F}(\underline{v}_1) + \underline{F}(\underline{v}_2)) + O(|\underline{v}_2 - \underline{v}_1|).$$

Replacing $\underline{v}_1$ and $\underline{v}_2$ with $\underline{u}_i^n$ and $\underline{u}_{i+1}^n$, respectively, in (4.3.23), we obtain $\underline{G}_{i+\frac{1}{2}}$ in (4.2.3),

$$\underline{G}_{i+\frac{1}{2}} = \tfrac{1}{2}(\underline{F}_i^n + \underline{F}_{i+1}^n) - \tfrac{1}{2}|\underline{A}(\tfrac{\underline{u}_i^n+\underline{u}_{i+1}^n}{2})|(\underline{u}_{i+1}^n - \underline{u}_i^n) + O(|\underline{u}_{i+1}^n - \underline{u}_i^n|)$$

$$= \tfrac{1}{2}(\underline{F}_i^n + \underline{F}_{i+1}^n) - \tfrac{k}{2}|\underline{A}(\tfrac{\underline{u}_i^n+\underline{u}_{i+1}^n}{2})|D_+\underline{u}_i^n + O(h).$$

Similarly,

$$\underline{G}_{i-\frac{1}{2}} = \frac{1}{2}(\underline{F}_{i-1}^n + \underline{F}_i^n) - \frac{k}{2}|\underline{A}(\frac{\underline{u}_{i-1}^n+\underline{u}_i^n}{2})|D_+\underline{u}_{i-1}^n + O(h).$$

Substituting $G_{i+\frac{1}{2}}$ and $G_{i-\frac{1}{2}}$ into (4.2.3) results in a first order upwind method

$$\underline{u}_i^{n+1} = \underline{u}_i^n - \lambda(\underline{G}_{i+\frac{1}{2}} - \underline{G}_{i-\frac{1}{2}})$$
$$= \underline{u}_i^n - \lambda D_0\underline{F}_i^n + \frac{\lambda h^2}{2}D_-(\underline{A}_{i+\frac{1}{2}}D_+\underline{u}_i^n),$$

where $\underline{A}_{i+\frac{1}{2}} = \underline{A}(\frac{\underline{u}_i^n+\underline{u}_{i+1}^n}{2})$. Observe that the last term on the right hand side of (4.3.24) is a diffusion term.

We now construct a second order accurate upwind difference due to Fromm [16]. To begin we shall consider the scalar equation

(4.3.25) $$\partial_t v + c\partial_x v = 0,$$

where c is a constant. Approximate the equation by the Lax-Wendroff method (3.5.3)

(4.3.26) $$u_i^{n+1} = u_i^n - ckD_0u_i^n + \frac{c^2k^2}{2}D_+D_-u_i^n.$$

It was shown in Section III.6 that the Lax-Wendroff method has a lagging phase error. In an effort to construct a finite difference with a zero phase error, Fromm constructed a finite difference method with second order accuracy that has a leading phase error. The idea of the construction is to project the exact solution at an advanced time with $|c|\lambda = 1$. The solution for $|c|\lambda < 1$ is obtained by Leith's backward differencing in time from projected solution. Let $\tilde{k} > k$ denote an alternate step and let $u^{n\tilde{+}1}$ represent the approxmate solution at time $nk + \tilde{k} > (n+1)k$. For $|c|\frac{\tilde{k}}{h} = 1$, the solution $u^{n\tilde{+}1}$ is obtained by following the characteristic back to $t = nk$. (See Section III.4.) The projected solution is

(4.3.27) $$u^{n\tilde{+}1} = u_{i-1}^n.$$

The projected solution shall be used to obtain an approximate solution u_i^{n+1}. Fromm used an implicit version of the Lax-Wendroff method (4.3.15)

(4.3.28) $$\frac{u_i^{n\tilde{+}1} - u_i^{n+1}}{\tilde{k}-k} = -cD_0u_i^{n\tilde{+}1} + \frac{c^2(\tilde{k}-k)}{2}D_+D_-u^{n\tilde{+}1} .$$

Replacing $u_i^{n\tilde{+}1}$ in (4.3.28) with its projected value u_{i-1}^n and solving for u_i^{n+1},

(4.3.29) $$u_i^{n+1} = u_{i-1}^n - (ck-h)D_0u_{i-1}^n - \frac{(ck-h)^2}{2}D_+D_-u_{i-1}^n.$$

This finite difference method has a leading phase error. The Lax-Wendroff method (4.3.26) and (4.3.19) are averaged to obtain a second order accurate method

(4.3.30) $$\begin{aligned} u_i^{n+1} &= u_i^n - \frac{ck}{2}D_0u_i^n - \frac{ck}{2}D_0u_{i-1}^n + \frac{c^2k^2}{4}D_+D_-u_i^n \\ &\quad + \frac{c^2k^2 - 2ckh}{4} D_+D_-u_{i-1}^n \\ &= u_i^n - \frac{c\lambda}{2}hD_0u_i^n - \frac{c\lambda}{2}hD_0u_{i-1}^n + \frac{c^2\lambda^2}{4}h^2D_+D_-u_i^n \\ &\quad + (\frac{c^2\lambda^2 - 2\lambda}{4})h^2D_+D_-u_{i-1}^n, \end{aligned}$$

which is stable provided the CFL conditions, $|c|\lambda \leq 1$, is satisfied. The averaged solution has an improved phase error. This method is known as Fromm's zero average phase method.

Fromm's zero average phase error method (4.3.30) is actually a second order accurate upwind difference method. This can be demonstrated by rewriting (4.3.30),

(4.3.30') $$u_i^{n+1} = \begin{cases} u_i^n - c\lambda hD_+u_i^n + \frac{c\lambda}{4}(1+c\lambda)h(D_+u_{i+1}^n - D_+u_{i-1}^n), & c < 0 \\ u_i^n - c\lambda hD_-u_i^n - \frac{c\lambda}{4}(1-c\lambda)h(D_-u_{i+1}^n - D_-u_{i-1}^n), & c > 0. \end{cases}$$

This method is not monotonicity preserving, however, van Leer [76] modified (4.3.30') resulting in a second order accurate upwind difference method that is monotonicity preserving. (This method will verify the conjecture made earlier in this section that the class of monotonicity preserving finite difference methods is larger than the class of monotone finite difference methods.) van Leer [76] introduces a "smoothness monitor" θ_i defined by

(4.3.31) $$\theta_i^n = \begin{cases} 1, & D_+u_i^n = D_-u_i^n = 0 \\ \dfrac{D_+u_i^n}{D_-u_i^n}, & \text{otherwise.} \end{cases}$$

Fromm's method (4.3.30') with $c > 0$ may be written as the average of the two methods

(4.3.32a) $$u_i^{n+1} = u_i^n - c\lambda hD_-u_i^n - \frac{c\lambda}{2}(1-c\lambda)h(D_-u_{i+1}^n - D_-u_i^n)$$

and

(4.3.32b) $$u_i^{n+1} = u_i^n - c\lambda hD_-u_i^n - \frac{c\lambda}{2}(1-c\lambda)h(D_-u_i^n - D_-u_{i-1}^n),$$

both of which are stable provided the CFL condition, $|c|\lambda \leq 1$, is satisfied. The case when $c < 0$ is completely analogous. The definition of monotonicity preserving can be expressed in the following fashion. With $c > 0$, the wave moves to the left and if the CFL condition is satisfied, Then u_i^{n+1} must lie between u_i^n and u_{i-1}^n (and with $c < 0$, the wave moves to the right and if the CFL condition is satisfied, then u_i^{n+1} lies between u_{i+1}^n and u_i^n) for every i. With this monotonicity preserving may be expressed by the inequality

$$(4.3.33) \qquad 0 < \frac{u_i^{n+1}-u_i^n}{u_i^n-u_{i-1}^n} < 1$$

for every grid point i. If the two finite difference methods are monotonicity preserving, then the average of these two methods is monotonicity preserving.

Consider the modification of (4.3.32)

$$(4.3.34a) \qquad u_i^{n+1} = u_i^n - c\lambda hD_-u_i^n - \frac{c\lambda}{2}(1-c\lambda)h(1-S(\theta_i^n))(D_-u_{i+1}^n - D_-u_i^n)$$

and

$$(4.3.34b) \qquad u_i^{n+1} = u_i^n - c\lambda hD_-u_i^n - \frac{c\lambda}{2}(1-c\lambda)h(1-S(\theta_{i-1}^n))(D_-u_i^n - D_-u_{i-1}^n),$$

where S is a function to be chosen that is independent of c and λ. The average method is

$$(4.3.35) \qquad u_i^{n+1} = u_i^n - c\lambda hD_-u_i^n - \frac{c\lambda}{4}(1-c\lambda)h(D_-u_{i+1}^n - D_-u_{i-1}^n)$$
$$+ \frac{c\lambda}{4}(1-c\lambda)[S(\theta_i^n)(D_-u_{i+1}^n - D_-u_i^n) - S(\theta_{i-1}^n)(D_-u_i^n - D_-u_{i-1}^n)].$$

The linearized stability analysis of (4.3.35) yields $|c|\lambda < 1$ (the CFL condition) and $S^2(\theta) < 1 + O(h)$. Suppose $D_-u_{i+1}^n = D_-u_{i-1}^n$ so that Fromm's method (4.3.30') reduces to the monotonicity preserving method

$$(4.3.36) \qquad u_i^{n+1} = u_i^n - c\lambda hD_-u_i^n.$$

In this case (4.3.35) should reduce to (4.3.36). This will occur provided

$$S(\theta_i^n) = -S(\theta_{i-1}^n).$$

However, if $D_-u_{i+1}^n = D_-u_{i-1}^n$,

$$\theta_i^n = \frac{D_+u_i^n}{D_-u_i^n} = \frac{D_-u_{i+1}^n}{D_-u_i^n} = \frac{D_-u_{i-1}^n}{D_-u_i^n} = \frac{D_-u_{i-1}^n}{D_+u_{i-1}^n} = \frac{1}{\theta_{i-1}^n}$$

and hence,

$$S(\theta_i^n) = -S\left(\frac{1}{\theta_i^n}\right).$$

For $0 < |c|\lambda < 1$, the monotonicity condition (4.3.33) must be satisfied. For (4.3.34a),

$$-\frac{u_i^{n+1}-u_i^n}{u_i^n-u_{i-1}^n} = c\lambda + \frac{c\lambda}{2}(1-c\lambda)(1-S(\theta_i^n))(\frac{D_-u_{i+1}^n - D_-u_i^n}{D_-u_i^n})$$
$$= c\lambda + \frac{c\lambda}{2}(1-c\lambda)(1-S(\theta^n))(\theta^n-1).$$

Condition (4.3.33) will hold provided $\partial_{c\lambda}(-\frac{u_i^{n+1}-u_i^n}{u_i^n-u_{i-1}^n}) > 0$ for $|c|\lambda = 0$ and 1, where $\partial_{c\lambda}$ denote differentiation with respect to $c\lambda$. For (4.3.34a) this gives rise to

$$1 - \frac{1}{2}(1-S(\theta_i^n))(1-\theta_i^n) > 0$$

for $c\lambda = 0$, and for $c\lambda = 1$

$$1 + \frac{1}{2}(1-S(\theta_i^n))(1-\theta_i^n) > 0.$$

Combining these two inequalities gives

$$|(1-S(\theta_i^n))(1-\theta_i^n)| < 2$$

or

(4.3.37) $$\frac{\theta_i^n - 3}{\theta_i^n - 1} > S(\theta_i^n) > \frac{\theta_i^n + 1}{\theta_i^n - 1} .$$

Similarly, for (4.3.34b)

$$|(1-S(\theta_i^n))(1-\frac{1}{\theta_i^n})| < 2$$

or

(4.3.38) $$\frac{\theta_{i+1}^n}{\theta_{i-1}^n} > S(\theta_i^n) > \frac{1-3\theta_i^n}{\theta_{i-1}^n} .$$

van Leer [76] chose

$$S(\theta_i^n) = \frac{|\theta_i^n| - 1}{|\theta_i^n| + 1} ,$$

which satisfies inequalities (4.3.37) and (4.3.38). This function is written in the more convenient form, using the definition of θ_i^n (4.3.31),

(4.3.39) $$S(\theta_i^n) = \begin{cases} \dfrac{|D_+u_i^n|-|D_-u_i^n|}{|D_+u_i^n|+|D_-u_i^n|} , & |D_+u_i^n| + |D_-u_i^n| > \varepsilon \\ 0 , & \text{otherwise,} \end{cases}$$

where $\varepsilon > 0$ is chosen as a measure of negligible variation in the grid function u.

The extension of this method to the case of variable coefficients is direct. However, as in the case of the first order upwind method, if $c(x)$ is of one sign, Fromm's method (4.3.30') is in conservation form. However, if $c(x)$ is not of one sign, Fromm's method is not in conservation form. Fromm's method can be modified so that it is in conservation form,

$$(4.3.40) \quad u_i^{n+1} = u_i^n - c\lambda h D_0 u_i^n + \frac{c^2\lambda^2}{2} h(D_- u_{i+1}^n - D_- u_i^n)$$
$$+ \frac{c\lambda}{8}(1-|c|\lambda)h(D_- u_{i+2}^n - D_- u_{i-1}^n - D_- u_i^n + D_- u_{i-1}^n)$$
$$- \frac{|c|\lambda}{8}(1-|c|\lambda)h(D_- u_{i+2}^n - 3D_- u_{i+1}^n + 3D_- u_i^n - D_- u_{i-1}^n).$$

Fromm's method (4.3.40) can be extended to a system of hyperbolic conservation laws (4.1.27). Let $\underline{A}(\underline{v})$ denote the Jacobian of $\underline{F}(\underline{v})$ and let $\mu_1(\underline{v}),\ldots,\mu_p(\underline{v})$ denote the eigenvalues of $A(v)$. There exists a nonsingular matrix $\underline{T}(\underline{v})$ such that $\underline{A}(\underline{v}) = \underline{T}(\underline{v})D(\underline{v})\underline{T}^{-1}(\underline{v})$, where $\underline{D}(\underline{v}) = \text{diag}(\mu_1(\underline{v}),\ldots,\mu_p(\underline{v}))$. Let $|\underline{D}(\underline{v})| = \text{diag}(|\mu_1(\underline{v})|,\ldots,|\mu_p(\underline{v})|)$. Then

$$|\underline{A}(\underline{v})| = \underline{T}(\underline{v})|\underline{D}(\underline{v})|\underline{T}^{-1}(\underline{v}),$$

which commutes with $\underline{A}(\underline{v})$. Let $\underline{u}_i^n$ denote the approximation to $\underline{v}(ih,nk)$. In (4.3.40) replace u_i^n with $\underline{u}_i^n$, $c\lambda$ with $\lambda\underline{A}(\underline{u}_i^n)$, $D_- u^n$ with $D_- F^n$, and $|c|$ with $|\underline{A}(\underline{u}_i^n)|$, that is

$$\underline{u}_i^{n+1} = \underline{u}_i^n - \lambda\underline{A}(\underline{u}_i^n)h\underline{D}_0\underline{u}_i^n + \frac{\lambda^2\underline{A}^2(\underline{u}_i^n)}{2}h(D_-\underline{F}_{i+1}^n - D_-\underline{F}_i^n)$$
$$+ \frac{\lambda\underline{A}(\underline{u}_i^n)}{8}(\underline{I}-|\underline{A}(\underline{u}_i^n)|\lambda)h(D_-\underline{F}_{i+2}^n - D_-\underline{F}_{i+1}^n - D_-\underline{F}_i^n + D_-\underline{F}_{i-1}^n)$$
$$- \frac{\lambda|\underline{A}(\underline{u}_i^n)|}{8}(\underline{I}-|\underline{A}(\underline{u}_i^n)|\lambda)h(D_-\underline{F}_{i+2}^n - 3D_-\underline{F}_{i+1}^n$$
$$+ 3D_-\underline{F}_i^n - D_-\underline{F}_{i-1}^n).$$

We present one final group of monotonicity preserving upstream methods, again due to van Leer [76]. These methods are based on the first order accurate methods due to Godunov [24] for equations of gas dynamics in Lagrangian coordinates. (Godunov's method will be discussed in greater detail in the next section.)

To begin we describe Godunov's method to approximate the scalar, linear equation with constant coefficients (4.3.25). Divide the x-axis into intervals of equal length h, where the i-th interval $I_i = [x_i, x_{i+1}]$ with center $x_{i+\frac{1}{2}} = x_i + \frac{h}{2}$. Consider the approximate solution for any x at time $t = nk$, denoted by $u^n(x)$ (we can at least begin, for $n = 0$, $u^0(x) = f(x)$). Project $u^n(x)$ onto the space of functions that are constant on each interval I_i by setting

$\bar{u}^n(x) = \bar{u}^n_{i+\frac{1}{2}}$ for $x \in I_i$ ($\bar{u}^n(x)$ denote the result of this projection), where

$$\bar{u}_{i+\frac{1}{2}} = \frac{1}{h}\int_{x_i}^{x_{i+1}} u^n(x)dx \tag{4.3.42}$$

(see Figure 4.10a). The bar "$^{-}$" over a quantity refers to an average of that quantity. The solution at the next time interval $t = (n+1)k$ is obtained exactly,

$$u^{n+1}(x) = \bar{u}^n(x-ck) \tag{4.3.43}$$

(see Figure 4.10b). If the CFL condition, $|c|\lambda \leq 1$, is satisfied then this translation of $\bar{u}^n(x)$ by an amount ck will not exceed h.

Again, projecting onto the space of piecewise constant functions, set $\bar{u}^{n+1}(x) = \bar{u}^{n+1}_{i+\frac{1}{2}}$ for $x \in I_i$, where

$$\bar{u}^{n+1}_{i+\frac{1}{2}} = \frac{1}{h}\int_{x_i}^{x_{i+1}} u^{n+1}(x)dx = \frac{1}{h}\int_{x_i}^{x_{i+1}} \bar{u}\ (x-ck)dx \tag{4.3.44}$$

(see Figure 4.10c).

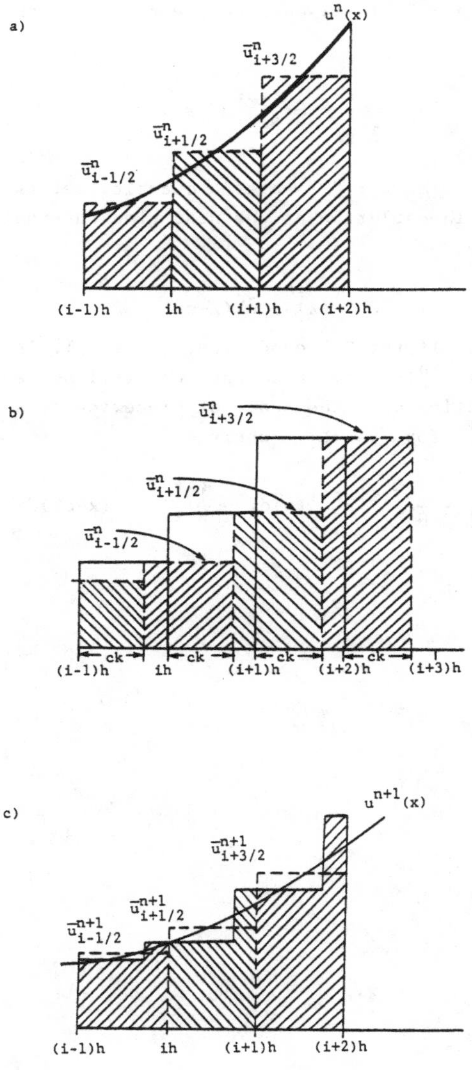

Figure 4.10 a) Projection of $u^n(x)$ onto Piecewise Constant Function $\bar{u}^n(x) = \bar{u}^n_{i+\frac{1}{2}}$, $x \in I_i$; b) Advection of Piecewise Constant $\bar{u}^n(x)$ on (4.3.43) for $c > 0$; c) Projection of $u^{n+1}(x)$ onto Piecewise Constant Function $\bar{u}^{n+1}(x) = \bar{u}^{n+1}_{i+\frac{1}{2}}$, $x \in I_i$.

For $c > 0$ and $x \in I_i$, by (4.3.43),

$$u^{n+1}(x) = \begin{cases} \bar{u}^n_{i-\frac{1}{2}}, & x_i < x < x_i + ck \\ \bar{u}^n_{i+\frac{1}{2}}, & x_i + ck < x < x_{i+1}. \end{cases}$$

Substituting into (4.3.44)

$$\bar{u}^{n+1}_{i+\frac{1}{2}} = \frac{1}{h}\int_{x_i}^{x_i+ck} \bar{u}^n_{i-\frac{1}{2}}\, dx + \frac{1}{h}\int_{x_i+ck}^{x_{i+1}} \bar{u}^n_{i+\frac{1}{2}}\, dx$$

$$= c\lambda\bar{u}^n_{i-\frac{1}{2}} + (1-c\lambda)\bar{u}^n_{i+\frac{1}{2}} = \bar{u}^n_{i+\frac{1}{2}} - c\lambda h D_0^{1/2}\bar{u}^n_i,$$

where $D_0^{1/2}\bar{u}^n_i = \dfrac{\bar{u}^n_{i+\frac{1}{2}} - \bar{u}^n_{i-\frac{1}{2}}}{h}$, that is, the centered dividend difference of $\bar{u}^n_i$ over the interval of length h rather than $2h$. Similarly, for $c < 0$

$$u^{n+1}_{i+\frac{1}{2}} = (1-|c|\lambda)\bar{u}^n_{i+\frac{1}{2}} + |c|\lambda\bar{u}^n_{i+\frac{3}{2}} = \bar{u}^n_{i+\frac{1}{2}} - |c|\lambda h D_0^{1/2} u_{i+1}.$$

Combining these two cases gives the first order upwind method (4.3.10) applied to the piecewise constant (grid averages) $\bar{u}^n_{i+\frac{1}{2}}$.

Rather than projecting $u^n(x)$ onto space of piecewise constant functions, suppose $u^n(x)$ is projected onto the space of piecewise linear functions, that is, functions that are linear on each I_i. Define

(4.3.45) $$\bar{u}^n(x) = \bar{u}^n_{i+\frac{1}{2}} + \overline{(\partial_x u^n(x))}_{i+\frac{1}{2}}(x-x_{i+\frac{1}{2}}), \quad x \in I_i,$$

where $\bar{u}^n_{i+\frac{1}{2}}$ is given by (4.3.42) and $\overline{(\partial_x u^n(x))}_{i+\frac{1}{2}}$ denotes the gradient of $u^n(x)$ at some point $\bar{x} \in I_i$. As in the piecewise constant case, the solution at the next time level is

$$u^{n+1}(x) = \bar{u}^n(x-ck)$$

and the projection onto the space of piecewise linear function, denoted by $\bar{u}^{n+1}(x)$, is

(4.3.46) $$\bar{u}^{n+1}(x) = \bar{u}^{n+1}_{i+\frac{1}{2}} + \overline{(\partial_x u^{n+1}(x))}_{i+\frac{1}{2}}(x-x_{i+\frac{1}{2}}), \quad x \in I_i,$$

where $\bar{u}^{n+1}$ is given by (4.3.44).

For $c > 0$ and $x \in I_i$, by (4.3.43) and (4.3.45),

$$u^{n+1}(x) = \begin{cases} u_{i-\frac{1}{2}} + \overline{(\partial_x u^n(x))}_{i-\frac{1}{2}}(x-ck-x_{i-\frac{1}{2}}) \ , & x_i < x < x_i + ck \\ u_{i+\frac{1}{2}} + \overline{(\partial_x u^n(x))}_{i+\frac{1}{2}}(x-ck-x_{i+\frac{1}{2}}) \ , & x_i + ck < x < x_{i+1} \end{cases}$$

Substituting into (4.3.44)

$$\text{(4.3.47a)} \quad \bar{u}^{n+1}_{i+\frac{1}{2}} = \frac{1}{h}\int_{x_i}^{x_i+ck} (\bar{u}^n_{i-\frac{1}{2}} + \overline{(\partial_x u^n(x))}_{i-\frac{1}{2}}(x-ck-x_{i-\frac{1}{2}})dx$$

$$+ \frac{1}{h}\int_{x_i+ck}^{x_{i+1}} (\bar{u}^n_{i+\frac{1}{2}} + \overline{(\partial_x u^n(x))}_{i+\frac{1}{2}}(x-ck-x_{i+\frac{1}{2}}))dx$$

$$= \bar{u}^n_{i+\frac{1}{2}} - c\lambda h D_0^{1/2}\bar{u}^n_i - \frac{c\lambda}{2}(1-c\lambda)h(\overline{(\partial_x u^n(x))}_{i+\frac{1}{2}} - \overline{(\partial_x u^n(x))}_{i+\frac{1}{2}} .$$

Similarly, for $c < 0$

$$\text{(4.3.47b)} \quad \bar{u}^{n+1}_{i+\frac{1}{2}} = \bar{u}^n_{i+\frac{1}{2}} - |c|\lambda h D_0^{1/2}\bar{u}^n_{i+1}$$

$$- \frac{|c|\lambda}{2}(1-|c|\lambda)h(\overline{(\partial_x u^n(x))}_{i+\frac{3}{2}} - \overline{(\partial_x u^n(x))}_{i+\frac{1}{2}})$$

Combining these two cases results in a second order upwind method applied to the piecewise grid averages $\bar{u}^n_{i+\frac{1}{2}}$.

Higher order accuracy can be achieved by projecting $u^n(x)$ onto the space of piecewise polynomials of higher degree. van Leer [78] also considers the projection onto the space of piecewise quadratic functions.

It remains to describe how $\overline{(\partial_x u^n(x))}_{i+\frac{1}{2}}$ is to be evaluated. There are several possibilities, of which, two will be discussed here. First, $(\partial_x u^n(x))_{i+\frac{1}{2}}$ can be approximated by a centered difference, that is, $D_0\bar{u}^n_{i+\frac{1}{2}}$ which is second order accurate. Substitution into (4.3.47) gives the second order upwind method (4.3.30') applied to the grid averages $\bar{u}^n_{i+\frac{1}{2}}$ rather than the function values at the grid points u^n_i.

Another possibility is to determine $\overline{(\partial_x u^n(x))}_{i+\frac{1}{2}}$ so that $u^n(x)$ so that $u^n(x)$ and $\bar{u}^n(x)$ have the same first moment, that is, by using (4.3.45)

$$\int_{x_i}^{x_{i+1}} \bar{u}^n(x)(x-x_{i+\frac{1}{2}})dx = \int_{x_i}^{x_{i+1}} (\bar{u}^n_{i+\frac{1}{2}} + \overline{(\partial_x u^n(x))}_{i+\frac{1}{2}}(x-x_{i+\frac{1}{2}}))(x-x_{i+\frac{1}{2}})dx$$

$$= \int_{x_i}^{x_{i+1}} u^n(x)(x-x_{i+\frac{1}{2}})dx$$

or

$$\overline{(\partial_x u^n(x))}_{i+\frac{1}{2}} = \frac{\int_{x_i}^{x_{i+1}} u^n(x)(x-x_{i+\frac{1}{2}})dx}{\int_{x_i}^{x_{i+1}} (x-x_{i+\frac{1}{2}})^2 dx} = \frac{12}{h^3}\int_{x_i}^{x_{i+1}} u^n(x)(x-x_{i+\frac{1}{2}})dx.$$

This gives rise to a recursive formula for $\overline{(\partial_x u^n(x)}_{i+\frac{1}{2}}$,

(4.3.48) $$\overline{(\partial_x u^{n+1}(x))}_{i+\frac{1}{2}} = \overline{(\partial_x u^n(x))}_{i+\frac{1}{2}} + (1-c\lambda)(1-2c\lambda-2c^2\lambda^2)\overline{(\partial_x u^n(x))}_{i+\frac{1}{2}}$$

$$- c\lambda(3-6c\lambda + 2c^2\lambda^2)\overline{(\partial_x u^n(x))}_{i-\frac{1}{2}} + 6c\lambda(1-c\lambda)hD_0^{1/2}\bar{u}^n_i.$$

To begin, set $n = 0$ and

$$\overline{(\partial_x u^0(x))}_{i+\frac{1}{2}} = \frac{12}{h^3}\int_{x_i}^{x_{i+1}} f(x)(x-x_{i+\frac{1}{2}})dx.$$

A complete step of the second order upwind method (4.3.47) with (4.3.48) is depicted in Figure 4.11.

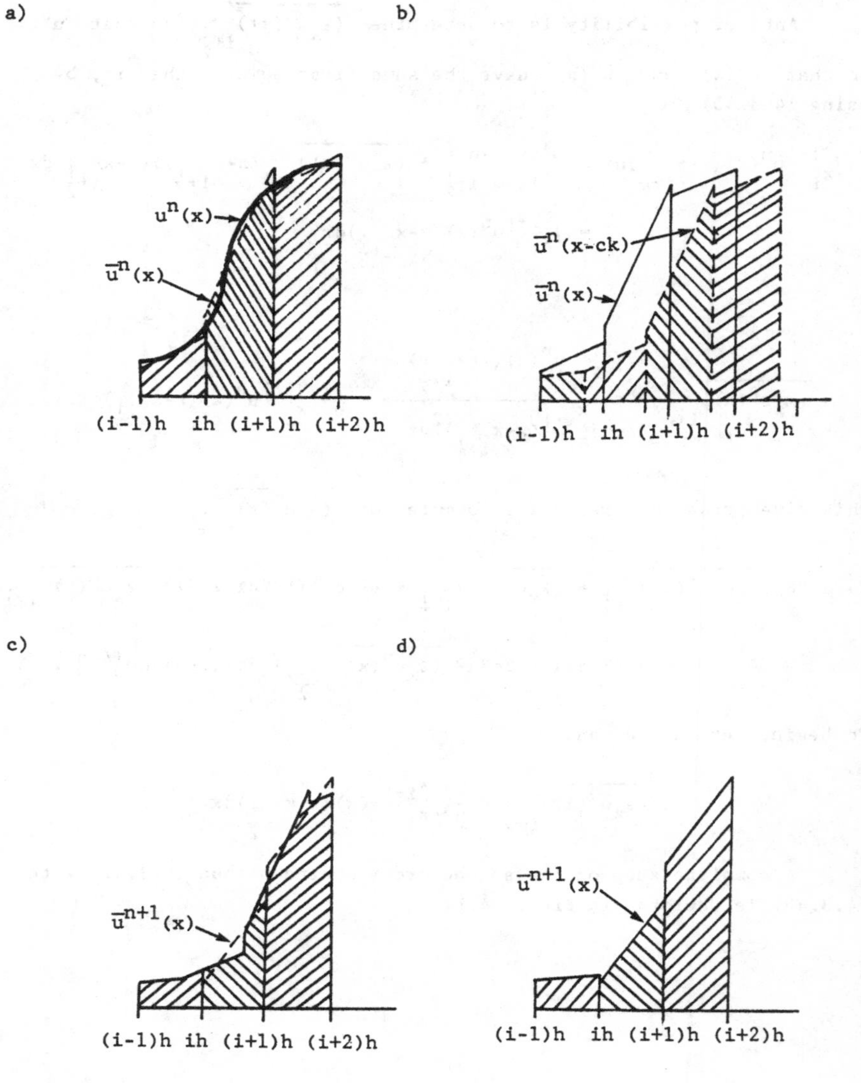

Figure 4.11 a) Projection of $u^n(x)$ onto Piecewise linear Function $\bar{u}^n(x)$ (4.3.45); b) Advection of Piecewise Linear Function $\bar{u}^n(x)$ on (4.3.43) for $c > 0$; c) Projection of $u^{n+1}(x)$ onto Piecewise Linear Function $\bar{u}^{n+1}(x)$ using (4.3.48); d) $\bar{u}^{n+1}(x)$ ready to begin next step.

Now we modify the second order upwind method (4.3.47) so that it is monotonicity preserving. A sufficient condition for a method satisfying the CFL condition to be monotonicity preserving is: if $\bar{u}^n_{i+\frac{1}{2}}$ lies between $\bar{u}^n_{i-\frac{1}{2}}$ and $\bar{u}^n_{i+\frac{3}{2}}$, then $\bar{u}^{n+1}_{i+\frac{1}{2}}$ must lie between $\bar{u}^n_{i-\frac{1}{2}}$ and $\bar{u}^n_{i+\frac{1}{2}}$, for $c > 0$ and between $\bar{u}^n_{i+\frac{1}{2}}$ and $\bar{u}^n_{i+\frac{3}{2}}$ for $c < 0$.

The term $\overline{(\partial_x u^n(x))}_{i+\frac{1}{2}}$ will be modified so that:

1) the linear function (4.3.45) on each interval I_i does not assume values outside the range spanned by the adjacent grid averages, that is, all values of $\bar{u}^n(x)$ for $x \in I_i$ in (4.3.45) lies between $\bar{u}^n_{i-\frac{1}{2}}$ and $\bar{u}^n_{i+\frac{3}{2}}$ (see Figure 4.12).

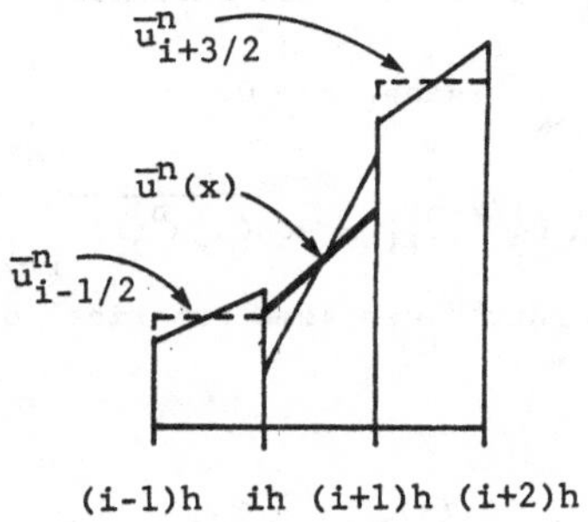

Figure 4.12. Slope of $\bar{u}^n(x)$ on I_i is modified so that all values of $\bar{u}^n(x)$ for $x \in I_i$ lie between $\bar{u}^n_{i-\frac{1}{2}}$ and $\bar{u}^n_{i+\frac{3}{2}}$.

2) If $\operatorname{sgn} D_0^{1/2}\bar{u}_i^n \neq \operatorname{sgn} D_0^{1/2}\bar{u}_{i+1}^n$, that is, if $\bar{u}_{i+\frac{1}{2}}^n$ attains a local extremum, then $\overline{(\partial_x u^n(x))}_{i+\frac{1}{2}}$ is set to 0 so that the extremum is not accentuated (see Figure 4.13).

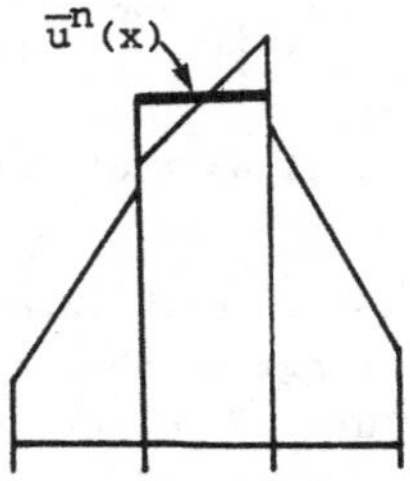

Figure 4.13. If $\bar{u}_{i+\frac{1}{2}}^n$ attains a local extremum, then slope $\overline{(\partial_x u^n(x))}_{i+\frac{1}{2}}$ is set to 0.

3) If $\operatorname{sgn} D_0^{1/2}\bar{u}_i^n = \operatorname{sgn} D_0^{1/2}\bar{u}_{i+1}^n \neq \operatorname{sqn}\overline{(\partial_x u^n(x))}_{i+\frac{1}{2}}$, then $\overline{(\partial_x u^n(x))}_{i+\frac{1}{2}}$ is set to 0. This introduces some numerical diffusion (see Figure 4.14).

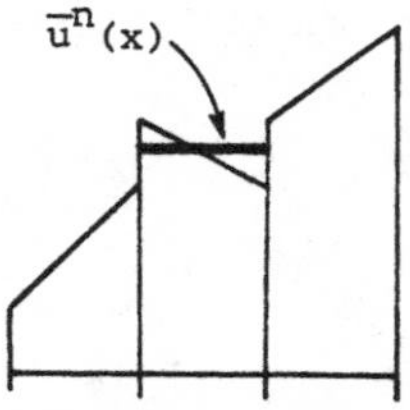

Figure 4.14. If the Slope $\overline{(\partial_x u^n(x))}_{i+\frac{1}{2}}$ does not agree with the trend in adjacent intervals, it is reduced to 0.

These requirements can be represented by replacing $\overline{(\partial_x u^n(x))}_{i+\frac{1}{2}}$ in (4.3.35) with

$$(4.3.49)\quad \overline{(\partial_x u^n(x))}^{mono}_{i+\frac{1}{2}} = \begin{cases} \min\{2|D_0^{1/2}\bar{u}_i^n|, |\overline{(\partial_x u^n(x))}_{i+\frac{1}{2}}|, 2|D_0^{1/2}\bar{u}_{i+1}^n|\} \\ \quad \overline{\operatorname{sgn}(\partial_x u^n(x))}_{i+1}, \text{ if } \operatorname{sgn} D_0^{1/2}\bar{u}_i^n = \\ \quad \operatorname{sgn} D_0^{1/2}\bar{u}_{i+1}^n = \overline{\operatorname{sgn}(\partial_x u^n(x))}_{i+\frac{1}{2}}, \\ 0, \text{ otherwise} \end{cases}$$

(see van Leer [78]).

IV.4. The Godunov and Random Choice Method

In this section we describe two methods based on the solution of a Riemann problem (see Section IV.1). In the method due to Godunov [24], the exact solution to a sequence of local Riemann problems is used to obtain a first order accurate upwind finite difference method that is monotonicity preserving.

Consider the Riemann problem defined by the hyperbolic system of conservation laws (4.1.27)

$$(4.4.1)\qquad \partial_t \underline{v} + \partial_x \underline{F}(\underline{v}) = 0$$

along with the step function initial condition

$$(4.4.2)\qquad \underline{v}(x,0) = \underline{f}(x) = \begin{cases} \underline{v}_L, & x < 0 \\ \underline{v}_R, & x > 0 . \end{cases}$$

The (weak) solution, called a similarity solution, which depends on the states $\underline{v}_L$ and $\underline{v}_R$ ad the ratio x/t, called the similarity variable, will be denoted by $\underline{R}(x/t, \underline{v}_L, \underline{v}_R)$.

Let $\mu_{min}(\underline{v}) = \min_{1\le j\le p} \mu_j(\underline{v})$ and $\mu_{max}(v) = \max_{1\le j\le p} \mu_j(\underline{v})$, which denotes the slowest and fastest signal speed of (4.1.27), respectively, then the solution of the Riemann problem (4.4.1)-(4.4.2) satisfies

$$\underline{R}(x/t, \underline{v}_L, \underline{v}_R) = \begin{cases} \underline{v}_L, & x/t < \mu_{min}(\underline{v}_L) \\ \underline{v}_R, & x/t > \mu_{max}(\underline{v}_R). \end{cases}$$

Suppose that u_i^n is given. Godunov [24] considered the piecewise constant function of x,

(4.4.3) $$\underline{u}^n(x) = \underline{u}_i^n, \quad x \in I_i^{1/2}.$$

where $I_i^{1/2} = [(i-\frac{1}{2})h, (i+\frac{1}{2})h]$. To advance the solution, from time nk to $(n+1)k$, consider the initial-value problem given by (4.4.1) and (4.4.3). On each interval $I_i = [ih, (i+1)h]$, this initial-value problem defines a Riemann problem. The initial-value problem (4.4.1) and (4.4.3) thus defines a sequence of Riemann problems. If the CFL condition,

(4.4.4) $$\max_{\substack{1 \le j \le p \\ \underline{v}}} |\mu_j(\underline{v})|\lambda \le \frac{1}{2}$$

is satisfied, the waves generated by the individual Riemann problems will not interact. In this case, these solutions to the local Riemann problems can be combined by superposition into a single exact solution $\underline{v}^e(x,t)$ for $nk \le t \le (n+1)k$,

(4.4.5) $$\underline{v}^e(x,t) = \underline{R}\left(\frac{x-(i+\frac{1}{2})h}{t-nk}, \underline{u}_i^n, \underline{u}_{i+1}^n\right), \quad x \in I_i.$$

To obtain $\underline{u}^{n+1}$, Godunov projected $\underline{v}^e(x,t)$ onto the space of piecewise constant functions by

(4.4.6) $$\underline{u}_i^{n+1} = \frac{1}{h}\int_{I_i^{1/2}} \underline{v}^e(x,(n+1)k)dx.$$

Since $\underline{v}^e(x,t)$ represents an exact solution to (4.4.1), using (4.1.29) over the rectangle $I_i^{1/2} \times [nk,(n+1)k]$

(4.4.7) $$\int_{I_i^{1/2}} \underline{v}^e(x,(n+1)k)dx - \int_{I_i^{1/2}} \underline{v}^e(x,nk)dx + \int_{nk}^{(n+1)k} F(v^e((i+\tfrac{1}{2})h,t)) - \int_{nk}^{(n+1)k} \underline{F}(v^e((i-\tfrac{1}{2})h,t))dt = 0.$$

Using (4.4.5), $\underline{v}^e((i-\frac{1}{2})h,t) = \underline{R}(0,\underline{u}_{i-1}^n,\underline{u}_i^n)$ and $\underline{v}^e((i+\frac{1}{2})h,t) = \underline{R}(0,\underline{u}_i^n,\underline{u}_{n+1}^n)$ an independent of t. Substituting into (4.4.7),

$$\int_{nk}^{(n+1)k} \underline{F}(\underline{v}^e((i+\tfrac{1}{2})h,t))dt = \underline{F}(\underline{R}(0,\underline{u}_i^n,\underline{u}_{i+1}^n))k$$

and

$$\int_{nk}^{(n+1)k} \underline{F}(\underline{v}^e((i-\tfrac{1}{2})h,t))dt = F(R(0,u_{i-1}^n,\underline{u}_i^n))k.$$

Relation (4.4.7) may be written, with the help of (4.4.6), in the form

(4.4.8) $$\underline{u}_i^{n+1} = \underline{u}_i^n - \lambda(\underline{F}(\underline{R}(0,\underline{u}_i^n,\underline{u}_{i+1}^n)) - \underline{F}(\underline{R}(0,\underline{u}_{i-1}^n,\underline{u}_i^n)))$$
$$= \underline{u}_i^n - \lambda(\underline{F}_{i+\frac{1}{2}}^n - \underline{F}_{i-\frac{1}{2}}^n),$$

where $\underline{u}_{i+\frac{1}{2}}^n = \underline{R}(0,\underline{u}_i^n,\underline{u}_{i+1}^n)$ and $\underline{F}_{i+\frac{1}{2}}^n = \underline{F}(\underline{u}_{i+\frac{1}{2}}^n)$. Clearly, Godunov's method is in conservation form.

The random choice method (also known as Glimm's method) follows the construction of the Godunov method through the solution of the Riemann problems and the construction of $\underline{v}^e(x,t)$ in (4.4.5). Where Godunov's method uses an averaging process, (4.4.6), to advance the solution to $t = (n+1)k$, the random choice method chooses a representative point value of the locally exact solution to obtain $\underline{u}_i^{n+1}$. This representative point value is achieved by sampling (4.4.5). Let ξ_n denote an equidistributed random variable in the interval $(-\frac{1}{2},\frac{1}{2})$ (see Section III.12). Sampling the locally exact solution (4.4.5),

(4.4.9) $$\underline{u}_i^{n+1} = \underline{v}^e((i+\xi_n)h,(n+1)k)$$
$$= \underline{R}(\frac{(\xi_n-\frac{1}{2})h}{k}, \underline{u}_{i-1}^n,\underline{u}_i^n),\ \xi_n < 0$$
$$\underline{R}(\frac{(\xi_n-\frac{1}{2})h}{k}, \underline{u}_i^n,\underline{u}_{i+1}^n),\ \xi_n > 0.$$

As described in Section III.12, one value of ξ_n is chosen per time step and used for each value of i.

The CFL condition given by (4.4.4) was necessary so that the waves generated at the midpoint of the interval $I_i^{1/2}$ by the individual Riemann problems would not propagate beyond their respective intervals in time k. Thus the waves from the different Riemann problems would not interact. By explicitly treating the wave interactions from the individual Riemann problems, Leveque [44] was able to choose a step size k larger than dictated by the CFL condition.

The only error estimates available for the approximation of a nonlinear conservation law by the random choice method were obtained by Colella [7] for the special case of Burgers' equation. Let

$$\varepsilon(h,t) = \sum_i |u_i^n - v(ih,nk)|h$$

denote the discrete L_1 error. For smooth regions

$$\varepsilon(h,t) \sim ch|\log h|$$

and for the interaction of a shock with a rarefaction wave

$$\varepsilon(h,t) \sim c_1(t-t_0)h^{1/2}|\log h| + c_2|\log h|,$$

where t_0 is the time at which the shock strikes the rarefaction. In the second error estimate it would appear that the error would grow in time. However, numerical studies performed by LaVita [36] demonstrate that there is little growth of the error with time for fixed h. LaVita's results indicate that the error behaves like $O(h)$, independent of time, even in the presencee of discontinuities.

The solution to the Riemann problem (4.4.1)-(4.4.2) consists of $p+1$ constant states $\underline{v}_0, \underline{v}_1, \ldots, \underline{v}_p$, where $\underline{v}_0 = \underline{v}_L$ and $\underline{v}_p = \underline{v}_R$, separated by p waves. As in Section IV.1, the k-th wave is a rarefaction if $\mu_k(\underline{v}_{k-1}) < \mu_k(\underline{v}_k)$, a shock with speed S of $\mu_k(\underline{v}_{k-1}) > S > \mu_k(\underline{v}_k)$, and, in the linearly degenerate case, a contact discontinuity propagation with speed $\mu_k(\underline{v}_{k-1}) = \mu_k(\underline{v}_k)$.

Both the Godunov method and the random choice do not make use of all of the information in $\underline{R}(x/t,\underline{v}_L,\underline{v}_R)$. However, the Riemann problem can be very costly to solve and in many cases is impossible to solve exactly. It has been proposed (Harten and Lax [31], Roe [54-55]) that the exact solution $\underline{R}(x/t,\underline{v}_L,\underline{v}_R)$ to the Riemann problem (4.4.1)-(4.4.2) be replaced with an approximate solution, denoted by $\underline{r}(x/t,\underline{v}_L,\underline{v}_R)$ which has a much less conplex structure. It is essential, however, that this approximation be conservative and satisfy the entropy inequality (4.1.36).

With the aid of (4.4.5), $\underline{u}_i^{n+1}$ in (4.4.6) may be written in the form

$$\begin{aligned}\underline{u}_i^{n+1} &= \frac{1}{h}\int_{(i-\frac{1}{2})h}^{ih} R\left(\frac{x-(i-\frac{1}{2})h}{k}, \underline{u}_{i-1}^n, \underline{u}_i^n\right)dx \\ &+ \frac{1}{h}\int_{ih}^{(i+\frac{1}{2})h} \underline{R}\left(\frac{x-(i+\frac{1}{2})h}{k}, \underline{u}_i^n, \underline{u}_{i+1}^n\right)dx \\ &= \frac{1}{h}\int^{h/2} \underline{R}(x/k, \underline{u}_{i-1}^n, \underline{u}_i^n)dx + \frac{1}{h}\int_{-\frac{h}{2}}^{0} \underline{R}(x/k, \underline{u}_i^n, \underline{u}_{i+1}^n)dx.\end{aligned}$$

Definition. Let $\underline{r}(x/t,\underline{v}_L,\underline{v}_R)$ denote an approximate solution to the Riemann problem (4.4.1)-(4.4.2). A finite difference method is of Godunov type if it can be written in the form

$$(4.4.11)\quad \underline{u}_i^{n+1} = \frac{1}{h}\int_0^{h/2}\underline{r}(x/k,\underline{u}_{i-1}^n,\underline{u}_i^n)dx + \frac{1}{h}\int_{-\frac{h}{2}}\underline{r}(x/k,\underline{u}_i^n,\underline{u}_{i+1}^n)dx.$$

Consider a three-point finite difference method in conservation form ((4.2.3) with $q = 1$)

$$(4.4.12)\qquad \underline{u}_i^{n+1} = \underline{u}_i^n - \lambda(\underline{G}_{i+\frac{1}{2}} - \underline{G}_{u-\frac{1}{2}}),$$

where

$$(4.4.13)\qquad \underline{G}_{i+\frac{1}{2}} = \underline{G}(\underline{u}_i^n,\underline{u}_{i+1}^n)$$

and $\underline{G}$ denotes the numerical flux (see Section IV.2).

We now present the discrete analogue of the entropy condition (4.1.36a).

Definition. A finite difference method (4.4.12) is consistent with the entropy condition (4.1.36a) if

$$(4.4.14)\qquad U_i^{n+1} < U_i^n - \lambda(E^n_{i+\frac{1}{2}} - E^n_{i-\frac{1}{2}}),$$

where $U_i^n = U(\underline{u}_i^n)$ and

$$(4.4.15)\qquad E^n_{i+\frac{1}{2}} = E(\underline{u}_i^n,\underline{u}_{i+1}^n).$$

In this case $E(\underline{w}_1,\underline{w}_2)$ is a numerical entropy flux that is consistent with the entropy flux F_E,

$$(4.4.16)\qquad E(\underline{v},\underline{v}) = F_E(\underline{v}).$$

Definition. An approximate solution $\underline{r}(x/t,\underline{v}_L,\underline{v}_R)$ of a Riemann problem (4.4.1)-(4.4.2) is consistent with the integral form of the conservation law if

$$(4.4.17)\qquad \int_{-h/2}^{h/2}\underline{r}(x/t,\underline{v}_L,\underline{v}_R)dx = \frac{h}{2}(\underline{v}_L + \underline{v}_R)-k(\underline{F}(\underline{v}_R) - F(\underline{v}_L)),$$

for h satisfying $\max_{\substack{1<j<p\\ \underline{v}}} |\mu_j(v)|\lambda < \frac{1}{2}$ (the CFL condition (4.4.4)).

Integrating the conservation law (4.4.1) over the rectangle $[-\frac{h}{2},\frac{h}{2}] \times [0,k]$,

(4.4.18) $$\int_{-h/2}^{h/2}\underline{v}(x,k)dx - \int_{-h/2}^{h/2}\underline{v}(x,0)dx + \int_0^k\underline{F}(\underline{v}(\tfrac{h}{2},t))dt - \int_0^k\underline{F}(\underline{v}(-\tfrac{h}{2},t))dt$$

However, if the CFL condition (4.4.4) is satisfied by k and h defining the rectangle, then

$$\int_0^k F(v(\tfrac{h}{2}, t))dt = \int_0^k F(\underline{v}_R)dt = k\underline{F}(\underline{v}_R)$$

and

$$\int_0^k F(v(-\tfrac{h}{2},t))dt = \int_0^k F(\underline{v}_L)dt = k\underline{F}(\underline{v}_L).$$

Also

$$\int_{-h/2}^{h/2}\underline{v}(x,0)dx = \frac{h}{2}(\underline{v}_L + \underline{v}_R).$$

Substituting these results into (4.1.18),

(4.4.19) $$\int_{-h/2}^{h/1}\underline{v}(x,t)dx = \frac{h}{2}(\underline{v}_L + \underline{v}_R) - k(\underline{F}(\underline{v}_R) - \underline{F}(\underline{v}_L)).$$

<u>Definition</u>. An approximate solution $\underline{r}(x/t,\underline{v}_L,\underline{v}_R)$ of a Riemann problem (4.4.1)-(4.4.2) is <u>consistent with the integral form of the entropy condition</u> if

(4.4.20) $$\int_{-\frac{h}{2}}^{h/2}U(\underline{r}(x/t,\underline{v}_L,\underline{v}_R))dx \leq \frac{h}{2}(U(\underline{v}_L) + U(\underline{v}_R)) - k(F_E(\underline{v}_R) - F_E(\underline{v}_L)).$$

This inequality is analogous to the entropy inequality (4.1.36b).

The following theorem is due to Harten and Lax [31].

<u>Theorem 4.9</u>. Let $\underline{r}(x/t,\underline{v}_L,\underline{v}_R)$ denote an approximate solution to a Riemann problem (4.4.1)-(4.4.2). If $\underline{r}$ satisfies (4.4.17) and (4.4.20), then the Godunov type method (4.4.11) using $\underline{r}$ is consistent with (4.4.1) and satisfies the entropy condition (4.4.14).

If the solution of a Godunov type method, which satisfies the hypothesis of Theorem 4.9, converges boundedly and a.e. to $\underline{v}(x,t)$ as $h \to 0$ for fixed λ, then $\underline{v}(x,t)$ is a weak solution of the conservation law (4.4.1) and satisfies the entropy condition. This follows from an extension of Theorem 4.6 (see Harten, Lax, and van Leer [32]).

In the remainder of this section we consider some particular approximate solutions $\underline{r}(x/t,\underline{v}_L,\underline{v}_R)$ to the Riemann problem (4.4.1)-(4.4.2). These methods are called Riemann solvers.

Roe's [54-56] method of approximating the solution to the Riemann problem (4.4.1)-(4.4.2) consists of computing the exact solution of the Riemann problem with equation (4.4.1) replaced by the locally linearized system

(4.4.21) $$\partial_t\underline{v} + \underline{A}(\underline{v}_L,\underline{v}_R)\partial_x\underline{v} = 0,$$

where the matrix $\underline{A}(\underline{v}_L,\underline{v}_R)$ denotes an approximation to the Jacobian of $\underline{F}(\underline{v})$,

(4.4.22) $$\underline{F}(\underline{v}_R) - \underline{F}(\underline{v}_L) = \underline{A}(\underline{v}_L,\underline{v}_R)(\underline{v}_R - \underline{v}_L)$$

Consider the scalar case with Burgers' equation. Using (4.4.22) and $F(v) = \frac{1}{2}v^2$,

$$A(v_L,v_R) = \frac{F(v_R)-F(v_L)}{v_R - v_L} = \frac{1}{2}(v_L + v_R) \equiv S.$$

The exact solution of the linearized Riemann problem gives the approximate solution to the nonlinear Riemann problem,

(4.2.23) $$\underline{r}(x/t,v_L,v_R) = f(x-St) = \begin{cases} v_L, & x/t < S \\ v_R, & x/t > S. \end{cases}$$

The Riemann solver (4.2.23) is exact if $v_L > v_R$, which corresponds to a shock. However, if $v_L < v_R$, which corresponds to a rarefaction wave, Roe's corresponding solution is an expansion shock. An expansion shock, which is nonphysical, is a discontinuity from which the characteristics emanate (see Figure 4.9).

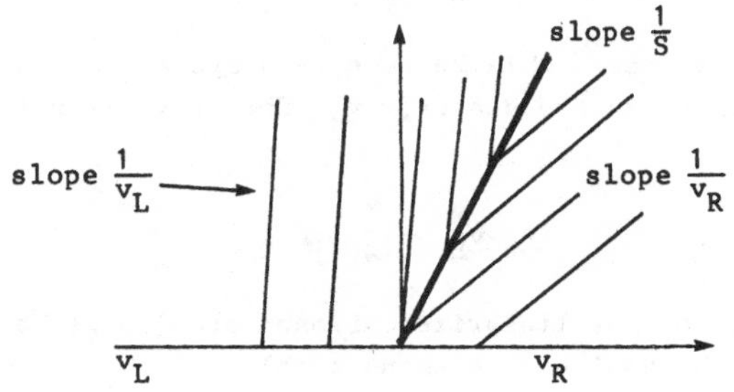

Figure 4.9. Roe's Riemann Solver Representing a Rarefaction Wave as an Expansion Shock.

Clearly, in this case the entropy condition i violated.

Consider the linearized Riemann problem corresponding to (4.4.1)-(4.4.2). In this case $\underline{A}(\underline{v}_L,\underline{v}_R)$ is a p×p matrix with real eigenvalues, which are assumed to be ordered, $\alpha_1 < \alpha_2 < \ldots < \alpha_p$. Let $\underline{R}_i(\underline{v}_L,\underline{v}_R)$ denote the right eigenvector of $\underline{A}(\underline{v}_L,\underline{v}_R)$ associated with eigenvalue $\alpha_i(\underline{v}_L,\underline{v}_R)$ for $i = 1,\ldots,p$. The solution to the Riemann problem consists of p+1 constant states $\underline{v}_0,\underline{v}_1,\ldots,\underline{v}_p$, where $\underline{v}_0 = \underline{v}_L$ and $\underline{v}_p = \underline{v}_R$, which is separated by a fan of p characteristics (see Figure 410).

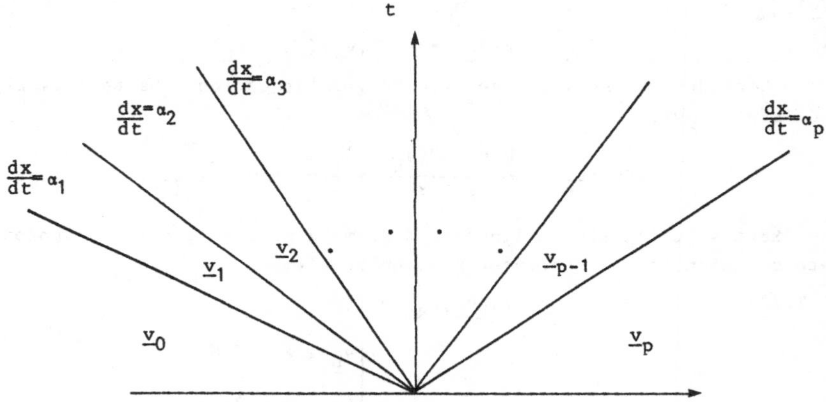

Figure 4.10. Solution of Linearized Riemann Problems.

$\underline{v}_R-\underline{v}_L$ may be written in terms of the p linearly independent eigenvectors R_i,

$$\underline{v}_R-\underline{v}_L = \sum_{j=1}^{p} \gamma_j\underline{R}_j, \tag{4.4.24}$$

where γ_j are constants. This represents a system of linear equations for the γ_j's. Intermediate states, $\underline{v}_k$ for $1 < k < p-1$ may be obtained from

$$\underline{v}_k-\underline{v}_L = \sum_{j=1}^{k} \gamma_j\underline{R}_j. \tag{4.4.25}$$

The exact solution of the linearized Riemann problem gives the approximate solution of the nonlinear Riemann problem

$$r(x/t,\underline{v}_R-\underline{v}_L) = v_k \ , \ \alpha_a < x/t < \alpha_{k+1},$$

for $k = 0,1,\ldots,p$, where $\alpha_0 = -\infty$ and $\alpha_{p+1} = +\infty$.

Harten and Hyman [29] have modified Roe's method to eliminate the nonphysical solution, that is, so that the Riemann solver satisfies the entropy condition. Define the matrix $\underline{T}(\underline{v}_L,\underline{v}_R)$, whose i-th column is $\underline{R}_i(\underline{v}_L,\underline{v}_R)$. Multiply equation (4.4.21) on the left by $T^{-1}(\underline{v}_L,\underline{v}_R)$. If $\underline{w} = \underline{T}^{-1}(\underline{v}_L,\underline{v}_R)\underline{v}$, then equation (4.4.21) reduces to

$$\partial_t \underline{w} + \underline{D}(\underline{v}_L,\underline{v}_R)\partial_x \underline{w} = 0,$$

where $\underline{D}(\underline{v}_L,\underline{v}_R) = \mathrm{diag}(\alpha_1(\underline{v}_L,\underline{v}_R),\ldots,\alpha_p(\underline{v}_L,\underline{v}_R))$, which represents p uncoupled scalar equations

(4.4.27) $$\partial_t w_j + \alpha_j(\underline{v}_L,\underline{v}_R)\partial_x w_j = 0.$$

The initial condition (4.4.2) becomes

$$\underline{w}(x,0) = \begin{cases} \underline{w}_L &, x < 0 \\ \underline{w}_R &, x > 0 \end{cases}$$

where $\underline{w}_L = \underline{T}^{-1}(\underline{v}_L,\underline{v}_R)\underline{v}_L$ and $\underline{w}_R = \underline{T}^{-1}(\underline{v}_L,\underline{v}_R)\underline{v}_R$. Thus the Riemann problem with p coupled waves has been reduced to p scalar Riemann problems given by (4.4.27) and

(4.4.28) $$\underline{w}_j(x,0) = \begin{cases} \underline{w}_{j,L}, & x < 0 \\ \underline{w}_{j,R}, & x > 0 \end{cases}$$

where $w_{j,L}$ and $w_{j,R}$ denote the j-th components of $\underline{w}_L$ and $\underline{w}_R$, respectively. The solution of each represents one of the p waves. As seen above, if $w_{j,L} < w_{j,R}$ the Roe's method violates the entropy condition. Harten and Hyman's modification consists of introducing an intermediate state w_j^*. The addition of the intermediate states adds diffusion. Let $\tilde{\alpha}_{j,L}$ and $\tilde{\alpha}_{j,R}$ denote the approximate speeds of the left and right ends of the rarefaction fan, respectively. Define

(4.4.29a) $$\alpha_{j,L} = \alpha_j(\underline{v}_L,\underline{v}_R) - (\alpha_j(\underline{v}_L,\underline{v}_R) - \tilde{\alpha}_{j,L})^+$$

and

(4.4.29b) $$\alpha_{j,R} = \alpha_j(\underline{v}_L,\underline{v}_R) + (\tilde{\alpha}_{j,R} - \alpha_j(\underline{v}_L,\underline{v}_R))^+$$

where the j-wave is a shock $(w_{j,L} > w_{j,R})$.

Harten and Hyman's modified approximate solution to the j-th Riemann problem is

$$(4.4.30) \qquad S_j(x/t, w_{j,L}, w_{j,R}) = \begin{cases} w_{j,L}, & x/t < \alpha_{j,L} \\ w_j^*, & \alpha_{j,L} < x/t < \alpha_{j,R} \\ w_{j,R}, & x/t > \alpha_{j,R} . \end{cases}$$

The value of w_j^* is chosen so that the consistency condition (4.4.17) holds. On the interval $[-\frac{h}{2}, \frac{h}{2}]$, using (4.4.30) with $t = k$,

$$S_j(x/k, w_{j,L}, w_{j,R}) = \begin{matrix} w_{j,L}, & -\frac{h}{2} < x < k\alpha_{j,L} \\ w_j^*, & k\alpha_{j,L} < x < h\alpha_{j,R} \\ w_{j,R}, & k\alpha_{j,R} < x < \frac{h}{2} . \end{matrix}$$

Substitution into the integral in (4.4.17) gives

$$(\frac{h}{2} + k\alpha_{j,L})w_{j,L} + k(\alpha_{j,R} - \alpha_{j,L})w_j^* + (\frac{h}{2} - k\alpha_{j,R})w_{j,R}$$

$$= \frac{h}{2}(w_{j,L} + w_{j,R}) - k\alpha_j(\underline{v}_L, \underline{v}_R)(w_{j,R} - w_{j,L}),$$

where the flux function is $\alpha_j(\underline{v}_L, \underline{v}_R)w_j$. This may be written as

$$(4.4.31) \qquad (\alpha_{j,R} - \alpha_{j,L})w_j^* = (\alpha_j(\underline{v}_L, \underline{v}_R) - \alpha_{j,L})w_{j,L}$$

$$+ (\alpha_{j,R} - \alpha_j(\underline{v}_L, \underline{v}_R))w_{j,R}.$$

The modified Riemann solver (in terms of $\underline{v}$), using $\underline{v} = \underline{T}\ \underline{w}$, becomes

$$(4.4.32) \qquad \underline{r}(x/t, \underline{v}_L, \underline{v}_R) = \sum_{j=1}^{p} S_j(x/t, w_{j,L}, w_{j,R})\underline{R}_j(\underline{v}_L, \underline{v}_R)$$

It remains to describe how the approximate speeds $\tilde{\alpha}_{j,L}$ and $\tilde{\alpha}_{j,R}$ are determined. Using Liu's generalization of Oleinik's entropy condition define

$$(4.4.33a) \qquad \tilde{\alpha}_{j,L} = \max_{0 \le \theta \le 1} \alpha_j(\underline{v}_L, \underline{v}(\theta))$$

and

$$(4.4.33b) \qquad \tilde{\alpha}_{j,R} = \max_{0 \le \theta \le 1} \alpha_j(\underline{v}(\theta), \underline{v}_R),$$

where $\underline{v}(\theta) = (1-\theta)\underline{v}_L + \theta\underline{v}_R$.

In the case where $\underline{F}(\underline{v})$ is convex, (4.4.33) reduces to

(4.4.34a) $$\tilde{\alpha}_{j,L} = \min(\alpha_j(\underline{v}_L,\underline{v}_L),\ \alpha_j(\underline{v}_L,\underline{v}_R))$$

and

(4.4.34b) $$\tilde{\alpha}_{j,R} = \max(\alpha_j(v_L,\underline{v}_R),\ \alpha_j(\underline{v}_R,\underline{v}_R)),$$

that is, check (4.4.33) at $\theta = 0$ and 1. In this case $\alpha_j(\underline{v}_L,\underline{v}_L)$ is the j-th eigenvalue of $\underline{A}(\underline{v}_L,\underline{v}_L) = \underline{A}(\underline{v}_L)$, where $\underline{A}(\underline{v}_L)$ denotes the Jacobian of $\underline{F}(\underline{v})$ evaluated at $\underline{v} = \underline{v}_L$. Similarly, $\alpha_j(\underline{v}_R,\underline{v}_R)$ is the j-th eigenvalue of $\underline{A}(\underline{v}_R,\underline{v}_R) = \underline{A}(\underline{v}_R)$.

As an example, consider the scalar conservation laws, Burgers' equation. In this case

$$\alpha(v_L,v_R) = A(v_L,v_R) = \frac{\frac{1}{2}v_R^2 - \frac{1}{2}v_L^2}{v_R - v_L} = \frac{v_R+v_L}{2}.$$

Replacing v_R with $v(\theta)$,

$$\alpha(v_L,v(\theta)) = \frac{(2-\theta)v_L + \theta v_R}{2}$$

and replacing v_L with $v(\theta)$,

$$\alpha(v(\theta),v_R) = \frac{(1-\theta)v_L + (1+\theta)v_R}{2}.$$

Setting $\theta = 0$ and 1 and substituting in (4.4.34)

$$\tilde{\alpha}_L = \min(v_L,S)$$

and

$$\tilde{\alpha}_R = \max(S,v_R),$$

where $S = \frac{v_L+v_R}{2}$. If $v_L < S < v_R$, corresponding to a rarefaction wave, $\tilde{\alpha}_L = v_L$ and $\tilde{\alpha}_R = v_R$ which are the exact speeds of the end-points of the rarefaction fan. Substituting into (4.4.29), $\alpha_L = v_L$ and $\alpha_R = v_R$. The intermediate state (4.4.31) is $v^* = S$ and the approximate solution to the Riemann problem is

$$r(x/t,v_L,v_R) = \begin{cases} v_L & ,\ x/t < v_L \\ \frac{v_L+v_R}{2} & ,\ v_L < x/t < v_R \\ v_R & ,\ x/t > v_R \end{cases}$$

If $v_L > S > v_R$, corresponding to a shock, $\tilde{\alpha}_L = \tilde{\alpha}_R = S$ and substituting into (4.4.29), $\alpha_L = \alpha_R = S$. In this case the intermediate state disappears and the approximate solution to the Riemann problem is

$$r(x/t,v_L,v_R) = \begin{cases} v_L\ ,\ x/t < S \\ v_R\ ,\ x/t > S. \end{cases}$$

Roe's Riemann solver includes all of the p-1 intermediate states. Can Riemann solvers be constructed satisfying the conditions of Theorem 4.9 while including fewer intermediate states? In Harten, Lax, and van Leer [32] (see also Harten and Lax [31]) Riemann solvers are constructed with one and two intermediate states. The Riemann solver with one intermediate state will be described.

Let μ^{min} denote a lower bound on $\mu_{min}(\underline{v})$ and μ^{max} denote an upper bound on $\mu_{max}(\underline{v})$. The approximate solution to the Riemann problem (4.4.1)-(4.4.2) is

$$\underline{r}(x/t,\underline{v}_L,\underline{v}_R) = \begin{cases} \underline{v}_L, & x/t < \mu^{min} \\ \underline{v}^*, & \mu^{min} < x/t < \mu^{max} \\ \underline{v}_R, & x/t > \mu^{max} \end{cases} \tag{4.4.35}$$

(see Figure 4.11). The value of $\underline{v}^*$ is chosen so that

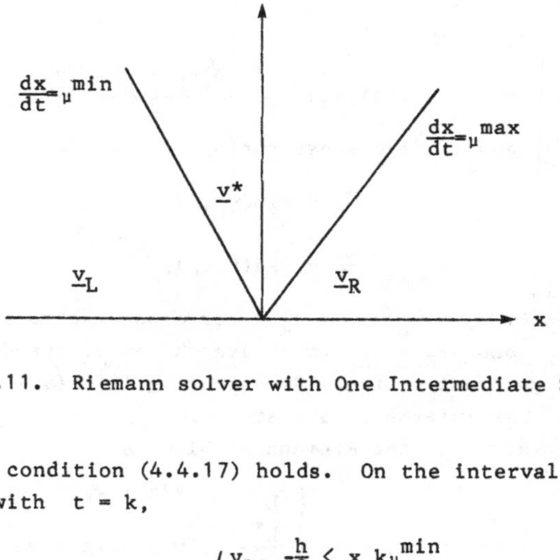

Figure 4.11. Riemann solver with One Intermediate State.

The consistency condition (4.4.17) holds. On the interval $[-\frac{h}{2},\frac{h}{2}]$, using (4.4.27) with $t = k$,

$$\underline{r}(x/k,\underline{v}_L,\underline{v}_R) = \begin{cases} \underline{v}_L, & -\frac{h}{2} < x\ k\mu^{min} \\ \underline{v}^*, & k\mu^{min} < x < h\mu^{max} \\ \underline{v}_R, & k\mu^{max} < x < \frac{h}{2}\ . \end{cases} \tag{4.4.27}$$

Substitution into the integral in (4.4.17) gives

$$(\tfrac{h}{2} + k\mu^{min})\underline{v}_L + k(\mu^{max}-\mu^{min})\underline{v}^* + (\tfrac{h}{2} - k\mu^{max})\underline{v}_R$$
$$= \tfrac{h}{2}(\underline{v}_L+\underline{v}_R) - k(\underline{F}(\underline{v}_R) - F(\underline{v}_L)),$$

or

(4.4.36) $$v^* = \frac{\mu^{max}\underline{v}_R - \mu^{min}\underline{v}_L}{\mu^{max} - \mu^{min}} - \frac{\underline{F}(\underline{v}_R) - \underline{F}(\underline{v}_L)}{\mu^{max} - \mu^{min}} .$$

It is shown in Harten, Lax, and van Leer [32] that (4.4.27)-(4.4.28) along with the associated numerical flux satisfy the entropy inequality (4.4.20).

Examles of other Riemann solvers are given in Engquist and Osher [14], Osher [49], and van Leer [78].

IV.5. Corrective and High Resolution Methods

In Section IV.3 it was shown that every monotone finite difference approximation to a hyperbolic conservation law was first order accurate and hence, was a second order approximation to a parabolic equation of the form (4.1.21). $F(v)$ denote the flux function of a scalar conservation law. If v_L and v_R denote two states and S denotes the speed of the discontinuity (given by the Rankine-Hugoniot condition (4.1.13)), then a flux function in a coordinate system moving with the discontinuity is given by (4.1.23)),

4.5.1) $$g(v) = F(v) - Sv.$$

If $g(v_L) = g(v_R) = C$, a constant, then the entropy condition becomes

(4.5.2) $$(g(v) - C)\text{sgn}(v_R - v_L) > 0$$

for $v \in (v_L, v_R)$. In Section IV.1 it was shown that for strict inequality in (4.5.2), which corresponds to a shock wave, a viscous profile exists and the transition width is given by

$$W(v_L, v_R) = \int_{v_L}^{v_R} \frac{\beta(w)}{g(w) - C}\, dw,$$

where $\beta(w)$ is the diffusion coefficient in (4.1.21). From this we conclude that, if a monotone finite difference method results in a discrete shock, then the spread (or width) is inversely proportional to $g(v) - C$.

For the case in which (4.5.2) is an equality, which corresponds to a contact discontinuity, the scalar conservation law reduces to a linear hyperbolic equation

(4.5.3) $$\partial_t v + S\partial_x v = 0.$$

Approximate this equation by the Lax-Friedrichs method (4.2.12), which to second order accuracy approximates the parabolic equation

$$\partial_t v + S\partial_x v = \beta\partial_x^2 v \tag{4.5.4}$$

where $\beta = \frac{1-\lambda^2 S^2}{2\lambda^2}$ and $\lambda = k/h$ (see Section IV.3). Introduce a change of variable $y = x - St$, which results in a coordinate system moving with speed S. Equation (4.5.4) reduces to

$$\partial_t v = \beta\partial_y^2 v. \tag{4.5.5a}$$

If the initial jump is at $x = 0$, the initial condition becomes

$$v(y,0) = \begin{cases} v_L, & y < 0 \\ v_R, & y > 0 \end{cases} \tag{4.5 5b}$$

the solution to (4.5.5) is

$$v(y,t) = \tfrac{1}{2}(v_L+v_R) + \tfrac{1}{2}(v_R-v_L)\mathrm{erf}\left(\frac{y}{\sqrt{4\beta t}}\right),$$

where erf denotes the error function,

$$\mathrm{erf}(z) = \frac{2}{\sqrt{\pi}}\int_0^z e^{-r^2}dv.$$

Writing the solution in x and by using the value of β,

$$v(x,t) = (v_L+v_R) + \tfrac{1}{2}(v_R-v_L)\mathrm{erf}\left(\frac{x-St}{\sqrt{2\lambda hn(1-\lambda^2S^2)}}\right).$$

Thus the width of the transition in a contact discontinuity for the first order Lax-Friedrichs method (or any first order accurate method) at time $t = nk$ is proportional to $n^{1/2}$, that is

$$W_1(v_L,v_R) = \text{const.}n^{1/2}$$

Similarly, Harten [26] showed that the width off the transition in a constant discontinuity for a q-th order method at time $t = nk$ is

$$W_q(v_L,v_R) = \text{const.}n^{1/q+1}.$$

We shall first describe corrective methods that can be used in conjunction with standard finite difference methods to prevent the smearing of shocks and contact discontinuities. The first of which is the artificial compression method (ACM) due to Harten [26-27].

Consider the scalar conservation law (4.1.1)

$$\partial_t v + \partial_x F(v) = 0$$

along with initial condition (4.1.6). Let the solution $v(x,t)$ contain a shock with speed $S(t)$, when $v_L(t)$ and $v_R(t)$ denote the states on the left and right of the shock, respectively. Let the shock be denoted by $(v_L(t), v_R(t), S(t))$. Assume that $v(x,t)$ does not take on any values vetween $v_L(t)$ and $v_R(t)$.

Consider a function $g(v,t)$ that satisfies the conditions

1) $g(v,t) \equiv 0$ for v outside the interval $(v_L(t), v_R(t))$

2) $g(v,t)\, \mathrm{sgn}(v_R(t) - v_L(t)) > 0$ for v in the interval $(v_L(t), v_R(t))$.

Such a function g is called an <u>artificial compression flux</u>.

The solution, $v(x,t)$, to (4.1.1) also satisfies the modified conservation law

$$\partial_t v + \partial_x (F(v) + g(v,t)) = 0. \tag{4.5.9}$$

$v(x,t)$ does not take on any values between $v_L(t)$ and $v_R(t)$ so that $v \notin (v_L(t), v_R(t))$. By property 1) of $g(v,t)$, for $v \notin (v_L(t), v_R(t))$, $g(v,t) \equiv 0$ and equation (4.5.9) reduces to (4.1.1).

Let $\tilde{F}(v) = F(v) + g(v,t)$ denote the modified flux function of (4.5.9). By property 1) of $g(v,t)$, $g(v_L(t),t) = g(v_R(t)) \equiv 0$ and, by the R-H condition (4.1.13),

$$\tilde{F}(v_R(t)) - \tilde{F}(v_L(t)) = (F(v_R(t) + g(v_R(t),t))$$

$$- (F(v_L(t)) + g(v_L(t),t))$$

$$= F(v_R(t) - F(v_L(t))$$

$$= S(t)(v_R(t) - v_L(t)).$$

Thus $(v_L(t), v_R(t), S(t))$ is a discontinuity for (4.5.9). It remains to show that it is admissible in the sense that the entropy condition (4.5.2) is satisfied. Define the flux funtion, $\tilde{g}(v)$, in a coordinate system moving with the discontinuity $(v_L(t), v_R(t), S(t))$,

$$\tilde{g}(v) = \tilde{F}(v) - S(t)v.$$

Again, using property 1) of $g(v,t)$, $\tilde{g}(v_L) = \tilde{F}(v_L) - S(t)v_L = F(v_L) - S(t)v_L = F(v_L) - S(t)v_L = C = F(v_R) - S(t)v_R = \tilde{F}(v_R) - S(t)v_R = \tilde{g}(v_R)$. The entropy conditin (4.5.2) holds for $g(v) = F(v) - S(t)v$. Using this and property 2) of $g(v,t)$,

$$[\tilde{g}(v)-C]\mathrm{sgn}(v_R(t)-v_L(t))$$

$$= [(F(v)+g(v,t)-S(t)v-C]\mathrm{sgn}(v_R(t)-v_L(t))$$

$$= [g(v)-C]\mathrm{sgn}(v_R(t)-v_L(t)) + g(v,t)\mathrm{sgn}(v_R(t)-v_L(t)) > 0.$$

Observe that by property 2) of $g(v,t)$, the entropy condition (4.5.10) is a strict inequality, so that if $(v_L(t),v_R(t),S(t))$ is a contact discontinuity for the original equation (4.1.1), with $[g(v)-C)\mathrm{sgn}(v_R(t)-v_L(t)) = 0$, $(v_L(t),v_R(t),S(t))$ is a shock for the modified equation (4.5.9). Also, if $(v_L(t),v_R(t),S(t))$ is a shock for the original equation (4.1.1), it remains a shock for the modified equation (4.5.9).

What is the purpose of introducing the artificial compression flux and the modified conservation law (4.5.9)? Using (4.5.10),

$$|\tilde{g}(v)-C| > |g(v)-C|$$

and so the spread of the shock for the modified system, which is inversely proportional to this quantity, is reduced.

The modified conservation law (4.5.9) could be approximated directly by monotone finite difference method in conservation form. However, the addition of the artificial compression flux function to the flux function of (4.1.1) results in a more restrictive CFL condition,

$$\max_v |\partial_v F + \partial_v g|\lambda < 1.$$

For this reason, as well as for ease of application, the artificial compression method is applied using the <u>split-flux</u> method. This is similar to the fractional step method. Consider the conservation law (4.1.1). Approximate (4.1.1) by the monotone finite difference method in conservation form

$$u_i^{n+1} = Qu_i^n = H(u_{i-1}^n, u_i, u_{i+1})$$
$$= u_i^n - \lambda(G_{i+\frac{1}{2}} - G_{i-\frac{1}{2}}),$$

where $G_{i+\frac{1}{2}} = G(u_i^n, u_{i+1}^n)$ and $G(v,v) = F(v)$ (see (4.3.3)). The CFL condition for (4.5.10) is

$$\max_v |a(v)|\lambda < 1,$$

where $a(v) = \partial_v F$.

Next, consider the conservation law

$$\partial_t v + \partial_x g(v,t) = 0, \tag{4.5.11}$$

where $g(v,t)$ is the artificial compression flux function. By Property 1) of $g(v,t)$, $g(v_L(t),t) = g(v_R(t),t) = 0$, or

$$\frac{g(v_R(t),t) - g(v_L(t),t)}{v_R(t) - v_L(t)} = 0.$$

Thus the discontinuity is a stationary shock. Approximate (4.5.11) by a finite difference method in conservation form

$$u_i^{n+1} = Q'u_i^n = u_i^n - \lambda'(G'_{i+\frac{1}{2}} - G'_{i-\frac{1}{2}}), \tag{4.5.12}$$

where $G'_{i+\frac{1}{2}} = G'(u_i^n, u_{i+1}^n)$ and $G'(v,v) = g(v,t)$. The monotone operator Q' is called an <u>artificial compressor</u>. The CFL condition for (4.5.12) is

$$\max_v |\partial_v g(v,t)|\lambda' < 1.$$

In general, $g(v,t)$ will only be known at the discrete points x_i. In this case, we replace the CFL condition with the discrete form of the CFL condition

$$\max_{u_{i+1}^n \neq u_i^n} \left|\frac{g_{i+1}^n - g_i^n}{u_{i+1}^n - u_i^n}\right|\lambda' < 1. \tag{4.5.13}$$

Combining the two methods (4.5.10) and (4.5.12) gives rise to the two-step method

$$u_i^{n+1} = Q'(Qu_i^n)$$

or

$$\tilde{u}_i^{n+1} = u_i^n - \lambda(G_{i+\frac{1}{2}} - G_{i-\frac{1}{2}}), \tag{4.5.14a}$$

$$u_i^{n+1} = u_i^n - \lambda(\tilde{G}_{i+\frac{1}{2}} - \tilde{G}_{i-\frac{1}{2}}), \tag{4.5.14b}$$

where $G'_{i+\frac{1}{2}} = G'(\tilde{u}_i^{n+1}, \tilde{u}_{i+1}^{n+1})$. The first step (4.5.14a) smears the discontinuity as it propagates it. In the second step (4.5.14b), called a corrective step, the smeared discontinuity is compressed while not being propagated. Since the application of Q' does not involve any motion, it does not alter the physical time in the solution obtained in the first step. Thus the timestep $k' = \lambda' h$ is regarded as a dummy timestep.

We now describe the artificial compresor Q'. Clearly, the finite difference method (4.5.14b) should have maximal resolution of a stationary shock. Harten [26] chose the upwind method in conservation form

$$u_i^{n+1} = u_i^n - \lambda'(G'_{i+\frac{1}{2}} - G'_{i-\frac{1}{2}}),$$

where

(44.5.15) $$G'_{i+\frac{1}{2}} = \tfrac{1}{2}(g_{i+1}^n + g_i^n) - |g_{i+1}^n - g_i^n|\,\mathrm{sgn}(u_{i+1}^n - u_i^n)$$

and $g_i^n = g(u_i^n, nk)$. To see that this is an upwind method define

(4.5.16a) $$\gamma_{i+\frac{1}{2}} = \begin{cases} \dfrac{g_{i+1}^n - g_i^n}{u_{i+1}^n - u_i^n}, & u_{i+1}^n \neq u_i^n \\ (\partial_v g)(u_i^n), & u_{i+1}^n = u_i^n \end{cases}$$

and

(4.5.16b) $$\gamma^+_{i+\frac{1}{2}} = \max(0, \gamma_{i+\frac{1}{2}}) = \tfrac{1}{2}(\gamma_{i+\frac{1}{2}} + |\gamma_{i+\frac{1}{2}}|),$$

(4.5.16b) $$\gamma^-_{i+\frac{1}{2}} = \min(0, \gamma_{i+\frac{1}{2}}) = \tfrac{1}{2}(\gamma_{i+\frac{1}{2}} - |\gamma_{i+\frac{1}{2}}|),$$

The artificial compressor can be written in the form

(4.5.17) $$u_i^{n+1} = u_i^n - k'\gamma^-_{i+\frac{1}{2}} D_+ u_i^n - k'\gamma^+_{i-\frac{1}{2}} D_-^n{}_i u.$$

The CFL condition for (4.5.17) is

(4.5.18) $$|\gamma_{i+\frac{1}{2}}|\lambda' \leq 1$$

(see Section IV.3). Harten [27] showed that if the CFL condition

(4.5.18) is satisfied, then the upwind method (4.5.15) (or (4.5.16)) is monotonicity preserving.

Theorem 4.10. Let $u_i^{n+1} = Q'u_i^n$, $n > 0$, where Q' is the artificial compression flux (4.5.14b) and (4.5.15) satisfying (4.5.18) and with u_i^0 given by

$$(4.5.19) \qquad u_i^0 = \begin{cases} v_L \ , & i < I_L \\ U_i \ , & I_L < i < I_R \\ v_R \ , & i < I_R \ , \end{cases}$$

where U_i is a monotone grid function. Then $u_i^n \to u_i$ pointwise as $n \to \infty$ where u_i is given by

$$(4.5.20) \qquad u_i = \begin{cases} v_L \ , & i < I \\ v_L + \alpha(v_R - v_L), & i = I \\ v_R \ , & i > I \ , \end{cases}$$

where $0 < \alpha < 1$ and $I_L < I < I_R$ are uniquely determined by the conservation relation

$$\sum_{i=I_L}^{I_R} u_i^0 = \sum_{i=I_L}^{I_R} u_i = (I-\alpha-I_L+1)v_L + (I_R-I+\alpha)v_R.$$

The final step is to describe the construction of the artificial compression flux $g(v,t)$. This will be defined as a grid function. A problem arises in the implementation of the artificial compression method, that is, the values $v_L(t)$ and $v_R(t)$, as well as the shock location, are not known. The methods must be modified so that this needed information is obtained from the values of u_i^n.

Construct an artificial compression flux $g(v,t)$ that satisfies the two conditions

(1) $g(v,t) = O(h)$,

$g(v(x+h,t) - g(v(x,t),t) = O(h^2)$ for $v \notin (v_L(t), v_R(t))$.

(2) $g(v,t)\text{sgn}|v_R(t)-v_L(t)) > 0$ for $v \in (v_L(t), v_R(t))$.

One possibility is $hD_+u_i^n = u_{i+1}^n - u_i^n$. For if v is smooth, $v((i+1)h,t)-v(ih,t) = O(h)$. Also $(hD_+u_i^n)\text{sgn}(u_{i+1}^n - u_i^n) = |u_{i+1}-u_i^n| > 0$.

Clearly a grid function is piecewise monotone. Divide the real line into a sequence of intervals $I_j < i < I_{j+1}$ for which u_i^n is monotone. Define

$$(4.5.22) \qquad g_i^n = \begin{cases} \operatorname{sgn}(u^n_{I_{j+1}} - u^n_{I_j})\min(|hD_+u_i^n|, |hD_-u_i^n|), & I_j < i < I_{j+1} \\ 0 \ , & \text{otherwise.} \end{cases}$$

For $u_{i+1}^n \neq u_i^n$, $\gamma_{i+\frac{1}{2}}$ defined by (4.5.16a) satisfies $|\gamma_{i+\frac{1}{2}}| < 1$. Thus the discrete CFL condition (4.5.18) reduces to

$$(4.5.23) \qquad \lambda' < 1.$$

With this choice of g_i^n, the second part of property 1) of the discrete artificial compression flux is satisfied,

$$(4.5.24) \qquad g_{i+1}^n - g_i^n = O(h^2),$$

provided that $\partial_x^2 v$ is bounded.

Due to the construction of g_i^n in (4.5.22), the artificial compressor must be redefined on each of the intervals $I_j < i < I_{j+1}$. Write

$$(4.5.25) \qquad Q' = {}_j Q'_j,$$

where Q'_j is defined by

$$(4.5.26) \qquad u_i^{n+1} = Q'u_i^n = \begin{cases} u_i^n - k'D_0u_i^n - \frac{k'}{2}\operatorname{sgn}(u^n_{I_{j+1}} - u^n_{I_j})(|D_+u_i^n| - |D_-u_i^n|), \\ \qquad I_j < i < I_{j+1} \\ u_i^n \ , \quad \text{otherwise.} \end{cases}$$

If the CFL condition (4.5.23) is satisfied, then u_i^{n+1} is monotone on each interval $I_j < i < I_{j+1}$.

The artificial compression method need only be applied to intervals $[I_j, I_{j+1}]$ which contain a discontinuity. If, however, the location of the discontinuity is not known, the artificial compression method may be applied to each interval. If the artificial compression method is applied on an interval where v is smooth the accuracy of the solution is not diminished. For, by (4.5.22) and (4.5.23), the artificial compression method is second order accurate in regions where

v is smooth while the solution obtained by the monotone finite difference method (4.5.10) is first order accurate.

Harten [27] shows that in practice the partition into intervals of monotonicity $[I_j, I_{j+1}]$ need not be found. This partition can be indirectly accomplished by replacing g_i^n in (4.5.20) with

$$g_i^n = S_{i+\frac{1}{2}}\max\{0,\min(|hD_+u_i^n|, S_{i+\frac{1}{2}}|hD_-u_i^n|)\},$$

where $S_{i+\frac{1}{2}} = \mathrm{sgn}(u_{i+1}^n - u_i^n)$. The artificial compression method is obtained by substitution of (4.5.27) into (4.5.14b) and (4.5.15).

The analogue of Theorem 4.10 holds where the artificial compression flux is replaced with the numerical artificial compression flux.

<u>Theorem 4.11</u>. Let $u_i^{n+1} = Q'u_i^n$, $n > 0$, where Q' is the artificial compressor defined by (4.5.14b), (4.5.15), and (4.5.25) with $\lambda' < 1$ and u_i^0 given by (4.5.19), where U_i is a <u>strictly</u> monotone function. Then $u_i^n \to u_i$ pointwise as $n \to \infty$, where u_i is given by (4.5.20)-(4.5.21).

The artificial compression method can be extended to systems of conservation laws. The method outlined for the scalar conservation law is applied componentwise to the system of conservation laws. However, the numerical artificial compression flux (4.5.27) (see Harten [27]),

(4.5.28) $$\underline{g}_i^n = 2h\alpha_i^n D_0 \underline{u}_i^n,$$

where α_i^n is a nonnegative scalar grid function defined by

(4.5.29) $$\alpha_i^n = \max[0, \min_{1<j<p}\{\frac{\min(hD_+u_{j,i}^n, (hD_-u_{j,i}^n)S_{j,i+\frac{1}{2}})}{|hD_+u_{j,i}^n| + |hD_-u_{j,i}^n|}\}],$$

where $S_{j,i+\frac{1}{2}} = \mathrm{sgn}(u_{j,i+1}^n - u_{j,i}^n)$ and $u_{j,i}^n$ denotes the j-th component of $\underline{u}_i^n$, for $j = 1,\dots,p$. (Observe that when $p = 1$, (4.5.28)-(4.5.29) reduces to (4.5.27).) This modification constructs the numerical artificial compression flux in the correct direction.

The flux-connected transport (FCT) method of Boris and Book [3] (see also Book, Boris, and Hain [2], Boris and Book [4], and Zalesak [83]) is a corrective or anti-diffusion type method. The method is motivated by the result that a first order finite difference method,

(4.5.30) $$\underline{u}_i^{n+1} = Q\underline{u}_i^n ,$$

that approximates the system of conservation laws (4.1.27)

$$\partial_t \underline{v} + \partial_x \underline{F}(\underline{v}) = 0$$

is also consistent with the diffusion equation

(4.5.31) $$\partial_t \underline{v} + \partial_x \underline{F}(\underline{v}) = k\partial_x(\beta(\underline{v},\lambda)\partial_x \underline{v})$$

where $\beta(v,\lambda) > 0$ is the coefficient of numerical diffusion. Let $r(\underline{v},\lambda)$ denote a positive function and consider the modified diffusion equation

(4.5.32) $$\partial_t \underline{v} + \partial_x \underline{F}(\underline{v}) = k\partial_x((\beta(\underline{v},\lambda)-r(\underline{v},\lambda))\partial_x \underline{v}).$$

Since $r(\underline{v},\lambda) > 0$, $\beta(v,\lambda)-r(v,\lambda) < \beta(v,\lambda)$ and as such there is less diffusion or anti-diffusion.

Rather than solve (4.5.32) directly, flux splitting is used to solve the system of conservation laws(4.1.27) (which also solves (4.5.31)) and then solve the diffusion equation

(4.5.33) $$\partial_t \underline{v} = -k\partial_x(r(\underline{v},\lambda)\partial_x \underline{v}).$$

Let

$$\underline{u}_i^{n+1} = D\underline{u}_i^n$$

approximate equation (4.5.33). This method alone is unstable since the equation it is approximating (4.5.3) is not well-posed (it represents the backward diffusion equation). The method used to approximate (4.5.32) is the two-step method

(4.5.34a) $$\tilde{\underline{u}}_i^{n+1} = Q\underline{u}_i^n$$

(4.5.34b) $$\underline{u}_i^{n+1} = D\tilde{\underline{u}}_i^{n+1} .$$

Combining the two steps gives rise to the method

$$\underline{u}_i^{n+1} = DQ\underline{u}_i^n,$$

which is stable provided the original finite difference method (4.5.30) is stable and provided that

(4.5.35) $$\beta(\underline{v},\lambda) - r(\underline{v},\lambda) > 0.$$

Condition (4.5.35) will place more of a restriction of λ than the CFL condition.

The first step (4.5.34a) which represents the solution to (4.1.27) results in smearing or diffusing discontinuities, while the second step (4.5.34b) acts to compress the smeared discontinuities.

As an example (see Boris and Book [3]), consider the two-step Lax-Wendroff method (4.2.26), which is second order accurate. However, by adding the zero-th order diffusion term $\frac{\eta h^2}{k} D_+ D_- \underline{u}_i^n$, where $\eta > 0$ (the diffusion/anti-diffusion coefficient), the method is first order accurate (Boris and Book [3] chose $\eta = \frac{1}{8}$). This can be added as a fractional step to the two-step Lax-Wendroff method, resulting in the three step method

(4.5.36a) $$\underline{u}_{i+\frac{1}{2}}^{n+\frac{1}{2}} = \frac{1}{2}(\underline{u}_{i+1}^n + \underline{u}_i^n) - \frac{\lambda}{2}(\underline{F}_{i+1}^n - \underline{F}_i^n),$$

(4.5.36b) $$\underline{u}_i^{n+1} = \underline{u}_i^n - \lambda(\underline{F}_{i+\frac{1}{2}}^{n+\frac{1}{2}} - \underline{F}_{i-\frac{1}{2}}^{n+\frac{1}{2}}),$$

(4.5.36c) $$\underline{u}_i^{n+1} = \underline{u}_i^{n+1} + \eta h^2 D_+ D_- \underline{u}_i^n .$$

This method is in conservation form and is monotone.

In the local linearization, $\underline{A}(\underline{v}) = \partial_{\underline{v}} \underline{F}(\underline{v})$ is taken to be locally constant, that is, $\underline{A} \equiv \underline{A}_0$. In this case, the three steps (4.5.36) can be combined,

(4.5.37) $$\underline{u}_i^{n+1} = \underline{u}_i^n - k\underline{A}_0 D_0 \underline{u}_i^n + (\frac{k^2}{2}\underline{A}_0^2 + \eta k^2) D_+ D_- \underline{u}_i^n .$$

As seen in Section III.8, the stability of (4.5.37) can be reduced to the stability of p scalar equations of the form

$$u_{j,i}^{n+1} = u_{j,i}^n - k\mu_j D_0 u_{j,i}^n + (\frac{k^2}{2}\mu_j^2 + \eta h^2) D_+ D_- u_{j,i}^n ,$$

where μ_j denotes the j-th eigenvalue of $\underline{A}_0$, for $j = 1,\ldots,p$. Taking the discrete Fourier transform of (4.5.38),

$$\hat{u}^{n+1}(\xi) = \rho(\xi)\hat{u}^n(\xi),$$

where the symbol

$$\rho(\xi) = 1 + (\mu_j^2\lambda^2 + 2\eta)(\cos\xi - 1) - i\mu_j\lambda\sin\xi$$

or

$$|\rho(\xi)|^2 = 1 - 4[2\eta - ((\mu_j^2\lambda^2 + 2\eta)^2 - \mu_j^2\lambda^2)\sin^2\frac{\xi}{2}]\sin^2\frac{\xi}{2}$$

If the CFL condition, $|\mu_j|\lambda \le 1$, is satisfied and $\mu_j^2\lambda^2 + 2\eta \le 1$, then $|\rho(\xi)| \le 1$ (the von Neumann condition) and hence the scalar method (4.5.38) is stable. This places an additional restriction on λ,

$$|\mu_j|\lambda < (1-2\eta)^{1/2}.$$

And in the nonlinear cases, for (4.5.36),

$$\max_{\substack{1 \le j \le p \\ \underline{v}}} |\mu_j(v)|\lambda < (1-2\eta)^{1/2} \tag{4.5.39}$$

The anti-diffusion equation (4.5.33) is approximated by

$$\underline{u}_i^{n+1} = D\underline{u}_i^n = \underline{u}_i^n - \eta h^2 D_+ D_- \underline{u}_i^n.$$

This anti-diffusion step is combined with (4.5.36) by adding it as a fourth step

$$\underline{u}_i^{n+1} = D\hat{\underline{u}}_i^{n+1} = \hat{\underline{u}}_i^{n+1} - \eta h^2 D_+ D_- \hat{\underline{u}}_i^{n+1}, \tag{4.5.36d}$$

where the term $\underline{u}_i^{n+1}$ in (4.5.36c) is replaced by $\hat{\underline{u}}_i^{n+1}$. The combined (four-step) method (4.5.36) is stable provided,

$$\max_{\substack{1 \le j \le p \\ \underline{v}}} |\mu_j(\underline{v})|\lambda < \frac{1}{2}\ .$$

The anti-diffusion method (4.5.36d) may be written in conservation form

$$\underline{u}_i^{n+1} = \hat{\underline{u}}_i^{n+1} - (\underline{f}_{i+\frac{1}{2}} - \underline{f}_{i-\frac{1}{2}}),$$

where the anti-diffusion flux

$$\underline{f}_{i+\frac{1}{2}} = \eta(\hat{\underline{u}}_{i+1}^{n+1} - \underline{u}_i^{n+1}).$$

The anti-diffusion method is not monotonicity preserving. This can be remedied by correcting the anti-diffusion fluxes so that 1) no new maxima or minima in the approximate solution are generated and 2) no existing maxima is accentuated (see Section IV.3). Boris and Book [ ] suggest the modification

$$f^c_{j,i+\frac{1}{2}} = S_{j,i+\frac{1}{2}} \max\{0, \min(S_{j,i+\frac{1}{2}} hD_- \hat{u}_{j,i}^{n+1}, \eta|hD_+\hat{u}_{j,i}^{n+1}|,\ S_{j,i+\frac{3}{2}} hD_+\hat{u}_{j,i+1}^{n+1})\}, \tag{4.5.37}$$

where $f^c_{j,i+\frac{1}{2}}$ and $\hat{u}^{n+1}_{j,i}$ denote the j-th components of $\underline{f}^c_{i+\frac{1}{2}}$ and $\hat{\underline{u}}^{n+1}_i$, respectively, for $j = 1,\ldots,p$. The anti-diffusion method in conservation form becomes

$$\underline{u}^{n+1}_i = \hat{\underline{u}}^{n+1}_i - (\underline{f}^c_{i+\frac{1}{2}} - \underline{f}^c_{i-\frac{1}{2}}),$$

where the corrected flux $\underline{f}^c_{i+\frac{1}{2}}$ is given by (4.5.37). Another procedure for modifying the anti-diffusion flux which is better suited to multi-dimensional conservation laws is presented in Zalesak [77].

The artificial compression method has two advantages over the flux-corrected transport method. First the ACM has as its stability requirement the CFL condition while the FCT has a more restrictive stability condition. Second, the ACM treats contact discontinuities as shocks and as such prevents the smearing of contact discontinuities. This is not the case with the FCT, where the smearing of the contact discontinuity is given by (4.5.7) and (4.5.8) where the proportionality constants may be small.

It would be desirable for a finite difference method in conservation form to be of high order accuracy in regions where the solution is smooth and to be monotonic (first order accurate) at a discontinuity. this would provide maximal accuracy without oscillations near a discontinuity. This is the motivation for the self-adjusting hybrid method of Harten and Zwas [25] (see also Harten [27]).

Consider a monotonic finite difference operator Q_1 and a q-th order accurate ($q \geq 2$) finite difference operator Q_q in conservation form

$$\underline{u}^{n+1}_i = Q_1\underline{u}^n_i = \underline{u}^n_i - \lambda(\underline{G}^{1,n}_{i+\frac{1}{2}} - \underline{G}^{1,n}_{i-\frac{1}{2}}) \tag{4.5.39}$$

and

$$\underline{u}^{n+1}_i = Q_q\underline{u}^n_i = \underline{u}^n_i - \lambda(\underline{G}^{q,n}_{i+\frac{1}{2}} - \underline{G}^{q,n}_{i-\frac{1}{2}}). \tag{4.5.40}$$

To construct the hybrid method, we shall take a convex combination of the two methods by introducing a dimensionless quantitu $\theta_{i+\frac{1}{2}}$, called a switch, where $0 \leq \theta_{i+\frac{1}{2}} \leq 1$. So as not to violate the conservation, we hybridize Q_1 and Q_q through their numerical fluxes. Define the hybrid finite difference operator Q by

$$\underline{u}^{n+1} = Q\underline{u}^n_i = \underline{u}^n_i - \lambda(\underline{G}^n_{i+\frac{1}{2}} - \underline{G}^n_{i-\frac{1}{2}}),$$

where

(4.5.42) $$G^n_{i+\frac{1}{2}} = \theta_{i+\frac{1}{2}}\underline{G}^{1,n}_{i+\frac{1}{2}} + (1-\theta_{i+\frac{1}{2}})G^{q,n}_{i+\frac{1}{2}}$$

The switch should have the properties

1) $\theta \approx 1$ at a discontinuity,

2) $\theta = O(h^r)$ in smooth regions,

where r is a positive integer large enough that the solution is q-th order accurate in smooth regions.

Substituting (4.5.42) into (4.5.41),

$$\underline{u}_i^{n+1} = Q\underline{u}_i^n = \underline{u}_i^n + \lambda(\theta_{i+\frac{1}{2}}(\underline{G}^{1,n}_{i+\frac{1}{2}} - \underline{G}^{q,n}_{i+\frac{1}{2}})$$
$$- \theta_{i-\frac{1}{2}}(\underline{G}^{1,n}_{i-\frac{1}{2}} - \underline{G}^{q,n}_{i-\frac{1}{2}})).$$

If in property 2) of θ, $r > q-1$, that is, in smooth regions $\theta = O(h^{q-1})$, then

$$\underline{u}_i^{n+1} = Q\underline{u}_i^n = Q_q\underline{u}_i^n + O(h^{q+1}).$$

Let $\underline{\rho}^1(\xi)$, $\underline{\rho}^q(\xi)$, and $\underline{\rho}(\xi)$ denote the amplification matrices of the locally linearized operators Q_1, Q_q, and Q, respectively. Suppose $\|\rho^1(\xi)\| \le 1 + C_1k$ and $\|\rho^q(\xi)\| \le 1 + C_qk$, so that the finite difference methods (4.5.39) and (4.5.40) are stable (see Theorem 2.1). Set θ equal to a constant with $0 \le \theta \le 1$, then by locally linearizing (4.5.41) and (4.5.42) and taking the discrete Fourier transform,

$$\underline{\rho}(\xi) = \theta\underline{\rho}^1(\xi) + (1-\theta)\underline{\rho}^q(\xi)$$

and

$$\begin{aligned}\|\underline{\rho}(\xi)\| &= \|\theta\underline{\rho}^1(\xi) + (1-\theta)\underline{\rho}^q(\xi)\| \\ &\le \theta\|\underline{\rho}^1(\xi)\| + (1-\theta)\|\underline{\rho}^q(\xi)\| \\ &= \theta(1+C_1k) + (1-\theta)(1+C_qk) \\ &= 1 + Ck,\end{aligned}$$

where $C = \theta C_1 + (1-\theta)C_q$. Stability of the linearization of (4.5.41)-(4.5.42) follows from Theorem 2.1.

As an example, let Q_q denote the two-step Lax-Wendroff method ($q = 2$) and Q_1 denote the monotonic method obtained by adding the diffusion term $\frac{\eta h^2}{k} D_+ D_- \underline{u}_i^n$ to the second step of the two-step Lax-Wendroff method (4.5.36a-c). The hybrid method is

(4.5.43a) $$\underline{u}_{i+\frac{1}{2}}^{n+\frac{1}{2}} = \tfrac{1}{2}(\underline{u}_{i+1}^n + \underline{u}_i^n) - \tfrac{\lambda}{2}(\underline{F}_{i+1}^n - \underline{F}_i^n)$$

(4.5.43b) $$\underline{u}_i^{n+1} = \underline{u}_i^n - \lambda(\underline{F}_{i+\frac{1}{2}}^{n+\frac{1}{2}} - \underline{F}_{i-\frac{1}{2}}^{n+\frac{1}{2}}) + \eta h^2 D_+(\theta_{i-\frac{1}{2}}^n D_- \underline{u}_i^n).$$

Harten [27] chosen $\eta = \frac{1}{8}$, so that for the first order method the linearized stability condition (4.5.39) is

(4.5.44) $$\max_{\substack{1 < j < p \\ \underline{v}}} |\mu_j(v)| \lambda < \frac{\sqrt{3}}{2}$$

Since the locally linearized Lax-Wendroff method is stable under condition (4.5.44), which is more restrictive than the CFL condition, the locally linearized hybrid method is stable under condition (4.5.44).

It remains to describe the construction of the switch $\theta_{i+\frac{1}{2}}^n$. Consider first the scalar conservation law. In this case, $\theta_{i+\frac{1}{2}}^n$ is chosen so that the hybrid method (4.5.43) is monotonicity preserving.

There are many possible choices, one of which is (see Harten [27])

(4.5.45) $$\theta_{i+\frac{1}{2}}^n = \max(\theta_i^n, \theta_{i+1}^n),$$

where

(4.5.46) $$\theta_i^n = \begin{cases} \dfrac{|hD_+u_i^n| - |hD_-u_i^n|}{|hD_+u_i^n| + |hD_-u_i^n|}, & |hD_+u_i^n| + |hD_-u_i^n| > \varepsilon \\ 0, & \text{otherwise,} \end{cases}$$

where $\varepsilon > 0$ is chosen so that a variation in u_i^n which is less than ε is negligible. Compare this with van Leer's monotonicity preserving method (4.3.34) with (4.3.49).

Clearly $0 \le \theta_i^n \le 1$, where there are only two cases in which $\theta_i^n = 1$; for $|hD_+u_i^n| = 0$ and $|h\phi_-u_i^n| > \varepsilon$ or $|hD_+u_i^n| > \varepsilon$ and $|hD_-u_i^n| = 0$.

Harten [27] proves that the hybrid method (4.5.43) with switch given by (4.5.45)-(4.5.46) is monotonicity preserving for

$$\max_v |a(v)|\lambda \le \frac{\sqrt{2}}{2},$$

where $a(v) = \partial_v F(v)$.

The switch given by (4.5.45)-(4.5.46) can be extended to systems of conservation laws. Two such ways are: 1) replace the absolute value in (4.5.46) with a suitably chosen norm, or 2) replace u_i^n in (4.5.46) with a scalar function of $\underline{u}_i^n$ which has a discontinuity when $\underline{u}_i^n$ has a discontinuity and is smooth otherwise. This second choice will be discussed further in Volume II.

Since in regions that contain discontinuities the hybrid method is essentially first order accurate, the artificial compression method may be used to sharpen the discontinuities. However, the artificial compression method must <u>not</u> be used in smooth regions. For this purpose the switch is used again. Modify the artificial compression method (4.5.12), (4.5.15), and (4.5.28)-(4.5.29) by

$$\underline{u}_i^{n+1} = \tilde{\underline{u}}_i^{n+1} - \frac{\lambda}{2}(\theta^n_{i+\frac{1}{2}}\underline{G}'^n_{i+\frac{1}{2}} - \theta^n_{i-\frac{1}{2}}\underline{G}'^n_{i-\frac{1}{2}}),$$

where $\tilde{\underline{u}}_i^{n+1}$ is the value of $\underline{u}_i^{n+1}$ obtained from the hybrid method (4.5.41).

<u>Definition</u>. The <u>total variation</u> (TV) of a grid function $u = \{u_i\}$

(4.5.47) $$TV(u) = \sum_j |u_{i+1} - u_i|.$$

<u>Definition</u>. A finite difference method in conservation form (4.2.6) is <u>total variation nonincreasing</u> (TVNI) if for all grid functions $u = \{u_i\}$ of bounded total variation, that is, $TV(u) < +\infty$,

(4.5.48) $$TV(Qu) \le TV(u).$$

The following theorem (Harten [28]) states the hierarchy of the properties (monotonic, monotonicity preserving, and total variation nonincreasing) of a finite difference method.

Theorem 4.12. a) A monotone finite difference method is a TVNI method. b) A TVNI finite difference method is monotonicity preserving.

A high resolution* method due to Harten [28] uses a first order accurate 3-point TVNI finite difference method to construct a second order accurate 5-point TVNI finite difference method. First the high resolution method will be described for a scalar conservation law (4.1.1).

A general 5-point finite difference method in conservation form is

$$u_i^{n+1} = u_i^n - \lambda(G_{i+\frac{1}{2}} - G_{i-\frac{1}{2}}),$$

where $G_{i+\frac{1}{2}} = G(u_{i-1}^n, u_i^n, u_{i+1}^n, u_{i+2}^n)$ with $G(v,v,v,v) = F(v)$. This may be rewritten in the form

$$u_i^{n+1} = u_i^n + C^+_{i+\frac{1}{2}} hD_+u_i^n + C^-_{i-\frac{1}{2}} hD_-u_i^n, \tag{4.5.49}$$

where

$$C^+_{i+\frac{1}{2}} = C^+(u_{i-1}^n, u_i^n, u_{i+1}^n, u_{i+2}^n) \tag{4.5.50a}$$

and

$$C^-_{i-\frac{1}{2}} = C^-(u_{i-2}^n, u_{i-1}^n, u_i^n, u_{i+1}^n). \tag{4.5.50b}$$

In the 3-point case, $C^+_{i+\frac{1}{2}} = C^+(u_i^n, u_{i+1}^n)$ and $C^-_{i-\frac{1}{2}} = C^-(u_{i-1}^n, u_i^n)$. For example, $C_\pm$ for a 3-point method is obtained by the mean value theorem,

$$\lambda(G(u_i^n, u_{i+1}^n) - G(u_i^n, u_i^n)) = -C^+(u_i^n, u_{i+1}^n)hD_+u_i^n$$

$$\lambda(G(u_{i-1}^n, u_i^n) - G(u_i^n, u_i^n)) = -C^-(u_{i-1}^n, u_i^n)hD_-u_i^n.$$

Lemma 1. Suppose the coefficients (4.5.50) of the method (4.5.49) satisfy

$$C^+_{i+\frac{1}{2}},\ C^-_{i+\frac{1}{2}} \geq 0 \tag{4.5.51a}$$

---

*High resolution refers to a narrow transition region between the two states of a shock.

and

(4.5.51a) $$C^+_{i+\frac{1}{2}} + C^-_{i+\frac{1}{2}} \leq 1,$$

then the finite difference method (4.5.49) is TVNI.

<u>Proof</u>: Replacing i with $i+1$ in (4.5.49) and subtracting (4.5.49) from this gives

$$hD_+u_i^{n+1} = C^+_{i+\frac{3}{2}} hD_+u_{i+1}^n + (1 - C^-_{i+\frac{1}{2}} - C^+_{i+\frac{1}{2}}) hD_+u_i^n$$
$$+ C^-_{i-\frac{1}{2}} D_+u_{i-1}^n,$$

where $hD_+u_i^n \equiv hD_-u_{i+1}^n$. Using inequalities (4.5.51), the coefficients of (4.5.52) are nonnegative and

$$|hD_+u_i^{n+1}| \leq C^+_{i+\frac{3}{2}} |hD_+u_{i+1}^n| + (1 - C^-_{i+\frac{1}{2}} - C^+_{i+\frac{1}{2}}) |hD_+u_i^n|$$
$$+ C^-_{i-\frac{1}{2}} |hD_+u_{i-1}^n|.$$

Summing over i,

$$TV(u^{n+1}) = \sum_i |hD_+u_i^{n+1}| \leq \sum_i C_{i+\frac{3}{2}} |hD_+u_{i+1}^n| + \sum_i (1 - C^+_{i+\frac{1}{2}} - C^-_{i+\frac{1}{2}})$$
$$|hD_+u_i^n| + \sum_i C^-_{i-\frac{1}{2}} |hD_+u_{i-1}^n|$$
$$= \sum_i C^+_{i+\frac{1}{2}} |hD_+u_i^n| + \sum_i (1 - C^+_{i+\frac{1}{2}} - C^-_{i+\frac{1}{2}}) |hD_+u_i^n|$$
$$+ \sum_i C^-_{i+\frac{1}{2}} |hD_+u_i^n|$$
$$= \sum_i |hD_+u_i^n| = TV(u^n).$$

Thus the condition (4.5.48) in the definition of TVNI is satisfied.

Consider a first order accurate 3-point TVNI finite difference method in the form (4.5.49)-(4.5.50) satisfying the assumption in Lemma 1. Since the leading term in the truncation error is $k\partial_x(\beta(v,\lambda)\partial_x v)$, this method approximates to second order accuracy the solution of the diffusion equation

$$\partial_t v + \partial_x F(v) = k(\partial_x \beta(v,\lambda)\partial_x v)$$

which may be written in the form

$$\partial_t v + \partial_x (F(v) - \tfrac{1}{\lambda} g(v,\lambda)) = 0, \tag{4.5.53}$$

where

$$g(v,\lambda) = k\beta(v,\lambda)\partial_x v. \tag{4.5.54}$$

With this in mind, suppose the first order 3-point TVNI method is used to solve the modified conservation law

$$\partial_t v + \partial_x (F(v) + \tfrac{1}{\lambda} g(v,\lambda)) = 0,$$

then it approximates to second order accuracy the solution to the original conservation law (4.1.1).

Define the numerical flux of the 3-point TVNI method by

$$\begin{aligned} G_{i+\frac{1}{2}} &= G(u_i^n, u_{i+1}^n) = \tfrac{1}{2}(F_{i+1}^n + F_i^n) - \tfrac{1}{\lambda} Q(\lambda \bar{a}^n_{i+\frac{1}{2}}) h D_+ u_i^n) \\ &= F_i^n - \tfrac{1}{2}[\bar{a}^n_{i+\frac{1}{2}} + \tfrac{1}{\lambda} Q(\lambda \bar{a}^n_{i+\frac{1}{2}})] h D_+ u_i^n , \end{aligned} \tag{4.5.55}$$

where

$$\bar{a}^n_{i+\frac{1}{2}} = \begin{cases} \dfrac{F_{i+1}^n - F_i^n}{u_{i+1}^n - u_i^n}, & u_{i+1}^n \neq u_i^n \\ a(u_i^n) & , u_{i+1}^n = u_i^n, \end{cases} \tag{4.5.56}$$

and $a(v) = \partial_v F(v)$.

<u>Lemma 2</u>. Suppose the function $Q(\eta)$ in (4.5.55) satisfies

$$|y| \leq Q(y) \leq 1 \quad \text{for} \quad 0 \leq |y| \leq \mu \leq 1. \tag{4.5.57}$$

The 3-point finite difference method in conservation form (4.2.6) where $G_{i+\frac{1}{2}}$ is given by (4.5.55) is TVNI provided

$$\lambda \max |\bar{a}^n_{i+\frac{1}{2}}| \leq \mu. \tag{4.5.58}$$

Observe that (4.5.58) is a CFL type condition.

Proof. Substituting the numerical flux (4.5.55) with (4.2.6)

(4.5.59)
$$u_i^{n+1} = u_i^n - \lambda(G_{i+\frac{1}{2}} - G_{i-\frac{1}{2}})$$
$$= u_i^n + \frac{1}{2}(Q(\lambda\bar{a}^n_{i+\frac{1}{2}}) - \lambda\bar{a}^n_{i+\frac{1}{2}})hD_+u_i^n$$
$$- \frac{1}{2}(Q(\lambda\bar{a}^n_{i-\frac{1}{2}}) - \lambda\bar{a}^n_{i-\frac{1}{2}})hD_-u_i^n$$
$$= u_i^n + C_{i+\frac{1}{2}}hD_+u_i^n - C_{i-\frac{1}{2}}hD_-u_i^n .$$

In this case, using (4.5.57) with $y = \lambda\bar{a}^n_{i+\frac{1}{2}}$,

$$C^{\pm}_{i+\frac{1}{2}} = \frac{1}{2}(Q(\lambda\bar{a}^n_{i+\frac{1}{2}}) \mp \lambda\bar{a}^n_{i+\frac{1}{2}}) > 0$$

and

$$C^{+}_{i+\frac{1}{2}} + C^{-}_{i+\frac{1}{2}} = Q(\lambda\bar{a}^n_{i+\frac{1}{2}}) < 1,$$

provided $0 < \lambda|\bar{a}^n_{i+\frac{1}{2}}| < \mu < 1$ for every i, or

$$\max_i|\bar{a}^n_{i+\frac{1}{2}}|\lambda < \mu.$$

Since $C^{\pm}_{i+\frac{1}{2}}$ satisfy the assumption of Lemma 1, the method is TVNI.

The high resolution method is based on replacing the flux $F(v)$ in (4.5.55) with $F(v) + \frac{1}{\lambda}g(v,\lambda)$. Define the modified flux $F^m(v) = F(v) + \frac{1}{\lambda}g(v,\lambda)$ at the grid points by

(4.5.60a)
$$F_i^{n,m} = F^m(u_i^n) = F_i^n + \frac{1}{\lambda}g_i^n,$$

(4.5.60b)
$$g_i^n = g(u_{i-1}^n, u_i^n, u_{i+1}^n)$$

(4.5.60c)
$$\bar{a}^{n,m}_{i+\frac{1}{2}} = \bar{a}^n_{i+\frac{1}{2}} + \frac{1}{\lambda}\gamma^n_{i+\frac{1}{2}}$$

(4.5.60d)
$$\gamma^n_{i+\frac{1}{2}} = \frac{g_{i+1}^n - g_i^n}{u_{i+1}^n - u_i^n} .$$

Substitution of the modified flux (4.5.60) in the numerical flux (4.5.55) gives rise to this modified numerical flux

(4.5.61) $$G^m_{i+\frac{1}{2}} = \frac{1}{2}(F^{n,m}_{i+1} + F^{n,m}_i - \frac{1}{\lambda}Q(\bar{\lambda}a^{n,m})hD_+u^n_i)$$
$$= \frac{1}{2}(F^n_{i+1} + F^n_i) + \frac{1}{2\lambda}(g^n_{i+1} + g^n_i - Q(\lambda\bar{a}^n_{i+\frac{1}{2}} + \gamma^n_{i+\frac{1}{2}})hD_+u^n_i).$$

Harten [28] shows that if $Q(y)$ is Lipschitz continuous and g^n_i satisfies

(4.5.62a) $$g^n_{i+1} + g^n_i = (Q(\lambda\bar{a}^n_{i+\frac{1}{2}}) - (\lambda\bar{a}^n_{i+\frac{1}{2}})^2)hD_+u^n_i + O(h^2)$$

and

(4.5.62b) $$g^n_{i+1} - g^n_i \equiv \gamma^n_{i+\frac{1}{2}}hD_+u^n_i = O(h^2),$$

then the modified numerical flux (4.5.61) may be written in the form

$$G^m_{i+\frac{1}{2}} = G^{LW}_{i+\frac{1}{2}} + O(h^2),$$

where $G^{LW}_{i+\frac{1}{2}}$ denotes the numerical flux for the second order accurate one-step Lax-Wendroff method,

$$G^{LW}_{i+\frac{1}{2}} = \frac{1}{2}(F^n_{i+1} + F^n_i) - \frac{k}{2}(\bar{a}^n_{i+\frac{1}{2}})^2D_+u^n_i$$
$$= \frac{1}{2}(F^n_{i+1} + F^n_i - \frac{1}{\lambda}(\lambda\bar{a}^n_{i+\frac{1}{2}})^2hD_+u^n_i).$$

Thus if g^n_i can be constructed so that conditions (4.5.62) are satisfied, then the modified (5-point) method is second order accurate. Also g^n_i should be constructed so that the resulting 5-point method is TVNI. Define

(4.5.63a) $$g^n_i = \begin{cases} S^n_{i+\frac{1}{2}}\max(0,\min(|\tilde{g}^n_{i+\frac{1}{2}}|,\tilde{g}^n_{i-\frac{1}{2}}\cdot S^n_{i+\frac{1}{2}})) \\ \quad = S^n_{i+\frac{1}{2}}\min(|\tilde{g}_{i+\frac{1}{2}}|,|\tilde{g}^n_{i-\frac{1}{2}}|),\ \tilde{g}^n_{i-\frac{1}{2}}\tilde{g}^n_{i+\frac{1}{2}} > 0 \\ 0\ ,\quad \text{otherwise,} \end{cases}$$

where

(4.5.63b) $$\tilde{g}^n_{i+\frac{1}{2}} = \frac{1}{2}(Q(\lambda\bar{a}^n_{i+\frac{1}{2}}) - (\lambda\bar{a}^n_{i+\frac{1}{2}})^2)hD_+u^n_i$$

and

$$S_{i+\frac{1}{2}} = \text{sgn}(g^n_{i+\frac{1}{2}}).$$

Harten [28] shows that $Q(y)$ satisfies (4.5.51) (the assumption of Lemma 1) with g^n_i defined by (4.5.63). So the finite difference method in conservation form with numerical flux given by (4.5.61) is TVNI provided the CFL type condition (4.5.58) is satisfied.

It remains to describe $Q(y)$. The numerical diffusion coefficient has the form

$$\beta(v,\lambda) = \frac{1}{2}(Q(\lambda a(v)) - \lambda^2 a^2(v))$$

(see Section IV.3). A natural choice for $Q(y)$, subject to the constraint (4.5.57) of Lemma 1, is $Q(y) = |y|$. Substitution of this value in the 3-point method (4.5.59) reduces it to the upstream method (4.3.13). However, when $y = 0$, which corresponds to $a(v) = 0$ giving rise to a steady shock, $\beta = 0$ and the entropy condition may be violated (see Harten [28]). A remedy is to define

$$Q(y) = \begin{cases} |y|, & |y| \geq 2\varepsilon \\ \frac{y^2}{4\varepsilon}+\varepsilon, & |y| < 2\varepsilon. \end{cases}$$

This choice of Q introduces more numerical diffusion for $|y| < 2\varepsilon$, however, $\beta(v,\lambda) > 0$ for $|y| = \lambda|a(v)| < 1$. The only value of y for which $\beta(v,\lambda) = 0$ is $y = 1$. This can also be remedied by choosing μ in (4.5.57) to be strictly less than 1.

To extend this method to systems of conservation laws (4.1.27), let $\underline{A}(\underline{v})$ denote the Jacobian of $\underline{F}(\underline{v})$. By the definition of hyperbolicity there exists a bounded, nonsingular matrix $\underline{T}(\underline{v})$ with bounded inverse such that $\underline{T}^{-1}(\underline{v})A(\underline{v})T(\underline{v}) = \underline{D}$, where $\underline{D}$ is a diagonal matrix whose elements are the eigenvalues of $\underline{A}(\underline{v})$. The columns of $T(v)$ are the right eigenvectors of $\underline{A}(\underline{v})$, $\underline{r}_j(\underline{v})$, $j = 1,\dots,p$ (see Section IV.1).

Define $\underline{u}^n_{i+\frac{1}{2}} = \frac{1}{2}(\underline{u}^n_i + \underline{u}^n_{i+1})$ and let $\alpha^n_{j,i+\frac{1}{2}}$ denote the component of $hD_+\underline{u}^n_i$ in the characteristic coordinate system $\{r_j(\underline{u}^n_{i+\frac{1}{2}})\}$,

$$hD_+\underline{u}^n_i = \sum_{j=1}^{p} \alpha^n_{j,i+\frac{1}{2}} r^n_{j,i+\frac{1}{2}},$$

where $r^n_{j,i+\frac{1}{2}} = r_j(\underline{u}^n_{i+\frac{1}{2}})$, so that

(4.5.65) $$\alpha^n_{j,i+\frac{1}{2}} = \underline{\ell}^n_{j,i+\frac{1}{2}} hD_+u^n_i,$$

where $\underline{\ell}^n_{j,i+\frac{1}{2}} = \ell_j(\underline{u}^n_{i+\frac{1}{2}})$ and $\underline{\ell}_j(\underline{v})$ denotes the j-th left eigenvector of $\underline{A}(\underline{v})$.

The high resolution method (4.5.60), (4.5.61), and (4.5.63) extended to systems of conservation laws consists of

(4.5.66a) $$\underline{u}^{n+1}_i = \underline{u}^n_i - (\underline{G}_{i+\frac{1}{2}} - \underline{G}_{i-\frac{1}{2}}),$$

where the numerical flux is

(4.5.66b) $$\underline{G}_{i+\frac{1}{2}} = \frac{1}{2}(\underline{F}^n_{i+1} + \underline{F}^n_i) + \frac{1}{2\lambda}\sum_{j=1}^{p} \underline{r}^n_{j,i+\frac{1}{2}}(g^n_{j,i}+g^n_{j,i+1} - Q_j(\lambda\mu^n_{j,i+\frac{1}{2}} + \gamma^n_{j,i+\frac{1}{2}})\alpha^n_{j,i+\frac{1}{2}},$$

(4.5.66c) $$g^n_{j,i} = s^n_{j,i+\frac{1}{2}}\max(0,\min(|\tilde{g}^n_{j,i+\frac{1}{2}}|,\tilde{g}^n_{j,i-\frac{1}{2}}s_{j,i+\frac{1}{2}})),$$

(4.5.66d) $$s^n_{j,i+\frac{1}{2}} = \mathrm{sgn}(\tilde{g}^n_{j,i+\frac{1}{2}}),$$

(4.5.66e) $$g^n_{j,i+\frac{1}{2}} = \frac{1}{2}(Q_j(\lambda\mu^n_{j,i+\frac{1}{2}}) - (\lambda\mu_{j,i+\frac{1}{2}})^2)\alpha^n_{j,i+\frac{1}{2}},$$

(4.5.66f) $$\gamma^n_{j,i+\frac{1}{2}} = \begin{cases} \dfrac{g^n_{j,i+1} - y^n_{j,i}}{\alpha^n_{j,i+\frac{1}{2}}}, & \alpha^n_{j,i+\frac{1}{2}} \neq 0 \\ 0 \;, & \alpha^n_{j,i+\frac{1}{2}} = 0, \end{cases}$$

and $\mu^n_{j,i+\frac{1}{2}} = \mu_j(\underline{u}^n_{i+\frac{1}{2}})$, the j-th eigenvalue of $\underline{A}(\underline{u}^n_{i+\frac{1}{2}})$.

IV.6. Numerical Examples

Here we present the results of a test problem given by the scalar conservation law, Burgers' equation

(4.6.1a) $$\partial_t v + \partial_x(\tfrac{1}{2}v^2) = 0, \quad -\infty < x < +\infty \;, \; t > 0$$

with the continuous initial condition

(4.6.1b) $$v(x,t = \begin{cases} 1 \;, & x < 0 \\ 1-x, & 0 < x < 1 \\ 0 \;, & x > 1. \end{cases}$$

The solution to this initial-value problem is a compression wave, which initially, consists of a fan for $0 < x < 1$. In constrast to an expansion (rarefaction) wave in which the fan expands with time, for the compression wave the fan closes (or compresses) with time until a shock is formed. In this example, the solution is a continuous wave that gets steeper until $t = 1$, when it becomes a shock. The exact solution for $t < 1$ is

$$(4.6.2a) \qquad v(x,t) = \begin{cases} 1, & x < t \\ (x-1)/(t-1), & t < x < 1 \\ 0, & x > 0 \end{cases}$$

and for $t > 1$

$$(4.6.2b) \qquad v(x,t) = \begin{cases} 1, & x > (t+1)/2 \\ 0, & x < (t+1)/2. \end{cases}$$

To avoid dealing with boundary conditions x and t are restricted to the intervals, $-1 < x < 3$ and $0 < t < 2.5$. In all calculations the grid spacing was taken to be $h = 0.02$. Figures 4.12-4.19 depict results at times $t = 0.5$, 1.0, and 2.5. At time $t = 0.5$ the wave is still continuous though it is half as wide as initially $(0.5 < x < 1.0)$. Time $t = 1.0$ corresponds to the formation of the shock with speed $S = 0.5$. And time $t = 2.5$ corresponds to the developed shock 0.75 units downstream from its initial position at $x = 1$.

Figure 4.12 depicts the results using the Lax-Friedrichs method. The corners at the endpoints of the compression wave are rounded. The shock is greatly smeared with the transition spanning twelve to fourteen grid points at $t = 2.5$.

Figure 4.13 depicts the results of the first order accurate upwind method. The corners of the endpoints of the compression wave are only slightly rounded. The shock is much less smeared with a transition spanning about four grid points.

Figure 4.14 depicts the results of the upward method with artificial compression. The approximation to the compression wave is identical to the results in Figure 4.13. This should be expected since the ACM is not applied in this region. The shock has a transition spanning only two grid points.

Figure 4.15 depicts the result of the two-step Lax-Wendroff method without artificial viscosity. The approximation to the compression wave is quite accurate. The corners at the endpoints of the compression wave are only slightly rounded. There is an overshoot at the shock at time $t = 2.5$ that goes beyond the range of the plot. This peak value is 1.39. The transition of the shock at time $t = 2.5$

spans two to three grid points. In Figure 4.16 the calculation is repeated using the artificial viscosity (4.2.36) with $\varepsilon = 0.9$. The overshoot is reduced having peak value 1.12 while the transition of the shock still spans two to three grid points.

Figure 4.17 depicts the results of the hybrid method of Harten and Zwas with ACM. In this case the high order method is the two-step Lax-Wendroff Method. The solution is free of oscillations. The transition of the shock at time $t = 2.5$ spans two to three grid points.

Figure 4.18 depicts the results of the anti-diffusion method of Boris and Book. The compression wave is very accurately approximated. There is a slight overshoot behind the shock which the transition at time $t = 2.5$ spans only one to two grid points.

Figure 4.19 depicts the results of the random choice method using the Van der Corput sampling procedure. The corners at the endpoints of the compression wave are perfectly sharp. The compression wave is not perfectly smooth. These fluctuations are due to the randomness. The shock is perfectly sharp, that is, the transition spans zero grid points. The shock, as represented in the Figure, is not a vertical segment. This is due to the coarseness of the grid. The location, however, at time $t = 2.5$ is one grid point ahead of the shock. On the average, the shock speed is correct.

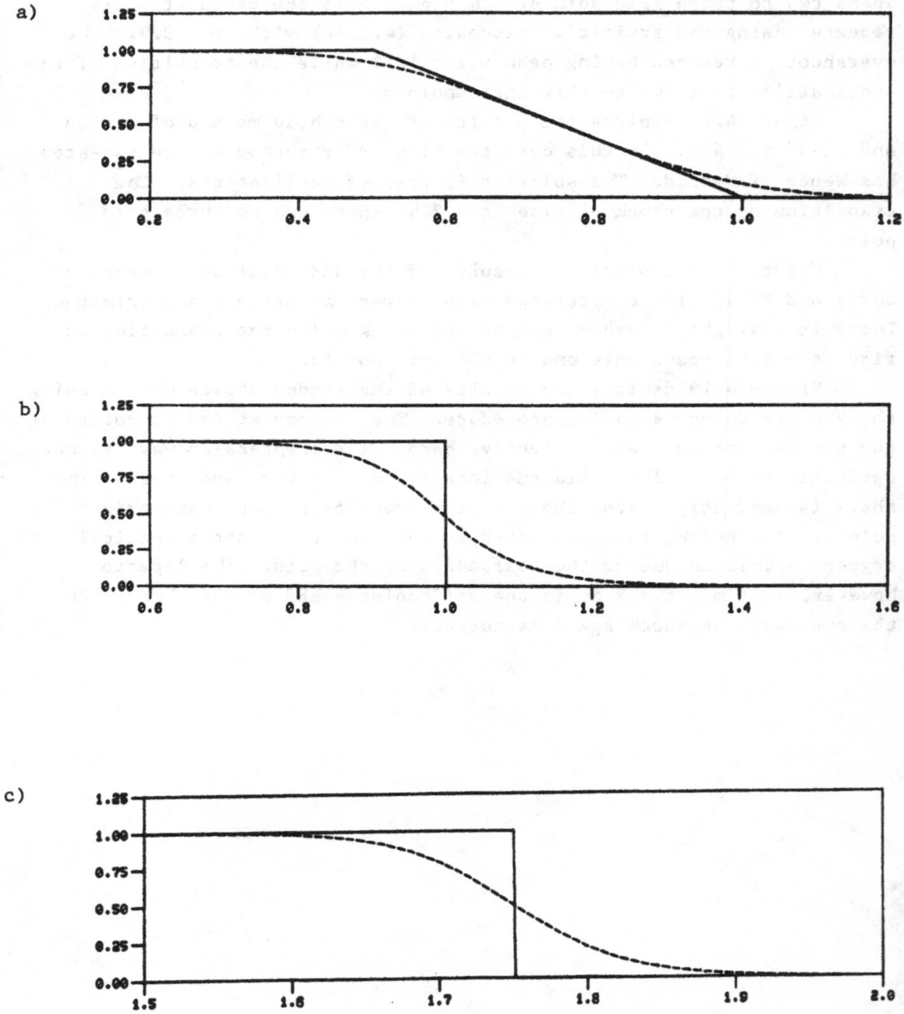

Figure 4.12. Lax-Friedrichs Method of Times
a) t = 0.5, b) t = 1.0, and c) t = 2.5.

a) b) c)

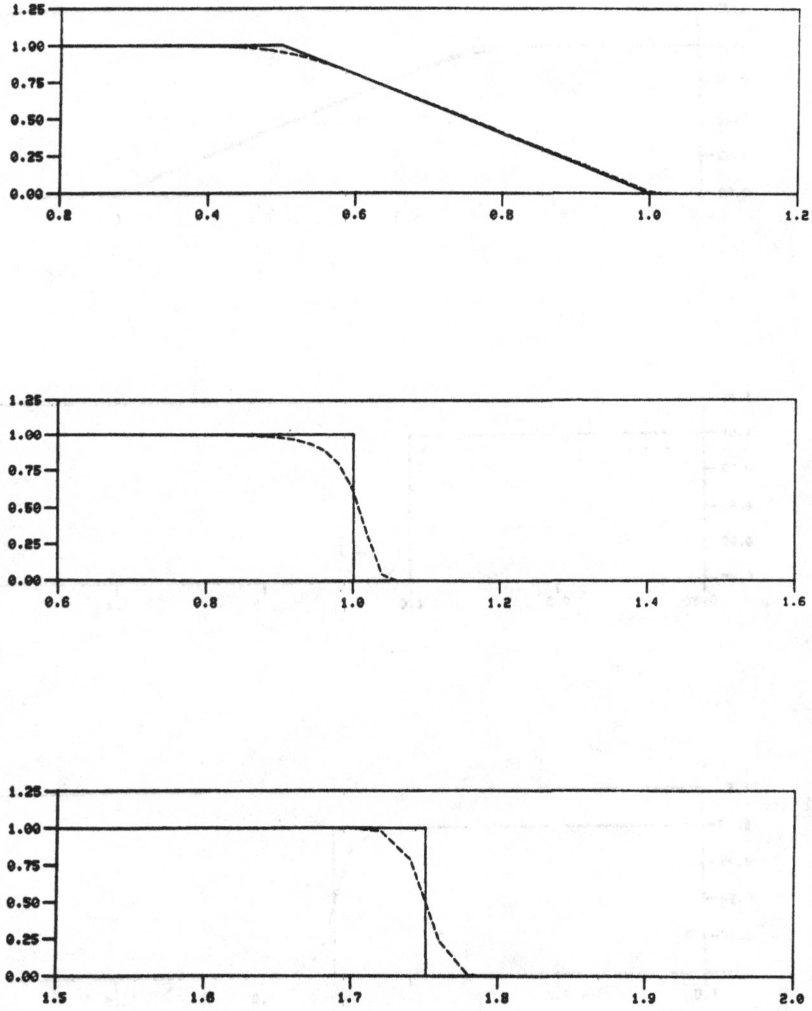

Figure 4.13. First Order Upwind Method at Times
a) t = 0.5, b) t = 1.0, and c) t = 2.5.

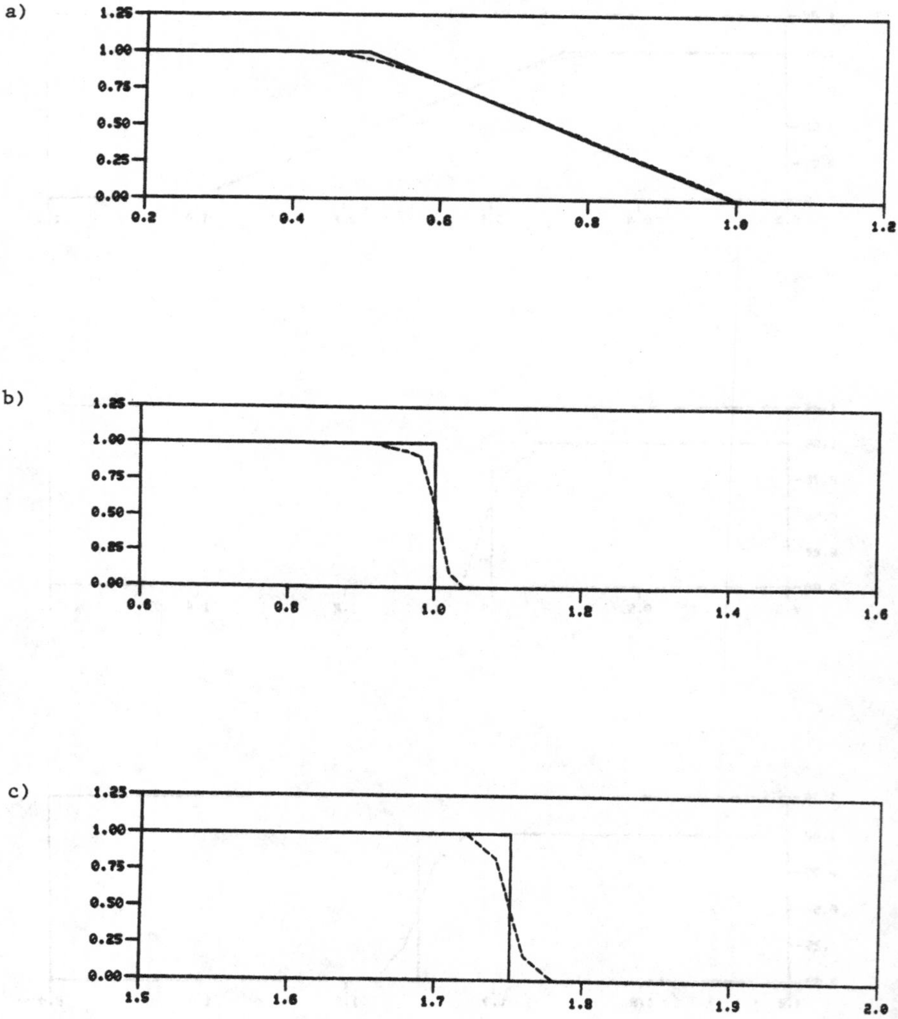

Figure 4.14. First Order Upwind Method with ACM at Times a) t = 0.5, b) t = 1.0, and c) t = 2.5.

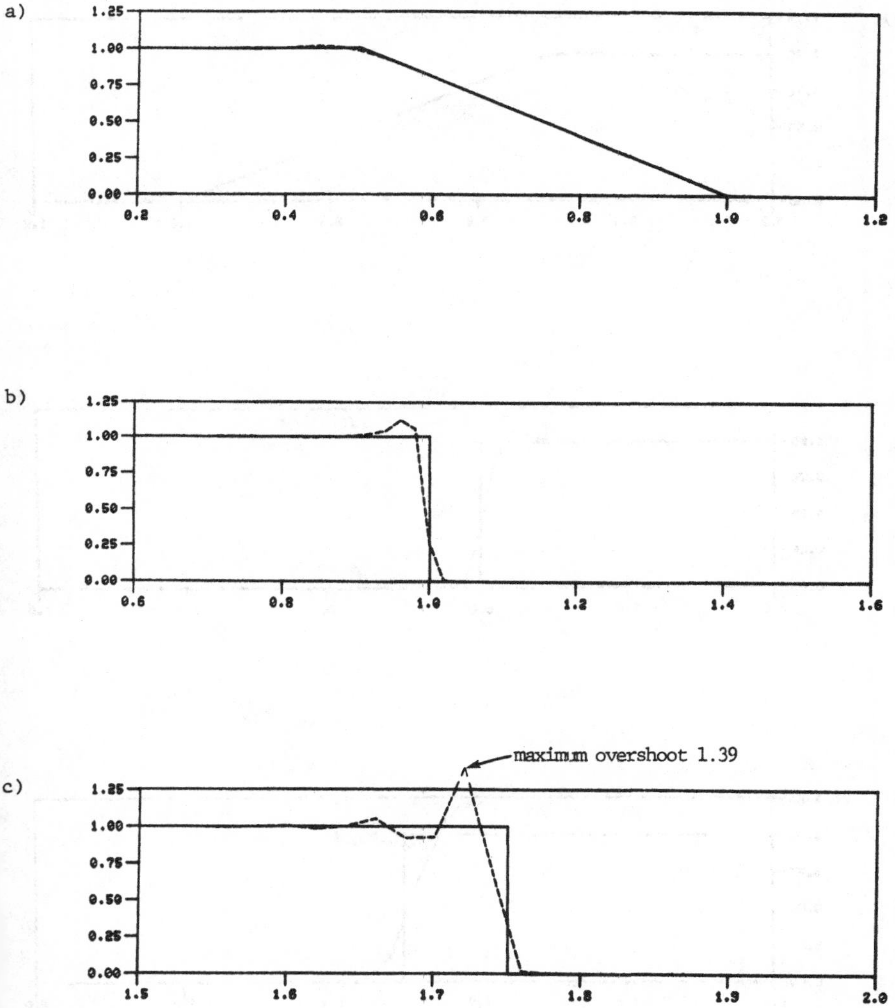

Figure 4.15. Two-Step Lax-Wendroff Method at Times a) $t = 0.5$, b) $t = 1.0$, and c) $t = 2.5$.

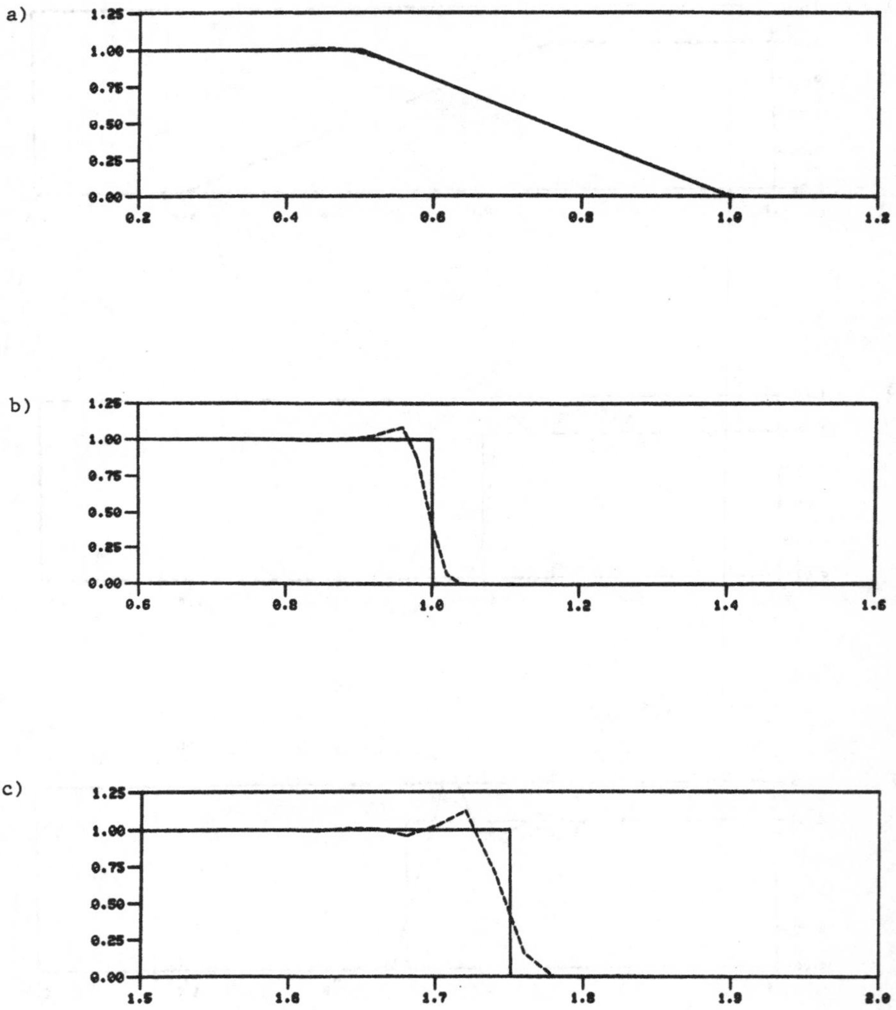

Figure 4.16. Two-Step Lax-Wendroff Method with Artificial Viscosity at Times a) t = 0.5, b) t = 1.0, and c) t = 2.5.

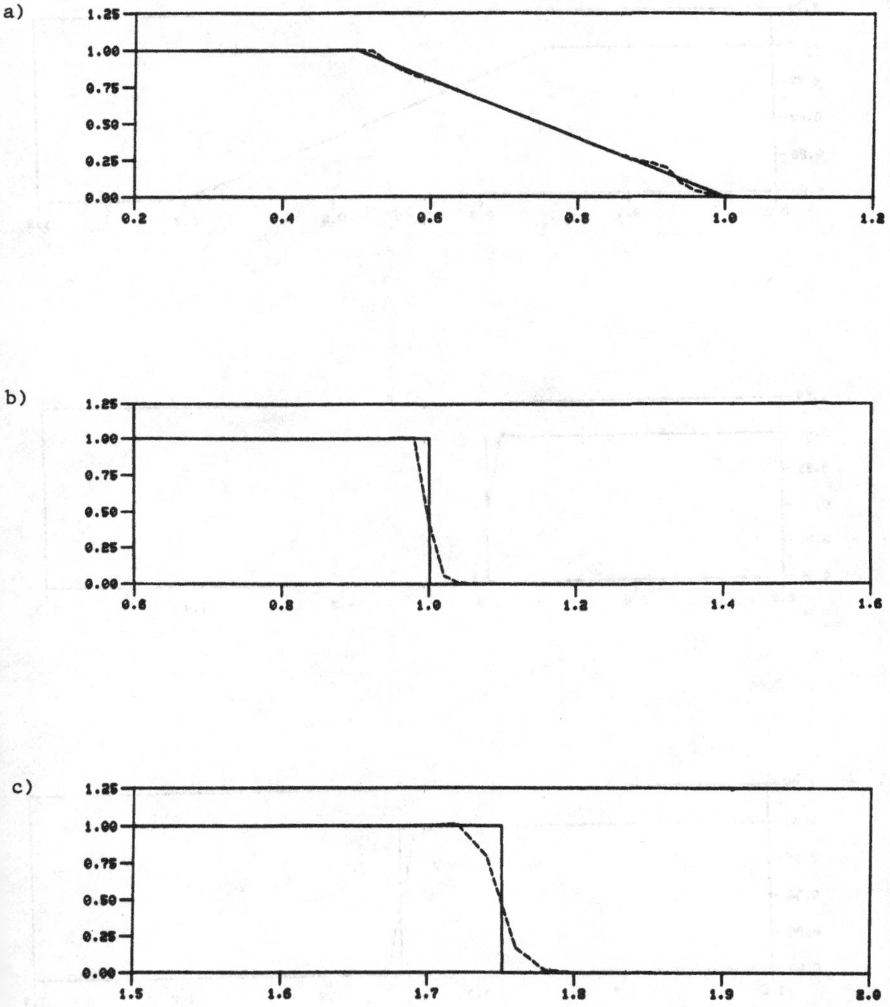

Figure 4.17. Hybrid Method of Harten and Zwas with ACM at Times a) $t = 0.5$, b) $t = 1.0$, and c) $t = 2.5$.

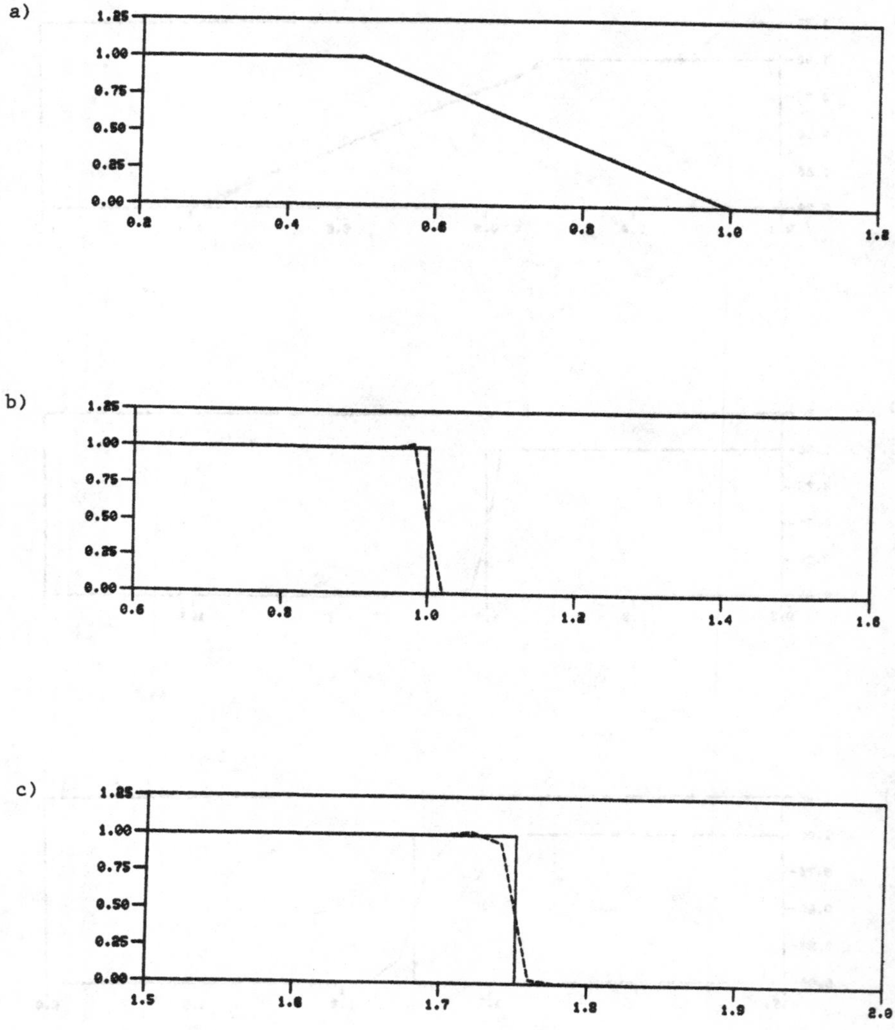

Figure 4.18. Anti-diffusion Method of Boris and Book at Times a) $t = 0.5$, b) $t = 1.0$, and c) $t = 2.5$.

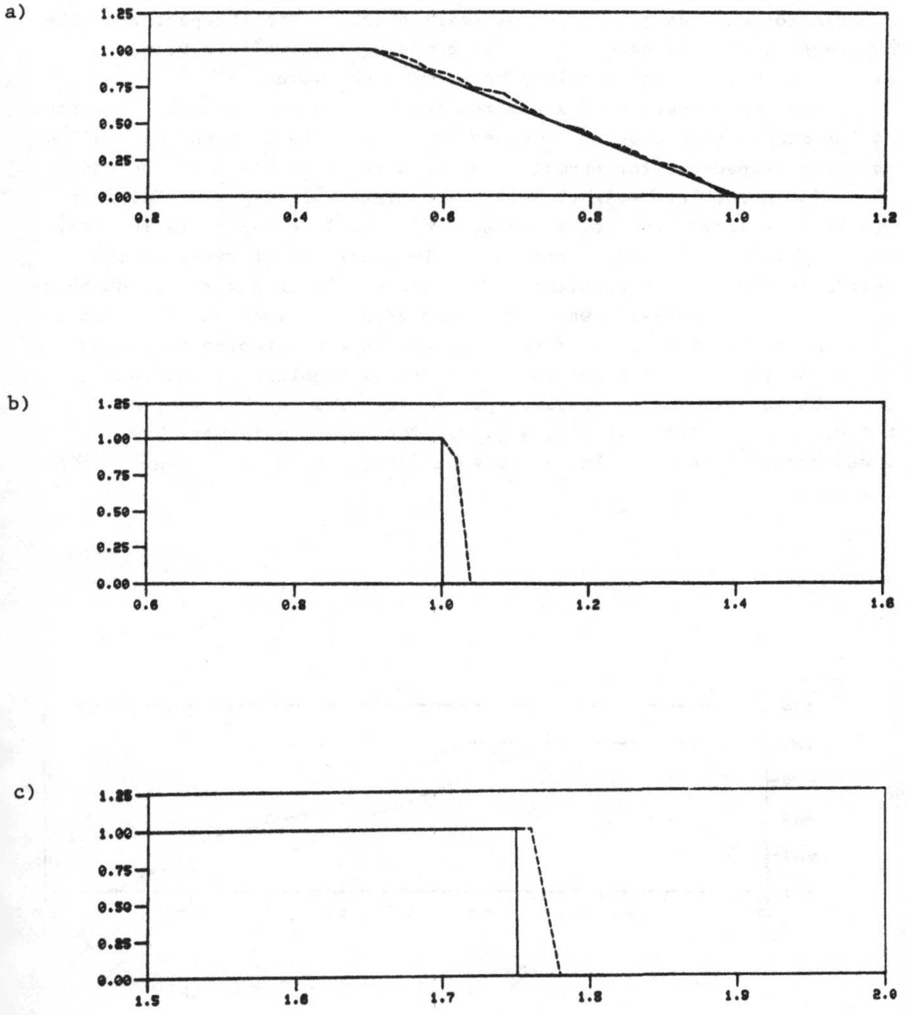

Figure 4.19. Random Choice Method with Van der Corput Sampling Procedure at Times a) t = 0.5, b) t = 1.0, and c) t = 2.5.

The random choice method, through the Riemann problems, models a compression wave as a sequence of small shocks. The steepening of the compression wave is accomplished by removing intermediate values (states in the Riemann problem) assumed by the wave.

The approximation of a compression wave is particularly sensitive to the sampling procedure. Suppose there is a fluctuation in this sampling procedure, for example, two or more consecutive random variables lie in the left (right) half of the interval $(-\frac{1}{2},\frac{1}{2})$. This can result in a large step being created which will be amplified in time. These artificially large steps result in incorrect (larger) shock speeds in the Riemann problems. This can result in a more rapid steepening of the compression wave and a premature development of a shock.

As an example, suppose the compression wave problem is solved using the random choice method with a random sampling procedure rather than the Van der Corput sampling procedure. The results at time $t = 0.5$ are depicted in Figure 4.20. The approximate wave has steepened into several shocks, in particular, there is a large jump of .16 at $x = .76$.

Figure 4.20. Random Choice Method with Random Sampling Procedure at Time $t = 0.5$.

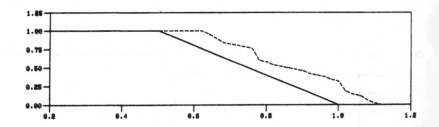

IV.7. References

1. Albright, N., P. Concus, and W. Proskurowski, Numerical Solution of the Multi-dimensional Buckley-Leverett Equatoins by a Sampling Method. SPE Paper No.7681 (1979).

2. Book, D. L., J. P. Boris, and K. Hain, Flux-Corrected Transport II: Generalizations of the Method, J. Comp. Phys., 18, 248 (1975).

3. Boris, J. and D. L. Book, Flux-Corrected Transport I. SHASTA, A Fluid Transport Algorithm That Works, J. Comp. Phys., 11, 38 (1973).

4. ________________________, Flux-Corrected Transport III. Minimal-Error FCT Algorithms, J. Comp. Phys., 20, 397 (1976).

5. Chorin, A. J., Random Choice Solution of Hyperbolic Systems, J. Comp. Phys., 22, 517 (1976).

6. _____________, Random Choice Methods with Applications to Reacting Gas Flow, J. Comp. Phys., 25, 253 (1977).

7. Colella, P., "The Effects of Sampling and Operator Splitting on the Accuracy of the Glimm Scheme," Ph.D. Thesis, University of California, Department of Mathematics, Berkeley (1978).

8. ___________, Glimm's Method for Gas Dynamics, SIAM J. Sci. Stat. Comp., 3, 76 (1982).

9. ___________, Approximate Solution of the Riemann Problem for Real Gases, LBL Report N., LBL-14442, Lawrence Berkeley Laboratory, Berkeley, (1982).

10. Colella, P. and H. Glaz, Numerical Modeling of Inviscid Shocked Flows of Real Gases, Proceedings of the Eighth International Conference in Fluid Dynamics, Lecture Notes in Physics, Springer-Verlag, New York, (1983).

11. Concus, P. and W. Proskurowski, Numerical Solution of a Non-linear Hyperbolic Equation by the Random Choice Method, J. Comp. Phys., 30, 153 (1979).

12. Courant, R., K. O. Friedrichs, and H. Lewy, Über die Partiellen Differenzengleichungen der Mathematischen Physic, Math. Ann., 100, 32 (1928). English Translation: AED Research and Development Report NYO-7689, New York University, New York (1956).

13. ___________, E. Isaacson, and M. Rees, On the Solution of Nonlinear Hyperbolic Differential Equations by Finite Differences, Comm. Pure Appl. Math., 5, 243 (1952).

14. Enquist, B. and S. Osher, Stable and Entropy Satisfying Approximations for Transonic Flow Calculations, Math. Comp., 34, 45 (1980).

15. Fok, S. K., Extension of Glimm's Method to the Problem of Gas Flow in a Duct of Variable Cross-Section, LBL Report No.LBL-12322, Lawrence Berkeley Laboratory, Berkeley, (1980).

16. Fromm, J. E., A Method for Reducing Dispersion in Convective Difference Schemes, J. Comp. Phys., 3, 176 (1968).

17. Glaz, H., "Development of Random Choice Numerical Methods for Blast Wave Models," NSWC Report NSWC/WDL TR 78-211, Silver Spring (1979).

18. __________, and T.-P. Liu, The Asymptotic Analysis of Wave Interactions and Numerical Calculations of Transonic Nozzle Flow, Adv. Appl. Math., 5, 111 (1984).

19. Glimm, J., Solutions in the Large for Nonlinear Hyperbolic Systems of Conservation Laws, Comm. Pure Appl. Math., 18, 697 (1965).

20. __________, J. Marchesin, and D. O. McBryan, Statistical Fluid Dynamics: Unstable Fingers, Comm. Math. Phys., 74, 1 (1980).

21. ______________________________, A Numerical Method for Two Phase Flow with an Unstable Interfact, J. Comp. Phys. 38, 179 (1981).

22. ______________________________, Unstable Fingers in Two Phase Flow, Comm. Pure Appl. Math., 24, 53 (1981).

23. Glimm, J., G. Marshall, and B. Plohr, A Generalized Riemann Problem for Quasi-One-Dimensional Gas Flows, Adv. Appl. Math., 5, 1 (1984).

24. Godunov, S. K., Finite Difference Method for Numerical Computation of Discontinuous Solutions of the Equation of Fluid Dynamics, Mat. Sbornik., 47, 271 (1959).

25. Harten, A. and G. Zwas, Self-Adjusting Hybrid Schemes for Shock Computations, J. Comp. Phys., 9, 568 (1972).

26. ________, The Artificial Compression Method for Computation of Shocks and Contact Discontinuities: I. Single Conservation Laws, Comm. Pure Appl. Math., 30, 611 (1977).

27. ________, The Artificial Compression Method for Computation of Shocks and Contact Discontinuities: III. Self-Adjusting Hybrid Methods, Math. Comp. 32, 363 (1978).

28. ________, High Resolution Schemes for Hyperbolic Conservation Laws, J. Comp. Phys., 49, 357 (1983).

29. Harten, A. and J. M. Hyman, Self-Adjusting Grid Methods for One-Dimensional Hyperbolic Conservation Laws, Los Alamos Scientific Laboratory Report No. LA-9105, (1983).

30. Harten, A., J. M. Hyman, and P. D. Lax, On Finite-Difference Approximations and Entropy Conditions for Shocks, Comm. Pure Appl. Math., 29, 297 (1976).

31. Harten, A. and P. D. Lax, A Random Choice Finite Difference Scheme for Hyperbolic Conservation Laws, SIAM J. Num. Anal., 18, 289 (1981).

32. Harten, A., P. D. Lax, and B. van Leer, On Upstream Differencing and Godunov-Type Schemes for Hyperbolic Conservation Laws, SIAM Review, 25, 35 (1983).

33. Jennings, G., Discrete Shocks, Comm. Pure Appl. Math., 27, 25 (1974).

34. John, F., On Integration of Parabolic Equations by Difference Methods, Comm. Pure Appl. Math., 5, 155 (1952).

35. Keyfitz, B., Solutions with Shocks, an Example of an L_1-Contractive Semigroup, Comm. Pure Appl. Math., 24, 125 (1971).

36. La Vita, J., Error Growth in Using the Random Choice Method for the Inviscid Burgers Equation, SIAM J. Sci. Stat. Comp., 1, 327 (1980).

37. Lax, P. D., Weak Solutions of Nonlinear Hyperbolic Equations and Their Numerical Computations, Comm. Pure Appl. Math., 7, 159 (1954).

38. ________, On the Stability of Difference Approximations to the Solution of Hyperbolic Equations with Variable Coefficients, Comm. Pure Appl. Math., 14, 497 (1967).

39. ________, Nonlinear Partial Differential EQuations and Computing, SIAM Review, 11, 7 (1969).

40. ________, Shock Waves and Entropy, Contributions to Nonlinear Functional Analysis, S. H. Zaratonell, ed., Academic Press, New York (1971).

41. Lax, P. D. and A. N. Milgram, Parabolic Equations, Contributions to the Theory of Partial Differential Equations, Annals of Math. Studies, 33, 167 (1954).

42. Lax, P. D. and B. Wendroff, Systems of Conservation Laws, Comm. Pure Appl. Math., 13, 217 (1960).

43. ________________, Difference Schemes for Hyperbolic Equations with High Order Accuracy, Comm. Pure. Appl. Math., 17, 381 (1964).

44. Leveque, R. J., Large Time Step Shock-Capturing Techniques for Scalar Conservation Laws, SIAM J. Num. Anal., 19, 1091 (1982).

45. Mac Cormack, R. W., The Effect of Viscosity in Hypervelocity Impact Cratering, AIAA Paper No. 69-354, (1969).

46. Marshall, G. and R. Méndez, Computational Aspects of the Random Choice Method for Shallow Water Equations, J. Comp. Phys., 39, 1 (1981).

47. Oleinik, O. A., Discontinuous Solutions of Nonlinear Differential Equations, Uspekhi Mat. Nauk., 12, 3 (1957), (Amer. Math. Soc. Transl. Ser. 2, 26, 95).

48. Osher, S., Shock Modeling in Aeronautics, Numerical Methods in Fluid Dymanics, M. Baines and K. W. Morton eds., Academic, New York (1983).

49. ________, Reimann Solvers, The Entropy Condition and Difference Approximation, SIAM J. Num. Anal., 21, 217 (1984).

50. Osher, S. and S. Chakravarthy, Upwind Schemes and Boundary Conditions with Applications to Euler Equations in General Geometrics, J. Comp. Phys., 50, 447 (1983).

51. ________________, High Resolution Schemes and Entropy Conditions, SIAM J. Num. Anal., 21, 955 (1984).

52. Richtmyer, R. D., A Survey of Difference Methods for Non-Steady Fluid Dymanics, NCAR Technical Notes 63-2, Boulder (1962).

53. Richtmyer, R. D. and K. W. morton, Difference Methods for Initial-Value Problems, 2nd. ed., Interscience, New York (1967).

54. Roe, P. L., The Use of the Riemann Problem in Finite Difference Schemes, Proceedings of the Seventh International Conference on Numerical Methods in Fluid Dynamics, Lecture Note in Physics, Vol.141, W.C. Reynolds and R.W. Mac Cormack eds., Springer-Verlag, New York, (1981).

55. _________, Approximate Riemann Solvers, Parameter Vectors, and Difference Schemes, J. Comp. Phys., 43, 357 (1981).

56. _________, Numerical Modeling of Shockwaves and Other Discontinuities, Numerical Methods in Aeronautical Fluid Dynamics, P. L. Roe ed., Academic Press, New York (1982).

57. Sod, G. A., A Numerical Study of a Converging Cylindrical Shock, J. Fluid Mech., 83, 785 (1977).

58. _________, A Study of Cylindrical Implosion, Lectures on Combustion Theory, Samuel Z. Burstein, Peter D. Lax, and Gary A. Sod, editors, Department of Energy Research and Development Report COO-3077-153, New York University, (1978).

59. _________, A Survey of Several Finite Difference Methods for Systems of Nonlinear Hyperbolic Conservation Laws, J. Comp. Phys., 27, 1 (1978).

60. _________, "A Hybrid Random Choice Method with Application to Internal Combustion Engines," Proceedings Sixth International Conference on Numerical Methods in Fluid Dynamics, Lecture Notes in Physics, Vol. 90, H. Cabannes, M. Holt, and V. Rusanov, eds. Springer-Verlag (1979).

61. _________, "Automotive Engine Modeling with a Hybrid Random Choice Method," SAE paper 790242, (1979).

62. _________, "A Hybrid Random Choice Method with Application to Automotive Engine Modeling," Proceedings of the Third IMACS International Symposium on Computer Methods for Partial Differential Equations, Bethlehem, Pennsylvania, (1979).

63. _________, Computational Fluid Dynamics with Stochastic Techniques, Proceedings of the von Karman Institute for Fluid Dynamics Lecture Series on Computational Fluid Dynamics, Brussels, (1980).

64. _________, "Automotive Engine Modeling with a Hybrid Random Choice Method, II," SAE Paper 800288, (1980).

65. _________, Numerical Modeling of Turbulent Combustion on Reciprocating Internal Combustion Engines, SAE Paper 820041, (1982).

66. _________, A Flame Dictionary Approach to Unsteady Combustion Phenomena, Mat. Aplic. Comp., 3, 157 (1984).

67. _________, A Random Choice Method with Application to Reaction-Diffusion Systems in Combustion, Comp. and Maths. Appl., II, (1985).

68. _________, A Numerical Study of Oxygen-Diffusion in a Spherical Cell with the Michaelis-Menten Oxygen Uptake Kinetics, to appear in Advances in Hyperbolic Partial Differential Equations, Vol.III, (1985).

69. Sod, G. A., A Flame Dictionary Approach to Unsteady Combustion Phenomena with Applicaiton to a Flame Interacting with a Cole Wall, Proceedings of the Ninth International Conference on Numerical Methods in Fluid Dynamics, Lecture Notes in Physics, to appear.

70. Steger, J. L. and R. F. Warming, Flux Vector Splitting of the Inviscid as Dynamics Equation with Application to Finite Difference Methods, J. Comp. Phys., 40, 263 (1981).

71. Strang, G., Accurate Partial Difference Methods II: Nonlnear Problems, Numerische Mathematik, 6, 37 (1964).

72. Tadmor, E., The Large-time Behavior of the Scalar, Genuinely Nonlinear Lax-Friedrichs Scheme, Math. Comp., 43, 353 (1984).

73. __________, Numerical Viscosity and the Entropy Condition for Conservative Difference Schemes, Math. Comp., 43, 369 (1984).

74. van Leer, B., Towards the Ultimate Conservative Difference Scheme I. The Quest for Monotonicity, Proceedings of the Third International Conference on Numerical Methods in Fluid Dynamics, Lecture Notes in Physics, Vol.18, H. Cabannes and R. Teuram eds., Springer-Verlag, New York (1973).

75. ____________, Towards the Ultimate Conservation Difference Scheme II. Monotonicity and Conservation Combined in a Second-Order Scheme, J. Comp. Phys., 14, 361 (1974).

76. ____________, Towards the Ultimate Conservation Difference Scheme III. Upstream-Centered Finite-Difference Scheme for Ideal Compressible Flow, J. Comp. Phys., 23, 263 (1977).

77. ____________, Towards the Ultimate Conservation Difference Scheme IV. A New Approach of Numerical Convection, J. Comp. Phys., 23, 276 (1977).

78. ____________, On the Relation Between the Upwind-Difference Schemes of Godunov, Enquist-Osher, and Roe, SIAM J. Sci. Stat. Comp., 5, 1 (1984).

79. von Neumann and R. D. Richtmyer, A Method for the Numerical Calculation of Hydrodynamic Shocks, J. Appl. Phys., 21, 232 (1950).

80. Woodward, P. and P. Colella, High Resolution Difference Schemes for Compressible Gas Dynamics, Proceedings Seventh International Conference on Numerical Methods in Fluid Dynamics, Lecture Notes in Physics, Vol.141, W. C. Reynolds and R. W. MacCormack eds., Springer-Verlag, New York, (1981).

81. ___________________________, The Numerical Simulation of Two-Dimensional Fluid Flow with Strong Shocks, to appear.

82. Yanenko, N. N., The Method of Fractional Steps, Springer-Verlag, New York (1971).

83. Zalesak, S. T., Fully Multidimensional Flux-Corrected Transport, J. Comp. Phys., 31, 335 (1979).

84. ______________, High Order "ZIP" Differencing of Convective Terms, J. Comp. Phys., 40, 497 (1981).

V. STABILITY IN THE PRESENCE OF BOUNDARIES

V.1. Introduction

Consider the system of differential equations

(5.1.1) $$\partial_t \underline{v} = P(\partial_x)\underline{v}, \quad t > 0, \quad a < x < b ,$$

with initial conditions

(5.1.2) $$\underline{v}(x,0) = \underline{F}(x)$$

and with boundary conditions

(5.1.3) $$\begin{aligned} \underline{\underline{A}}\underline{v}(a,t) &= 0, \\ \underline{\underline{B}}\underline{v}(b,t) &= 0, \end{aligned}$$

where $\underline{v}$ denotes a vector with p components, P is a polynomial in ∂_x whose coefficients are matrices, and $\underline{A}$ and $\underline{B}$ are matrices of rank $< p$.

Approximate equation (5.1.1) by the finite difference method

(5.1.4) $$\underline{u}^{n+1} = Q\underline{u}^{u} = \sum_j \underline{Q}_j S_+^j u^n,$$

where the $\underline{Q}_j$'s are $p\times p$ matrices. At the center of the grid, away from the boundaries, everything will be in order, at least at the outset. However, the boundaries are going to require special consideration.

In general, the finite difference approximation requires more boundary conditions than the differential equation requires if it is to lead to a well posed problem.

For example, consider the equation $\partial_t v = \partial_x v$, $0 < x < 1$, $t > 0$ with initial data $v(x,0) = f(x)$. The solution is constant along the characteristics $x + t =$ constant, that is, $v(x,t) = f(x+t)$, as depicted in Figure 5.1.

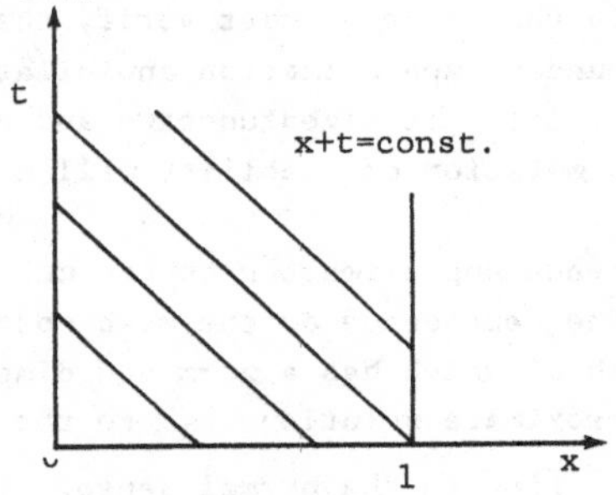

Figure 5.1.

A boundary condition can be imposed on the right boundary $x = 1$, but no boundary condition can be imposed on the left boundary.

Suppose the equation is approximated by the Lax-Wendroff method (3.5.3)

$$u_i^{n+1} = u_i^n + kD_0u_i^n + \frac{k^2}{2}D_+D_-u_i^n$$
$$= a_1u_{i-1}^n + a_2u_i^n + a_3u_{i+1}^n ,$$

where a_1, a_2, and a_3 are constants. Here boundary conditions are required on both right and left. The floating boundary condition on the left is going to be imposed to preserve stability and accuracy. It is this floating (or artificial) boundary condition that is going to be the source of many problems.

Stability for the pure initial-value problem has been discussed using the Fourier method. However, the Fourier argument in its previous form does not apply because $\sum_{k=-\infty}^{\infty} u_k e^{ih\xi}$ does not exist for the initial boundary-value problem.

If $p = 1$, then the finite difference method can be written in the form

$$\underline{u}^{n+1} = \underline{Q}_N\underline{u}^n ,$$

where $\underline{Q}_N$ is a matrix of $(N-1)\times(N-1)$ with $h = 1/N$. We want to find the eigenvectors of this matrix $\underline{Q}_N$. Observe that if $u_j = e^{ijx}$, then $S_+u_j = e^{ijh}u_j$ and so $u = \{u_j\}$ is an eigenfunction of the shift operator S_+. Can we construct eigenvectors out of the shift operators? If $\underline{Q}_N$ is a polynomial in S_+ (and S_-) as in Chapter I, then when $\underline{Q}_N$ is applied, $u = \{u_j\}$ is multiplied by ρ an eigenvalue of $\underline{Q}_N$. If $|\rho| < 1 + Ck$, where C is a constant, we have boundedness of this this particular solution. If all are bounded, then we have the presumption of stability.

For any number a, real or complex, and $u_j = e^{aj}$, $S_+u_j = e^a u_j$. If there are no boundaries, then $\|u_j\| = +\infty$ and it is not an eigenfunction of S_+. However, if we are considering a half plane problem, $x > 0$ with a left boundary at $x = 0$, then e^{aj} with $\text{Re}\, a < 0$ becomes an acceptable eigenfunction* of S_+ which becomes multiplied by an eigenvalue ρ. In this case we must verify that either $|\rho| < 1 + Ck$ or the boundary approximation annihilates the eigenfunction. A better definition of eigenfunction and eigenvalue is required and another formulation of stability will have to be considered.

The finite difference approximation $u_h(ih,nk)$ (where the subscript h denotes the dependence on the mesh spacing), lies in a sequence of spaces, each of which has a norm and dimension all its own. Convergence of the approximate solution u_h to the exact solution v

*This is not an eigenfunction in the normal sense.

should mean that there exists a sequence of spaces $S_1, S_2, S_3, \ldots$, a sequence of norms $\| \ \|^{(1)}, \| \ \|^{(2)}, \| \ \|^{(3)}, \ldots$, with $\| \ \|^{(m)}$ defined on S_m, and a sequence of vectors $u_{h_1}, u_{h_2}, u_{h_3}, \ldots$ with $u_{h_m} \in S_m$ such that $\|v - u_{h_m}\|^{(m)} \to 0$ as $m \to \infty$. Then v must belong to all of the spaces S_m. Stability should mean that there exists a bound $C(t)$ independent of the space and independent of m such that $\|u_{h_m}\|^{(m)} \leq C(t)$. These ideas will be more definitely established in the next two sections.

V.2. Ryabenki-Godunov Theory.

Consider the initial boundary-value problem (5.1.1)-(5.1.3). Let $h = (b-a)/N$ denote the mesh spacing of the grid with which the interval $a \leq x \leq b$ is partitioned. Consider the finite difference method (5.1.4) written in the form

$$\underline{u}_h^{n+1} = \underline{Q}_h \underline{u}_h^n , \tag{5.2.1}$$

where $\underline{u}_h^n = (u_1^n, u_2^n, \ldots, u_{N-1}^n)^T$ and $\underline{Q}_h$ is the block matrix of order $p(N-1) \times p(N-1)$ whose row elements are the element matrices $\underline{Q}_j$ used in defining $\underline{Q}$ in (5.1.4). The dimension of $\underline{Q}_h$ grows as $h \to 0$, that is, as the mesh is refined.

We are now able to state the key definition of the Ryabenki-Godunov theory.

Definition 5.1. Consider a sequence of spaces $S_1, S_2, S_3, \ldots, S_m, \ldots$ with a corresponding sequence of norms $\| \ \|^{(1)}, \| \ \|^{(2)}, \| \ \|^{(3)}, \ldots, \| \ \|^{(m)}, \ldots$ and a one-parameter family of operators $\{\underline{Q}_h(k)\}$ (where there is a relation between h and m). Suppose there exists a sequence of vectors $\underline{u}_{h_1} \in S_1$, $\underline{u}_{h_2} \in S_2$, $\underline{u}_{h_3} \in S_3, \ldots, \underline{u}_{h_m} \in S_m, \ldots$, and a sequence of complex numbers $\lambda_{h_1}, \lambda_{h_2}, \lambda_{h_3}, \ldots, \lambda_{h_m}, \ldots$ such that
(1) $\|\underline{u}_{h_m}\|^{(m)} = 1$,
(2) $k_m \to 0$ as $m \to \infty$, where k_m is the time step corresponding to mesh spacing h_m,
(3) $\|\underline{Q}_{h_m} \underline{u}_{h_m} - \lambda_{h_m} \underline{u}_{h_m}\|^{(m)} h_m^{-q} \to 0$ as $m \to \infty$ for all q,
(4) and $\lambda_{h_m} \to \lambda$ as $m \to \infty$,
then λ is said to belong to the spectrum of the family $\{\underline{Q}_h\}$.

As a result of condition (3) of the definition, $\underline{u}_{h_m}$ and λ_{h_m} are approximate eigenvectors and eigenvalues of $\underline{Q}_{h_m}$. However, the purpose of weakening this in (3) is to enable us to consider one boundary without the other in discussing stability. This point will be clarified as we continue. The values of λ in the spectrum are like the range of the symbol $\underline{Q}$.

Consider a problem on a finite domain as in problem (5.1.1)-(5.1.3). Using the definition of the spectrum, the boundary condition on the left will not affect the boundary condition on the right.

Similarly, the boundary condition on the right will not affect the boundary condition on the left. To check stability of a finite domain problem as above, we have to consider the stability of three cases:

1) the pure initial-value problem,
2) the left-quarter-plane problem, that is, boundary on the right,
3) the right-quarter-plane problem, that is, boundary on the left.

It is shown in the monograph by Godunov and Ryabenki [1] that the spectrum of the original problem is the sum of the spectrums of the above three cases.

Formally, suppose the finite difference method is of the form

$$u^{n+1} = (\sum_j a_j S_+^j)u^n \,, \tag{5.2.2}$$

with the symbol given by

$$\rho(\xi) = \sum_j a_j e^{-ij\xi} \,.$$

The symbol ρ can be written in the form $\rho(\xi) = \Gamma(e^{i\xi})$, where $\Gamma(z) = \sum_j a_j z^{-j}$. Hence the von Neumann condition takes the form $|\Gamma(z)| \leq 1 + Ck$ for $|z| = 1$. These values of the symbol can be viewed as limits of approximate eigenvalues of the family of finite difference operators corresponding to the approximate eigenvectors

$$(u_{h_i})_\ell = z^\ell, \ |z| = 1 \,. \tag{5.2.3}$$

If a boundary on the left, we have to determine if there exist approximate eigenvectors of the form

$$(u_{h_i})_\ell = z^\ell, \ |z| < 1 \,, \tag{5.2.4}$$

where $|z| < 1$ is required in order that u_{h_i} has finite norm. Similarly, if a boundary is on the right, we have to determine if there exist approximate eigenvectors of the form

$$(u_{h_i})_\ell = z^\ell, \ |z| > 1 \,,$$

where $|z| > 1$ is required in order that u_{h_i} has finite norm. This leads to the following theorem.

<u>Theorem 5.1</u>. (a) (Ryabenki-Godunov Condition). A necessary condition for stability is $|\lambda| \leq 1$, where λ is in the spectrum of the family of difference operators.

(b) If the spectrum of the family of difference operators is not bounded by 1, the scheme is not stable.

<u>Proof</u>: Let λ be the point in the spectrum of $\{\underline{Q}_h\}$. There exists an approximate eigenvector $\underline{u}_{h_m}$ and an approximate eigenvalue λ_{h_m} such that

$$|\lambda_{h_m} - \lambda| = \delta \quad \text{and} \quad (\underline{Q}_{h_m}\underline{u}_{h_m} - \lambda_{h_m}\underline{u}_{h_m})h_m^{-q} = \underline{\varepsilon}_{h_m}, \text{ where } \|\underline{\varepsilon}_{h_m}\|^{(m)} \to 0$$

faster than h_m^q for any q. Take $\underline{u}_{h_m}$ as $\underline{u}_{h_m}^0$, then

$$\underline{u}_{h_m}^1 = \underline{Q}_{h_m}\underline{u}_{h_m}^0 = \lambda_{h_m}\underline{u}_{h_m} + \underline{\varepsilon}_{h_m}$$

$$\underline{u}_{h_m}^2 = \underline{Q}_{h_m}\underline{u}_{h_m}^1 = \underline{Q}_{h_m}(\lambda_{h_m}\underline{u}_{h_m} + \underline{\varepsilon}_{h_m})$$

$$= \lambda_{h_m}^2\underline{u}_{h_m} + \lambda_{h_m}\underline{\varepsilon}_{h_m} + \underline{\varepsilon}_{h_m}$$

$$\underline{u}_{h_m}^n = \underline{Q}_{h_m}\underline{u}_{h_m}^{n-1} = \lambda_{h_m}^n\underline{u}_{h_m} + \sum_{i=0}^{n-1}\lambda_{h_m}^i\underline{\varepsilon}_{h_m} .$$

(a) If $|\lambda| < 1$, then for some h_m, $|\lambda_{h_m}| < 1$ and $\|\lambda_{h_n}^n\underline{u}_{h_n}\|^{(m)} < \|\underline{u}_{h_m}\|^{(m)}$. Letting $n = \frac{1}{k}$, we see that $\|\lambda_{h_m}^i\underline{\varepsilon}_{h_m}\|^{(m)}$ is bounded.
(b) If $|\lambda| > 1$, then for some h_m, $\lambda_{h_m} > 1$ and we have exponential growth and hence no stability.

If every λ in the spectrum is such that $|\lambda| < 1$, we have only a presumption of stability because

1) $(\underline{Q}_h)$ is not a norm,

2) the Ryabenki-Godunov condition can be weaker than the von Neumann condition for the pure initial-value problem. To see this, suppose $\lambda_{h_m} = 1 + C\sqrt{k_m}$, where k_m is the time step associated with the difference operator $\underline{Q}_{h_m}$. This value of λ_{h_m} is an acceptable approximate eigenvalue with $\lambda = 1$ as the limit, so that the Ryabenki-Godunov condition is satisfied. However, $\lambda_{h_m} = 1 + C\sqrt{k_m}$ is not allowed by the von Neumann condition. Thus the Ryabenki-Godunov condition is necessary but not sufficient for stability. An example demonstrating this will be given below. However, the Ryabenki-Godunov condition is stronger than the von Neumann condition in the sense that a larger class of normal modes is considered, for near a boundary there may be modes which decay away from the boundary and affect the stability of the difference method.

We now present some examples of computing the spectrum. As a first example consider $\partial_t x = \partial_x^2 v$, $t > 0$, $0 < x < 1$, that is, equation (5.1.1) with $P(z) = z^2$ and $p = 1$, where $v(x,0) = f(x)$, and $v(0,t) = v(1,t) = 0$. Approximate this by

(5.2.6) $$u_i^{n+1} = Qu_i^n = u_i^n + kD_+D_-u_i^n ,$$

with initial condition

(5.2.7) $$u_i^0 = f(ih) ,$$

and boundary conditions

(5.2.8) $$u_0^n = u_N^n = 0,$$

where $h = 1/N$. Letting $r = k/h^2$, the finite difference method may be written in matrix form as in the definition, where

$$\underline{Q}_h = \begin{pmatrix} 1-2r & r & & & \\ r & 1-2r & r & & \\ & \ddots & \ddots & \ddots & \\ & & & & r \\ & & & r & 1-2r \end{pmatrix}.$$

This represents a family of matrix equations, one member for each grid size h. The eigenvalues of this matrix are

$$\lambda_{h_j} = 1 - r(\cos \frac{2j\pi}{N} - 1), \quad j = 1,2,\ldots,N-1$$

and the associated eigenvectors are $\underline{u}_h = \{u_\ell\}$, where the ℓ-th component is $u_\ell = \text{Im}(\exp \frac{2\pi i j \ell}{N})$. We see that $\|\underline{Q}_h \underline{u}_h - \lambda_h \underline{u}_h\| = 0$ and $\underline{u}_h$ satisfies the two boundary conditions.

For the pure initial-value problem we have the symbol $\rho(\xi) = 1 + 2r(\cos\xi - 1)$. And for all values of ξ in the domain of ρ, there exists a sequence of approximate eigenvalues $\{\lambda_h\}$ such that $\lambda_h \to \rho(\xi)$. In fact, in this case the spectrum of the family $\{\underline{Q}_h\}$ is just the set of values assumed by the symbol.

This example was rigged because zero or periodic boundary conditions ensure that the von Neumann condtion is sufficient for stability.

Consider the equation $\partial_t v = \partial_x^2 v$, $t > 0$, $0 < x < 1$, with $v(x,0) = f(x)$, $v(0,t) = 1$, and $v(1,t) = 0$. Approximate the equation and the initial condition by (5.2.6)-(5.2.7) and approximate the boundary conditions by

$$u_0^n = 1 \quad \text{and} \quad u_N^n = 0 ,$$

where $k = 1/N$.

Look for solutions of the finite difference method of the form $u_i^n = C\Gamma^n z^i$, where C is a normalization constant, such that $\Gamma \to \lambda$, with λ in the spectrum. For $|z| = 1$, consider the pure initial-value problem. For $|z| < 1$, disregard the right boundary condition and consider only the left boundary conditions. For $|z| > 1$ disregard the left boundary condition and consider only the right boundary condition.

Consider the case where $|z| = 1$, that is, the pure initial value problem. In this case the symbol is, as in the first example,

$$\rho(\xi) = 1 + 2r(\cos\xi - 1),$$

where $r = \frac{k}{h^2}$. $|\rho(\xi)| < 1$, provided that $r < \frac{1}{2}$.

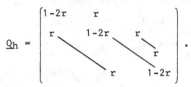

Consider the right half-plane problem where the boundary is on the left. Using the boundary coundition, $u_0^n = C\Gamma^n z^0 = C\Gamma^n = 1$ for all n. This is impossible, however, for if $|\Gamma| < 1$, then Γ^n is changing for each n. Hence the boundary condition on the left eliminates any contribution to the spectrum.

Consider the left half-plane problem, where the boundary is on the right. Using the boundary condition, $u_N^n = C\Gamma^n z^N = 0$ for all n. For $|z| > 1$, this is impossible. Hence the boundary condition on the right also eliminates any contribution to the spectrum.

So again in this case stability is governed by the pure initial-value problem.

Again, consider $\partial_t v = \partial_x^2 v$, $t > 0$, $0 < x < 1$, with $v(x,0) = f(x)$, $v(1,t) = 0$, and some boundary condition on the left. Approximate the equation and the initial condition by (5.2.6)-(5.2.7) and approximate the boundary conditions by

$$\alpha u_0^n + \beta u_1^n = 0$$
$$u_N^n = 0,$$

where $h = 1/N$ and α and β are constants. That is, the left boundary condition is approximated by $\alpha u_0^n + \beta u_1^n = 0$. For example, if the boundary condition on the left is $av(0,t) + bv_x(0,t) = 0$ approxi- approximated by $au_0^n + b(\frac{u_1^n - u_0^n}{h}) = 0$. This is in the form $\alpha u^n{}_0 + \beta u_1^n = 0$, where $\alpha = a - b/h$ and $\beta = b/h$.

Look for approximations of the form $u_j^n = C\Gamma^n z^j$, where C is a normalization constant such that $\Gamma \to \lambda$ with λ in the spectrum of the family of difference operators:

$$\begin{aligned} u_j^{n+1} &= Qu_j^n \\ &= Q(C\Gamma^n z^j) \\ &= \Gamma(C\Gamma^n z^j) \\ &= C\Gamma^{n+1} z^j. \end{aligned}$$

We have three cases to consider:

1) The case where $|z| = 1$, which corresponds to the pure initial-value problem as in the preceeding two examples. Let $z = e^{i\omega h}$, then $z^j = e^{i\omega jh} = e^{i\omega x}$, where $x = jh$. In this case $\Gamma = \rho(-\omega h)$, where ρ denotes the symbol of the above difference operator for the pure initial-value problem. Γ is the approximate eigenvalue, which for stability, by the Ryabenki-Godunov condition, $|\Gamma| \leq 1$ is required. This condition is satisfied provided that $r = k/h^2 \leq \frac{1}{2}$, the same as for the pure initial-value problem.

2) Consider the case where $|z| > 1$, which corresponds to a left quarter plane problem. The boundary condition on the right yields $u_N^n = 0$ or $C\Gamma^n z^N = 0$. This has no nontrivial solution. Hence the problem does not contribute anything to the spectrum.

3) Consider the case where $|z| < 1$, which corresponds to a right quarter plane problem. The left boundary condition $\alpha u_0^n + \beta u_1^n = 0$ becomes $\alpha C\Gamma^n z^0 + \beta C\Gamma^n z^1 = 0$ or $\alpha + \beta z = 0$, which

gives $z = -\frac{\alpha}{\beta}$. Examine $Qu_j^n = \Gamma u_j^n$ which is equivalent to

$$ru_{j+1}^n + (1 - 2r)u_j^n + ru_{j-1}^n = u_j^n$$

or

$$ru_{j+1}^n + (1 - 2r - \Gamma)u_j^n + ru_{j-1}^n = 0 .$$

By substitution of the assumed form of $u_j^n = C\Gamma^n z^j$,

$$rz^2 + (1 - 2r - \Gamma)z + r = 0 .$$

Solving for Γ, with $z = -\frac{\alpha}{\beta}$,

$$\Gamma = 1 - 2r - \frac{\alpha^2+\beta^2}{\alpha\beta} r .$$

And so an approximate eigenvalue Γ is obtained that depends on the boundary condition,

$$\Gamma \to \lambda = \lim_{h\to 0}(1 - 2r - \frac{\alpha^2+\beta^2}{\alpha\beta} r).$$

If the left boundary condition is $av(0,t) + b\partial_x v(0,t)$, as above, then $\alpha = a - b/h$ and $\beta = b/h$. This gives

$$\frac{\alpha^2 + \beta^2}{\alpha\beta} = \frac{(ah-b)^2 + b^2}{(ah-b)b} \to -2 \text{ as } h \to 0$$

from which $\lambda = 1 - 2r + 2r = 1$, so that $|\lambda| \leq 1$ and the Ryabenki-Godunov condition is satisfied.

We shall now give an example of a boundary generated instability. In the first example it was shown that if $v(0,t) = 0$ was approximated by $u_0^n = 0$, there was no trouble. Suppose this boundary condition is approximated by $2u_0^n - u_1 = 0$ which comes from $h^2D_+D_-u_0^n = 0$*. Here $\alpha = 2$ and $\beta = -1$, so $\lambda = \lim_{h\to 0}(1 - 2r - \frac{\alpha^2+\beta^2}{\alpha\beta}r) = 1 + \frac{r}{2} > 1$ for all $r = k/h^2 > 0$. Hence the Ryabenki-Godunov condition is violated. Thus the method is unstable, where the instability has been brought about by the boundary condition.

As a demonstration, we use this approximation to the initial-boundary value problem (5.2.6)-(5.2.8) with $f(x) = \sin(\pi x)$, where $N = 20$ and $r = 1/2$. The growth in the numerical solution at time $t = 0.0625$, 0.125, 0.1875, and 0.25 depicted in Figure 5.2 represent the boundary generated instability.

---

*This is a reasonable approximation since $h^2D_+D_-v = h^2\partial_x^2 v + O(h^4) = 0 + O(h^2)$, which is a second order approximation to $v = 0$.

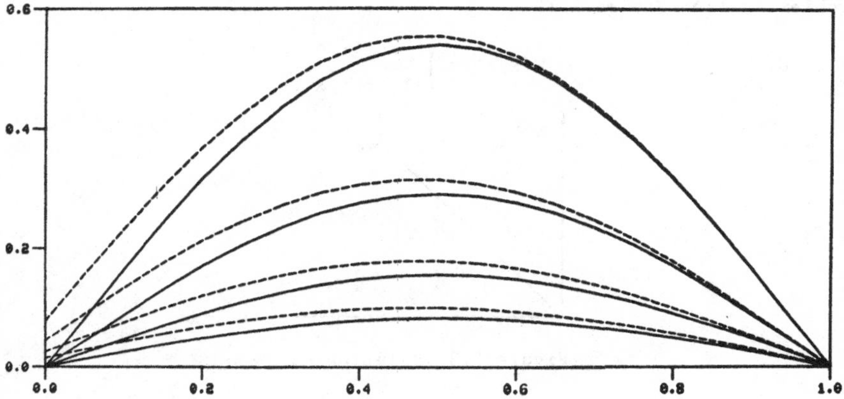

Figure 5.2. Boundary Generated Instability

This next example will be one in which the Ryabenki-Godunov conditions is satisfied yet there are oscillations propagated by the boundary (see Kreiss and Oliger [11]). Consider $\partial_t v = \partial_x v$, $t > 0$, $x > 0$ with $v(x,0) = 1$ and $v(1,t) = 1$. Approximate this equation by the leap-frog method (3.10.1)

$$u_i^{n+1} = u_i^{n-1} + r(u_{i+1}^n - u_{i-1}^n) ,$$

where $r = k/h$. The characteristics are $x + t =$ constant so that boundary conditions cannot be imposed at $x = 0$. However, the finite difference method will require an artificial boundary condition at $x = 0$, $u_0^n = 0$, for example.

With this initial data, $v(x,t) = 1$. Make a change of variable and define w_i^n by

(5.2.9) $$u_i^n = 1 + (-1)^i w_i^n .$$

Then by substitution, the difference equation for w_i^n becomes

$$w_i^{n+1} = w_i^{n-1} - r(w_{i+1}^n - w_{i-1}^n) ,$$

which approximates the equation

$$\partial_t W = -\partial_x W, \; t > 0, \; x > 0$$
$$W(x,0) = 0,$$
$$W(0,t) = -1 .$$

For this problem the characteristics are $x - t = \text{constant}$, that is, the characteristics eminate from the boundary $x = 0$. Hence this problem allows a boundary condition at $x = 0$. The solution of this problem is (see Figure 5.3)

(5.2.10) $$W(x,t) = \begin{cases} 0, & t < x \\ -1, & t > x . \end{cases}$$

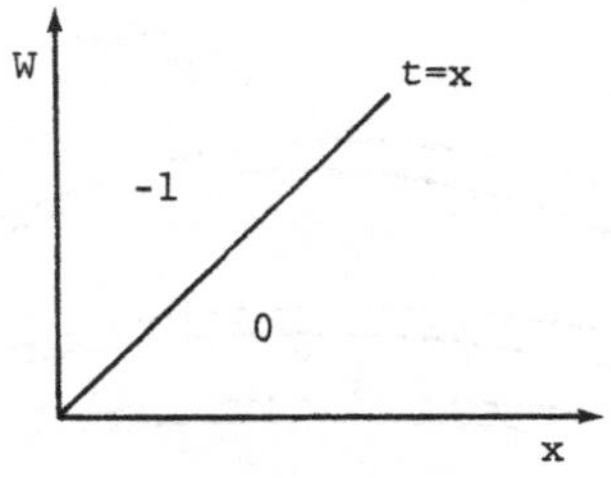

Figure 5.3. Solution W.

Therefore the approximate solution u_i^n (obtained by substitution of (5.2.10) in (5.2.9) is given by (see Figure 5.4)

$$u_i^n = \begin{cases} 1, & nk < ih \\ 1 + (-1)^i, & nk > ih . \end{cases}$$

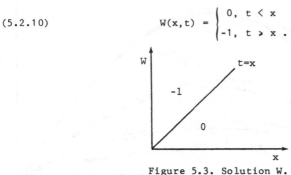

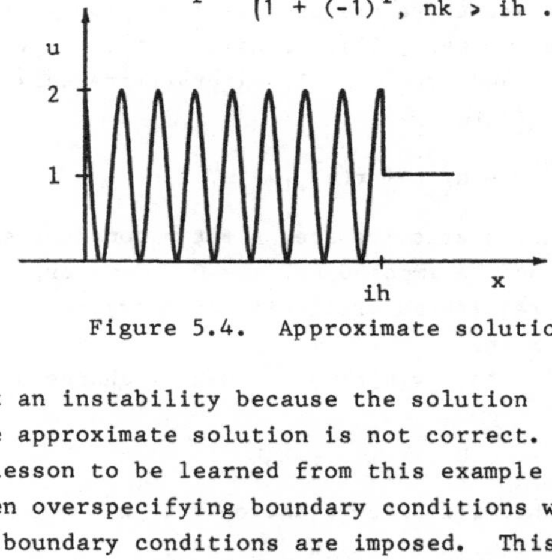

Figure 5.4. Approximate solution u_i^n .

This is not an instability because the solution u_i^n is bounded, but clearly the approximate solution is not correct.

The lesson to be learned from this example is that one must be careful when overspecifying boundary conditions which is the case when artificial boundary conditions are imposed. This is a nondissipative method. It will be seen in the next section that nondissipative methods are most prone to boundary condition instability. Dissipative methods will damp some of the oscillations; however, near the boundary there will still be trouble.

A similar example due to Gustafsson shows how these oscillations can be made even worse in a finite domain problem. Consider the same problem above where $0 < x < 1$ and the additional boundary condition

$v(1,t) = 0$. Approximate this equation by the leap-frog method as in the last example. The boundary condition on the left is approximated by first-order extrapolation, that is, $u_0^n = u_1^n$ and the boundary condition on the right is approximated by $u_N^n = 0$.

Under these conditions it is shown that if the approximate solution has the form $u_i^n = Cr^n z^i$, it grows like

$$\|u^n\|_2 = \text{constant } N^{t/2}.$$

Thus at the boundary $x = 0$, a wave is generated that grows like $N^{t/2}$. This wave, if not damped, will be reflected at the opposite boundary $x = 1$. When this reflected wave reaches the left boundary, $x = 0$, it is again increased by another factor of $N^{1/2}$. This process is continued.

The oscillations get worse if a higher order extrapolation method is used to approximate the left boundary condition. Suppose, for example, the left boundary condition is approximated by $D_+^j u_0^n = 0$. It can be shown that the wave grows like $\|u^n\|_2 = \text{constant } N^{(j-1/2)t}$.

In the final example, (Richtmyer [16]), consider the wave equation (3.1.1) written as a system

$$\partial_t v = \partial_x w$$

$$\partial_t w = \partial_x v \ , \quad t > 0, \ x > 0$$

with initial conditions

$$v(x,0) = f(x),$$

$$w(x,0) = g(x),$$

and a given boundary condition at $x = 0$. Approximate this by the fractional step method

$$w_{i+\frac{1}{2}}^{n+\frac{1}{2}} - w_{i-\frac{1}{2}}^{n-\frac{1}{2}} = r(v_{i+1}^n - v_i^n) \ , \tag{5.2.11}$$

$$v_i^{n+1} - v_i^n = r(w_{i+\frac{1}{2}}^{n+\frac{1}{2}} - w_{i-\frac{1}{2}}^{n+\frac{1}{2}}) \ , \tag{5.2.12}$$

where $r = k/h$. Suppose that the boundary condition at $x = 0$ is approximated at $i = 0$ by

$$v_0^n + q w_{1/2}^{n+\frac{1}{2}} = 0 \ , \tag{5.2.13}$$

where q is a constant. We shall see that this approximation of the boundary condition at $x = 0$ cannot be imposed for an arbitrary value of q. Clearly, by the CFL condition, a necessary condition for stability is $r \le 1$.

Consider situations of the form

(5.2.14) $$v_i^n = Vr^n z^i$$

and

(5.2.15) $$w_{i+\frac{1}{2}}^{n+\frac{1}{2}} = Wr^{n+\frac{1}{2}} z^{i+\frac{1}{2}},$$

where V and W are constants. For $|z| = 1$ we consider the pure initial-value problem, which is stable for $r \leq 1$.

Consider the right quarter plane problem, where $|z| < 1$ is required. Substituting the solution (5.2.14)-(5.2.15) into finite difference equations (5.2.11) and (5.2.12),

$$Wr^{n+\frac{1}{2}} z^{i+\frac{1}{2}} - Wr^{n-\frac{1}{2}} z^{i+\frac{1}{2}} - r(Vr^n z^{i+1} - Vr^n z^i) = 0$$

and

$$Vr^{n+1} z^i - Vr^n z^i - r(Wr^{n+\frac{1}{2}} z^{i+\frac{1}{2}} - Wr^{n+\frac{1}{2}} z^{i-\frac{1}{2}}) = 0,$$

or

(5.2.16) $$\begin{pmatrix} z^{1/2}(r-1) & -rr^{1/2}(z-1) \\ -rr^{1/2}(z-1) & z^{1/2}(r-1) \end{pmatrix} \begin{pmatrix} W \\ V \end{pmatrix} = \begin{pmatrix} 0 \\ 0 \end{pmatrix}.$$

In order to have a nontrivial solution, the determinant must vanish, that is,

$$z(r-1)^2 - r^2 r(z-1)^2 = 0$$

or

$$\frac{r}{(r-1)^2} r^2 = \frac{z}{(z-1)^2},$$

from which we see that if $|z| < 1$ and $r \leq 1$, then $|r| < 1$.

Multiplying the first equation in (5.2.16) by W, multiplying the second equation in (5.2.16) by V and subtracting yields

$$z^{1/2}(r-1)W^2 - z^{1/2}(r-1)V^2 = 0$$

or

(5.2.17) $$|V| = |W|.$$

Substituting the solution (5.2.14)-(5.2.15) into the boundary condition approximation (5.2.13)

$$Vr^n + qWr^{n+\frac{1}{2}} z^{1/2} = 0.$$

Solving for q,

$$q = -\frac{V}{W}(rz)^{-1/2},$$

from which it follows, using (5.2.17), that

$$|q| = \frac{|V|}{|W|}|rz|^{-1/2} = |rz|^{-1/2} > 1,$$

for $|z| < 1$ and $|r| < 1$. Thus the Ryabenki-Godunov condition is satisfied if $|q| > 1$.

V.3. Stability of Initial Boundary-Value Problems for Hyperbolic Equations

How does one prescribe boundary conditions for hyperbolic systems? Consider first the scalar problem

$$\partial_t v = a\partial_x v, \quad t > 0, \quad x > 0$$
$$v(x,0) = f(x)$$

where a is a real constant. The solution is $v(x,t) = f(x + at)$, that is, the solution is constant along the characteristics $x + at =$ constant. There are three cases corresponding to $a < 0$, $a = 0$, and $a > 0$.

For the case where $a < 0$, a boundary condition must be specified at $x = 0$, namely $v(0,t) = g(t)$. In this case the characteristics are emanating from the line $x = 0$. The characteristics are incoming, as depicted in Figure 5.5. To ensure continuity

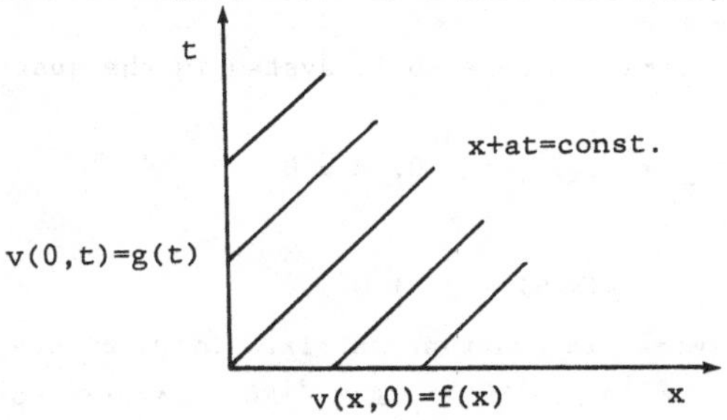

Figure 5.5. Characteristics for $a < 0$

require that $g(0) = v(0,0) = f(0)$. This is a compatibility condition.

For the case where $a = 0$, the equation reduces to an ordinary differential equation $\frac{dv}{dt} = 0$. The characteristics are vertical lines. and $v(x,t) = f(x)$, as depicted in Figure 5.6.

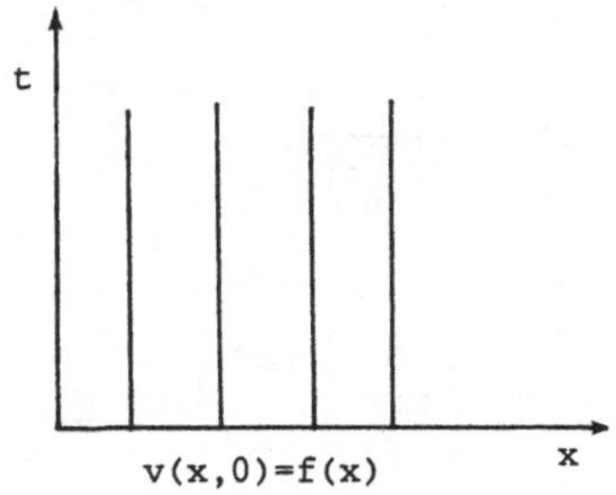

Figure 5.6. Characteristics for $a = 0$.

Finally, for the case where $a > 0$, characteristics are entering entering the boundary $x = 0$, that is, the characteristics are outgoing, as depicted in Figure 5.7. Boundary conditions cannot be arbitrarily prescribed at $x = 0$, for, since the solution is $v(x,t) = f(x + at)$, $v(0,t) = f(at)$, and prescribing any boundary condition other than $f(at)$ would be a contradiction.

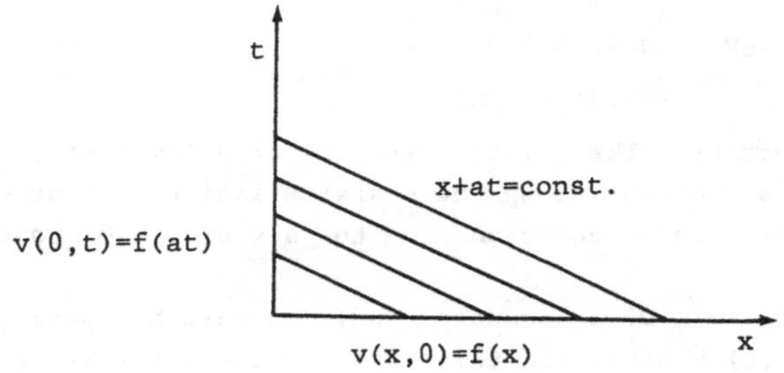

Figure 5.7. Characteristics for $a > 0$.

Next consider the symmetric hyperbolic system in the quarter plane

(5.3.1) $$\partial_t \underline{v} = \underline{A}\partial_x \underline{v}\ , \quad t > 0,\ x > 0$$

with initial condition

$$\underline{v}(x,0) = \underline{f}(x)$$

where $\underline{A}$ is a $p \times p$ symmetric constant matrix. There exists a matrix $\underline{T}$ such that $\underline{A} = T^{-1}\underline{\Lambda}\,\underline{T}$, where $\underline{\Lambda} = \mathrm{diag}(\lambda_1,\lambda_2,\dots,\lambda_p)$ is a diagonal matrix with the eigenvalues of $\underline{A}$. If $\underline{T}$ is multiplied by appropriate permutation matrices, the elements of $\underline{\Lambda}$ are ordered, that is,

(5.3.2) $$\lambda_1 \le \lambda_2 \le \dots \le \lambda_r < 0 < \lambda_{r+1} \le \lambda_{r+2} \le \dots \le \lambda_p .$$

Let $\underline{\Lambda}^{I} = \mathrm{diag}(-\lambda_1,-\lambda_2,\dots,-\lambda_r)$ and $\underline{\Lambda}^{II} = \mathrm{diag}(\lambda_{r+1},\lambda_{r+2},\dots,\lambda_p)$ so that

(5.3.3) $$\underline{\Lambda} = \left(\begin{array}{c|c} -\Lambda^{I} & 0 \\ \hline 0 & \Lambda^{II} \end{array}\right).$$

Define $\underline{w} = \underline{T}^{-1}\underline{v}$, so that

(5.3.4) $$\underline{w}_t = \begin{pmatrix} \underline{w}^{I} \\ \underline{w}^{II} \end{pmatrix}_t = \left(\begin{array}{c|c} -\underline{\Lambda}^{I} & 0 \\ \hline 0 & \underline{\Lambda}^{II} \end{array}\right) \begin{pmatrix} \underline{w}^{I} \\ \underline{w}^{II} \end{pmatrix}_x \equiv \underline{\Lambda w}_x .$$

Assume for now that $\underline{A}$ has no zero eigenvalues.

The solutions $\underline{w}^I$ and $\underline{w}^{II}$ are of the form $f(x + \lambda t)$. The values of $\underline{w}^{II}$, (called outgoing variables), at the boundary $x = 0$, since $\lambda > 0$, as in the scalar case, are determined by the initial conditions, and so boundary conditions for $\underline{w}^{II}$ cannot be prescribed. However, for $\underline{w}^I$ (called incoming variables) since $\lambda < 0$, as in the scalar case, values must be prescribed at the boundary $x = 0$. One choice would be $\underline{w}^I = \underline{g}^I(t)$. However, $\underline{w}^{II}$ is known at the boundary so a more general boundary condition may be considered,

(5.3.5) $$\underline{w}^I = \underline{S}^I\underline{w}^{II} + \underline{g}^I(t)$$

where $\underline{S}^I$ is an rx(p-r)-matrix.

Suppose now we add a boundary on the right, at $x = 1$. In this case the incoming variablesat the left boundary are outgoing variables at the right boundary. Similarly, the outgoing variables at the left boundary are incoming variables at the right boundary.

At the right boundary, values of $\underline{w}^I$ are determined by the initial conditions and cannot be prescribed, whereas values of $\underline{w}^{II}$ must be prescribed. One choice of boundary condition is $\underline{w}^{II} = \underline{g}^{II}(t)$. In this case, however, $\underline{w}^I$ is known at the boundary, so a more general boundary condition may be considered,

(5.3.6) $$\underline{w}^{II} = \underline{S}^{II}\underline{w}^I + \underline{g}^{II}(t) ,$$

where $\underline{S}^{II}$ is a $(p-r) \times r$ matrix (see Figure 5.8).

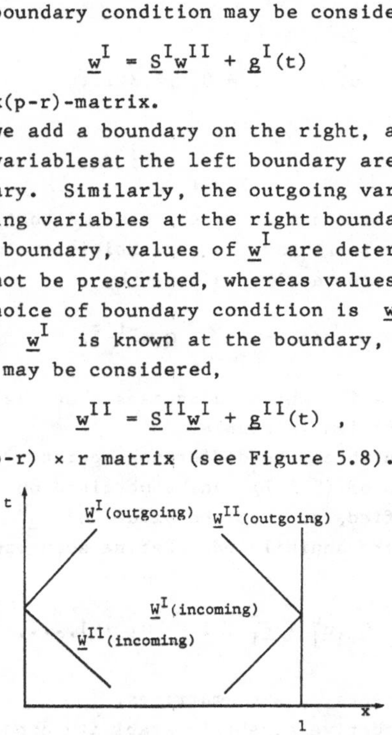

Figure 5.8.Incoming and outgoing variables on right and left boundaries

Consider $\lambda = 0$ being an eigenvalue of $\underline{A}$. Let $\underline{w}^0$ correspond to $\lambda = 0$. At the left boundary $(x = 0)$, $\underline{w}^0$ is known, $\underline{w}^0(0,t) = \underline{f}(0)$ for every t as in the scalar case. Similarly, at the right boundary $(x = 0)$, $\underline{w}^0$ is known, $\underline{w}^0(1,t) = \underline{f}(1)$ for every t, as in the scalar case. So a reasonable choice is to treat $\underline{w}^0$ as an outgoing variable at both boundaries. Thus the decomposition of $\underline{w}$ will be different at the two boundaries. At the left boundary

$$\begin{pmatrix} \underline{w}^I \\ \underline{w}^{II} \end{pmatrix}_t = \left(\begin{array}{c|c} -\underline{\Lambda}^I & 0 \\ \hline 0 & {}^0\underline{\Lambda}^{II} \end{array}\right) \begin{pmatrix} \underline{w}^I \\ \underline{w}^{II} \end{pmatrix}_x$$

and at the right boundary

$$\begin{pmatrix} \underline{w}^I \\ \underline{w}^{II} \end{pmatrix}_t = \left(\begin{array}{c|c} -\underline{\Lambda}^I{}_0 & 0 \\ \hline 0 & \underline{\Lambda}^{II} \end{array}\right) \begin{pmatrix} \underline{w}^I \\ \underline{w}^{II} \end{pmatrix}_x$$

Again, consider the quarter-plane problem (5.3.1), where Λ, as in (5.3.3), is nonsingular, with boundary conditions (5.3.5) and $\underline{f}(x) = 0$. Approximate the transformed equation (5.3.4) by

(5.3.7)
$$\underline{u}^{n+1} = \underline{Q}\underline{u}^n ,$$
$$u_i^0 = 0,\ i = 0,1,2,3,\ldots,$$

where

(5.3.8)
$$\underline{Q} = \sum_{j=-m_1}^{m_2} \underline{Q}_j S_+^j ,$$

and $\underline{Q}_j$ are $p \times p$ constant matrices. Assume that $m_2 > 1$ and $\underline{Q}_{-m_1}$ and $\underline{Q}_{m_2}$ are nonsingular. For the pure initial-value problem, the symbol (or amplification matrix) is given by

(5.3.9)
$$\underline{G}(\xi) = \sum_{j=-m_1}^{m_2} \underline{Q}_j e^{-ij\xi}.$$

Assume that $\|G(\xi)\|_2 < 1$, which guarantees that the method for the pure initial-value problem is stable.

From the way in which the difference operator $\underline{Q}$ is defined in (5.3.8), the solution of (5.3.7) can be obtained only if boundary conditions are specified, so that the values of $\underline{u}_i^n$ for $i = -m_1+1,\ldots,-1,0$ are annihilated. Define such boundary conditions as follows

(5.3.10)
$$\underline{u}_i^n = \sum_{j=1}^{m} \underline{C}_{ij}\underline{u}_j^n + \underline{F}_i^n ,\ i = -m_1 + 1,\ldots,-1,0,$$

where $\underline{C}_{ij}$ are constants $p \times p$ matrices.

Our goal is to derive algebraic stability conditions. In an early paper by Kreiss [8], algebraic stability conditions were established for the above initial boundary-value problem under the additional assumption that the difference method was dissipative of order $2s$, (see Sections III.6 and III.8). With this assumption the high frequency terms would have no influence.

In a more recent paper by Gustafsson, Kreiss, and Sundstrom [6] (see also Kreiss and Oliger [11]), by reformulating the notation of stability algebraic stability conditions were obtained for nondissipative methods, as well as, for the case where $\underline{A}$ is a function of x and t. These results will now be presented.

Introduce two discrete norms; the first is the discrete ℓ_2-norm over time rather than space defined by

$$\|\underline{F}\|_k^2 = \sum_{n=0}^{\infty} \|\underline{F}^n\|^2 k,$$

where $\underline{v} = (v_1^2 + v_2^2 + \ldots + v_p^2)^{1/2}$. The second discrete norm is the discrete ℓ_2-norm over space and time defined by

$$\|\underline{u}\|_{hk}^2 = \sum_{n=0}^{\infty} \|\underline{u}^n\|_2^2 k.$$

With this stability for the initial boundary-value problem may be defined.

Definition. The finite difference method (5.3.7) and (5.3.10) is stable if there exist constants α_0 and K_0 such that for all $\underline{F}_j$ with $\|\underline{F}_j\|_k < \infty$ and all $\alpha > \alpha_0$ the following estimate holds

(5.3.11) $$(\alpha - \alpha_0)\|e^{-nk}\underline{u}_i^n\|_{hk}^2 \leq K_0^2 \sum_{j=-m_1+1} \|e^{-\alpha nk}\underline{F}_j\|_k^2 .$$

This is not the only choice for definition of stability (see Gustafsson et al. [6]). However, this definition prohibits the unwanted behavior in (5.2.9). Also the algebraic conditions can be generalized to a finite domain problem and variable coefficients.

The question of stability can be reduced to an estimate of the solution to a resolvent equation. Consider the resolvent equation

(5.3.12) $$(\underline{Q} - r\underline{I})\underline{z}_j = 0, \; j = 1,2,\ldots,$$

with $\|\underline{z}\|_2 < \infty$ and

(5.3.13) $$\underline{z}_i^n = \sum_{j=0}^{m} \underline{C}_{ij}\underline{z}_j^n + \underline{G}_i, \quad i = -m_1+1,\ldots,-1,0.$$

Equation (5.3.13) is an ordinary differential equation with constant coefficients and its most general solution can be written in the form (using continuity arguments)

(5.3.14) $$\underline{z}_i = \sum_{|\kappa_j|<1} \underline{P}_j(i)\kappa_j^i ,$$

where $\underline{P}_j$ are polynomials in i with vector coefficients and κ_j with $\|\kappa_j\| < 1$ are solutions of the characteristic equation

(5.3.15) $$\mathrm{Det}\Big|\sum_{j=-m_1}^{m_2} \underline{Q}_j\kappa_j - r\underline{I}\Big| = 0.$$

The coefficients of the polynomials $\underline{P}_j$ can be obtained by substituting (5.3.14) into the characteristic equation (5.3.15).

The following lemma is due to Gustafsson et al. [6] (see also Kreiss and Oliger [11]):

Lemma. For $|\Gamma| > 1$ the characteristic equation (5.3.15 has no solution κ_j with $\|\underline{\kappa}_j\| = 1$ and the number of roots κ_j such that $|\kappa_j| < 1$ is equal to the number of boundary conditions (5.3.13) (counted according to their multiplicity).

We now state the main theorem for the quarter plane problem with constant coefficients.

Theorem 5.2. The Ryabenki-Godunov condition* is satisfied if and only if the resolvent equation (5.3.12) and (5.3.13) has no eigenvalue Γ with $|\Gamma| > 1$. The finite difference method is stable in the sense of the above definition if and only if there is, in addition to the Ryabenki-Godunov condition, a constant $K > 0$ such that for all Γ, with $|\Gamma| > 1$, and all $\underline{G}_j$

$$(|\Gamma| - 1)\|\underline{z}\| < K \sum_{j=-m_1+1}^{0} |\underline{G}_j|^2.$$

This result can be generalized to the case with two boundaries, Gustafsson et al. [5]:

Theorem 5.3. Consider the finite difference method for $t > 0$ and $0 < x < 1$ and assume that the corresponding left and right quarter plane problems are stable in the sense of (5.3.11), then the original problem is stable.

Furthermore the equation $\partial_t v = \underline{A}\partial_x \underline{v} + \underline{B}\ \underline{v}$ can be treated as a perturbation of the original equation (5.3.1) and the stability of the finite difference method is given by the following theorem:

Theorem 5.4. Assume that the finite difference method (5.3.7) and (5.3.9) is stable in the sense of (5.3.11). If the approximation is perturbed by adding to the difference operators $\underline{Q}_j$ in (5.3.8) in terms of order k, then the resulting finite difference method is stable in the same sense.

Finally we consider the variable coefficient case with the equation is

$$\partial_t \underline{v} = \underline{A}(x)\partial_x \underline{v}\ ,$$

which is approximated by (5.3.10) where

---

*In this setting the Ryabenki-Godunov condition is satisfied if and only if the resolvent equations (5.3.12) and (5.3.13) with $\underline{G}_j = 0$ have no eigenvalues with $|\lambda| > 1$.

$$\underline{Q} = \sum_{j=-m_1}^{m_2} \underline{Q}_j(x) S_+^j ,$$

and where we assume that $\underline{Q}_j(x)$ is twice continuously differentiable. Under this and the assumptions made in the constant coefficient case the Lax-Nirenberg Theorem (Theorem 3.9) guarantees stability of the pure initial-value problem. Stability of the initial boundary-value problem will follow provided the left quarter plane problem with constant coefficient obtained by replacing $\underline{Q}_j(x)$ with $\underline{Q}_j(1)$ and the right quarter plane problem with constant coefficients obtained by replacing $\underline{Q}_j(x)$ with $\underline{Q}_j(0)$ are stable. See Kreiss [9].

We shall now present some finite difference methods for initial boundary-value problems for hyperbolic equations. Two types of methods will be considered: dissipative and nondissipative.

Consider the quarter plane problem (5.3.1) where without loss of generality we assume that $\underline{A}$ is diagonal and the boundary condition is of the form $\underline{v}^I = \underline{S}^I \underline{v}^{II}$, that is, $g(t) = 0$.

For the dissipative type method, consider the Lax-Wendroff method

(5.3.17) $$\underline{u}^{n+1} = \underline{u}^n + \underline{A} D_0 \underline{u}^n + \frac{k^2}{2} A^2 D_+ D_- \underline{u}^n,$$

along with the boundary condition

(5.3.18) $$\underline{u}_0^I = \underline{S}^I \underline{u}_0^{II}.$$

We shall consider ways of specifying $\underline{u}_0^{II}$. Three possible choices are:

(5.3.19) $$(hD_+)^j \underline{u}_0^{II} = 0, \quad j \text{ is a positive integer}$$

(5.3.20) $$(\underline{u}_0^{II})^{n+1} = (\underline{u}_0^{II})^n + k\underline{A}^{II} D_+ (\underline{u}_0^{II})^n,$$

where $\underline{A}^{II} = \underline{\Lambda}^{II}$ in (5.3.3) since $\underline{A}$ is diagonal, and

(5.3.21) $$(\underline{u}_0^{II} + \underline{u}_1^{II})^{n+1} - k\underline{A}^{II} D_+ (\underline{u}_0^{II})^{n+1} = (\underline{u}_0^{II} + \underline{u}_1^{II})^n - k\underline{A}^{II} D_+ (\underline{u}_0^{II})^n.$$

It is shown in Gustafsson et al [5] that the Lax-Wendroff method (3.5.3) is stable in the sense of (5.3.11) with the boundary condition (5.3.18) approximated by any of the three choices (5.3.19)-(5.3.21).

For the nondissipative type method, consider the leap-frog method (3.10.1)

(5.3.22) $$\underline{u}^{n+1} - \underline{u}^{n-1} = 2k\underline{A} D_0 \underline{u}^n.$$

It is shown in Gustafsson et al [12] that the leap-frog method (5.3.22) is stable in the sense of (5.3.11) with the boundary condition (5.3.18)

approximated by (5.3.20) or (5.3.21) but is not stable with the boundary condition (5.3.18) approximated by (5.3.19). A direct verification that the boundary condition approximated by (5.3.19) results in an unstable method is given in Kreiss and Oliger [11].

Here we see that the stability of the approximation to an initial boundary-value problem depends on the finite difference method and for the pure initial-value problem.

We shall now present a useful result of Goldberg and Tadmor [2]. This gives a sufficient stability condition for explicit dissipative difference approximations to initial boundary-value problems in the quarter plane. This condition is independent of the finite difference method is used to solve the pure initial-value problem and is given in terms of the out flow boundary conditions.

Consider the quarter plane problem (5.3.1) where, without loss of generality, $\underline{A}$ is assumed to be diagonal with boundary conditions given by (5.3.5). Approximate this equation by (5.3.7) where $\underline{Q}$ defined by (5.3.8) is dissipative of order $2s$.

From the way in which the difference operator $\underline{Q}$ is defined in (5.3.8), at each time step m_1 boundary values of $\underline{u}_i^n$ for $i = -m_1 + 1,\dots,-1,0$ are required. For the outflow variables the boundary condition is approximated by

(5.3.23) $$\sum_{j=0}^{q} \underline{C}_{ij}^{(1)} (\underline{u}_{i+j}^{II})^{n+1} = \sum_{j=0}^{q} \underline{C}_{ij}^{(0)} (\underline{u}_{i+j}^{II})^{n}, \quad i = -m_1+1,\dots,-1,0$$

where $\underline{C}_{ij}^{(1)}$ and $\underline{C}_{ij}^{(0)}$ are constant diagonal matrices of order $(p-r) \times (p-r)$. For the inflow variables the physical boundary condition is

(5.3.24) $$(\underline{u}_0^{I})^n = \underline{S}^{I}(\underline{u}_0^{II})^n + \underline{g}(t),$$

with the m_1-1 additional conditions of the form

(5.3.25) $$(\underline{u}_i^{I})^n = \sum_{j=0}^{m} \underline{D}_{ij}(\underline{u}_{i+j}^{II})^n + \underline{g} \ , \quad i = -m+1,\dots,-1,0$$

where $\underline{D}_{ij}$ are constant $r \times (p-r)$ matrices and $\underline{g}_i$ are vectors depending on h and $g(t)$.

The following theorem simplifies the stability conditions, where stability is in the sense of (5.3.11).

<u>Theorem 5.5</u>. The finite difference approximation (5.3.7), (5.3.8), (5.3.23)-(5.3.25) is stable if and only if the $p-r$ scalar components of the outflow variable are stable.

<u>Proof</u>. Since the matrices $\underline{Q}_j$ in (5.3.8) are diagonal divide (5.3.7) in terms of the inflow and outflows variables,

(5.3.26) $$(\underline{u}_i^I)^{n+1} = \underline{Q}^I(\underline{u}_i^I)^n$$

and

(5.3.27) $$(\underline{u}_i^{II}) = \underline{Q}^{II}(\underline{u}_i^{II})^n,$$

where

$$\underline{Q}^I = \sum_{j=-m_1}^{m_2} \underline{Q}_j^I S_+^j,$$

$$\underline{Q}^{II} = \sum_{j=-m_1}^{m_2} \underline{Q}_j^{II} S_+^j,$$

and

$$\underline{Q}_j = \begin{matrix} \underline{Q}^I & \underline{0} \\ \underline{0} & \underline{Q}^{II} \end{matrix} .$$

As a result of (5.3.27) and the boundary condition (5.3.23), the outflow problem is self-contained. However, the inflow problem given by (5.3.26) with boundary conditions (5.3.24) and (5.3.25) depends on the outflow variables as inhomogeneous boundary values in the boundary conditions (5.3.24) and (5.3.25). The stability of the original finite difference method is equivalent to the stability of two separate problems:

1) the inflow problem (5.3.26) with inhomogeneous boundary values (5.3.24) and (5.3.25)

2) the outflow problem (5.3.27) with boundary conditions (5.3.23).

The estimate (5.3.22) as the definition of stability gives bounds for inhomogeneous boundary values. Hence it suffices to consider the homogeneous boundary values for the inflow problems. Since $\underline{Q}_j^I$ are diagonal, the inflow problem (5.3.26) separates into r independent dissipative approximations with homogeneous boundary conditions. These were shown by Kreiss [7] to the stable independent of the finite difference method for the pure initial-value problem.

Now consider the outflow problem (5.3.27). Similarly, since $\underline{Q}_j^{II}$, $\underline{C}_{ij}$, and $\underline{C}'_{ij}$ are diagonal, the outflow problem separates into $p-r$ independent approximations.

Thus the problem has been reduced to the following scalar problem

(5.3.28) $$v_t = av_x, \quad t > 0, \quad x > 0$$
$$v(x,0) = f(x)$$

where $a > 0$. The finite difference method (5.3.7) and (5.3.8) takes the form

(5.3.29) $$u^{n+1} = Qu^n ,$$
$$u_i^0 = f(ih), \quad i = 0,1,\ldots,$$

where

where

$$Q = \sum_{j=-m_1}^{m_2} q_j S_+^j .$$

The boundary conditions (5.3.23) reduce to

(5.3.30) $$\sum_{j=0}^{q} c_{ij}^{(1)} u_{i+j}^{n+1} = \sum_{j=0}^{q} c_{ij}^{(0)} u_{i+j}^{n}, \; i = -m_1+1,\ldots,-1,0$$

where $c_{ij}^{(1)}$ and $c_{ij}^{(0)}$ are constants.

A special case of the boundary condition (5.3.31) called <u>translatory</u> is when constants c_{ij} and c'_{ij} are independent of i, that is,

(5.3.31) $$\sum_{j=0}^{q} c_j u_{i+j}^{n+1} = \sum_{j=0}^{q} c'_j u_{i+j}^{n}, \; i = -m_1+1,\ldots,-1,0.$$

In this case, define the boundary method $T^{(\alpha)}$, $\alpha = 0,1$ by

$$T^{(1)} u_i^{n+1} = T^{(0)} u_i^n, \; i = 0,\pm1,\pm2,\ldots$$

(5.3.32) $$T^{(\alpha)} = \sum_{j=0}^{q} c_j^{(\alpha)} S_+^j, \; \alpha = 0,1 .$$

This boundary method generates the boundary condition (5.3.31) when i is restricted to $-m_1+1,\ldots,-1,0$.

The main result of Goldberg and Tadmer [2] provides an easily verifiable condition that is sufficient for stability, which depends entirely on the boundary conditions.

<u>Theorem 5.6</u>. The finite difference approximation (5.3.29) and (5.3.31) is stable if the boundary method (5.3.32) is stable and if

(5.3.33) $$T^{(1)}(\kappa) \equiv \sum_{j=0}^{q} C_j^{(1)} \kappa^j \neq 0 \quad \text{for every} \quad |\kappa| \leq 1.$$

The stability condition is independent of the finite difference method for the pure initial-value problem. In addition, if the boundary method is known to be stable, then stability of the entire finite difference approximation depends on the condition (5.3.33).

In the case of explicit boundary conditions of the form

(5.3.34) $$u_j^{n+1} = \sum_{j=0}^{q} C_j u_j^n, \; i = -m_1+1,\ldots,-1,0,$$

then $T^{(1)}(\kappa) \equiv 1$ and Theorem 5.6 is automatically satisfied.

Consider the boundary approximation given by the explicit method

$$u_i^{n+1} = u_i^n + a(u_{i+1}^n - u_i^n), \; i = -m_1+1,\ldots,-1,0$$

where $\lambda = k/h$. This method is stable and satisfies the CFL condition provided $\lambda \leq 1/a$. Since this approximation is stable and of the form

(5.3.34), we see that the approximation to the initial value problem with the above boundary approximation is stable.

Next consider the boundary approximation given by the implicit method

$$u_i^{n+1} - a(u_{i+1}^{n+1} - u_i^{n+1}) = u_i^n, \quad i = -m_1+1,\dots,-1,0.$$

This method is unconditionally stable. Since this approximation is not of the form (5.3.34), $T^{(1)}(\kappa)$ must be evaluated,

$$T^{(1)}(\kappa) = 1 - \lambda a(\kappa-1) = 1 + \lambda a(1-\kappa).$$

Since $\lambda a > 0$, for $|\kappa| < 1$

$$\mathrm{Re}\, T^{(1)}(\kappa) = 1 + \lambda a(1 - \mathrm{Re}\,\kappa) > 1.$$

From which the stability of the approximation to the pure initial-value problem with the above boundary approximation is stable.

V.4. References

1. Godunov, S. K. and V. S. Ryabenki, The Theory of Difference Schemes - An Introduction, North Holland Publishing Co., Amsterdam (1964).

2. Goldberg, M., "On a Boundary Extrapolation Theorem by Kreiss," Math. Comp. 31, 469 (1977).

3. ___________ and E. Tadmor, "Scheme-Independent Stability Criteria for Difference Approximations of Hyperbolic Initial-Boundary Value Problems I," Math. Comp., 32, 1097 (1978).

4. Gustafsson, B., "On the Convergence Rate for Difference Approximations to Mixed Initial Boundary Value Problems," Report #33, Department of Computer Science, Uppsala University, Uppsalay Sweden (1971).

5. ___________, H. O. Kreiss, and A. Sundstrom, "Stability Theory of Difference Approximations for Mixed Initial Boundary-Value Problems, II," Math. Comp. 26, 649 (1972).

6. Kreiss, H. O., "Stability Theory for Difference Approximations of Mixed Initial Bounday-Value Problems, I, "Math. Comp. 22, 703 (1968).

7. ___________, "Initial Boundary Value Problems for Hyperbolic Systems," Comm. Pure Appl. Math., 23, 277 (1970).

8. ___________, "Difference Approximations for Initial Boundary-Value Problems," Proc. Roy. Soc. Lond. A., 323, 255 (1971).

9. ___________, "Initial Boundary-Value Problems for Hyperbolic Partial Differential Equations," Proceedings of the Fourth International Conference on Numerical Methods in Fluid Dynamics, Ed. R. D. Richtmeyer, vol.35, Lecture Notes in Physics, Springer-Verlag, Berlin (1974).

). Kreiss, H. O. and E. Lundguist, "On Difference Approximations with Wrong Boundary Values," Math. Comp., 22, 1 (1968).

11. ________ and J. Oliger, Methods for the Approximate Solutions of Time Dependent Problems, GARP Publication Series No.10, 1973.

12. Osher, S., "Stability of Difference Approximations of Dissipative Type for Mixed Initial Boundary-Value Problems, I," Math. Comp., 23, B35 (1969).

13. ________, "On Systems of Difference Equations with Wrong Boundary Values," Math. Comp., 23, 567 (1969).

14. ________, "Initial Boundary-Value Problems for Hyperbolic Equations in Regions with Corners, I," Trans. Amer. Math. Soc., 176, 141 (1973).

15. Kreiss, H. O., "Stability of Parabolic Difference Approximation to Certain Mixed Limited Boundary-Value Problems," Math. Comp., 26, 13 (1972).

16. Richtmyer, R. D., The Stability Criterion of Godunov and Ryabenki for Difference Schemes, AEC Computing and Applied Mathematics Center Report NYO-1480-4, Courant Institute of Mathematical Sciences, New York University, New York (1964).

17. Varah, J. M., "Maximum Norm Stability of Difference Approximations to the Mixed Initial Boundary-Value Problem for the Heat Equation," Math. Comp., 24, 31 (1970).

Appendix A - The Kreiss Matrix Theorem and Its Consequences

In the vector case we saw that something more than the von Neumann condition is required for stability. Some of the additional conditions sufficient for stability were considered in Section II.7. In this appendix some of the more technical conditions sufficient for stability are considered.

Consider the finite difference method

$$(A.1) \qquad Q_1\underline{u}^n = Q_2\underline{u}^{n-1} \ , \ n > 1$$

where $\underline{u} = (u_1,\ldots,u_p)^T$, Q_1 and Q_2 are $p \times p$ matrices of the form

$$Q_1 = \sum_{j=-m_3}^{m_4} Q_{1j}S_+^j \ ,$$

$$Q_2 = \sum_{j=-m_1}^{m_2} Q_{2j}S_+^j \ ,$$

with the Q_{1j}'s and the Q_{2j}'s $p \times p$ constant matrices.

Apply the discrete Fourier transform to (A.1),

$$(A.2) \qquad \underline{u}^n(\xi) = \underline{G}(\xi)\underline{u}^{n-1}(\xi), \ n > 1,$$

where $\underline{G}(\xi)$ is the amplification matrix

$$\underline{G}(\xi) = \left(\sum_{j=-m_3}^{m_4} Q_{1j}e^{-ij\xi}\right)^{-1} \left(\sum_{j=-m_1}^{m_2} Q_{2j}e^{-ij\xi}\right) .$$

By repeated applications of (A.2),

$$\hat{\underline{u}}^n(\xi) = \underline{G}^n(\xi)\hat{\underline{u}}^0(\xi) .$$

From Theorem 2.1, we see that stability of the finite difference method is equivalent to $\underline{G}(\xi)$ being uniformly bounded for all powers of n for $nk < 1$. An equivalent form of this condition is that there exists a constant $\alpha > 0$ such that the matrices $\underline{A} = e^{-\alpha k}\,\underline{G}$ satisfy the condition

$$\|A^n\| < K$$

for all positive integers n, where K is a constant independent of ξ.

Definition. Let F be a family of matrices depending on a parameter ξ. The family F is called stable (A) if for every $A \in F$ and all $n > 0$,

$$\|\underline{A}^n\| < K ,$$

where K is a constant independent of $\underline{A}$.

A necessary condition for statement (A) to hold is that all eigenvalues κ_i of $\underline{A}$ lie inside or on the unit circle. Hence it is necessary for the eigenvalues λ_i of $\underline{G}$ to satsify $|\lambda_i| < e^{\alpha k}$ for some constant α. This is equivalent to the von Neuamnn condition. We shall now state the Kreiss matrix theorem [4].

Theorem A.1. Let F be a family of square complex matrices $\underline{A}$ of order p. The following are equivalent to the family F being stable.

(R) There exists a constant C_R such that for all $\underline{A} \in F$ and all complex numbers z with $|z| > 1$, $(\underline{A}-z\underline{I})^{-1}$ exists and

$$|(z\underline{I}-\underline{A})^{-1}| < C_R(|z|-1)^{-1} \tag{A.3}$$

(S) There exists constants C_S and C_B and to each $\underline{A} \in F$ a non-singular matrix S such that

(i) $\max(\|S\|,\|S^{-1}\|) < C_S$

and

(ii) $B = \underline{S}\,\underline{A}\,\underline{S}^{-1}$ is upper triangular and its off-diagonal elements satisfy

$$|b_{ij}| < C_B(|-|\kappa_j|) ,$$

when κ_j are the diagonal elements of B_j, i.e., the eigenvalues of A.

(H) There is a constant $C_H > 0$ and for each $A \in F$ a positive definite hermitian matrix $\underline{H}$ such that

(i) $C_H^{-1}\underline{I} < \underline{H} < C_H\underline{I}$,

(ii) $\underline{A}^*\underline{H}\,\underline{A} < H$.

Remark. (R) The resolvent condition is used in theoretical proofs. This contains the von Neumann condition. It tells you that as you get far from 1 in the complex plane the blow up cannot be too much.

(S) It is observed that the larger the diagonal elements are the smaller the off-diagonal term must be. This is called the algebraic (or off-diagonal) condition.

(H) Two norms $\|\ \|_a$ and $\|\ \|_b$ are called equivalent if and only if there exists a constant $C_H > 0$ such that

$$\frac{1}{\sqrt{C_H}}\|\underline{v}\|_a < \|\underline{v}\|_b < \sqrt{C_H}\,\|\underline{v}\|_a .$$

(H) defines a norm. Condition (Hi) states that the new norm is equivalent to the ℓ_2-norm. Condition (Hii) implies that if we have stability in the new norm, we have stability in the old norm

(<u>equivalent norm condition</u>).

We shall present a proof that stability is equivalent to the resolvent condition (R). For a complete proof see Morton and Scheckler [6]. The most difficult part is to show that the resolvent condition implies stability.

The proof used here is a modification of a theorem due to Laptev [5] first found by Gilbert Strang and pointed out to the author by Gideon Zwas.

<u>Proof</u>: Stability $\Longrightarrow$ (R). Due to stability all of the eigenvalues κ_i of $\underline{A} \in F$ lie inside the closed unit disc. Therefore, for $|z| > 1$, $(z\underline{I}-\underline{A})^{-1}$ exists. By expanding in Taylor series

$$\|(z\underline{I}-\underline{A})^{-1}\| = \|\sum_{n=0}^{\infty} \underline{A}^n z^{-(n+1)}\|$$
$$< \sum_{n=0}^{\infty} \|\underline{A}^n\| |z^{-(n+1)}|$$
$$< C_R \sum_{n=0}^{\infty} |z^{-(n+1)}|,$$

where C_R is a constant independent of $\underline{A}$, using conditions (A).

$$= \frac{C_R}{|z|-1} .$$

Hence the resolvent condition is proved.

In order to show that the resolvent condition implies stability we make use of a contour formula in Theorem 10 of Dunford and Schwartz [2] to write

$$\underline{A}^n = \frac{1}{2\pi i} \int_\Gamma z^n (z\underline{I}-\underline{A})^{-1}\, dz \, , \tag{A.4}$$

where the eigenvalues of $\underline{A}$ lie inside or on the unit circle. Let Γ be a circle of radius $r > 1$ centered at the origin, so that $z = re^{i\theta}$ and (A.4) becomes

$$\underline{A}^n = \frac{r^{n+1}}{2\pi} \int_0^{2\pi} e^{i(n+1)\theta} (re^{i\theta}\underline{I}-\underline{A})^{-1}\, d\theta \, . \tag{A.5}$$

Let $\underline{B} = (re^{i\theta}\underline{I}-\underline{A})^{-1}$. Using Cramer's rule, consider a typical element of B,

$$B_{\ell m} = \frac{\sum_{j=0}^{p-1} u_j e^{ij\theta}}{\sum_{j=0}^{p} v_j e^{ij\theta}} \, , \tag{A.6}$$

where the coefficients u_j and v_j are independent of θ. Multiply (A.6) by the complex conjugate of the denominator gives $Re(B_{\ell m})$ and $Im(B_{\ell m})$ as rational trigonometric functions with a nonvanishing denominator and numerator of degree $2p-1$. The denominator is non-vanishing since $\rho(\underline{A}) < 1$ and $|z| > 1$ so that $\det(z\underline{I}-\underline{A}) \neq 0$. Therefore,

$$\phi(\theta) \equiv \mathrm{Re}(B_{\ell m}) = \frac{\sum_{j=0}^{2p-1} a_j \cos j\theta + b_j \sin j\theta}{\sum_{j=0}^{2p} c_j \cos j\theta + d_j \sin j\theta}$$

and

$$\psi(\theta) \equiv \mathrm{Im}(B_{\ell m}) = \frac{\sum_{j=0}^{2p-1} \alpha_j \cos j\theta + \beta_j \sin j\theta}{\sum_{j=0}^{2p} c_j \cos j\theta + d_j \sin j\theta} .$$

Both ϕ and ψ do not contain singularities on $[0,2\pi]$, since the denominator does not vanish. Using the property $\cos n\theta \cos m\theta = \frac{1}{2}(\cos(m+n)\theta + \cos(m-n)\theta)$, we see by direct calculation that $\phi'(\theta)$ and $\psi'(0)$ are real rational trigonometric functions with numerators of degree $4p-1$. Since a trigonometric polynomial of degree $q(= 4p-1)$ has no more than $2q(8p-2)$ roots in $[0,2\pi]$, $\phi(\theta)$ and $\psi(\theta)$ has less than 8p regions of monotonicity in $[0,2\pi]$. Using the second mean value theorem of integral calculus on a typical interval $[\theta_1, \theta_2]$ where ϕ is monotone

$$\int_{\theta_1}^{\theta_2} [\cos(n+1)\theta + \sin(n+1)\theta]\phi(\theta)d\theta$$

$$= \int_{\theta_1}^{\theta_2} \cos(n+1)\theta\phi(\theta)d\theta + \int_{\theta_1}^{\theta_2} \sin(n+1)\theta\phi(\theta)d\theta = I_1 + I_2$$

$$= \phi(\theta_2)\left[\frac{(\sin(n+1)\theta_2 - \cos(n+1)\theta_2) - (\sin(n+1)\xi - \cos(n+1)\xi)}{n+1}\right]$$

$$+ \phi(\theta_1)\left[\frac{(\sin(n+1)\xi - \cos(n+1)\xi) - (\sin(n+1)\theta_1 - \cos(n+1)\theta_1)}{n+1}\right]$$

where $\xi \in (\theta_1, \theta_2)$. For $\psi(\theta)$ over a typical interval $[\bar{\theta}_1, \bar{\theta}_2]$ of monotonicity, similar integrals I_3 and I_4 are obtained. The integrals I_1, I_2, I_3, and I_4 satisfy

$$|I_1| , |I_2| < \frac{4}{n+1} \max_{[\theta_1, \theta_2]} |\phi(\theta)|$$

$$|I_3| , |I_4| < \frac{4}{n+1} \max_{[\bar{\theta}_1, \bar{\theta}_2]} |\psi(\theta)| .$$

By substitution of these inequalities into (A.5),

$$|(\underline{A}^n)_\ell| \leq |\mathrm{Re}(\underline{A}^n)_{\ell m}| + |\mathrm{Im}(\underline{A}^n)_{\ell m}|$$

$$\leq \frac{r^{n+1}}{2\pi} \sum_{[\theta_j,\theta_{j+1}]} |I_1| + |I_2| + \sum_{[\bar\theta_j,\bar\theta_{j+1}]} |I_3| + |I_4|$$

$$\leq \frac{r^{n+1}}{2\pi} \cdot \frac{4}{n+1} \quad 2 \sum_{[\theta_j,\theta_{j+1}]} \max|\phi| + 2 \sum_{[\bar\theta_j,\bar\theta_{j+1}]} \max|\psi| ,$$

where the sums are over all regions of monotonicity. Since each sum has less than $8p$ terms,

$$|(\underline{A}^n)_{\ell m}| < \frac{r^{n+1}}{2\pi}\frac{4}{n+1}(4\cdot 8p)\max_\theta |B_{\ell m}|$$

$$= \frac{64r^{n+1}p}{\pi(n+1)} \max_\theta |B_\ell| \leq \frac{64r^{n+1}p}{\pi(n+1} \cdot \frac{C_R}{r-1} ,$$

where the resolvent condition has been used. Choose $r = 1 + \frac{1}{n+1}$, then

$$|(\underline{A}^n)_{\ell m}| \leq \frac{64epC_R}{\pi} .$$

Finally, since $\|M\|_2 \leq \sqrt{\sum_{i,j} |M_{ij}|^2} \leq p \operatorname{Max}_{i,j} |M_{ij}|$,

$$\|\underline{A}^n\| \leq \frac{64eC_Rp^2}{\pi} ,$$

for all $n > 0$ and $\underline{A} \in F$. This gives an extremely good estimate of $\|A^n\|$.

We shall now prove Theroem 3.2 iv).

<u>Definition</u>. A sequence of complex numbers $c_1,\dots,c_p$ is said to be <u>nested</u> with <u>nesting constant</u> K if and only if

$$|c_r-c_s| \leq K|c_\ell-c_m| \quad \text{for} \quad \ell \leq r \leq s \leq m ,$$

see Morton and Schechter [6].

It can be seen that any set of p numbers can be ordered so that it is nested with $K \leq 2^p$. To see this, taking any one as the first and choosing the number closest to it as the second, the number (in the remaining $p-2$ numbers) closest to the second as the third, and so on,

$$|c_{i-1}-c_i| \leq |c_{i-1}-c_j|$$

for $j \geq i$. Thus for $\ell \leq r \leq s \leq m$

$$|c_r-c_s| \leq |c_r-c_m| + |c_s-c_m|$$

and

$$|c_r-c_m| \leq |c_{r-1}-c_r| + |c_{r-1}-c_m| , \quad \text{if} \quad r > 1$$

$$\leq 2|c_{r-1}-c_m)| \leq \dots \leq 2^{r-\ell}|c_\ell-c_m| .$$

Similarly, $|c_s - c_m| < 2^{s-\ell}|c_\ell - c_m|$. Thus

$$|c_r - c_s| < (2^{r-\ell} + 2^{s-\ell})|c_r - c_m| < 2P|c_s - c_m| .$$

The following theorem is due to Buchanan [1].

Theorem A.2. Let F be a family of matrices $\underline{A}$ which are in upper triangular form with their eigenvalues nested with constant K. Then F is stable if and only if the von Neumann condition $|\kappa_i| < 1$, on the eigenvalues is satisfied and the off-diagonal elements satisfy

$$|A_{ij}| < \overline{K} \max(1 - |\kappa_i|,\ 1 - |\kappa_j|,\ \kappa_i - \kappa_j|) , \tag{A.8}$$

when $\overline{K}$ is a constant independent of $\underline{A}$.

The following proposition gives sufficient conditions for stability in special cases.

Theorem 2.3.iv). Consider the finite difference method (A.1) whose amplification matrix $\underline{G}(\xi)$ satisfies the von Neumann condition. If all elements of $\underline{G}(\xi)$ are bounded, all but one of the eigenvalues of $\underline{G}(\xi)$ lie in a circle inside of the unit circle; that is, all but one eigenvalues λ_i are such that $|\lambda_i| < r < 1$ (where r is the radius of the circle inside the unit circle), then the finite difference method is stable.

Proof: $\underline{G}$ can be put into nested triangular form by means of a unitary transformation. Let M denote the bound on the element of $\underline{G}$. Then the elements of the transformed matrix are still bounded by M. Without impairing the nesting, take the eigenvalue which does not lie inside a disc contained in the unit disc to be λ_1 (the first diagonal element). Hence, the inequality (A.8) in the Buchanan theorem is satisfied with $\bar{K} < \frac{M}{1-r}$. The von Neumann condition is simply a restriction on the exceptional eigenvalues. λ_1, namely $|\lambda_1| < 1 + O(k)$. Stability follows from the Buchanan theorem.

REFERENCES

1. Buchanan, M. L., A Necessary and Sufficient Condition for Stability of Difference Schemes for Initial Value Problems, SIAM J., 11, 919 (1963).

2. Dunford, N. and J. T. Schwartz, Linear Operators, Part I, Interscience, New York (1957).

3. Gershgorin, S., Über die Abgrenzung der Eigenwerle einer Matrix, Izvestiya Akad, Nauk, USSR, Zer. Math., 7, 749 (1931).

4. Kreiss, H. O., Über die Stabilitatsdefinition für Differenzieg-leichungen die partielle Differentialgleichungen approximeren, Nordisk Tidskr Informations-Behandling, 2, 153 (1962).

5. Lapfev, G. I., Conditions for the Uniform Well-Posedness of the Cauchy Problem for Systems of Equations, Soviet Math. Dokl., 16 65 (1975).

6. Morton, K. W. and S. Scheckter, On the Stability of Finite Difference Matrices, SIAM J. Num. Anal., Ser. B. 2, 119 (1965).

Appendix B. Kreiss's Stability Theorems for Dissipative Methods

Consider the first-order hyperbolic system

(B.1) $$\partial_t \underline{v} = \underline{A}\partial_x \underline{v} \ , \quad -\infty < x < +\infty \ , \ t > 0 \ ,$$

where $\underline{v} = (v_1, \ldots, v_p)^T$ and $\underline{A}$ is a $p \times p$ constant matrix. Approximate equation (B.1) by the finite difference method

(B.2) $$\underline{u}_i^{n+1} = Q \, \underline{u}_i^n \ ,$$

where

(B.3) $$Q = \sum_{j=-m_1}^{m_2} \underline{A}_j S_+^j \ ,$$

and $\underline{A}_j$ are $p \times p$ matrices that depend on h and k only. The amplification matrix $\underline{G}(\xi)$ of the finite difference method (B.2) is

(B.4) $$\underline{G}(\xi) = \sum_{j=-m_1}^{m_2} \underline{A}_j e^{-ij\xi} \ .$$

Definition. A finite difference method (B.2) is dissipative of order 2s, where s is a positive integer, if there exists a constant $\delta > 0$ such that

$$|\lambda_j(\xi)| \leq 1 - \delta |\xi|^{2s}$$

with $|\xi| \leq \pi$, where $\lambda_j(\xi)$ are the eigenvalues of the amplication matrix $\underline{G}(\xi)$ on (B.2).

The importance of the idea of dissipation can be seen from the following theorem due to Kreiss [3] and simplified by Parlett [6]. The proof presented here is due to Parlett.

Theorem B.1. Suppose the finite difference method (B.2) approximating equation (B.1) is

i) accurate of order $2s - 1$, and

ii) dissipative of order $2s$,

then it is stable.

Proof. In order to prove this we shall use a lemma due to Parlett [6].

Lemma 1. Let $\underline{L}$, $\underline{D}$, $\underline{U}$ denote strictly lower triangular diagonal and upper triangular matrices, respectively. If $\underline{L} + \underline{D} + \underline{U}$ is a normal $p \times p$ matrix, then

$$\|\underline{U}\| \leq C\|\underline{L}\| \ ,$$

where C is a constant that depends on p.

Proof. The Frobenius norm

$$\|\underline{A}\|^2 = \sum_{i,j=1}^{p} |a_{ij}|^2$$

shall be used. Let the Euclidean length of the i-th column of $\underline{L}$ be ℓ_i, of the i-th row of $\underline{U}$ be u_i.

Since the corresponding rows and columns of a normal matrix have equal Euclidean length

$$u_j^2 + |\ell_{j1}|^2 + \dots + |\ell_{j,j-1}|^2 = \ell_j^2 + |u_{ij}|^2 + \dots + |u_{j-1,j}|^2$$

for $j = 1,\dots,p-1$. Summing this result for $j = 1,\dots,k$ and omitting the $|\ell_{ij}|^2$ on the left-hand side,

$$\sum_{j=1}^{k} u_j^2 < \sum_{j=1}^{k} \ell_j^2 + \sum_{j=1}^{k} \sum_{m=1}^{j-1} u_{mj}^2 .$$

Summing the second term on the right by rows,

$$\sum_{j=1}^{k} u_j^2 < \sum_{j=1}^{k} \ell_j^2 + \sum_{j=1}^{k-1} \sum_{m=j+1}^{k} u_{jm}^2 .$$

for $k = 1,\dots,p-1$. Using this recursively for $k = p-1, p-2,\dots,1$

$$\sum_{j=1}^{p-1} u_j^2 < \sum_{k=1}^{p-1} \sum_{j=1}^{k} \ell_j^2 < (p-1) \sum_{j=1}^{p-1} \ell_j^2 .$$

Hence, $\|\underline{U}\| < \sqrt{p-1}\, \|\underline{L}\|$.

In Theorem 3.4, if (B.2) is a (2s - 1)-th order accurate approximation to (B.1), then the amplification matrix $\underline{G}(\xi)$ of (B.2) can be written as

$$\underline{G}(\xi) = e^{-i\lambda\underline{A}\xi} + O(|\xi|^{2s}) , \tag{B.5}$$

where $\lambda = k/h$, or

$$\underline{G}(\xi) = e^{-i\lambda\underline{A}\xi} + \underline{S}_{2s} ,$$

where $\|\underline{S}_{2s}\| < C|\xi|^{2s}$. Using the definition of hyperbolicity, there exists a nonsingular matrix $\underline{T}(\xi)$ with $\|\underline{T}\|$, $\|\underline{T}^{-1}\| < K_1$ and such that $\underline{TAT}^{-1} = \underline{D}$, where $\underline{D}$ is a diagonal matrix with real elements. Hence,

$$\underline{G}'(\xi) = \underline{TGT}^{-1} = e^{-i\lambda\underline{D}\xi} + \underline{S}'_{2r} ,$$

where $\|\underline{S}'_{2r}\| < C'|\xi|^{2r}$. Let $\underline{U}(\xi)$ be a unitary matrix that triangulates G', that is,

$$\underline{U}^*\underline{G}'\underline{U} = \underline{U}^* e^{-i\lambda\underline{D}\xi}\, \underline{U} + \underline{U}^*\underline{S}'_{2r}\, \underline{U} = \underline{D}' + \underline{N} ,$$

where $\underline{D}'$ is diagonal and $\underline{D}' + \underline{N}$ is upper-triangular.

Clearly the strictly lower-triangular elements of $\underline{D}'$ and $\underline{N}$ must cancel and hence, have the same norm. However, $\underline{U}^* e^{-\bar{i}\lambda \underline{D}\xi}\underline{U}$ is unitary and so by the above lemma its upper-triangular part has norm less than a constant times the norm of its lower-triangular part. The norm of the upper triangular part of $\underline{U}^* e^{-i\lambda \underline{D}\xi}\underline{U}$ and hence of $\underline{U}^*\underline{G}'\underline{U}$ is $O(|\xi|^{2r})$.

For the upper triangular matrix $\underline{D}' + \underline{N}$ a positive definite matrix $\hat{\underline{H}}$ will be constructed so that condition (H) of the Kreiss matrix (Theorem A.1) is satisfied. For any diagonal matrix $\underline{D}$, $\underline{D}(\underline{D}' + \underline{N})\underline{D}^{-1} = \underline{D}' + \underline{DND}^{-1}$, we have only to choose $\underline{D}$ so that $\|DND^{-1}\|$ is sufficiently small.

Following the argument of Kreiss [3], take $\underline{D} = \mathrm{diag}(d, d^2, .., d^n)$ with $d > 1$ and define $\underline{N}_D = \underline{DND}^{-1}$, then

$$\|\underline{N}_D\| = \|\underline{DND}^{-1}\| \leq d^{-1}\|N\| \leq \tilde{K}d^{-1}|\xi|^{2r} ,$$

where K is some constant. Choose d such that $Kd^{-1} \leq \frac{\delta}{3}$ (where δ is from the definition of dissipation). By condition ii), since $\underline{D}'$ is diagonal with the eigenvalues of $\underline{G}'$, and hence of $\underline{G}$, $\|\underline{D}'^*\underline{D}'\| \leq 1 - 2\delta|\xi|^{2r}$ and

$$\begin{aligned}\|(\underline{D}' + \underline{N}_D)^*(\underline{D}' + \underline{N}_D)\| &\leq \|\underline{D}'^*\underline{D}'\| + 2\|\underline{N}_D\| + \|\underline{N}_D\|^2 \\ &\leq 1 - \delta|\xi|^{2r} , \quad |\xi| \leq \pi .\end{aligned}$$

Hence,

$$(\underline{D}' + \underline{N})^*\underline{D}^2(\underline{D}' + \underline{N}) \leq (1 - \delta|\xi|^{2r})\underline{D}^2 , \quad |\xi| \leq \pi ,$$

and

$$\underline{G}^*\hat{\underline{H}}\underline{G} \leq (1 - \delta|\xi|^{2r})\hat{\underline{H}} \leq \hat{\underline{H}} , \quad |\xi| \leq \pi ,$$

with

$$\hat{\underline{H}} = \underline{T}^*\underline{UD}^2\underline{U}^*\underline{T} .$$

Furthermore,

$$\frac{d^2}{\tilde{K}} \underline{I} \leq \hat{\underline{H}} \leq \tilde{K}d^{2n}\underline{I} .$$

Therefore the two norms are equivalent and by condition (H) of the Kreiss matrix theorem (Theorem A.1) the finite difference method (B.2) is stable.

A finite difference method may be dissipative only to an even order, so the above theorem is of use only for schemes which are accurate of odd order, that is, 1, 3, ... We now consider a theorem for finite difference methods that are accurate to an even order (see Kreiss [3]).

Theorem B.2. Let the finite difference method (B.2) be dissipative of order 2r and assume that for $|\xi| \le \pi$ the amplification matrix $\underline{G}(\xi)$ (B.4) can be written in the form

$$\underline{G}(\xi) = \underline{I} + \sum_{\nu=1}^{2r-2} \frac{1}{\nu!}(-i\lambda\underline{A}\xi)^{\nu} + \underline{G}_{2r}(\xi) ,$$

where $\|\underline{G}_{2r}(\xi)\| = O(|\xi|^{2r})$ and $\lambda = \frac{k}{h}$; then it is stable.

The proof of this theorem is a direct consequence of two lemmas.

Lemma 2. Let $\underline{G}(\xi)$ be the amplification matrix of (B.2). Then (B.2) is stable if there exists a constant C such that for all ξ with $|\xi| \le \pi$ and all complex s with $|s| < 1$, the matrix $(s\underline{G}(\xi) - \underline{I})^{-1}$ exists and

(B.6) $$|(s\underline{G}(\xi) - \underline{I})^{-1}| \le C(1 - |s|)^{-1} .$$

Proof. This follows immediately from the resolvent condition (R) of the Kreiss matrix theorem (Theorem A.1).

Lemma 3. Let (B.2) be dissipative of order 2r where r is a positive integer. Then (B.2) is stable if

i) $\underline{G}(\xi)$ can be written in the form

(B.7) $$\underline{G}(\xi) = \underline{I} + \underline{G}_1(\xi) + \underline{G}_{2r}(\xi) , \quad |\xi| \le \pi ,$$

with $\|\underline{G}_{2r}(\xi)\| = O(|\xi|^{2r})$, and

ii) there exists a constant K and for every ξ a nonsingular matrix $\underline{R}$ with

(B.8) $$\max(\|\underline{R}\|, \|\underline{R}^{-1}\|) \le K$$

such that

(B.9) $$\underline{R}(\underline{I} + \underline{G}_1(\xi))\underline{R}^{-1} = \underline{D} = \mathrm{diag}(\kappa_1,\dots,\kappa_n) .$$

Proof. Without a loss of generality (due to (B.8) and (B.9)) assume that $\underline{I} + \underline{G}_1(\xi)$ is diagonal. Using Gershgorin's [1] estimates of the eigenvalues of a matrix, the eigenvalue λ_i of $\underline{G}(\xi)$ can be ordered in such a way that

(B.10) $$|\lambda_i - \kappa_i| \le \text{const.}\ |\xi|^{2r} .$$

Hence,

(B.11) $$s\underline{G}(\xi) - \underline{I} = \begin{pmatrix} s\lambda_1+sq_1-1 & sq_{12} & \cdots & sq_{1p} \\ sq_{21} & s\lambda_2+sq_2-1 & & sq_{2p} \\ \vdots & & \ddots & \vdots \\ sq_{p1} & & & s\lambda_p+sq_{p-1} \end{pmatrix} ,$$

where $|g_j| + |g_{jk}| \leq \text{const.}\ |\xi|^{2r}$ for j , $k = 1,2,\dots,p$.

Let $c_{\alpha,\beta}$ denote the elements of the matrix $(s\underline{G} - \underline{I})^{-1}$ and by $(s\underline{G} - \underline{I})_{\alpha\beta}$ denote the matrix obtained from $(s\underline{G} - \underline{I})$ by omitting the α-th row and the β-th column. By Cramer's rule we obtain estimates for $|c_{\alpha\beta}|$,

$$\text{(B.12)} \qquad |c_{\alpha\beta}| = \left|\frac{\det(s\underline{K} - \underline{I})_{\alpha\beta}}{\det(s\underline{G} - \underline{I})}\right|$$

$$\leq \text{const.}\ \frac{|\xi|^{2r(p-1)} + \sum_{\ell=1}^{p-1} \sum_{j} \left(\prod_{\nu=1}^{\ell} |s\lambda_{j\nu} - 1|\right)\cdot|\xi|^{2r(p-1-\ell)}}{\prod_{\nu=1}^{p} |s\lambda_\nu - 1|} .$$

Since (B.2) is dissipative of order 2r, $|\lambda_i| \leq 1 - \delta|\xi|^{2r}$; hence

$$|s\lambda_j - 1| \geq 1 - |s||\lambda_j| \geq \delta|\xi|^{2r}$$

and

$$|s\lambda_j - 1| \geq 1 - |s| .$$

Therefore, by substitution into (B.12)

$$|c_{\alpha\beta}| \leq \text{const.}\ (1 - |s|)^{-1}[\delta^{1-p} + \sum_{\ell=1}^{p-1} \binom{p-1}{\ell}\delta^{\ell+1-p}]$$

$$\leq \text{const.}\ \delta^{1-p}(1 - |s|)^{-1} ,$$

from which stability follows by the first lemma.

Theorem B.2 now follows directly from the definition of hyperbolicity and Lemma 3.

It should be noted that the first theorem (Theorem B.1) can be proved using the Lemmas 2 and 3 in this last theorem (Theorem B.2).

Consider the first order system of hyperbolic equations in ℓ space dimensions

$$\text{(B.13)} \qquad \partial_t\underline{v} = \underline{P}(\underline{x},\partial_{\underline{x}})\underline{v} \ , \ -\infty < x_j < +\infty \ , \ j = 1,\dots,\ell \ , \quad t > 0 \ ,$$

where

$$\text{(B.14)} \qquad \underline{P}(\underline{x},\partial_{\underline{x}}) = \sum_{j=1}^{\ell} \underline{B}_j(\underline{x})\partial_{x_j} \ ,$$

$\underline{x} = (x_1,\dots,x_\ell)$, and $\underline{B}_j(x)$ are $p \times p$ matrices whose elements are functions of $\underline{x}$.

And consider the constant coefficient first-order system of hyperbolic equations in ℓ space dimensions

$$\text{(B.15)} \qquad \partial_t\underline{v} = \underline{P}(\partial_{\underline{x}})\underline{v} \ , \quad -\infty < x_j < +\infty \ , \ j = 1,\dots,\ell \ , \ t > 0 \ ,$$

where

(B.16) $$\underline{P}(\partial_{\underline{x}}) = \sum_{j=1}^{\ell} \underline{C}_j \, \partial_{x_j}$$

and $\underline{C}_j$ are $p \times p$ constant matrices.

Let $\hat{\underline{v}}(\underline{\omega},1)$ denote the Fourier transform of $\underline{v}(\underline{x},t)$,

$$\hat{\underline{v}}(\underline{\omega},t) = \frac{1}{(2\pi)^{\ell/2}} \int_{-\infty}^{+\infty} e^{i(\omega o x)} \, \underline{v}(\underline{x},t)d\underline{x} \; ,$$

then by taking the Fourier transform of equation (B.15) we obtain the system of ordinary differential equations in t

$$\partial_t v(\hat{\underline{\omega}},t) = \underline{P}(i\underline{\omega})\hat{\underline{v}}(\underline{\omega},t) \; ,$$

where $\underline{P}(i\underline{\omega})$, called the symbol of $\underline{P}(\partial_{\underline{x}})$, is the $p \times p$ matrix

(B.17) $$\underline{P}(i\underline{\omega}) = i \sum_{j=1}^{\ell} \underline{C}_j \omega_j \; .$$

Approximate equation (B.12) by the finite difference method

(B.18) $$\underline{u}^{n+1} = \underline{Q} \, \underline{u}^n \; ,$$

where

(B.19) $$\underline{Q} = \sum_{j=-m_1}^{m_2} \underline{A}_j(\underline{x})S_+^{\,j} ,$$

and $\underline{A}_j(\underline{x})$ are $p \times p$ matrices whose elements depend on $\underline{x}$, h , and k. The amplification matrix of (B.18) is denoted by $\underline{G}(\underline{x},\underline{\xi})$, where

(B.20) $$\underline{G}(\underline{x},\underline{\xi}) = \sum_{j=-m_1}^{m_2} \underline{A}_j(\underline{x})e^{-ij\underline{\xi}} \; ,$$

and $\underline{\xi} = (\xi_1,\ldots,\xi_\ell)$.

__Definition__. A finite difference method (B.18) is __dissipative of order 2s__, where s is a positive integer, if there exists a constant $\delta > 0$ such that

$$|\lambda_j(\underline{x},\underline{\xi})| \le 1 - \delta \|\underline{\xi}\|^{2s} \; ,$$

for each x and each ξ , with $\|\xi\| \le \pi$, where λ_j are the eigenvalues of $\underline{G}(\underline{x},\xi)$.

Kreiss [3] proved a variable coefficient version of Theorem B.1.

__Theorem B.3__. Suppose that equation (B.13) and the finite difference method (B.14) have hermitian coefficient matrices that are Lipschitz continuous and uniformly bounded. If (B.14) is

i) accurate of order $2r - 1$, and

ii) dissipative of order $2r$,

then it is stable.

The idea of the proof is similar to that of the proof of Theorem B.1. However, $\hat{\underline{H}}$ (does not have the one property necessary for the extension to variable coefficients; that is, we need as $\|\underline{\xi}\| \to 0$, $\hat{\underline{H}} = \underline{I} + O(\|\xi\|^{2r-1})$. $\hat{\underline{H}}$ will be reconstructed so that it satisfies this additional requirement in a neighborhood of $\underline{\xi} = 0$.

As a first step we shall prove the following theorem for the case of constant coefficients (due to Parlett [6]).

Theorem B.4. Let equation (B.1) and the finite difference method (B.2) have constant hermitian coefficients. If for some positive integer r, (B.2) is

i) accurate of degree $2r - 1$,

ii) dissipative of order $2r$,

then there exists an hermitian matrix $\hat{\underline{H}}(\underline{\xi})$ satisfying

iii) $\beta^{-1}\underline{I} < \hat{\underline{H}} < \beta\underline{I}$ for some $\beta > 1$,

iv) $\underline{G}^*\hat{\underline{H}}\underline{G} < (1 - \delta\|\underline{\xi}\|^{2r})\hat{\underline{H}}$, $\|\underline{\xi}\| < \pi$,

v) $\hat{\underline{H}} = \underline{I} + O(\|\underline{\xi}\|^{2r-1})$ as $\|\underline{\xi}\| \to 0$,

where δ is the constant in the definition of dissipation.

Without a loss of generality asssume that $\underline{P}$ has been disgonalized by a unitary transformation, with eigenvalues μ_k. Let $\underline{\xi} = \sigma\underline{\xi}_0$, where $\underline{\xi}_0$ is one fixed unit vector and write $\underline{G}(\underline{\xi})$ in the form

(B.21) $$\underline{G}(\underline{\xi}) = \underline{M} + \sigma^{2r}\underline{Q} + O(\sigma^{2r}),$$

where $\underline{M} = e^{-i\underline{I}}$ is a diagonal matrix and $\underline{Q} = \underline{Q}(\underline{\xi}_0)$ is a hermitian matrix.

We shall require a technical lemma to proceed with the proof.

Lemma 4. For a dissipative finite difference method whose symbol $\underline{G}$ has the form (B.21), if $Q_{ik} = 0$ whenever $\mu_i = \mu_k$ for $i \neq k$, then

(B.22) $$Q_{ii} < -\delta + O(1) \quad \text{as } \delta \to 0 .$$

Proof. We shall show that the eigenvalues of $\underline{G}$ equal its diagonal elements up to order σ^{2r}. This will follow a stronger form of Gerschgorin's theorem (see Gerschgorin [1]) which states that if ℓ of the Gerschgorin's circles of a matrix $\underline{L}$ are disjoint from the rest then their union contains precisely ℓ eigenvalues of $\underline{L}$. This follows directly from the fact that the eigenvalues of a matrix depend continuously on its elements.

Let μ_k be a simple eigenvalue of $\hat{\underline{P}}$ and multiply the k-th row of $\underline{G}$ by $\sqrt{\sigma}$ which divides the k-th column of $\underline{G}$ by $\sqrt{\sigma}$. This similarity transformation on $\underline{G}$ leaves the eigenvalues λ_i of $\underline{G}$ unchanged. However, it makes the radius of the k-th Gerschgorin circle

$O(\sigma^{2r+1/2})$ and those of the remainder $O(\sigma^{2r-1/2})$. Since the centers of the circles are at G_{ii}, where

$$G_{ii} = e^{i\sigma\mu_i} + \sigma^{2r} Q_{ii} + O(\sigma^{2r+1}) \tag{B.23}$$

and $\mu_k \neq \mu_i$ for $i \neq k$, the k-th Gerschgorin circle is disjoint from the rest for sufficiently small r. Hence, there is an eigenvalue of $\underline{G}$ such that

$$\lambda = e^{i\sigma\mu_k} + \sigma^{2r} Q_{kk} + O(\sigma^{2r}) .$$

By the definition of dissipative $|\lambda| < 1 - \delta\sigma^{2r}$, so that

$$1 - 2\delta\sigma^{2r} > 1 + 2\sigma^{2r} Q_{kk} + O(\sigma^{2r})$$

or

$$Q_{kk} < -\delta + O(1) \quad \text{as} \quad \sigma \to 0 .$$

If μ_k is not a simple eigenvalue of $\underline{\hat{P}}$, all the rows and columns corresponding to μ_k are treated as above. Since $Q_{ik} = 0$ for $\mu_i = \mu_k$ with $i \neq k$, the radius of the Gerschgorin circles are still $O(\sigma^{2r+1/2})$. Therefore, at least one circle contains an eigenvalue of $\underline{G}$ so that (B.22) holds for that element.

If for some of the elements $Q_{ii} > -\delta + O(1)$, these would become disjoint from those with $Q_{ii} > -\delta + O(1)$ as $\sigma \to 0$ and we would have a contradiction for then they would contain an eigenvalue of $\underline{G}$.

If two eigenvalues μ_i and μ_k of $\underline{\hat{P}}$ are equal, then since $\underline{Q}$ is hermitian, with an appropriate choice of unitary transformation (used to diagonalize $\underline{\hat{P}}$) we can ensure that the elements Q_{i_k} are zero.

Define a hermitian matrix $\underline{\hat{H}_0}$ by

$$\hat{H}_{0_{ik}} = \begin{cases} 0 , & \text{if } \mu_i = \mu_k \\ \dfrac{2iQ_{ik}}{\mu_i - \mu_k} , & \text{if } \mu_i \neq \mu_k \end{cases} \tag{B.24}$$

and from this the hermitian matrix

$$\underline{\hat{H}} = \underline{I} + \sigma^{2r-1} \underline{\hat{H}_0} . \tag{B.25}$$

Expanding $\underline{M}$ (in equation (B.21)) in powers of σ and using (iv)

$$\underline{G^*\hat{H}G} = \underline{\hat{H}} + \sigma^{2r}[2\underline{Q} + i\underline{\hat{H}_0}\underline{\hat{P}} - i\underline{\hat{P}}\underline{\hat{H}_0}] + O(\sigma^{2r}) . \tag{B.26}$$

B- the construction of $\underline{\hat{H}}$, the off-diagonal elements of the expressio in the bracket of (B.26) are zero.

For $\underline{\xi} = \sigma\underline{\xi}_0$, by Lemma 4

(B.27) $$2\underline{Q} + i(\hat{\underline{H}}_0\hat{\underline{P}} - \hat{\underline{P}}\hat{\underline{H}}_0) < -2\delta\underline{I} + O(1) ,$$

where $\hat{\underline{H}}_0 = \hat{\underline{H}}_0(\underline{\xi}_0)$ is bounded (not uniformly in $\underline{\xi}_0$). Hence there is an $\sigma_0(\underline{\xi}_0)$, such that for $0 < \sigma < \sigma_0(\underline{\xi}_0)$, (iv) and (v) are satisfied in the direction ξ_0 , that is, by (B.26) and (B.27)

(B.28) $$\begin{aligned}\underline{G}^*\hat{\underline{H}}\underline{G} &< \hat{\underline{H}} - 2\delta\sigma^{2r}\underline{I} + O(\sigma^{2r})\\ &< (1 - \delta\sigma^{2r})\hat{\underline{H}} \quad \text{for} \quad \sigma < \sigma_0 .\end{aligned}$$

By the continuity of $\underline{P}$ and $\underline{Q}$ as a function of ξ there exists an $\varepsilon > 0$ such that $\|\underline{\xi}_0 - \underline{\xi}_1\| < \varepsilon$ where $\|\underline{\xi}_1\| = 1$ and

$$2\underline{Q}(\underline{\xi}_1) + i(\hat{\underline{H}}_0(\underline{\xi}_0)\hat{\underline{P}}(\underline{\xi}_1) - \hat{\underline{P}}(\underline{\xi}_1)\underline{H}_0(\underline{\xi}_0)) < -\delta\underline{I} .$$

Using the same reasoning that leads to (B.28), (iv) and (v) are satisfied for $\underline{\xi} = \sigma\underline{\xi}_1$ for $0 < \sigma < \sigma_1(\underline{\xi}_0)$. By the Heine-Borel theorem, the sphere $|\underline{\xi}| = 1$ can be covered by a finite number of such neighborhoods, in each of which $\hat{\underline{H}}_0$ can be taken constant. Hence, we can construct an $\hat{\underline{H}}(\underline{\xi})$ in a finite neighborhood of $|\underline{\xi}| < \sigma_1$ of the orgin satisfying condition (iii), (iv), and (v). Outside this neighborhood condition (v) has no effect so $\hat{\underline{H}}$ can be defined as in the proof of Theorem B.1.

We can now proceed to prove Kreiss' theorem (see Richtmyer and Morton [7]).

Consider a neighborhood of an arbitrary point, without loss of generality, taken to be the origin. Consider functions $\underline{u}$ such that

$$\underline{u}(\underline{x}) \equiv 0 \quad \text{for} \quad \|\underline{x}\| > \zeta > 0 .$$

Let $\underline{Q}_0$ denote the finite difference operator $\underline{Q}$ in (B.18) at $\underline{x} = 0$, that is,

(B.29) $$\underline{Q}_0 = \Sigma\, \underline{A}_j(0)S_+^j$$

and the corresponding symbol of $\underline{Q}_0$ given by (B.20) at $\underline{x} = 0$, or

(B.30) $$\underline{G}_0 = \underline{G}(0,\underline{\xi}) = \Sigma\, \underline{A}_j(0)e^{-ij\underline{\xi}} .$$

In the Theorem B.3 there exists an hermitian matrix $\hat{\underline{H}}(0,\underline{\xi})$ such that

(B.31) $$\hat{\underline{H}} = \underline{I} + \hat{\underline{H}}_{2r-1}(0,\underline{\xi})$$

where

(B.32) $$\|\hat{\underline{H}}_{2r-1}\| < K\|\underline{\xi}\|^{2r-1}$$

and K a constant. $\hat{\underline{H}}$ satisfies conditions (iv) and (v) of that theorem.

Define the operator $\underline{H}$ by

(B.33) $$(\underline{Hu})(\underline{x}) = (2\pi)^{p/2} \int \underline{\hat{H}}(0,\underline{\omega}h)\underline{\hat{u}}(\omega)e^{i\underline{\omega}\cdot\underline{x}}d\underline{\omega} ,$$

where $\underline{\hat{u}}$ denotes the Fourier transform of $\underline{u}$. Similarly, the operator $\underline{H}_{2r-1}$ is defined by replacing $\underline{\hat{H}}(0,\underline{\omega}h)$ in (B.33) with $\underline{\hat{H}}_{2r-1}(0,\omega h)$. Because of the properties $\underline{\hat{H}}$ there is a norm induced by the operator $\underline{H}$, namely

$$\|\underline{u}\|_H = \sqrt{(\underline{u},\underline{Hu})} .$$

Using this norm

$$\|\underline{Gu}\|_H^2 = (\underline{Gu},\underline{HGu}) = (\underline{Gu},(\underline{I} + \underline{H}_{2r-1})\underline{Gu}) = E + F$$

where $E = \|\underline{Gu}\|^2$ and $F = (\underline{Gu},\underline{H}_{2r-1}\underline{Gu})$. This representation of E and F can be decomposed further, namely,

(B.34) $$\|\underline{Gu}\|_H^2 = (E_0 + F_0) + (E - E_0) + (F - F_0) ,$$

where $E_0 = \|\underline{G}_0\underline{u}\|_H^2$ and $F_0 = (\underline{G}_0\underline{u},\underline{H}_{2r-1}\underline{G}_0\underline{u})$.

The task at hand is to obtain estimates for the three terms on the right hand side of (B.34).

Define the norm

(B.35) $$\|\underline{u}\|_\rho^2 = \int (2 \sin(\tfrac{1}{2} \underline{\xi}))^{2\rho}\|\underline{\hat{u}}\|^2 \, d\underline{k} .$$

Since $\hat{D}_{\pm j} = 2ie^{\pm i\xi_j/2} \sin(\frac{1}{2} \xi_j)$, by substitution into (B.35),

(B.36) $$\|u\|_\rho^2 = \sum_{j=1}^{p} \|D^\rho_{+j}\underline{u}\|^2 = \sum_{j=1}^{p} \|D^\rho_{-j}\underline{u}\|^2 .$$

From which $\|\underline{u}\|_p < 2^{p-q}\|\underline{u}\|_q$ for $p > q$.

We can readily obtain a bound on the first term in (B.34),

(B.37) $$\begin{aligned} E_0 + F_0 &= \|\underline{G}_0\underline{u}\|_H^2 = \int \underline{\hat{u}}^*\underline{G}_0^*\underline{\hat{H}}\underline{G}_0\underline{\hat{u}}d\underline{\omega} \\ &< \int \underline{\hat{u}}^*(1 - \tfrac{1}{2} \delta|\underline{\xi}|^{2r})\underline{\hat{H}}\underline{\hat{u}}d\underline{\omega} \\ &< \int \underline{\hat{u}}^*(1 - \tfrac{1}{2} \delta \sum_{j=1}^{p} (2 \sin(\tfrac{1}{2} \xi_j))^{2r})\underline{\hat{H}}\underline{\hat{u}}d\underline{\omega} \\ &= \|\underline{u}\|_H^2 - \tfrac{1}{2} \delta\hat{K}^{-1} \|u\|_r^2 , \end{aligned}$$

where K is a constant.

The finite difference operator $\underline{Q}$ may be written in the form (using the fact that it is accurate of order $2r - 1$)

(B.38) $$\underline{Q} = \sum_{m=0}^{2r-1} \frac{1}{m!} \underline{P}_D^m + \underline{Q}_D ,$$

where

$$\underline{P}_D = \sum_{j=1}^{p} \underline{B}_j(\underline{x}) D_j^{(2r-1)}$$

and

(B.39) $$D_j^{(2r-1)} = \sum_{m=0}^{r-1} (-1)^m \gamma_{2m+1} D_{0j}^{2m+1} ,$$

that is, a truncated expansion of $h\partial_{x_j}$ in power of the difference operator D_{0j} with γ_{2m+1} being real constants. $\underline{Q}_D$ represents the remainder of order greater than or equal to $2r$ but whose form is not needed.

The following technical lemma will be used whose proof can be found in Kreiss [3].

Lemma 5. Let $\underline{A}(\underline{x})$ denote a uniformly bounded and uniformly Lipschitz continuous matrix, with Lipschitz constant L. Then for $\underline{v}$, $\underline{u} \in L_2(\underline{x})$ we have

$$(\underline{u}, \underline{A}hD_{0j}\underline{v}) = -(hD_{0j}\underline{u}, \underline{Av}) + \phi_0(u,v) ,$$

$$(\underline{u}, \underline{A}hD_{+j}\underline{v}) = -(hD_{-j}\underline{u}, \underline{Av}) + \phi_1(u,v) ,$$

where ϕ_0 and ϕ_1 are functions such that

$$|\phi_i(u,v)| < hL\|u\| \|v\|$$

for $i = 0$ and 1.

We may rewrite $\underline{Q}$ in (B.34) in the form

$$\underline{Q} = e^{\underline{P}_D} - \sum_{m=2r}^{\infty} \frac{1}{m!} \underline{P}_D^m ,$$

so that, using Lemma 5,

$$\|(\sum_{m=0}^{2r-1} \frac{1}{m!} \underline{P}_D^m)\underline{u}\|^2 = (\underline{u}, [e^{-\underline{P}_D} - \sum_{m=2r+1}^{\infty} \frac{1}{m!} (-\underline{P}_D)^m][e^{\underline{P}_D} - \sum_{m=2r+1}^{\infty} \frac{1}{m!} \underline{P}_D^m]\underline{u}) + \phi(\underline{u},\underline{u})$$

$$= \|\underline{u}\|^2 + (\underline{u}, \underline{Ru}) + \phi(\underline{u},\underline{u}) ,$$

where $\underline{R}$ is a finite sum of terms $\underline{B}_j(\underline{x})\underline{D}_j^p$ with $p > 2r$ and ϕ is a bounded function as in the above lemma. The terms $(\underline{u}, \underline{Ru})$ can be changed by reordering the terms $\underline{B}_j$ and $\underline{D}$ to obtain (see Richtmyer and Morton [7])

$$(\underline{u}, \underline{Ru}) = (\underline{D}^r\underline{u}, \underline{DS}^r\underline{u}) + \bar{\phi}(\underline{u},\underline{u}) ,$$

where $\underline{S}$ is a bounded difference operator of order $2r$ less than that of $\underline{R}$, $\bar{\phi}$ is a bounded function as in the above lemma, and D^r is the product of r factors of the form $D_{\pm j}$. Thus

$$(B.40) \qquad \|\underline{G}\underline{u}\|^2 = \|(\sum_{m=0}^{2r-1} \frac{1}{m!} \underline{P}_D^m)\underline{u}\|^2 + 2\,\mathrm{Re}(\underline{Q}_D\underline{u}, (\sum_{m=0}^{2r-1} \frac{1}{m!} \underline{P}_D^m)\underline{u}) + \|\underline{Q}_D\underline{u}\|^2$$
$$= \|\underline{u}\|^2 + (D^r\underline{u}, \underline{\bar{S}}D^r\underline{u}) + \bar{\phi}(\underline{u},\underline{u}) ,$$

where $\underline{\bar{S}}$ is a bounded difference operator like $\underline{S}$ and ϕ is a bounded function as in the above lemma.

However, $E = E_0 = \|\underline{G}\underline{u}\|^2 - \|\underline{G}_0\underline{u}\|^2$ and by (B.40) consists maily of a term in $\underline{\bar{S}} = \underline{\bar{S}}_0$, where $\underline{\bar{S}}_0$ corresponds to $\underline{\bar{S}}$ at $\underline{x} = 0$. Since the support of $\underline{u}$ is limited to a sphere of radius ζ, replace $\underline{G}$ by $\underline{G}_0 + (\underline{G} - \underline{G}_0)g(\underline{x})$, where $g(\underline{x})$ is a smooth function such that $g(x) \equiv 0$ for $\|\underline{x}\| > 2\zeta$ and $g(\underline{x}) \equiv 1$ for $\|\underline{x}\| < \frac{3}{2}\zeta$, where $\underline{x}$ is sufficiently small. Thus the integration in $E - E_0$ need only be extended to a sphere of radius 2. Since the $\underline{B}_j(\underline{x})$'s are hermitian (see Lemma 10 in Kreiss [3])

$$(B.41) \qquad |E - E_0| \le M_1(\zeta)\|u\|_r^2 + C_1 h\|\underline{u}\|^2 ,$$

where $M_1(\zeta) \to 0$ as $\zeta \to 0$ and C_1 in a constant.

For the last term in (B.30)

$$|F - F_0| = |\mathrm{Re}((\underline{G} + \underline{G}_0)\underline{u}, \underline{H}_{2r-1}(\underline{G} - \underline{G}_0)\underline{u})|$$
$$\le \text{const. } (\underline{u}, \underline{H}_{2r-1}(\underline{G} - \underline{G}_0)\underline{u}) ,$$

since $\underline{G}$ is a bounded difference operator.

Factorize $\underline{\hat{H}}_{2r-1}$ as $\underline{\hat{H}}_{2r-1} = \underline{\hat{H}}_1\underline{\hat{H}}_2$, where

$$\underline{\hat{H}}_1 = (\sum_{j=1}^{p} (2 \sin \tfrac{1}{2}\xi_j)^2)^{r/2}\underline{I} ,$$

$$\underline{\hat{H}}_2 = \underline{\hat{H}}_1^{-1}\, \underline{\hat{H}}_{2r-1} ,$$

and "^" denotes the Fourier transform. It can be seen that

$$\|\underline{\xi}\|^r\underline{I} \ge H_1 \ge (\tfrac{2}{\pi})^r\|\xi\|^r\underline{I}$$

and

$$\|\underline{\hat{H}}_2\| \le (\tfrac{\pi}{2})^r\, C_2\|\xi\|^{r-1}.$$

where C_2 is a constant. Thus by substitution, since $\underline{H}_{2r-1}$ is hermitian,

$$|(\underline{u}, \underline{H}_{2r-1}(\underline{G} - \underline{G}_0)\underline{u})| \le |(\underline{H}_1\underline{u}, \underline{H}_2(\underline{G} - \underline{G}_0)\underline{u})|$$
$$\le \|\underline{H}_1\underline{u}\|\|\underline{H}_2(\underline{G} - \underline{G}_0)\underline{u}\|$$
$$\le \text{const. } \|\underline{u}\|_r\|(\underline{G} - \underline{G}_0)\underline{u}\|_{r-1} .$$

However, by the definition

$$\text{(B.42)}\qquad \|(\underline{G} - \underline{G}_0)\underline{u}\|_{r-1} \le \sum_{j=1}^{p} \|D_{+j}^{r-1}(\underline{G} - \underline{G}_0)\underline{u}\|$$

$$\le \sum_{j=1}^{p} \|\underline{G} - \underline{G}_0)D_{+j}^{r-1}\underline{u}\| + \text{const. } h\|\underline{u}\|$$

$$\le M_2(\zeta)\|\underline{u}\|_r + C_3 h\|\underline{u}\| ,$$

where $M_2(\zeta) \to 0$ as $\zeta \to 0$, using the fact that $\underline{G} - \underline{G}_0$ is at least a first order difference operator, so $(\underline{G} - \underline{G}_0)D_{+j}^{r-1}$ is at least of order r. Combining this result with (B.41)

$$\text{(B.43)}\qquad |E - E_0| + |F - F_0| \le M(\xi)\|\underline{u}\|_r^2 + C_4 h\|\underline{u}\|^2 ,$$

where $M(\zeta) \to 0$ as $\zeta \to 0$ and C_4 is a constant.

For some $\zeta_0 > 0$, by taking $\zeta < \zeta_0$ the first term on right in (B.43) dominated by the negative term in (B.37) so that

$$\text{(B.44)}\qquad \|\underline{u}^{n+1}\|_H^2 = \|\underline{Gu}^n\|_H^2 \le \|\underline{u}^n\|_H^2 + (C_1 + C_3)h\|u^n\|^2$$

$$\le (1 + \alpha k)\|\underline{u}^n\|_H^2 ,$$

where $\alpha = \frac{h}{k}(C_1 + C_3)$, which implies local stability.

In order to obtain the result for a general $\underline{u}^n$, let $\sum_i d_i^2(\underline{x}) \equiv 1$ be a smooth Gording type (see Hörmander [2]) partition of unity in which each $d_i(\underline{x})$ is a smooth real-valued function of $\underline{x}$ which vanishes outside some sphere S_j. At each fixed-point $\underline{x}$ only a uniformly bounded number of the $d_i(\underline{x})$'s are nonzero. Each sphere S_i is chosen so that (B.44) holds for all $\underline{u}^n$ whose support is restricted to S_i. Hence the radius of S_i is taken less than ζ_0. Define $\underline{H}_i$ at the center of the sphere S_i as above and let

$$\|\underline{u}\|_H^2 = \sum_i \|d_i\underline{u}\|_{H_i}^2 = \sum_i (d_i\underline{u}, \underline{H}_i d_i\underline{u}) .$$

This norm satisfies the inequalities common to all $\underline{H}_i$.

Since, in interchanging d_i with $\underline{G}$ only a finite number of d_i's contribute at each $\underline{x}$,

$$\|\underline{u}^{n+1}\|_H^2 = \sum_i (d_i\underline{Gu}^n, \underline{H}_i d_i \underline{Gu}^n)$$

$$\le \sum (\underline{Gd_i u}^n, \underline{H}_i \underline{Gd_i u}^n) + \text{const. } h\|\underline{u}^n\|^2$$

$$\le \sum \|d_i\underline{u}^n\|_{H_i}^2 (1 + \alpha k) + \text{const. } h\|\underline{u}^n\|^2$$

$$\le (1 + O(k))\|\underline{u}^n\|_H^2 .$$

It has yet to be shown that in general the condition of accuracy to order 2r - 1 cannot be overcome by these methods.

However, Parlett [6] has proved a theorem under slightly stronger conditions whose statement will be given.

Definition. A system of differential equation (B.13) is regular hyperbolic if it is hyperbolic and P has distinct eigenvalues.

Theorem B.5. Let (B.13) be a regular hyperbolic system with variable coefficients. Let the coefficients in (B.13) and (B.18) be hermitian, Lipschitz continuous and uniformly bounded. If for some integer r greater than one, (B.18) is

i) accurate of order 2r - 2 ,

ii) dissipative of order 2r ,

then (B.18) is stable.

References

1. Gershgorin, S., Über die Abgrenzung der Eigenwerte einer Matrix, Izvestiya Akad, Nauk, USSR, Ser. Math., 7, 749 (1931).

2. Hörmander, L., Linear Differential Operators, Springer-Verlag, Berlin (1969).

3. Kreiss, H. O., On Difference Approximations of the Dissipative Type for Hyperbolic Differential Equations, Comm. Pure. Appl. Math., 17, 335 (1964).

4. Lax, P. D., On the Stability of Difference Approximations to Solutions of Hyperbolic Equations with Variable Coefficients, Comm. Pure Appl. Math., 14, 497 (1961).

5. ________, and L. Nirenberg, On Stability of Difference Schemes: A Sharp Form of Garding's Inequality, Comm. Pure Appl. Math., 19, 473 (1966).

6. Parlett, B. N., Accuracy and Dissipation in Difference Schemes, Comm. Pure Appl. Math., 19, 111 (1966).

7. Richtmyer, R. D., and K. W. Morton, Difference Methods for Initial Value Problems, 2nd ed., Interscience, New York (1967).

Appendix C. The Lax-Nirenberg Theorem and a Special Case

As an introduction to the main theorem (the Lax-Nirenberg Theorem), we shall prove a weaker theorem due to Lax [2] for a single scalar equation

$$\partial_t v = a(x)\partial_x v, \quad t > 0, \quad -\infty < x < +\infty. \tag{C.1}$$

Approximate equation (C.1) by the finite difference method

$$u_i^{n+1} = Qu_i^n, \tag{C.2}$$

where

$$Q = \sum_{j=-m_1}^{m_1} C_j(x) S_-^j . \tag{C.3}$$

The symbol of (C.3) is given by

$$\rho(x, \xi) = \sum_{j=-m_1}^{m_2} C_j(x) e^{-ij\xi}. \tag{C.4}$$

Theorem C.1 (Lax). The finite difference method (C.2) is stable if the symbol (C.4) satisfies the following conditions

(i) Lipschitz continuous in x,

(ii) $|\rho(x, \xi)| < 1$ if $\xi \neq 0$,

(iii) $\rho(x,\xi) = 1 - Q(x)\xi^{2p} + O(\xi^{2p+1})$, when ξ is near 0 and $Q(x) > 0$ for all x.

Proof: Condition (iii) implies that a certain amount of numerical viscosity is present.

By condition (ii), for $\xi \neq 0$,

$$1 - \rho\rho^* > 0. \tag{C.5}$$

Furthermore, by condition (iii),

$$1 - \rho\rho^* = Q(x)\xi^{2p} + O(|\xi|^{2p+1}) \tag{C.6}$$

Using (C.5) and (C.6), we can write

$$1 - \rho\rho^* = g(x, \xi)(e^{i\xi}-1)^{2p}, \tag{C.7}$$

where g is a positive analytic function and is different from zero for ξ real. Define

$$d = \sqrt{g}(e^{i\xi}-1)^p.$$

which is analytic in ξ and Lipschitz continuous in x. Substituting d into (C.7)

(C.8) $$1 = \rho\rho^* + dd^*.$$

(Conditions (ii) and (iii) were so designed that we could write (C.8).) Construct the kernel

(C.9) $$K(x,\xi,\eta) = \rho(x,\xi)\rho^*(x,\eta) + d(x,\xi)d^*(x,\eta).$$

It can be seen that K satisfies the Lipschitz condition with respect to the following norm

(C.10) $$\|K(x)-K(x')\|_5 < \text{const.}|x-x^1|,$$

where $\|\cdot\|_5$ denotes the sum of the L_1-norms of the partial derivatives of K up to the second order in ξ and third order in η. Furthermore, by (C.8)

(C.11) $$K(x,\xi,\xi) \equiv 1.$$

It follows from the structure of K in (C.9) that K is given as the sum of two degenerate positive kernels. Therefore, for any $u \in L_2$ and for every x

(C.12) $$|\int\rho(x,\xi)u(\xi)d\xi|^2 < \int\int u^*(\xi)K(x,\xi,\eta)u(\mu)d\xi d\eta.$$

Expand K into a Fourier series with respect to η and ξ

(C.13) $$K(x,\xi,\eta) = \sum_{\ell,m} K_{\ell,m}(x)e^{i(m\eta-\ell\xi)}.$$

We can express the three properties of K, (C.10)-(C.12), in terms of the Fourier coefficient. Properties given by (C.10) give rise to

(C.10') $$|K_{\ell,m}(x)-K_{\ell,m}(x')| < \frac{\text{const.}(x-x')}{|\ell|^2|m|^3},$$

which is obtained by performing, in the integral representation of the Fourier coefficient of $K(x) - K(x')$, two integrations by parts with respect to ξ and three integrations by parts with respect to η and using (C.10) to estimate.

Expanding u into a Fourier series,

$$u(\xi) = \sum_j \omega_j e^{-ij\xi}$$

and using the representation of ρ in (C.4), by substitution into (C.11), for every x and every sequence $\{\omega_j\}$ in L_2

(C.11') $$|\sum_j c_j(x)\omega_j|^2 < \sum_{\ell,m} \omega_\ell K_{\ell,m}(x)\omega_m.$$

In (C.12) expanding both sides into a Fourier series and equating coefficients gives the relation

$$\sum_j K_{j-\ell,j-m}(x) = \delta_{\ell,m},$$

where $\delta_{\ell,m}$ denotes the Kronecker delta.

Let

$$v_i = Qu_i = \sum_j c_j(x) u_{i-j}^n ,$$

then we see $v_i = u_i^{n+1}$. Applying (C.11') to $\omega_\ell = u_{i-\ell}$ a "local energy estimate" is obtained,

$$(C.14) \qquad |v_i| < \sum_{\ell,m} u_{i-\ell}^* K_{\ell,m}(ih) u_{i-m}.$$

Let $r = i-\ell$ and $s = i-m$ and sum (C.14) over i

$$(C.15) \qquad \sum_i |v_i|^2 < \sum_{i,r,s} u_r^* K_{i-r,i-s}(ih) u_s .$$

The right hand side of (C.15) can be written as

$$(C.16) \qquad \sum_{k,r,s} u_r^* K_{i-r,i-s}(rh) u_s + \sum_{i,r,s} u_r^* (K_{i-r.i-s}(ih) - K_{i-r,i-s}(rh)) u_s .$$

We want to obtain bounds on the two sums in (C.16). Consider the first sum; using (C.12'), by substitution into (C.15)

$$(C.17) \qquad \sum_i |v_i|^2 < \sum_r |u_r|^2 + \text{second term.}$$

Using (C.10') applied to the second term in (C.16),

$$\sum_{r,s} u_r^* (K_{i-r,i-s}(ih) - K_{i-r,i-s}(rh)) u_s < O(h) \sum \frac{|u_r||u_s|}{(i-r)^2 (i-s)^2}$$

and by the Schwartz inequality,

$$(C.18) \qquad \sum_{r,s} u_r^* (K_{i-r,i-s}(ih) - K_{i-r,i-s}(rh)) u_s < O(h) \sum \frac{(|u_r|^2 + |u_s|^2)}{(i-r)^2 (i-s)^2} .$$

In the term containing $|u_r|$ carry out the summations first with respect to i and s, in the term containing $|u_s|$ carry out the summation first with respect to i and r. Using $i-r = \ell$ and $i-s = m$ as new indices

$$O(h) \sum |u_r|^2$$

as an upper bound on the second term in (C.16). Thus (C.15) can be written in the form

$$(C.19) \qquad \sum_i |v_i|^2 < (1 + O(h)) \sum_r |u_r|^2 .$$

Letting $\bar{x}$ denote any point between 0 and h, inequality (C.19) holds. Then integrating this inequality with respect to $\bar{x}$ from 0 to h,

$$\| u^{n+1} \|^2 < (1 + O(h)) \; u^n \|^2 ,$$

from which stability follows.

Consider the first order system of hyperbolic equations in ℓ space dimensions

(C.20) $$\partial_t \underline{v} = \underline{P}(\underline{x}, \partial_{\underline{x}})\underline{v}, \quad t > 0, \quad -\infty < x_j < +\infty, \quad j = 1,\ldots,\ell$$

where

(C.21) $$\underline{P}(\underline{x}, \partial_{\underline{x}}) = \sum_{j=1}^{\ell} \underline{B}_j(x)\partial_{x_j},$$

$\underline{x} = (x_1,\ldots,x_\ell)$ and $\underline{B}_j(x)$ are $p\times p$ matrices whose elements are functions of $\underline{x}$. Approximate equation (C.20) by the finite difference method

(C.22) $$\underline{u}^{n+1} = Q\underline{u}^n,$$

where

(C.23) $$\underline{Q} = \sum_{j=-m_1}^{m_2} \underline{A}_j(\underline{x})S_+^j.$$

and $\underline{A}_j(\underline{x})$ are $p\times p$ matrices whose elements are functions of $\underline{x}$, h, and k. The amplification matrix of (C.22) is

(C.24) $$\underline{G}(\underline{x},\underline{\xi}) = \sum_{j=-m_1}^{m_2} \underline{A}_j(\underline{x})e^{-ij\underline{\xi}},$$

where $\underline{\xi} = (\xi_1,\ldots,\xi_\ell)$.

We shall now state the main result, the Lax-Nirenberg theorem (see Lax and Nirenberg [2] and Richtmyer and Morton [4]).

Theorem C.2 (Lax-Nirenberg). Suppose in the finite difference method (C.22) the coefficient matrices $\underline{A}_j(\underline{x})$ are real and symmetric independent of t and have bounded second derivatives. Suppose the amplification matrix $\underline{G}(\underline{x},\underline{\xi})$ in (C.24) satisfies

$$\|\underline{G}(\underline{x},\underline{\xi})\|_2 \le 1$$

for every $\underline{x}$ and $\underline{\xi}$, then the method (C.22) for real vector functions $\underline{u}^n$ is stable.

This theorem requires much less information about the structure of the finite difference method (C.22) than Kreiss' theorem (see Appendix B).

There is no way of viewing the difference between the frozen coefficient case and the variable coefficient as a small perturbation. However, suppose that $\underline{u}^n$ has small support, then the difference operator Q (in the variable coefficient case) resembles the frozen coefficient case. So we want to write $\underline{u}^n = \sum_\ell \underline{u}^n$, where $\underline{u}^n$ have small support, but not too small.

Proof: Assume $\frac{k}{h} = O(1)$. Consider the hermitian matrix $\underline{R}$ defined by

(C.25) $$\underline{R} = \underline{I} - \underline{G}^*(\underline{x},\underline{\xi})G(\underline{x},\underline{\xi}).$$

Since $\|\underline{G}(\underline{x},\underline{\xi})\| < 1$, it follows that $\underline{R} > 0$ for all $\underline{x}$ and $\underline{\xi}$. Let $\underline{R}_\Delta$ denote the difference operator whose symbol is $\underline{R}(\underline{x},\underline{\xi})$. Since the $\underline{A}_j(x)$'s are differentiable, they are Lipschitz continuous and

(C.26) $$\underline{R}_\Delta = \underline{I} - \underline{Q}^*\underline{Q} + O(k).$$

<u>Claim</u>. $\|\underline{Q}\| < 1 + O(k)$ is equivalent to the inequality

(C.27) $$(\underline{v},\underline{R}_\Delta\underline{v}) > -Mk\|\underline{v}\|^2$$

for some vector function $\underline{v}$. To see this, suppose (C.27) is true, then by substitution of (C.26) into (C.27),

(C.28) $$\begin{aligned}(\underline{v},(I-\underline{Q}^*\underline{Q})\underline{v}) &= \|\underline{v}\|^2 - (\underline{v},\underline{Q}^*\underline{Q}\underline{v})\\ &= \|\underline{v}\|^2 - \|Qv\|^2 > -Mk\|\underline{v}\|^2.\end{aligned}$$

From which it follows that

$$\|\underline{Q}\underline{v}\|^2 < (1+Mk)\|\underline{v}\|^2$$

and

$$\|Q\|^2 \equiv \max_{\|\underline{v}\|\neq 0} \frac{\|\underline{Q}\underline{v}\|^2}{\|\underline{v}\|^2} < 1 + O(k).$$

Thus $\|\underline{Q}\| < 1 + O(k)$. Conversely, using (C.28),

$$\begin{aligned}(\underline{v},\underline{R}_\Delta\underline{v}) &= \|\underline{v}\|^2 - \|\underline{Q}\underline{v}\|^2\\ &> \|\underline{v}\|^2 - \|\underline{Q}\|^2\|\underline{v}\|^2 = (1-\|Q\|^2)\|\underline{v}\|^2\\ &> (1-(1+O(k)))\|\underline{v}\|^2\\ &= -Mk\|\underline{v}\|^2,\end{aligned}$$

where M is some constant.

With this claim to show stability, it suffices to prove the inequality (C.27) holds.

Let $\Omega_1, \Omega_2, \ldots \Omega_k \ldots$ be open sets and let K be a compact set such that $K \subset \bigcup_j \Omega_j$, then there exists funtion $\phi_j(x) \in C_0^\infty(\Omega_j)$ such that $\phi_j(x) > 0$, $\sum_j \phi_j < 1$, $\sum_j \phi_j = 1$ in K. This is called a <u>smooth partition of unity</u> (see Hörmander [1]). Write $\phi_j = d_j^2$ (since $\phi_j > 0$), then $\sum_j d_j^2 \equiv 1$. Scale the partition of unity by defining

$$e_j(\underline{x}) = d_j\left(\frac{\underline{x}}{\sqrt{k}}\right).$$

Then $\sum_j e_j^2 = 1$, having support $O(\sqrt{k})$, and satisfies a Lipschitz condition

$$|e_j(\underline{x}) - e_j(\underline{x}')| < \frac{C}{\sqrt{k}}|\underline{x}-\underline{x}'|$$

We want to show that

(C.29) $$|(\underline{u},\underline{R}_\Delta\underline{u}) - \Sigma(e_j\underline{u},\underline{R}_\Delta e_j\underline{u})| < \text{const. } k\|\underline{u}\|^2.$$

Let $\underline{R}_\Delta = \sum_j \underline{R}_j S_+^j$ and consider a typical term $\underline{R}_i S_+^i$, then by substitution into the left hand side of (C.29),

(C.30) $$(\underline{u},\underline{R}_i S_+^i \underline{u}) - \sum_j (e_j\underline{u}, R_i S_+^i e_j\underline{u}).$$

By the definition of the inner product,

$$(\underline{u},\underline{R}_i S_+^i \underline{u}) = \sum_\ell \underline{u}_\ell^T(\underline{R}_i S_+^i u)_\ell h,$$

$$(e_j\underline{u},\underline{R}_i S_+^i e_j\underline{u}) = \sum_\ell (e_j\underline{u})^T(\underline{R}_i S_+^i e,\underline{u})_\ell h,$$

so then, by substitution into (C.30), the left hand term in (C.29) becomes

(C.31) $$\sum_\ell (\underline{u}_\ell^T\underline{R}_i(\ell h) S_+^i \underline{u}_\ell)h(1-e_j S_+^i e_j)_\ell.$$

However,

$$\sum_j \frac{1}{2}(e_j(\underline{x}) - S_+^i e_j(\underline{x}))^2 = \frac{1}{2}(\sum_j e_j^2 - 2\sum_j e_j S_+^i e_j + \sum_j (S_+^i e_j)^2)$$

$$= \frac{1}{2}(1 - 2\sum_j e_j S_+^i e_j + 1)$$

$$= 1 - \sum_j e_j S_+^i e_j,$$

and by substitution into (C.31), we obtain the term

$$\frac{1}{2}\sum_\ell (\underline{u}_\ell^T\underline{R}_i(\ell h) S_+^i \underline{u}_\ell)h(e_j - S_+^i e_j)^2.$$

Since e_j is Lipschitz continuous,

$$|(e_j - S_+^i e_j)| < k\cdot(k)^{-1/2} = \sqrt{k}$$

and hence

$$(e_j - S_+^i e_j)^2 < k.$$

Since the $\underline{A}_j$'s are bounded, it follows that the $\underline{R}_j$'s are bounded

$$\frac{1}{2}\sum_\ell (\underline{u}_\ell^T\underline{R}_i(\ell h) S_+^i \underline{u}_\ell)h(e_j - S_+^i e_j)^2| < \text{const. } k\,\Sigma(\underline{u}_\ell^T u_\ell)h$$
$$= \text{const. } k\|\underline{u}\|^2.$$

Hence, we have proven the inequality (C.29), from which we can conclude that inequality (C.27) is proved if we can prove that the following inequality

(C.33) $$(\underline{u}_\ell,\underline{R}_\Delta\underline{u}_\ell) > -Mk\|\underline{u}_\ell\|^2,$$

where $\underline{u}_\ell = e_\ell\underline{u}$.

Take the origin as a point in the support of $\underline{u}_\ell$, and expand $\underline{R}$ in a Taylor series in space about the origin

$$\underline{R}(\underline{x},\underline{\xi}) = \underline{R}(0,\underline{\xi}) + \sum_{j=1}^{p} x_j \frac{\partial \underline{R}(\underline{x},\underline{\xi})}{\partial x_j}\bigg|_{\underline{x}=0} + O(\underline{x}^2).$$

This expansion is valid since the $\underline{A}_j(\underline{x})$'s have bounded second derivatives. Let $\underline{R}^{(0)}(\xi) = \underline{R}(0,\underline{\xi})$ and $\underline{R}^{(j)}(\xi) = \frac{\partial \underline{R}}{\partial x_j}(\underline{x},\underline{\xi})\big|_{\underline{x}=0}$. Then

(C.34) $$\tilde{\underline{R}}(\underline{x},\underline{\xi}) = \underline{R}^{(0)}(\underline{\xi}) + \sum_j x_j \underline{R}^{(j)}(\underline{\xi}) + O(k),$$

where $O(\underline{x}^2) = O(k)$, hence we may neglect this term. So define

$$\tilde{\underline{R}}(\underline{x},\underline{\xi}) = \underline{R}^{(0)}(\xi) + \sum_j x_j \underline{R}^{(j)}(\xi).$$

Let $\tilde{\underline{R}}_\Delta$ denote the finite difference operator whose symbol is $\tilde{\underline{R}}(\underline{x},\underline{\xi})$. Hence, in order to show the inequality in (C.33), we can reduce this to proving the following inequality

(C.35) $$(\underline{u}_\ell, \tilde{\underline{R}}_\Delta \underline{u}_\ell) \geq -Mk\|\underline{u}_\ell\|^2,$$

that is

(C.35') $$(\underline{u}_\ell, \underline{R}_\Delta^{(0)} \underline{u}_\ell) + \sum_j (\underline{v}_{j\ell}, \underline{R}^{(j)} \underline{u}_\ell) \geq -Mk\|\underline{u}_\ell\|^2.$$

where $\underline{v}_j = x_j \underline{u}$.

Since $\underline{R}(\underline{x},\underline{\xi}) \geq 0$, there exists a constant $\overline{M}$ such that

$$\underline{R}^{(0)}(\underline{\xi}) + x_j \underline{R}^{(j)}(\xi) + \overline{M}k\underline{I} \geq 0$$

for $|x_j| \leq \sqrt{k}$.

In order to obtain the desired estimate we shall need two lemmas. The first lemma shall be stated and proved now.

<u>Lemma 1</u>. Let $\overline{A}$ and $\overline{B}$ be hermitian matrices such that $\overline{A} + \overline{B} \geq 0$ and $\underline{A} - \underline{B} \geq 0$, then for any real number M and two vectors $\underline{v}_1$ and $\underline{v}_2$,

$$|(\underline{v}_1, \overline{B}\underline{v}_2)| \quad \tfrac{1}{2}M^2(\underline{v}_1, \overline{A}\underline{v}_1) + \tfrac{1}{2}M^{-2}(\underline{v}_2, \overline{A}\underline{v}_2).$$

<u>Proof</u>: For any vectors $\underline{w}_1$ and $\underline{w}_2$ we have

$$((\underline{w}_1+\underline{w}_2), (\overline{A}+\overline{B})(\underline{w}_1+\underline{w}_2)) + ((\underline{w}_1-\underline{w}_2), (\overline{A}-\overline{B})(\underline{w}_1-\underline{w}_2))$$

$$= 2(\underline{w}_1, \overline{A}\underline{w}_1) + 2(\underline{w}_2, \overline{A}\underline{w}_2) + 2(\underline{w}_1, \overline{B}\underline{w}_2) + 2(\underline{w}_2, \underline{B}\underline{w}_1) \geq 0.$$

So, by substitution of $\underline{w}_1 = M\underline{v}_1$ and $\underline{w}_2 = M^{-1}e^{i\theta}\underline{v}_2$, where θ is suitably chosen, we obtain the desired result.

Let $\overline{A} = \underline{R}^{(0)}(\xi) + M_1 k\underline{I}$ and $\overline{B} = \sqrt{k}\underline{R}^{(j)}(\xi)$. It can readily be seen that $\overline{A}$ and $\overline{B}$ satisfy the conditions of Lemma 1, so that

$$\sqrt{k}\,|(\sigma\underline{v}_{j\ell},\underline{R}_\Delta^{(j)}\underline{u}_\ell)| < \frac{M_2}{2}((\underline{u}_\ell,\underline{R}_\Delta^{(0)}\underline{u}_\ell) + M_1 k\|\underline{u}_\ell\|^2) + \frac{1}{2M_2}(\underline{v}_{j\ell},\underline{R}_\Delta^{0}\underline{v}_{j\ell}) + M_1 k\|\underline{v}_{j\ell}\|^2).$$

By the definition of $\underline{v}_{j\ell}$, $\|\underline{v}_{j\ell}\|^2 < O(k)\|\underline{u}_\ell\|^2$. With this and by taking $M_2 = \frac{\sqrt{k}}{p}$, the above inequality becomes

$$|(\underline{v}_{j\ell},\underline{R}_\Delta^{(j)}\underline{u}_\ell)| < \frac{1}{2p}(\underline{u}_\ell,\underline{R}_\Delta^{(0)}\underline{u}_\ell) + \frac{p}{2k}(\underline{v}_{j\ell},\underline{R}_\Delta^{0}\underline{v}_{j\ell}) + O(k)\|\underline{u}_\ell\|^2.$$

Using this estimate for the second term of (C.35'), we can obtain a lower bound for it

$$(\underline{u}_\ell,\tilde{\underline{R}}_\Delta\underline{u}_\ell) > (\underline{u}_\ell,\underline{R}_\Delta^{(0)}\underline{u}_\ell) - \sum_j(\frac{1}{2p}(\underline{u}_\ell,\underline{R}_\Delta^{(0)}\underline{u}_\ell) + \frac{p}{2k}(\underline{v}_{j\ell},\underline{R}_\Delta^{(0)}\underline{v}_{j\ell})) = (\underline{u}_\ell,\underline{R}_\Delta^{(0)}\underline{u}_\ell) - \frac{p}{2k}\sum_j(\underline{v}_{j\ell},\underline{R}_\Delta^{(0)}\underline{v}_{j\ell}), \tag{C.37}$$

so that it remains to prove

$$\frac{1}{2}(\underline{u}_\ell,\underline{R}_\Delta^{(0)}\underline{u}_\ell) - \frac{p}{2k}\sum_j(\underline{v}_{j\ell},\underline{R}_\Delta^{(0)}\underline{v}_{j\ell}) > -O(k)\|\underline{u}_\ell\|^2. \tag{C.38}$$

We shall need the second lemma now.

<u>Lemma 2</u>. If the symbol $\underline{R}$ is hermitian and $\phi(\underline{x})$ is a smooth real scalar function such that $\phi(\underline{x}) < 1$, then for $\underline{u}$ real

$$|(\phi\underline{u},\underline{R}_\Delta\phi\underline{u}) - (\phi\underline{u},\phi\underline{R}_\Delta\underline{u})| < K(L+L^2)(k)^2\|\underline{u}\|^2, \tag{C.39}$$

where L is the Lipschitz constant for ϕ and K is a constant which depends on $\underline{R}$ only.

<u>Proof</u>: Since $\underline{R}$ is hermitian and the difference operator $\underline{R}_\Delta$ whose symbol is $\underline{R}$ is of the form $\underline{R}_\Delta = \underline{R}_j S_+^j$, then the $\underline{R}_j(\underline{x})$'s are hermitian. Consider one such term $\underline{R}_j S_+^j$, then

$$\begin{aligned}\underline{I}_j &= (\phi\underline{u},\underline{R}_j S_+^j\phi\underline{u}) - (\phi\underline{u},\phi\underline{R}_j S_+^j\underline{u}) \\ &= \sum_\ell[(\phi(\ell h)u_\ell^T\underline{R}(\ell h)\phi((\ell+j)h)\underline{u}_{\ell+j}h) \\ &\quad - (\phi(\ell h)\underline{u}_\ell^T\phi(\ell h)\underline{R}_j(\ell h)\underline{u}_{\ell+j}h)] \\ &= \sum(\underline{u}_\ell^T\underline{R}_j\underline{u}_{\ell+j})\phi(\ell h)(\phi((\ell+j)h) - \phi(\ell h))h.\end{aligned} \tag{C.40}$$

Replacing ℓ by $\ell-j$ and j by $-j$, by (C.40),

(C.41) $$\underline{I}_j = \sum_\ell (\underline{u}^T_{\ell+j}\underline{R}_{-j}\underline{u}_\ell)\phi((\ell+j)h)(\phi(\ell h)-\phi((\ell+j)h))h.$$

Combining (C.40) and (C.41),

$$\underline{I}_j = \frac{1}{2}\sum_\ell(\phi(\ell+j)h) - \phi(\ell h))(\phi(\ell h)u_\ell\underline{R}, \underline{u}_{\ell+j} - \phi((\ell+j)h)\underline{u}^T_{\ell+j}\underline{R}_{-j}\underline{u}_j h$$

and

(C.42) $$\underline{I}_j < L|j|k\sum_\ell|(\phi(\ell h)-\phi((\ell+j)h))\underline{u}_\ell\underline{R}_j\underline{u}_{\ell+j} + \phi((\ell\pm j)h)(\underline{u}^T\underline{R}_j\underline{u}_{\ell+j} - \underline{u}^T_{\ell+j}\underline{R}_{-j}\underline{u}_p)|h .$$

We see that

$$|(\phi(\ell h)-\phi((\ell+j)h)\underline{u}^T_\ell\underline{R}\ \underline{u}_{\ell+j}|h < L|j|k\|\underline{R}_j\|\|\underline{u}\|^2$$

and

$$|\phi((\ell+j)h)(\underline{u}^T_\ell\underline{R}_j\underline{u}_{\ell+j} - \underline{u}^T_{\ell+j}\underline{R}_{-j}\underline{u}_\ell)|h < K|j|k\|\underline{u}\|^2,$$

where K is the Lipschitz constant for $\underline{R}$. Combining gives

$$|\underline{I}_j| < Lj^2k^2(L\|\underline{R}_j\|+K)\|\underline{u}\|^2,$$

from which the inequality follows by summing from $j=1$ to p.

Let ϕ in Lemma 2 be set equal to x; then

$$|(\underline{u}_{j\ell}, \underline{R}^{(0)}_\Delta\ \underline{u}_{j\ell}) - (x^2_j u_\ell, \underline{R}^{(0)}_\Delta\ \underline{u}_\ell)| < K(k)^2\|\underline{u}_\ell\|^2,$$

where K depends on $\underline{R}^{(0)}$ only, or

(C.43) $$(\underline{u}_{j\ell}, \underline{R}^{(0)}_\Delta\ \underline{u}_{j\ell}) = (x^2_j u_\ell, \underline{R}^{(0)}_\Delta\ \underline{u}_\ell) + O(k^2)\|\underline{u}_\ell\|^2.$$

By substitution of this result (C.43) into (C.42), setting $\sum_j x^2_j = r^2$

(C.44) $$(\underline{u}_\ell, \tilde{\underline{R}}_\Delta\underline{u}_\ell) > \frac{1}{2}(\underline{u}_\ell, \underline{R}^{(0)}_\Delta\ \underline{u}_\ell) - \frac{1}{2}(\frac{pr^2}{k}\underline{u}_\ell, \underline{R}^{(0)}_\Delta\ \underline{u}_\ell) \pm O(k)\|\underline{u}_\ell\|^2$$

$$= \frac{1}{2}((1 - \frac{pr^2}{k})\underline{u}_\ell, \underline{R}^{(0)}_\Delta\ \underline{u}_\ell) + O(k)\|\underline{u}_\ell\|^2.$$

Let $\underline{u}_\ell$ have support such that $\underline{u}_\ell = 0$ for $r > (\frac{k}{2p})^{1/2}$ and introduce the auxiliary function ϕ defined by

$$\phi(r) = \begin{cases} 1 - \frac{pr^2}{k}, & r < (\frac{k}{2p})^{1/2} \\ \frac{1}{2}, & r > (\frac{k}{2p})^{1/2} \end{cases}.$$

Rewriting (C.44) in terms of the function ϕ gives

(C.45) $$(\underline{u}_\ell, \underline{R}_\Delta\underline{u}_\ell) > \frac{1}{2}(\phi\underline{u}_\ell, \underline{R}^{(0)}_\Delta\ \underline{u}_\ell) + O(k)\|\underline{u}_\ell\|^2.$$

ϕ satisfies the Lipschitz condition where there is $\sqrt{p/k}$, so by Lemma 2

$$(\phi^2 \underline{u}_\ell, \underline{R}_\Delta^{(0)} \underline{u}_\ell) = (\phi \underline{u}_\ell, \underline{R}_\Delta^{(0)} \phi \underline{u}_\ell) + O(K^2 k^2) \|\underline{u}_\ell\|^2. \tag{C.46}$$

From the Fourier representation of the first term on the right in (C.46)

$$(\phi \underline{u}_\ell, \underline{R}_\Delta^{(0)} \phi \underline{u}_\ell) > 0$$

Since $O(K^2k^2) = O(k)$, by substitution of (C.46) into (C.45),

$$(\underline{u}_\ell, \tilde{\underline{R}}_\Delta \underline{u}_\ell) > O(k) \|\underline{u}_\ell\|^2.$$

We see that in the proof of the last claim the problem had been reduced to one of constant coefficients.

References

1. Hörmander, L., Linear Differential Operators, Springer-Verlag, Berlin (1969).

2. Lax, P. D., On the Stability of Difference Approximations to Solutions of Hyperbolic Equations with Variable Coefficients, Comm. Pure Appl. Math., 14, 497 (1961).

3. Lax, P. D. and L. Nirenberg, On Stability of Difference Schemes: A Sharp Form of Garding's Inequality, Comm. Pure. Appl. Math., 19, 473 (1966).

4. Richtmyer, R. D., and K. W. Morton, Difference Methods for Initial Value Problems, 2nd Ed., Interscience, New York (1967).

Appendix D: Hyperbolic Equations with Discontinuous Initial Data

Consider the hyperbolic equation

(D.1) $$\partial_t v + \beta \partial_x v = 0, \quad t > 0, -\infty < x < +\infty$$

with β a constant different from zero. The solution is uniquely determined by the initial condition

(D.2) $$v(x,0) = f(x).$$

Approximation equation (D.1) and (D.2) by

(D.3) $$u_i^{n+1} = Qu_i^n, \quad n > 0, \quad -\infty < i < +\infty,$$

where $u_i^0 = f(ih)$. Assume that Q is a polynomial of the form $Q = Q(S_+,S_-)$ (and accurate of order p). Let $\rho(\xi) = Q(e^{-i\xi},e^{i\xi})$ denote the symbol of Q.

We shall illustrate the solutions of u_i^n of (D.3) for the case where $f(x)$ is a step function $T(0)$, and

(D.4) $$T(y_0) = \begin{cases} 1, & y_0 < x < y_1, \ y_1 >> 0 \\ 0, & x < y_0 \\ g(x), & x > y_1 \ , \end{cases}$$

where $g(x)$ is arbitrary but sufficiently smooth such that $T \in L_2(\mathbb{R})$.

Consider the characteristic line $x + t = 0$ of Equation (D.1) and (D.2). The discontinuity in the initial value of (D.1) and (D.2) follows the characteristic line. However, in the numerical solution of (D.3) the discontinuity has spread out over a region surrounding the characteristic line (see Figure D.1) due to the numerical dissipation.

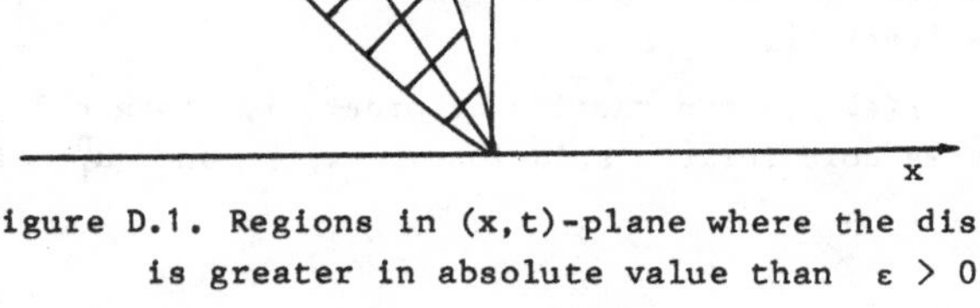

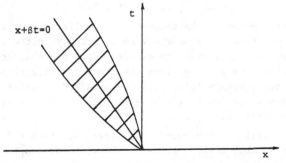

Figure D.1. Regions in (x,t)-plane where the disturbance is greater in absolute value than $\varepsilon > 0$.

The solution u_i^n of (D.3) has the feature that when p grows there appear parasitic waves, which is Gibb's phenomenon (see Dym and McKean [1]) from higher order terms of Q as displayed in Figure D.2.

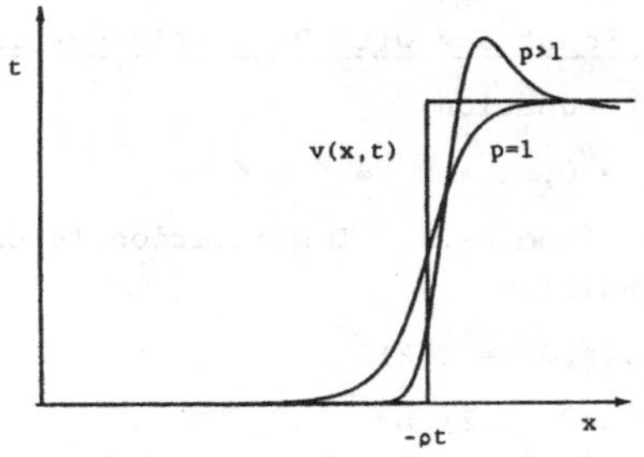

Figure D.2. u_i^n for p=1 and p>1.

We must choose methods that control these waves. It turns out that dissipation is useful in contolling these waves.

We shall now prove the main theorem of this section which is due to Apelkrans [9]. We first introduce the notion of contraction.

<u>Definition</u>. Let α and ξ be real. We say that ρ is <u>contractive of order</u> $\tau > 1$, if a function $R(\xi)$ exists such that for all $0 < \alpha < \alpha_0$, and for all ξ,

$$\rho(\xi - i\alpha) = (e^{\alpha\lambda\beta + \sigma\alpha^\tau})R(\xi),$$

with $|R(\xi)| < 1$ and $\lambda = k/h$.

Consider (D.1) and (D.2) with initial values

$$v(x,0) = \begin{cases} 1, & \text{for } x = 0, \\ 0, & \text{for } x \neq 0, \end{cases}$$

then the generalized solution of (D.1) has the form

$$v(x,\tau) = \begin{cases} 1, & x + \beta t = 0, \\ 0, & x + \beta t \neq 0. \end{cases}$$

The estimates given in the theorem and corollary below show that the influence upon the error from the jump continuity in the initial data decrease exponentially with the distance from the characteristic through the origin. The <u>distance</u> from the characteristic through the origin is given by $d(x,t) = |g(x,t)|$, where $g(x,t)$ satisfies equation $\partial_t g = \beta(x,t)\partial_x g$ with $g(x,0) = x$. In the case of Equation (D.1), $d(x,t) = |(\beta t + x)|$.

<u>Theorem D.1</u>. If $\rho(\xi)$ is contractive of order τ, then the following estimates hold for the solutions $v(x,t)$ and u_i^n of (D.1)

and (D.2), respectively, where the initial values are given by (D.4)

(D.6) $$|v(x,t) - u_i^n| < c(t)e^{-V(x,t)},$$

where $c(t)$ is a function of $t = nk$ and independent of h, and $V(x,t) = h^{-q}|\beta t+x|$, where $x = ih$ and $g = \frac{\tau-1}{\tau}$.

Proof. Let $\alpha = O(h^r)$, $0 < r < 1$ and α real. The exponent r will be specified later. Make the following transformation

$$w_i^n = e^{-\alpha i}u_i^n = e^{-\alpha x/h}u_i^n,$$

then w_i^n satisfies the difference equation

$$w_i^{n+1} = Q'w_i^n$$

If $\rho'(\xi)$ is the symbol of Q', then $\rho'(\xi) = \rho(\xi-i\alpha)$. Using the stability of Q and the contractivity of ρ

(D.7) $$\|w_i^n\|_2 < e^{\alpha\beta t/h+\gamma}\|w_i^0\|_2,$$

where $\gamma = \sigma\alpha^t h^{-1/\lambda}$. $w_i^0 = u_i^0$ since $u_i^0 \neq 0$ only for $i = 0$, hence $\|w_i^0\|_2 = \|u_i^0\|_2 = \sqrt{h}$. Using th definition of the ℓ_2-norm

(D.8) $$\|w_i^n\|_2^2 > h|w_i^n|^2,$$

for every i, from which it follows that, by substitution of $\|w_i^0\| = \sqrt{h}$ and inequality (D.8) and (D.7),

$$\sqrt{h}|w_i^n| < \sqrt{h}e^{\alpha\beta t/h+\gamma},$$

or

$$|w_i^n| < e^{\alpha\beta t/h+\gamma}.$$

However, $|w_i^n| = e^{-\alpha i}|u_i^n|$ and

(D.9) $$|u_i^n| < e^{\alpha i}e^{\alpha\beta t/h+\alpha}.$$

Choose $\alpha = -\text{sign}(\beta t+x)h^{1/\tau}$. The substitution of α into (D.7) gives

$$|u_i^n| < c_0(t)e^{-\text{sign}(\beta t+x)h^{-q}(\beta t+x)} = c_0(t)e^{-V(x,t)},$$

where $c_0(t) = \exp\{(-\text{sign}(\beta t+x)^\tau t\sigma/\lambda\}$, $q = \frac{\tau-1}{\tau}$, and $V(x,t) = h^{-q}|\beta t+x|$.

If $\alpha = 0$ (that is, $w_i^n = u_i^n$), then by the above argument, $|u_i^n| < 1$ for $x + \beta t = 0$ and $|v(x,t) - u_i^n| < 2$, where $v(x,t) = 1$ on $x + \beta t = 0$. Choose

$$C(t) = \begin{cases} C_0(t), & x + \beta t \neq 0 \\ 2, & x + \beta t = 0, \end{cases}$$

which is the desired result.

The following corollary will be considered which deals with the case in which the initial values are given by a step function (D.4)

Corollary. If $u_i^{n+1} = Qu_i^n$ is accurate of order p and ρ is contractive of order τ, then

$$|u_i^n - T(-\beta t)| \leq C(t)h^{(q-1)/2}e^{-V(x,t)},$$

$|x+\beta t|_0 < y$, where $V(x,t) = h^{-q}|\ t+x|$, u_i^n is the solution of (D.3) with $u_i = T(0)$, $C(t)$ is independent of h and $q = \frac{\tau-1}{\tau}$, $x = ih$, and $t = nk$.

Proof. First, for $\alpha > 0$,

$$\|e^{-\alpha x/h}T(0)\|_2 \leq \|T(0)\|_{max}(\sum_i he^{-2\alpha i})^{1/2}$$

$$= \|T(0)\|_{max}\sqrt{h}(\sum_i (e^{-2\alpha})^i)^{1/2}$$

$$= \frac{\|T(0)\|_{max}\sqrt{h}}{\sqrt{1-e^{-2\alpha}}} = O(\alpha^{-1/2}h^{1/2}).$$

To the left of the characteristic line $x + \beta t = 0$, since $\alpha > 0$, we obtain the desired estimate. This follows from Theorem D.1 with the right hand side of inequality (D.6) given by $\alpha^{-1/2} = h^{-1/2\tau} = h^{(q-1/2}$.

Let $T'(0)$ be the step function which is the reflection of $T(0)$ across the line $x = 0$ (t-axis). Then for $\alpha < 0$, as above

$$\|e^{-\alpha x/h}T'(0)\|_2 \equiv O(|\alpha|^{-1/2}h^{1/2}).$$

Thus the result holds to the right of the characteristic line as above. Combine $T(0)$ and $T'(0)$ to obtain the smooth function for every x.

$$\bar{T}(0) = T(0) + T'(0).$$

Let $\bar{v}(x,t)$ denote solution to (D.1) and (D.2) with $f(x) = \bar{T}(0)$ and $\bar{u}_i^n$ denote solution to corresponding finite difference method (D.3). Then since Q is accurate of order p, $\|\bar{v}(x,t)-\bar{u}_i^n\| = O(h^p)$, provided that $T(0)$ is sufficiently smooth. For $\alpha < 0$, write $T(0) = \bar{T}(0) - T'(0)$. The estimate follows by use of the triangle inequality.

We shall now prove a theorem which will determine the best possible $\tau > 1$, that is, the largest order of contractivity. We first need to state a technical lemma (see Section III.3 and Lax [4]).

Lemma 1. If the finite difference method (D.2) is accurate of order p, then

$$\rho(\xi) = e^{i\beta\lambda\xi+U(\xi)},$$

where $U(\xi)/(i\xi)^{p+1}$ is an analytic function of ξ.

Observe that since $U(\xi)/(i\xi)^{p+1}$ is an analytic function of ξ, $U(\xi)$ may be expanded in a Taylor series of the form

(D.10) $$U(\xi) = C_0(i\xi)^{p+1} + C_1(i\xi)^{p+2} + \dots .$$

Theorem D.2. Suppose that the finite difference method (D.3) is accurate of order p and dissipative of order $2s$. Let $r = 2s - p$. Then there exists a constant $\alpha_0 > 0$, real function $R(\xi)$, and a constant $\sigma > 0$ such that

$$\rho(\xi - i\alpha) = e^{\alpha\lambda\beta + \sigma\alpha^\tau} R(\xi), \quad \text{for} \quad |\alpha| < \alpha_0,$$

where $|R(\xi)| \leq 1$ for $|\xi| \leq \pi$. Furthermore, τ is given by

$$\tau = \begin{cases} 4s/(r+1) \text{ or } 8s(r+3), \text{ for } p = 2j-1 \\ 2s/r, \text{ for } C_0\alpha = 0,\ p = 4h-2 \\ \qquad\qquad \text{or} \qquad\qquad\qquad\qquad h = 1,2,\dots \\ \qquad C_0\alpha > 0,\ p = 4h \qquad\qquad\qquad\qquad\qquad \text{and } p = 2j \\ 6s/(r+2), \text{ for } C_0\alpha > 0,\ p = 4h-2 \\ \qquad\qquad\qquad\qquad\qquad\qquad h = 1,2 \\ \qquad C_0\alpha < 0,\ p = 4h \end{cases}$$

where $j = 1,2,\dots$.

Proof. Since the finite difference method (D.3) is dissipative of order $2s$ (s a natural number) there exists a constant $\delta_1 > 0$ such that

$$|\rho(\xi)| \leq 1 - \delta_1 \xi^{2s}, \quad \text{for} \quad |\xi| \leq \pi.$$

From the above lemma

$$|\rho(\xi)| = e^{\mathrm{Re}(u)} \leq e^{-\delta_1 \xi^{2s}}.$$

Now we write

$$\rho(\xi - i\alpha) = e^{\alpha\beta\lambda + i\beta\lambda\xi + U(\xi - i)}.$$

By expanding $U(\xi - i\alpha)$ in a Taylor series

$$U(\xi - i\alpha) = U(\xi) + \alpha U'(\xi) + \dots,$$

(D.11) $$|\rho(\xi - i\alpha)| = e^{\alpha\beta\lambda + \mathrm{Re}(U) + \alpha \mathrm{Re}(U') + \dots} \leq e^{\alpha\beta\lambda - \delta_1\xi^{2s} + H(\alpha,\xi,p)}$$

where

$$H(\alpha,\xi,p) = \mathrm{Re}\{\alpha U'(\xi) + \dots\}.$$

The leading terms of $H(\alpha,\xi,p)$ are

(D.12) $$H(\alpha,\xi,p) = \mathrm{Re}\{C_0(\alpha^{p+1} + a_1\alpha(i\xi)^p + a_2\alpha^2(i\xi)^{p-1} + \dots) + \dots\},$$

where $a_j = \binom{p+1}{j}$, for $j = 1,2,\dots$. Compare terms of the form $C\alpha^{j+1}\xi^{p-j}$, where C is a constant, and $j = 0,1,2,\dots,p-1$ with the term $-\delta^{2s}$ in (D.11). Applying a version of Hölder's inequality* to the terms $C\alpha^{j+1}\xi^{p-j}$ we obtain

$$|\alpha|^{j+1}|\xi|^{p-1} < d_j|\alpha|^{2s(j+1)/(r+1)} + \varepsilon\xi^{2s},$$

where $r = 2s-p$. The exponent $\tau_j = 2s(j+1)/(r+1)$ is a strictly monotonically increasing function of j for $j > 0$ and $r > 1$. We are interested in the sign of the terms in the expansion in (D.12). All imaginary terms have dropped out. Consider the case where p is an odd integer,

$$\text{(D.13)} \qquad H(\alpha,\xi,p) = C_0(\alpha^{p+1} + (-1)^{(p-1)/2}a_2\alpha^2\xi^{p-1} + \ldots).$$

From inequality (D.11) the smallest exponent of $|\alpha|$ is either $4s/(r+1)$ or $8s/(r+3)$ depending on the sign of C_0 and of p. If the negative terms in (D.13) are neglected,

$$\text{(D.14)} \qquad H(\alpha,\xi,p) < e_1\alpha^\tau + e_2\xi^{2s} + \ldots,$$

where $\tau = 4s/(r+1)$ or $\tau = 8s/(r+3)$ and $e_2 = O(\varepsilon)$, as $\varepsilon \to 0$. Similarly, for the case where p is even (D.14) holds with

$$\tau = 2s/r \quad \text{for} \quad C_0\alpha < 0,\ p = 4h-2 \quad \text{or} \quad C_0\alpha > 0,\ p = 4h$$

and

$$\tau = 6s/(r+2) \text{ for } C_0\alpha < 0,\ p = 4h-2 \text{ or } C_0\alpha > 0,\ p = 4h,$$

for $h = 1,2,\ldots$.

Now by substitution into (D.11)

$$|\rho(\xi-i\alpha)| < e^{\alpha\beta\lambda-(\delta_1-e_2)\xi^{2s}+e_1\alpha^\tau+\ldots},$$

from which, for ξ and α sufficiently small, we can choose $\varepsilon = \varepsilon_0 > 0$ so small that

$$\xi^{2s}(\delta_1-e_2+\ldots) > \tfrac{1}{2}\delta_1\xi^{2s}$$

where the dots refer to higher order terms in ξ and α. Hence there is a constant $\sigma > 0$ such that

$$\text{(D.15)} \qquad |(e_1+\ldots)\alpha^\tau| < \sigma\alpha^\tau$$

for $\varepsilon > \varepsilon_0$ since e_1 is uniformly bounded by the above lemma. $\rho(\xi-i\alpha)$ can be written in the form

$$\text{(D.16)} \qquad \rho(\xi-i\alpha) = e^{\alpha\beta\lambda+\sigma\alpha^\tau}R(\xi),$$

---

*<u>Hölder's inequality</u>. Let $A > 0$, $B > 0$, and $\varepsilon > 0$ be given, then

$$AB < \mu A^r + \varepsilon B^q,$$

where $r = q/(q-1)$, $q > 1$, and $\mu > 0$ uniformly bounded for $\varepsilon_0 > 0$, with $\varepsilon_0 > 0$.

where $|R(\xi)| < 1$ for ξ and α sufficiently small.

Suppose that in order to obtain (D.16) we had to choose $|\xi| < \xi_0$ (for some ξ_0). For the case where $|\xi| > \xi_0$, the term $e^{-\delta_1 \xi^{2s}}$ is available. By the definition of H, $H(\alpha,\xi,p) \to 0$ as $|\alpha| \to 0$ and

$$|\rho(\xi - i\alpha)| = e^{\alpha\beta\lambda\ \delta_1 \xi^{2s} + \sigma\alpha^\tau - \sigma\alpha^\tau + H(\alpha,\xi,p)}$$

$$< e^{\alpha\beta\lambda + \sigma\alpha^\tau - \frac{1}{2}\delta_1 \xi^{2s}}$$

for $|\alpha|$ sufficiently small. Hence there exists an $\alpha_0 > 0$ such that

$$\rho(\xi - i\alpha) = e^{\alpha\beta\lambda + \sigma\alpha^\tau} R(\xi), \quad \text{for } |\alpha| < \alpha_0$$

where $|R(\xi) < 1$ for $|\xi| < \pi$.

We now consider systems of hyperbolic equations with constant coefficients. In regions where the exact solution is known to be smooth, one might expect to obtain better rates of convergence by the use of higher order methods. In the theorem due to Majda and Osher [8] a model differential equation and corresponding general finite difference method with discontinuous initial data is considered. It is proven that the rate of convergence between the characteristics emanating from the origin is fixed independent of the order of the difference method.

The model differential equation has the form

$$\text{(D.17)} \qquad L\underline{v} \equiv \partial_t \underline{v} - \underline{A}\partial_x \underline{v} + \underline{B}\underline{v} = 0$$

where

$$\underline{v} = \begin{pmatrix} v_1 \\ v_2 \end{pmatrix}, \quad \underline{A} = \begin{pmatrix} 0 & 0 \\ 0 & -1 \end{pmatrix}, \quad \underline{B} = \begin{pmatrix} 0 & 1 \\ -1 & 0 \end{pmatrix}.$$

The initial values are given by

$$\underline{v}(x,0) = \begin{pmatrix} \phi(x) \\ 0 \end{pmatrix} = \begin{pmatrix} H(x) \\ 0 \end{pmatrix},$$

where $H(x)$ is the Heaviside function

$$H(x) = \begin{cases} 1, & x > 0 \\ 0, & x < 0 \end{cases}.$$

Discretize in space only. Consider the finite difference method given by

$$\text{(D.18)} \qquad L_i \underline{u}_i = \partial_t \underline{u}_i - (\underline{A}_i \underline{u}_i + \underline{B}\underline{u}_i)$$

where

$$\underline{u}_i = \begin{pmatrix} u_{1,i} \\ u_{2,i} \end{pmatrix} \quad \text{and} \quad \underline{A}_i = \begin{pmatrix} P_{1,i} & 0 \\ 0 & P_{2,i} \end{pmatrix}.$$

The initial values are given by

$$\underline{u}_i(0) = \begin{pmatrix} \phi(ih) \\ 0 \end{pmatrix}.$$

It is critical that use the average value of H(x) is used at the grid point corresponding to $x = 0$, that is, $i = 0$. Define

$$\phi(0) = \frac{1}{2}(\phi(0^+) + \phi(0^-)) = \frac{1}{2}.$$

The difference operators $P_{1,i}$ and $P_{2,i}$ are finite difference approximations to ∂_x and $-\partial_x$, respectively, which are accurate of order r and dissipative of order $2s > r+1$.

Let $\rho_1(\xi)$ and $\rho_2(\xi)$ denote the symbols of $P_{1,i}$ and $P_{2,i}$, then $P_{1,i} - P_{2,i}$ is said to satisfy the <u>ellipticity condition</u> if

$$\text{(D.19)} \qquad \frac{\rho_1(\xi)-\rho_2(\xi)}{h} = 2i\xi(1+\phi(\xi)),$$

where $(1+\phi(\xi)) \neq 0$ for $|\xi| \leq \pi$ and ϕ is an analytic function.

If $\underline{B} \equiv 0$ and the initial data for both components is the Heaviside function, the model Equation (D.17) reduces to two scalar equations of the form (D.1). In particular, near the characteristics $x \pm t = 0$, a Gibbs phenomenon takes place, so that

$$\overline{\lim_{h\to 0}} \max_{0\leq t<T_0} |\underline{v}(x,t)-\underline{u}_i(t)| \geq c > 0$$

for some constant c. However, in the region R_δ defined for any $\delta > 0$ by

$$R_\delta = \{(x,t) \mid \frac{|x-t|}{t} > \delta\ ,\ \frac{|x+t|}{t} > \delta,\ \delta \leq t \leq T_0\},$$

there is an optimal rate of convergence given by

$$\max_{(x,t)\in R_\delta} |\underline{v}(x,t)-\underline{u}_i(t)| \leq c_\delta h^r,$$

where c is constant which depends on δ.

When $\underline{B} \neq 0$, $\underline{B}$ is a dispersive coupling term and the solution is smooth everywhere except for a jump continuity in v and its derivative along with characteristics $x \pm t = 0$. Under these conditions Majda and Osher [5] proved the following theorem.

<u>Theorem D.3</u>. Suppose that the finite difference operators $P_{1,i}$ and $P_{2,i}$ are accurate of order r and dissipative. Assume $P_{1,i} - P_{2,i}$ satisfies the ellipticity condition (D.19). Then in the region

$$\text{(D.20)} \qquad R^1_\delta = R_\delta \cap \{(x,t): \|x\| < t\}$$

we have the estimate

$$\max_{(ih,t)\in R'_\delta} \|\underline{u}_i-\underline{v}\| \leq c_\delta h^2$$

The estimate in (D.20) is sharp in the following sense: If $r > 2$ and R is any region with compact closure in R_δ, then

$$\lim_{h\to 0} \frac{\max_R \|\underline{u}_i(t)-\underline{v}(x,t)\|}{h^2} > 0$$

where $x = ih$. In the region $R_\delta^2 = R_\delta \cap \{(x,t)\|x\| > t\}$, we obtain the estimate

$$\max_{(ih,t)\in R^2} \|\underline{u}_i-\underline{v}\| < c_\delta h^r,$$

where c_δ is a constant that depends on δ. Then the rate of convergence is optimal in R_δ^2. If $r > 2$, $\underline{u}_i - \underline{v}$ has an asymptotic error expansion in R_δ^1 given by

$$\underline{u}_i - \underline{v} = h^2 c_0(\underline{w}(x,t)) + O(h^3)$$

where $c_0 \neq 0$ is a universal constant and $\underline{w}(x,t)$ solves the Cauchy problem $L\underline{w} = 0$ with initial data

$$\underline{w}(x,0) = \begin{matrix} \delta'(0) \\ 0 \end{matrix},$$

where δ' is the derivative of the Dirac measure.

It should be observed that Richardson extrapolation can be used to obtain $O(h^3)$ in R_δ^1 for the above finite difference method which is accurate of order $r > 2$. However, this is not the best that can be obtained.

A better estimate can be achieved if the initial data is "preprocessed." To this end, consider the Dirichlet kernel given by

$$D_N(x) = \frac{\sin((N+\frac{1}{2})x)}{\sin(\frac{1}{2}x)} = \sum_{|\ell|<N} e^{i\ell x}.$$

We define the modified initial data for the semidiscrete (discretization in space only) operator (D.18) by the convolution integral

$$\phi_i^D = \frac{1}{2\pi}\int_{-\pi}^{\pi} D\,(x_i-y)\phi(y)dy \tag{D.21}$$

where $x_i = ih$ and $(2N+1)h = 2\pi$. This is a type of smoothing operation on the initial data.

For this choice of N and for x_i with $|i| < N$, a periodic function ϕ_i defined on these grid points, the ℓ-th finite Fourier coefficient is given by

$$\tilde{\phi}^D(\ell) = \frac{h}{2}\sum_{j=-N}^{N} \phi_j^D\, e^{-i\ell jh}, \quad |\ell| < N.$$

With this the solution to the semi-discrete difference method in (D.18) with the modified initial data (D.21) is given by

$$\underline{u}_j^D(t) = \sum_{\ell=-N}^{N} e^{i\ell x_j} \exp((Q(i\ell h)/h + B)t)\tilde{\phi}^D(\ell).$$

By modifying the initial data as in (D.21), in region R^1 an optimal rate of convergence is attained, given by the estimate

$$\max_{(x_i,t)\in R^1_\delta} \|\underline{u}_i^D(t)-\underline{v}(x,t)\| < Ch^r,$$

where C is a positive constant.

Consider the hyperbolic equation with variable coefficients.

(D.22) $$\partial_t v + \beta(x)\partial_x v = 0, \quad t > 0, \quad -\infty < x < +\infty,$$

where β is a real valued function. Its generalized solutions are uniquely determined by the initial condition

(D.23) $$v(x,0) = f(x).$$

Approximate equations (D.22) and (D.23) by

$$u_i^{n+1} = Qu_i^n,$$

when $Q = \sum_j Q_j(x)S_+^j$. Let $\rho(x,\xi) = \sum_j Q_j(x)e^{-ij\xi}$ denote the symbol of Q.

Let $g(x,t) = c$ denote a family of characteristics for equation (D.22) that satisfies $g(x,0) = x$. Consider the initial condition

$$v(x,0) = \begin{cases} 1, & \text{for } x = 0 \\ 0, & \text{for } x \neq 0, \end{cases}$$

then the generalized solution has the form

$$v(x,t) = \begin{cases} 1, & \text{for } g(x,t) = 0, \\ 0, & \text{for } g(x,t) \neq 0. \end{cases}$$

If $f(x)$ is sufficiently smooth, then equation (D.22) can be very closely approximated by stable finite difference methods and realistic error bounds can be achieved (e.g. Lax [4]). However, suppose f(x) has a discontinuity, this discontinuity propagates in the solution to equation (D.22) along a characteristic. This propagation is disturbed in the solution of the corresponding finite difference method (D.24). A theorem due to Apelkraus [1] gives a sharp bound for the error and indicates how the error depends on the order of accuracy of the difference method. This theorem will make use of the Lax-Nirenberg theorem (Theorem 3.8) in Section III.11.

We shall need to state some technical lemmas and introduce the idea of uniform contractivity.

<u>Lemma 2</u>. Assume that c_0 (in Theorem D.2) is uniformly bounded in x. Then under the assumptions of Theorem D.2,

$$R(\xi) = \rho(\xi) + \alpha R_0(\xi), \quad a \to 0,$$

where $\alpha = h^r$ for $0 < r < 1$, and $R_0(\xi)$ is uniformly bounded in x for $|\xi| < \pi$.

<u>Proof</u>: From Lemma 1 and from (D.13),

$$e^{\alpha\beta\lambda+i\xi\beta+U(\xi-i\alpha)} = e^{\alpha\beta\lambda+\sigma\alpha^\tau}R(\xi)$$

Hence

$$R(\xi) = e^{i\xi\beta\lambda-\sigma\alpha^\tau+U(\xi)+U'(\xi)+\ldots},$$

or

$$R(\xi) = \rho(\xi)e^{-\sigma\alpha^\tau+\alpha U'(\xi)+\ldots}$$

Using the definition of dissipative, choose ε_0 for every x in the in the proof of Theorem D.2. Since c_0 is uniformly bounded in x, there exists a $\sigma > 0$ in (D.12) which holds for every x. This is due to the fact that e_1 essentially depends on c_0 and ε_0; from which the lemma follows.

Definition. If Lemma 2 holds for a symbol ρ, then ρ is uniformly contractive of order τ.

Lemma 3. Suppose that the coefficient of (D.22), $\beta(x) \in C_0^\infty$, and the symbol (D.24) is contractive of order τ for every x, then it is uniformly contractive of order τ, providing the finite difference method is accurate of order p.

Proof: By expanding the terms in the finite difference method (D.24) in a Taylor series we obtain from Equation (D.22) that c_0 is a polynomial in $\beta(x)$ and its derivatives of order p or less. Since $\beta(x) \in C_0^\infty$, we conclude that c_0 is uniformly bounded in x and the result follows by application of Lemma 2.

We shall now state the main theorem.

Theorem D.4. Consider the initial value problem (D.22) and (D.25) approximated by the finite difference method (D.24) which is accurate of order p. Suppose that the symbol $\rho(x,\xi)$ of (D.24) is contractive of order τ for every x and $\beta(x) \in C_0^\infty$. Then there exists a function $c(t)$ independent of x and h such that

$$|v(x,t)-u_i^n| < c(t)e^{-v(x,t)},$$

where $v(x,t) = h^{-q}|g(x,t)|$, $q = (\tau-1)/\tau$, $x = ih$, and $t = nk$.

Proof: Since $\rho(x,\xi)$ is contractive of order τ, it follows that

$$|\rho(x,\xi)| = |R(x,\xi)_{\alpha=0}| < 1$$

for every x and ξ. As in the proof of Lemma 3, by expanding the terms of the difference method (D.24) in a Taylor series (see e.g. Section III.3 and Lax [6]) that $\sum_j Q_j(x)j^\ell$ are polynomials in $\beta(x)$ and its derivatives of order up to $\ell-1$, where $\ell=1,2,\ldots,p$. Since $\beta(x) \in C_0$, the conditions of the Lax-Nirenberg theorem hold.

Make the transformation

$$\bar{v}(x,t) = e^{-\alpha g(x,t)/h} v(x,t) \tag{D.26}$$

then $\bar{v}(x,t)$ satisfies (D.22) with $\bar{v}(x,0) = v(x,0)$ since $v(x,0) \neq 0$ only for $x = 0$. Similarly, make the transformation

$$w_i^n = e^{-\alpha g(x,t)/h} u_i^n, \tag{D.27}$$

where w_i^n satisfies the finite difference method

$$w_i^{n+1} = Q' w_i^n = e^{\alpha g(x,t+k)/h} Q e^{-\alpha g(x,t)/h} u_i^n. \tag{D.28}$$

Since $\beta(x) \in C_0^\infty$, it follows that g and its first and second derivatives are uniformly bounded in x and t. So we may write

$$\begin{aligned}\alpha/h(g(x,t) - g(x,t+k)) &= -\alpha\lambda g_t(x,t) + O(\alpha h)\\ &= -\alpha\lambda\beta(x) g_x(x,t) + O(\alpha h)\end{aligned}$$

as $h \to 0$, since $\partial_t g(x,t) = \beta(x)\partial_x g(x,t)$. We may now write the symbol $\rho'(x,\xi)$ of Q' in terms of the symbol of Q, namely

$$\rho'(x,\xi) = e^{-\alpha\lambda\beta g_x + O(\alpha h)} \rho(x,\xi - i\alpha\partial_x g(x,t)) + O(\alpha h)$$

as $h \to 0$, uniformly in x and t (since $\partial_t^2 g$ and $\partial_x^2 g$ are uniformly bounded in x and t). From Lemma 3 it follows that ρ is uniformly contractive of order τ, so that

$$\begin{aligned}\rho'(x,\xi) &= e^{-\alpha\lambda\beta\partial_x g + \alpha\beta\lambda\partial_x g + (\alpha\partial_x g)^\tau + O(\alpha h)} R(x,\xi) + O(\alpha h)\\ &= \rho(x,\xi) + O(\alpha^\tau + \alpha h),\end{aligned}$$

as $h \to 0$, uniformly in x and τ, since $R(x,\xi) = \rho(x,\xi) + O(\alpha^\tau)$.

Taking $\alpha = -(\text{sign } g(x,t))h^{1/\tau}$, we can conclude that there exists a function $R_1(x,\xi)$ and a constant $M > 0$ such that

$$\rho'(x,\xi) + \rho(x,\xi) + hR_1(x,\xi),$$

where $|R_1(x,\xi)| \leq M_1$ for all x and ξ. Hence the difference operator Q' can be divided into two difference operators

$$Q' = Q + h\bar{Q}.$$

By the Lax-Nirenberg theorem, the difference operator Q is stable and since $\|\bar{Q}\|_2 < \infty$ for h sufficiently small, the finite difference method (D.28) is stable, that is, there exists a function $c_0(t)$ independent of x and h such that

$$\|w_i^n\|_2 \leq c_0(t)\|w_i^0\|_2.$$

Using (D.29) and following the proof of Theorem D.1, we obtain the result

$$|u_i^n| < c_0(t)e^{-v(x,t)},$$

where $v(x,t) = h^{-q}|g(x,t)|$, and $q = \frac{\tau-1}{\tau}$.

The proof is completed by observing that $u_i^n \neq 0$ only when $g(x,t) = 0$ (where $x = ih$ and $t = nk$). Hence there exists a function $c(t)$ such that

$$|v(x,t)-u_i^n| < c(t)e^{-v(x,t)}.$$

A recent pointwise estimate that extends the results of Majda and Osher has been made by Mock and Lax [6] for the case of variable coefficients. They show that for a linear hyperbolic equation with variable coefficients in any number of dimensions, if the initial data is preprocessed in an appropriate manner near the discontinuities, then for a finite difference method of order q, the moments of $\underline{u}$ approximate the moments of $\underline{v}$ to order q. Furthermore, if the approximate solution is suitably postprocessed, then the approximate solutions (as well as its derivatives) will differ from the exact solution, by $O(h^{q-\delta})$ where δ is as small as one wishes.

The preprocessing of the initial data is accomplished through the use of the result from numerical integration.

Lemma 4. Let f be any C^∞ function on $\mathbb{R}^+$, with bounded support. Given any positive integer q, there exists a quadrature formula accurate of order q of the form

$$\int_0^\infty f(x)dx = \sum_{j=0}^{\infty} w_j f(jh)h + O(h^q),$$

where the weights w_j depend on q and $w_j = 1$ for $j > q$.

This lemma may be applied twice to obtain the following result:

Lemma 5. Let f be a piecewise C^α function with the discontinuity located at $x = 0$. Given any positive integer q,

$$\int_{-\infty}^{+\infty} f(x)dx = \sum_{j=-\infty}^{\infty} w_j'(f(jh)h + O(h^q),$$

where $w_0' = 2w_0$, $w_j' = w_{|j|}$ for $j \neq 0$, and $f(0) = \frac{f(0^-)+f(0^+)}{2}$. This result can be extended to several variables and more than one discontinuity.

We shall consider an operator L, in one space dimension for simplicity, of the form

(D.30) $$L = \partial_t - P(x,\partial_x) = \partial_t - B(x)\partial_x,$$

and the equation

(D.31) $$Lv = 0,\ t > 0,\ -\infty < x < +\infty,$$

where B is a real C^∞ function. We shall denote the L_2 inner product by

$$(u,v)_{L_2} = \int u(x)v(x)dx.$$

Also we shall denote the adjoint of L by L* and let v and v* satisfy $Lv = 0$ and $L^*v^* = 0$, respectively. Furthermore, assume that either v or v* vanishes for $|x|$ large. Using Green's theorem, for the strip $0 < t < T$,

(D.32) $$(v(T),v^*(T))_{L_2} = (v(0),v^*(0))_{L_2}.$$

Approximate equation (D.31) by a forward difference method accurate of order q, written in the form

(D.33) $$u_i^{n+1} = Qu_i^n = \sum_j Q_j(ih)u_{i-j}^n.$$

Let A* denote a backward difference method as in (D.33) approximating the adjoint equation, written in the form

(D.34) $$u^{*n}_i = Q^*u_i^n = \sum_j Q_j^*((i+j)h)u^{*n+1}_{i+j}.$$

Multiplying (D.32) by u^{*n+1}_i and summing

$$\begin{aligned}\sum_i u^{*n+1}_i u_i^{n+1} &= \sum_i u^{*n+1}_i \sum_j Q_j(ih)u_{i-j}^n\\ &= \sum_i u_i^n \sum_j Q_j^*((i+j)h)u^{*n}_{i+j}\\ &= \sum_i u_i^n u^{*n}_i .\end{aligned}$$

Hence, by multiplying both sides by h

(D.35) $$(u^{n+1},u^{*n+1}) = (u^n,u^{*n}).$$

For all integers N, by summing (D.35) from 0 to N-1,

(D.36) $$(u^N,u^{*N}) = (u^0,u^{*0}).$$

We now begin preprocessing the initial data. Assume v is a piecewise C^∞ solution of $Lv = 0$, where the initial data contains a single discontinuity at $x = 0$. Let the initial data for the finite difference method (D.33) be defined by

(D.37) $$\begin{aligned}u_j^0 &= w_j'v(jh,0)\ ,\ j \neq 0\\ u_0^0 &= w_0(v(0^-,0) + v(0^+,0)),\end{aligned}$$

where the weights w_j' are defined as above.

Let $\phi(x)$ be an arbitrary test function in C_0^∞ and let v^* denote the solution of

(D.38)
$$L^*v^* = 0$$
$$v^*(x,T) = \phi(x).$$

Let u_j^{*n} satisfy (D.34) the finite difference method for the adjoint equation, with

(D.39)
$$u_i^{*N} = \phi(ih),$$

where $Nk = T$. It can be seen that the finite difference method (D.34) for the adjoint equation is also accurate of order q. Assuming Q^* is a stable difference operator,

(D.40)
$$u_i^{*0} = v^*(ih,0) + O(h^q).$$

Making use of the above quadrature formula, where $f(x) = v(x,0)v^*(x,0)$ (a piecewise C^∞ function with a discontinuity at $x = 0$),

$$(v(0),v^*(0))_{L_2} = \int_{-\infty}^{+\infty} f(x)dx = \sum_j w_j' v(jh,0)v^*(ih,0)h + O(h^q).$$

Using the estimate in (D.40) along with (D.37),

$$\begin{aligned}(v(0),v^*(0))_{L_2} &= \sum_j u_j^0(u_j^{*0} + (h^q))h + O(h^q)\\ &= \sum_j u_j^0 u_j^{*0} h + O(h^q)\\ &\quad u^0 u^{*0} h + O(h^q).\end{aligned}$$

Using (D.32) and (D.36)

$$(v(T),v^*(T))_{L_2} = (u^N,u^{*N}) + O(h^q)$$

or, using (D.38) and (D.39)

(D.41)
$$(v(T),\phi)_{L_2} = (u^N,\phi) + O(h^q)$$

which establishes the desired estimate for the moments of u. Using (D.40) we can obtain the dependence of error term in (D.41) on ϕ. Rewrite (D.41) as

$$(v(T),\phi)_{L_2} = (u^N,\phi) + O((\max \frac{\partial^{q+1}\phi}{\partial x^{q+1}})h^q).$$

We shall now post-process the approximate solution to obtain a pointwise estimate from the weak estimate (D.42).

Let $s(x)$ be a function that satisfies the conditions

$$\int s(x)dx = 1, \ \int x^{\ell}s(x)dx = 0, \ \ell = 1,2,\ldots,p-1,$$

where p is an arbitrary integer whose support is contained in the interval (0,1). Define a specific test function ϕ by

$$\phi(x) = \frac{1}{\varepsilon}s(\frac{x-y}{\varepsilon}),$$

where $\varepsilon = h^{q/(q+p+2)}$. Using the fact that if $g \in C^{\infty}(y,y+\varepsilon)$,

$$g(y) = \int g(x) \ (x)dx + O(\varepsilon P),$$

we obtain (provided the interval $(y,y+\varepsilon)$ contains no discontinuities of u at time T)

$$v(y,T) + O(\varepsilon P) = \int v(x,T)\phi(x)dx$$

$$= (v(T),\phi)_{L_2} .$$

Substituting into (D.42),

$$v(y,T) = (u^N,\phi) + O(h^q(\max \frac{\partial^{q+1}\phi}{\partial x^{q+1}})) + O(\varepsilon^P).$$

However, using the definition of ϕ and ε

$$\frac{\partial^{q+1}\phi}{\partial x^{q+1}} = O(\frac{1}{\varepsilon^{q+2}})$$

and

$$\frac{h^q}{\varepsilon^{q+2}} = \varepsilon^P = h^{qp/(q+p+2)}.$$

This gives rise to the pointwise estimate

(D.43) $$v(y,T) = (u^N,\phi) + O(h^{qp/(q+p+2)}).$$

A similar estimate is obtained on the interval $(Y-\varepsilon,y)$. Estimate (D.43) shows that by post-processing the approximate solution u, u differs from v by an amount as close to $O(h^q)$ as we wish by choosing p sufficiently large.

These one-sided estimates make it possible to obtain accurate pointwise estimates up to the discontinuity.

References

1. Apelkrans, Mats Y. T., On Difference Schemes for Hyperbolic Equations with Discontinuous Initial Values, Math. Comp., 22, 525 (1968).

2. Davis, P. and P. Rabinowitz, Methods of Numerical Integration, Academic Press, New York (1974).

3. Dym, H. and H. P. McKean, Fourier Series and Integrals, Academic Press, New York, (1972).

4. Lax, P. D., On the Stability of Difference Approximations to the Solution of Hyperbolic Equations with Variable Coefficients, Comm. Pure Appl. Math., 14, 497 (1961).

5. Majda, A. and S. Osher, Propagation of Error with Regions of Smoothness for accurate Difference Approximations to Hyperbolic Equations, Comm. Pure Appl. Math., 30, 671 (1977).

6. Mock, M. S. and P. D. Lax, The Computation of Discontinuous Solutions of Linear Hyperbolic Equations, Comm. Pure Appl. Math., 31, (1978).

INDEX